DYNAMICS ON SURFACES

THE JERUSALEM SYMPOSIA ON
QUANTUM CHEMISTRY AND BIOCHEMISTRY

Published by the Israel Academy of Sciences and Humanities,
distributed by Academic Press (N.Y.)

1st JERUSALEM SYMPOSIUM: *The Physicochemical Aspects of Carcinogenesis* (October 1968)

2nd JERUSALEM SYMPOSIUM: *Quantum Aspects of Heterocyclic Compounds in Chemistry and Biochemistry* (April 1969)

3rd JERUSALEM SYMPOSIUM: *Aromaticity, Pseudo-Aromaticity, Antiaromaticity* (April 1970)

4th JERUSALEM SYMPOSIUM: *The Purines: Theory and Experiment* (April 1971)

5th JERUSALEM SYMPOSIUM: *The Conformation of Biological Molecules and Polymers* (April 1972)

Published by the Israel Academy of Sciences and Humanities,
distributed by D. Reidel Publishing Company (Dordrecht, Boston and Lancaster)

6th JERUSALEM SYMPOSIUM: *Chemical and Biochemical Reactivity* (April 1973)

Published and distributed by D. Reidel Publishing Company
(Dordrecht, Boston and Lancaster)

7th JERUSALEM SYMPOSIUM: *Molecular and Quantum Pharmacology* (March/April 1974)

8th JERUSALEM SYMPOSIUM: *Environmental Effects on Molecular Structure and Properties* (April 1975)

9th JERUSALEM SYMPOSIUM: *Metal-Ligand Interactions in Organic Chemistry and Biochemistry* (April 1976)

10th JERUSALEM SYMPOSIUM: *Excited States in Organic Chemistry and Biochemistry* (March 1977)

11th JERUSALEM SYMPOSIUM: *Nuclear Magnetic Resonance Spectroscopy in Molecular Biology* (April 1978)

12th JERUSALEM SYMPOSIUM: *Catalysis in Chemistry and Biochemistry Theory and Experiment* (April 1979)

13th JERUSALEM SYMPOSIUM: *Carcinogenesis: Fundamental Mechanisms and Environmental Effects* (April/May 1980)

14th JERUSALEM SYMPOSIUM: *Intermolecular Forces* (April 1981)

15th JERUSALEM SYMPOSIUM: *Intermolecular Dynamics* (October 1982)

16th JERUSALEM SYMPOSIUM: *Nucleic Acids: The Vectors of Life* (May 1983)

VOLUME 17

DYNAMICS
ON SURFACES

PROCEEDINGS OF THE SEVENTEENTH JERUSALEM SYMPOSIUM ON
QUANTUM CHEMISTRY AND BIOCHEMISTRY HELD IN
JERUSALEM, ISRAEL, 30 APRIL – 3 MAY, 1984

Edited by

BERNARD PULLMAN

*Institut de Biologie Physico-Chimique
(Fondation Edmond de Rothschild), Paris, France*

JOSHUA JORTNER and ABRAHAM NITZAN

Department of Chemistry, University of Tel Aviv, Israel

and

BENJAMIN GERBER

*Department of Physical Chemistry,
The Hebrew University, Jerusalem, Israel*

D. REIDEL PUBLISHING COMPANY

A MEMBER OF THE KLUWER ACADEMIC PUBLISHERS GROUP

DORDRECHT / BOSTON / LANCASTER

CHEMISTRY

Library of Congress Cataloging in Publication Data

Jerusalem Symposium on Quantum Chemistry and Biochemistry (17th: 1984)
  Dynamics on surfaces.

  (The Jerusalem symposia on quantum chemistry and biochemistry; v. 17)
  Includes index.
  1. Surface chemistry—Congresses. I. Pullman, Bernard, 1919-
II. Title. III. Series.
QD506.A1J47   1984   541.3'453   84-15145
ISBN 90-277-1830-X

Published by D. Reidel Publishing Company,
P.O. Box 17, 3300 AA Dordrecht, Holland.

Sold and distributed in the U.S.A. and Canada
by Kluwer Academic Publishers,
190 Old Derby Street, Hingham, MA 02043, U.S.A.

In all other countries, sold and distributed
by Kluwer Academic Publishers Group,
P.O. Box 322, 3300 AH Dordrecht, Holland.

TABLE OF CONTENTS

PREFACE

The Seventeenth Jerusalem Symposium focused on dynamics on surfaces, a subject of interest in the areas of chemistry, physics and material sciences. The Symposium reflected the high standeds of these distinguished scientific meetings, which convene yearly at the Israel Academy of Sciences and Humanities to discuss a specific topic in the broad area of quantum chemistry and biochemistry.

Molecule-surface dynamics constitutes a fast developing, current, research area. The main theme of the Symposium was built around a conceptual framework for the elucidation of the chemical and physical aspects of atomic and molecular scattering off surfaces, surface transport, surface reaction dynamics and radiative interactions on surfaces. The interdisciplinary nature of this research area was emphasized by intensive and extensive interactions between scientists from different disciplines and between theory and experiment.

Held under the auspices of the Israel Academy of Sciences and Humanities and the Hebrew University of Jerusalem, the Seventeenth Jerusalem Symposium was sponsored by the Institut de Biologie Physico-Chimique (Fondation Edmond de Rothschild) of Paris. We wish to express our deep thanks to Baron Edmond de Rothschild for his continuous and generous support, which makes him a true partner in this important endeavour. We would also like to express our gratitude to the Administrative Staff of the Israel Academy, and in particular to Mrs. Avigail Hyam, for the efficiency and excellency of the local arrangements.

Joshua Jortner
Bernard Pullman
Benjamin Gerber
Abraham Nitzan

THEORY OF ONE-PHONON ASSISTED ADSORPTION AND DESORPTION OF He ATOMS
FROM A LiF(001) SINGLE CRYSTAL SURFACE

Dieter Eichenauer and J. Peter Toennies

Max-Planck-Institut für Strömungsforschung, Göttingen, and
Sonderforschungsbereich 126 Göttingen/Clausthal
Federal Republic of Germany

ABSTRACT

    In connection with recent high resolution He atom inelastic scat-
tering experiments the probability for resonant one-phonon scattering
is calculated within the framework of the distorted wave Born approximation. In this approach the inelastic scattering is treated as a perturbation of the atom-static surface potential eigenstates due to the
dynamic atom-phonon coupling. Both the static and the dynamic potentials
are derived from semi ab initio atom-ion pair potentials. The calculation takes full account of bound state resonances, the surface corrugation and the crystal lattice dynamics. In comparison with experimental
data the theoretical results confirm the recent interpretation of resonant features with respect to the mechanisms of adsorption into and desorption out of specified surface bound states. They also reveal the
contributions of Rayleigh and bulk phonon creation and annihilation
processes to the inelastic scattering.

1. INTRODUCTION

    The mechanisms by which atoms or molecules are adsorbed on or de-
sorbed from surfaces are of great importance for understanding surface
catalyzed reactions. Up to recently most of the experimental data on
these processes stems from kinetic rate measurements in which sticking
probabilities are determined by monitoring the coverage of the surface
with gas atoms by LEED or some other technique [1]. Desorption rates
are typically measured by rapidly heating a crystal and observing the
gas evolution (thermal flash desorption). This work has led to the development of rather detailed models for describing the formation of a
chemisorbed overlayer on a crystal surface. Several elementary steps are
postulated. The initial step involves the trapping of the atom or molecule in a physisorbed state sometimes referred to as a precursor state
[2]. In a subsequent step after diffusion along the surface the adsorbed
molecule passes into a more tightly bound chemisorbed state. In desorption the same processes occur in reverse order. Thus the trapping into

1

B. Pullman et al. (eds.), Dynamics on Surfaces, 1–14.
© *1984 by D. Reidel Publishing Company.*

and desorption out of physisorbed states are fundamental rate determining processes in catalytic surface reactions and for this reason much
theoretical work has been devoted to their understanding [3,4,5].

Recent high resolution molecular beam experiments have opened up
the possibility to study these processes in complete microscopic detail.
In work in our laboratory a number of small resonance maxima were observed between the orders of magnitude larger diffraction maxima in the
scattering of monoenergetic ($\Delta v/v \approx 1\%$) He atom beams of about 20 meV
energy from LiF(001) crystals [6]. The small maxima were attributed to
diffraction into bound states of the atom-surface potential and subsequent ejection following an interaction with phonons at the surface. In
the bound states the atoms are constrained to move along the surface
with a translational energy equal to their initial plus bound state
energy. From the angular half widths the lifetimes in the various bound
states were found to increase with quantum number v ranging from
$1 \cdot 10^{-12}$ sec for v=0 to about $10 \cdot 10^{-12}$ sec for v=3 [7]. In subsequent experiments atoms were injected into a specific bound state by choosing
the appropriate resonant incident angles. Then the distribution of final
angles and the time of flight spectra of the atoms leaving these bound
states were measured [8]. The results indicated that a substantial fraction of the atoms appeared to leave the surface in grazing collisions after receiving sufficient energy by annihilating a single phonon. This
so-called "Scylla" process is shown schematically in Fig.1. From Fig.1
we recognize that in a Scylla process we are observing directly the elementary process of single phonon assisted desorption. Experimentally it

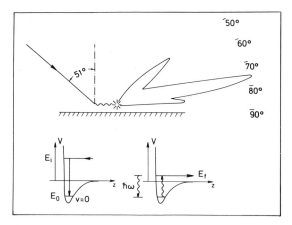

Fig.1: Schematic diagram showing a classical interpretation of
the experimental results for inelastic scattering of monoenergetic He atoms with 17.75 meV scattered from LiF along the
<110> azimuth. The atoms with incident angle 51° are trapped
by diffraction via a (10) reciprocal lattice vector into the
v=0 bound state. The final angle distribution of atoms leaving
the surface after annihilating a single phonon at the surface
is shown at the right. This process is called a Scylla resonance.

was also possible to identify a related scattering process by utilizing
a kinematic coincidence by which all atoms in the v=0 bound state are
elastically diffracted into a specific final angle nearly independent
of their kinetic energy in that bound state. Thus measured incident
angle distributions and the time of flight distributions could be interpreted to indicate that the atoms had been trapped into the bound state
by creating a single surface phonon. This process, called "Charybdis",
is the time-reverse of the Scylla process and thus represents a possible
elementary process of one-phonon assisted adsorption. Other resonance
processes in which the trapped atoms undergo transitions from one bound
state to another have also been identified [9]. We note that in all
these processes the kinematics are fully defined: the initial and final
atom velocities and scattering angles, the crystallographic directions,
the bound states and the phonon energies and momenta are all uniquely
specified in the experiments.

So far it has not been possible to identify all of the many observed resonances nor determine the relative contributions from surface
localized phonons and bulk phonons at the surface. To understand these
and other details we have embarked on a theoretical program to calculate
cross sections for Scylla and Charybdis processes. We feel that the
He-LiF system is an excellent prototype system which should be understood before approaching metal surfaces. From extensive experimental
studies it is well established that LiF surfaces are nearly perfect,
clean and have a high coherence length. Diffraction [11] and bound state
energies [12] are precisely known and the surface phonon dynamics have
been confirmed by inelastic time of flight experiments [13]. For metal
surfaces on the other hand recent He atom scattering experiments [10]
have shown a number of anomalies in the phonons and surface structures,
which will clearly affect adsorption and desorption processes.

The present paper reports on first results from this undertaking.
The next section (2.) describes the distorted wave Born approximation
formalism used. The following section (3.) describes briefly the semi ab
initio two-body potential model used to derive the potential between the
atom and the rigid crystal as well as the phonon coupling potential. In
section 4 we briefly describe the calculations of phonon dynamics on
which our results are based. Finally in section 5 we compare results for
a time of flight spectrum and an angular distribution for specific Scylla processes with experimental results. The paper closes with a brief
discussion.

2. SCATTERING THEORY

The discrete nature of the bound states and the fact that the de
Broglie wavelength ($\sim 1\text{Å}$) is comparable to the lattice spacing (Li^+-F^-
distance = 2.01 Å) indicate that a fully quantum mechanical dynamical
theory must be applied. Fortunately a formalism first developed by
Manson and *Celli* [14] is available for efficiently dealing with this
problem. In this theory the treatment is greatly simplified by using the
distorted wave Born approximation which is justified even in the presence of resonances since the inelastic scattering probabilities are

never more than 1% of the diffraction intensities. The accuracy of the
distorted wave Born approximation for calculating phonon inelastic He-
LiF(001) scattering has been demonstrated in several theoretical studies
for approximate interaction potential models [15,16]. In the following
we outline the theory only briefly.

The fundamental assumption at the root of all available rigorous
quantum theories of inelastic surface scattering is that the gas atom-
surface potential can be described as a superposition of two-body interactions. This assumption makes it possible to ultimately reduce the ma-
ny-body dynamics to a sum of two-particle dynamical couplings. This two
-body additivity assumption appears to be a priori well justified for
He-LiF. There is some evidence that this approximation also works for
metal surfaces [17]. This has been attributed to the fact that the scattering turning points are very far from the electron cores (typically 2
to 3 Å) where the electron density is so low that three-body forces have
only a small effect. In this approximation the total potential for the
interaction of the gas atom with the surface is given by

$$V(\vec{r},t) = \sum_{j,s} v^{(s)}\left(\vec{r} - \vec{r}_{jo}^{(s)} - \vec{u}_j^{(s)}(t)\right) \quad, \tag{1}$$

where $\vec{r}$ and $\vec{r}_j^{(s)} = \vec{r}_{jo}^{(s)} + \vec{u}_j^{(s)}(t)$ are the vector positions of the
gas and lattice atoms respectively; j and s refer to a surface unit
cell and lattice atom in the cell ($s=1:Li^+$; $s=2:F^-$). $v^{(s)}$ is a two-body potential and $\vec{u}_j^{(s)}(t)$ is the time dependent displacement of the
lattice atoms produced by the lattice phonon waves.

The relevant scattering cross section is described by the differential inelastic reflection coefficient (probability for the atom to be
scattered with final energy in the inverval (E_f, E_f+dE_f) and solid angle
$d\Omega_f$ centered about the polar angle Θ_f with respect to the surface
normal and the azimuthal angle ϕ_f). In quantum scattering theory it
can be written without approximation as [14]

$$\frac{dR}{dE_f\, d\Omega_f} = \frac{L^4 m^2 k_f}{4\pi^2\hbar^4 |k_{iz}|} \sum_{\{n_i\}} p(\{n_i\})$$

$$\cdot \sum_{\{n_f\}} |T_{fi}(t)|^2\, \delta\left(E_f + E_{\{n_f\}} - E_i - E_{\{n_i\}}\right) \quad, \tag{2}$$

where the prefactor is for normalization purposes (L is a quantization
length, m the gas atom mass, k_f the final momentum and k_{iz} the
normal component of the incident momentum ($|k_{iz}|=k_i \cos \Theta_i$)), $p(\{n_i\})$
is the temperature dependent occupation number distribution of the initial phonon states, $T_{fi}(t)$ is the time dependent transition matrix,
$E_f(=(\hbar^2/2m)k_f^2)$ and $E_i(=(\hbar^2/2m)k_i^2)$ are the final and initial kinetic
energies of the atom and $E_{\{n_f\}}$ and $E_{\{n_i\}}$ are the final and initial
vibrational energies of the crystal.

The standard two potential formalism, introduced by *Gell-Mann* and
Goldberger and others [18] is introduced in order to apply the Born approximation:

$$V(\vec{r},t) \;=\; V_{st}(\vec{r}) + v_{dyn}(\vec{r},t) \quad, \tag{3}$$

where V_{st} describes the dominant static interaction between the gas atom and the lattice atoms at their equilibrium positions, $\vec{r}_{jo}^{(s)}$:

$$V_{st}(\vec{r}) \;=\; \sum_{j,s} v^{(s)}(\vec{r} - \vec{r}_{jo}^{(s)}) \quad. \tag{4}$$

Strictly speaking V_{st} is the thermal average of the static potential implying an attenuation of the elastic intensity by the Debye-Waller factor. This is small in the present calculations and has been neglected. V_{st} is much larger than the second term v_{dyn} which describes the dynamic coupling of the gas atom to the lattice atom displacements.

This separation of the potential makes it possible to introduce the distorted wave Born approximation in which the term $V_{st}(\vec{r})$ is responsible for the elastic deflection of the atoms and the term $v_{dyn}(\vec{r},t)$ acts as a time dependent perturbation coupling the eigenstates of $V_{st}(\vec{r})$. Even with this simplification the solution of Eq.(2) is still a very formidable task because of the large number of phonon states which have to be accounted for. This can be circumvented by introducing a very elegant formalism first introduced by *van Hove* [19]. In the *van Hove* transformation the reflection coefficient is changed into an integral over the time correlation T-matrix averaged over the lattice vibrations. After considerable further calculations the reflection coefficient can be transformed to the following formula which is used in the actual calculations:

$$\frac{dR}{dE_f \, d\Omega_f} \;=\; \frac{m^2 \, k_f}{16\pi^5 \hbar^4 |k_{iz}| A_{uc}} \cdot n(\omega)$$

$$\sum_{\vec{Q}} \sum_{s\alpha} \sum_{s'\beta} \tilde{Z}^*_{s\alpha} \big(s\alpha \, |\,\mathrm{Im}\ \tilde{g}\ (\vec{Q},\omega^2)\,|\ s'\beta\big)\ \tilde{Z}_{s'\beta} \quad, \tag{5}$$

where $\omega=(E_f-E_i)/\hbar$, A_{uc} is the area of the surface unit cell and the Bose factor is $n(\omega)=(\exp(\hbar\omega/k_B T)-1)^{-1}$ for $\omega>0$ (phonon annihilation) and $n(\omega)=(\exp(\hbar\omega/k_B T)-1)^{-1}+1$ for $\omega<0$ (phonon creation). k_B is the Boltzmann constant. The sum on the right of Eq.(5) is over all phonon momenta $\vec{Q}$, as well as over the lattice ions $s=1$ and $s=2$, and the three degrees of freedom of each ion with the subscripts α and β referring to x, y or z. The $\tilde{Z}$'s contain all the information on the atom-surface dynamic coupling.

The time dependence of lattice displacements is contained in the matrix over the imaginary part of the surface projected Green's function for the phonons $(s\alpha\,|\,\mathrm{Im}\ \tilde{g}(\vec{Q},\omega^2)|\ s'\beta)$. The coupling between the atom motion and the lattice dynamics leads to the complicated sums in Eq.(5). We shall briefly describe in Section 4 how the matrix elements of $\mathrm{Im}\ \tilde{g}(\vec{Q},\omega^2)$ are calculated and discuss the calculation of the $\tilde{Z}$'s below.

The $\tilde{Z}$'s are given by

$$\tilde{Z}_{s\alpha}{}^{\cdot} = L \int_{-\infty}^{+\infty} dz \sum_{\Delta\vec{K}} \sum_{\vec{G}} \sum_{\vec{G}'} \sum_{\vec{G}''} \varphi_{\vec{G}'',f}^{(-)*}(z)$$

$$\cdot \{e^{-i\Delta\vec{K}\cdot\vec{S}_s} \int_{\infty}^{\infty} d\Delta k_z \, \Delta k_\alpha \, e^{i\Delta k_z(z-z_s)} L^3 v^{(s)}(\Delta\vec{k})\} \qquad (6)$$

$$\cdot \varphi_{\vec{G}',i}^{(+)}(z) \, \delta_{\Delta\vec{K}+\vec{Q},\vec{G}} \, \delta_{\Delta\vec{K},\vec{K}_f-\vec{K}_i+\vec{G}''-\vec{G}'} \quad ,$$

where $v^{(s)}(\Delta\vec{k})$ is the Fourier transform of the two-body potential

$$v^{(s)}(\Delta\vec{k}) = \frac{1}{L^3} \int_V d^3\rho \, v^{(s)}(\vec{\rho}) \, e^{-i\Delta\vec{k}\cdot\vec{\rho}} \qquad (7)$$

$V(=L^3)$ is the volume of quantization. $\vec{S}_s$ and z_s in Eq.(6) are the parallel and perpendicular components of the position vector of the s-th ion with respect to the center of the unit cell (note that $\Delta\vec{k} = (\Delta\vec{K}, \Delta k_z) = (\Delta k_x, \Delta k_y, \Delta k_z)$). The integral over Δk_z in Eq.(6) transforms the Fourier transform of the two-body potential into a z-dependent effective potential. The δ-symbols assure conservation of parallel momentum.

The wave functions $\varphi_{\vec{G},f}^{(-)}(z)$ and $\varphi_{\vec{G},i}^{(+)}(z)$ in Eq.(6) are obtained from the time independent atom scattering wave functions $\Psi_f^{(-)}$ and $\Psi_i^{(+)}$, which are much more complicated than those used in previous applications of the theory. Because of our interest in resonance processes these atom scattering wave functions must include all diffraction channels in order that selective adsorption into all bound states is properly accounted for. $\Psi_f^{(-)}$ and $\Psi_i^{(+)}$ are solutions of the following time independent Schrödinger equation [20]

$$[-\frac{\hbar^2}{2m} \nabla^2 + V_{st}(\vec{r})] \, \Psi(\vec{r}) = E \, \Psi(\vec{r}) \qquad (8)$$

The solution of Eq.(8) is facilitated by expanding the static potential in a Fourier expansion:

$$V_{st}(\vec{r}) = \sum_{\vec{G}} v_{\vec{G}}(z) \, e^{i\vec{G}\cdot\vec{R}} \quad . \qquad (9)$$

Recall that $\vec{r}=(\vec{R},z)$ and note that the sum is over all reciprocal lattice vectors of the two-dimensional surface lattice. According to Bloch's theorem $\Psi(\vec{r})$ can also be expanded:

$$\Psi(\vec{r}) = \frac{1}{\sqrt{A_{uc}}} \sum_{\vec{G}} \varphi_{\vec{G}}(z) \, e^{i(\vec{K}_i+\vec{G})\cdot\vec{R}} \quad , \qquad (10)$$

where $\vec{K}_i$ is the projection of $\vec{k}_i$ ($|\vec{K}_i|=k_i \sin \Theta_i$) onto the surface.
Insertion of Eq.(9) and (10) into the Schrödinger equation (8) leads
finally to a set of close coupling equations which have to be solved
numerically for the $\varphi_{\vec{G}}(z)$ appearing in Eq.(6). From the asymptotic
boundary conditions we get

$$\varphi_{\vec{G}}^{(+)}(z) \xrightarrow[z\to\infty]{} e^{-ik_{\vec{G}z}z} \delta_{\vec{G},\vec{0}} + A_{\vec{G}} e^{ik_{\vec{G}z}z} , \qquad (11)$$

where the time reversed solutions $\varphi_{\vec{G}}^{(-)}$ are obtained by changing the
signs of the arguments in the two exponents. The $A_{\vec{G}}$ are the scattering amplitudes.

Eq.(5) has to be solved numerically for each set of initial and
final conditions given by the scattering angles Θ_i, ϕ_i, Θ_f, ϕ_f and
beam energies E_i, E_f. Here E_i, ϕ_i and ϕ_f are fixed and thus E_f, Θ_i
and Θ_f were varied. The numerical solution is rather computer time
consuming since because of the relatively large corrugation it was necessary to include reciprocal lattice vectors up to 4th order. The total number of coupled channels was typically 25 and the computing time
2 minutes per value of $dR/dE_f d\Omega_f$ (UNIVAC 1100). Finally we wish to
point out that for discrete surface modes Eq.(5) can be greatly simplified [21,22].

3. POTENTIAL MODEL

The He-LiF(001) system is particularly suitable for developing and
testing the feasibility of describing both the static and dynamical potentials with a single two-body potential model. This approach has not
been tried previously, in the past either only the static or the dynamical part were derived from an empirical two-body potential. Being an
almost perfect ionic crystal the assumption of an additive potential
appears to be better justified than for almost any other crystal. Moreover because of their small number of electrons the He-Li$^+$ and He-F$^-$
potentials are amenable to a full ab initio quantum chemical calculation. With extensive experimental data on the static part of the potential it has been possible to carry out a large number of critical comparisons of this new two-body potential [23] before using it here to
describe the more complicated Scylla process.

In the previous section we left out a small potential term for reasons of clarity. In actual fact $V_{st}(\vec{r})$ in Eqs.(4) and (9) is given by

$$V_{st}(\vec{r}) = \sum_{j,s} v^{(s)}(\vec{r} - \vec{r}_{jo}^{(s)}) + V_{induced\ dipole} \qquad (12)$$

The second term on the right is due to a collective interaction of the
electric field $\vec{E}(\vec{r})$ produced by all lattice ions with the polarizability α_{He} of the helium atom:

$$V_{induced\ dipole} = -\frac{1}{2} \alpha_{He} |\vec{E}(\vec{r})|^2 \qquad (13)$$

Since this term is small it is a good approximation to neglect its effect on the phonon coupling.

This global treatment of the induction term, which dominates the attractive ion-atom gas phase interaction, prevents us from adopting semi-empirical potentials for He-Li$^+$ and He-F$^-$. Rather we seek a two-body potential which takes account of all first order overlap and electrostatic effects, but not the second order induced dipole term already included in Eq.(12). Also the model must take account of the long range dispersion interaction and its correct damping at small distances. A simple model which has been extensively tested for gas phase atom-atom interactions and recently for ion-metal surface [24] interactions is the new potential formula of *Tang* and *Toennies* [25]:

$$v(\rho) = A\ e^{-b\rho} - \sum_{n=3}^{\infty} \left(1 - \sum_{k=0}^{2n} \frac{(b\rho)^k}{k!}\ e^{-b\rho}\right) \frac{C_{2n}}{\rho^{2n}}, \qquad (14)$$

where the first term is obtained from a full SCF calculation and the second term takes account of the 2nd order dispersion interaction, of which each term is dampened separately. This potential predicts that the damping depends only on the range of the repulsive potential. In the present application we have found it sufficient to keep only the leading dispersion term and thus $v^{(s)}$ is approximated by

$$v^{(s)}(\rho) = A_{(1)}^{(s)}\ e^{-b_{(1)}^{(s)}\rho} - \left(1 - \sum_{k=0}^{6} \frac{(b_{(1)}^{(s)}\rho)^k}{k!}\ e^{-b_{(1)}^{(s)}\rho}\right) \frac{C_6^{(s)}}{\rho^6}. \quad (15)$$

The first order SCF parameters $A_{(1)}^{(s)}$ and $b_{(1)}^{(s)}$ were kindly calculated for us by *R. Ahlrichs* and *H.-J. Böhm* (He-Li$^+$: $A_{(1)}^{(1)}$=650.9 eV, $b_{(1)}^{(1)}$ = 5.092 Å^{-1}, He-F$^-$: $A_{(1)}^{(2)}$=4408 eV, $b_{(1)}^{(2)}$=4.435 Å^{-1} [26] and $C_6^{(1)}$ for He-Li$^+$ was taken from the literature ($C_6^{(1)}$=120 meV Å^6) [27]. Unfortunately $C_6^{(2)}$ for He-F$^-$ was not available from ab initio calculations and since semi-empirical formulae gave rather widely differing results this term had to be determined empirically from the bound state energies. The expression Eq.(12) was used and summed to convergence for a range of $C_6^{(2)}$ values. The bound state energies for the resulting potentials were calculated for each value of $C_6^{(2)}$. Fortunately a good fit with all bound state energies within the small experimental errors was found for one unique value of $C_6^{(2)}$=4425 meV Å^6. With this value for $C_6^{(2)}$ we have further tested the accuracy of this model by extensive calculations of diffraction intensities for comparison with the measurements of *Boato et al.* [11]. This study, to be published elsewhere [23], reveals that the ab initio model performed better than previous standard models for treating the rigid surface corrugation. Thus the model appears to describe very well the atom-rigid lattice interaction.

In summary we wish to point out that probably the success of the present model is due to a fortunate cancellation of many small additional neglected higher order terms. For example we have neglected explicit consideration of surface relaxation and charge transfer between surface lattice atoms suggested by a small anomaly in the Rayleigh surface phonon dispersion curves [13]. Also we have not considered quadrupole deformation of the surface atoms [28]. Moreover we have neglected higher order dispersion terms in Eq.(15) which are probably partly cancelled by largely unknown higher order repulsion terms neglected in the *Tang-Toennies* potential model [25]. Finally we should mention the neglect of a small three-body contribution to $V_{st}(\vec{r})$ [29] and the dielectric damping of the potential from deeper crystal layers. All of these effects are presumably accounted for in the empirical adjustment of $C_6^{(2)}$. Probably this is also the reason why our value of C_3=137 meV Å^3 for the long range crystal-atom attraction is considerably larger than the calculated value of 92 meV Å^3 [30].

4. PHONON DYNAMICS

For the reflection coefficient calculations it is necessary to know both the dispersion curves for the discrete surface localized phonon modes such as the Rayleigh mode and the density of states of both the surface localized and bulk phonons projected onto the surface plane. These data were obtained from a Green's function lattice dynamical program similar to one described in the literature [31] and kindly supplied to us by *G. Benedek*. The calculations are based on the breathing-shell model in the version proposed by *Schröder* and *Nüsslein* [32]. In its original form this potential showed a small (11%) discrepancy for the Rayleigh mode at the zone boundary. This has been largely removed by using newly adjusted parameters based on a fit both of the neutron data and the new He-data for the Rayleigh mode dispersion curve [13]. Thus the new calculations use the same ionic polarizabilities at the surface as in the bulk, also the lattice spacing is the same throughout the crystal.

In the Green's function method the information on the phonon eigensolutions is contained in the imaginary part of the surface projected Green's function of the semi-infinite lattice · Im $\tilde{g}(\vec{Q},\omega^2)$. The matrix elements of Im $\tilde{g}(\vec{Q},\omega^2)$, which enter into the calculation of the reflection coefficient, are related to the polarization vectors $\vec{e}_s(\vec{Q},\nu)$ [33]. If the wave vector $\vec{Q}$ is along one of the high symmetry directions on the crystal surface (<100> or <110> direction) as in the experiment then the phonon modes at the surface can be divided into two classes with respect to their polarization vectors $\vec{e}_s(\vec{Q},\nu)$ [34]. The first, which is the important one for our calculations, is characterized by sagittal polarization in the scattering plane. The second type of polarization is called shear horizontal and does not couple to the atom in the distorted wave Born approximation.

Fig.2 shows the most important, namely the s=s'=2, $\alpha=\beta=z$, component of the matrix Im $\tilde{g}(\vec{Q},\omega^2)$ along the <100> direction in the form of a contour diagram. This is the major contribution to the density of phonon states, which are involved in the scattering. It is seen that although the atoms can couple to a wide range of phonons the largest contributions are expected from Rayleigh phonons (S_1 in Fig.2) and bulk phonons near the origin. Note that the strong weighting of modes with low frequencies is further enhanced by the Bose factor (see Eq.(5)).

In the calculations reported here the resolution was ΔQ=0.05 Å^{-1} and $\Delta\omega$=1.25·10^{12} rad/sec, which is better than the experimental resolu-

tion. The sharp δ-functions occurring at the Rayleigh mode have been replaced by an equivalent rectangular area having a width equal to Δω.

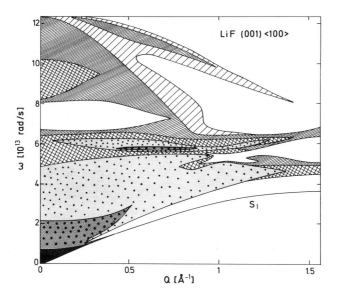

<u>Fig.2:</u> Diagram representing the matrix element $(s\alpha\,|\,\text{Im}\,\tilde{g}(\vec{Q},\omega^2)\,|\,s'\beta)$ of
the imaginary part of the surface projected Green's function in the
Q-ω space for s=s'=2 (specifying the F⁻ion) and α=β=z. The data are given for the LiF(001) surface with $\vec{Q}$ along the <100> crystal direction.
The low-lying single curve denotes the discrete Rayleigh phonon branch
while the areas above represent the continuous bulk bands. The shading
indicates the magnitude of the Green's function matrix element in units
of $10^{-4}\,\text{erg}^{-1}\,\text{cm}^2$: blank: 0...0.003; weakly hatched: 0.003...0.01;
strongly hatched: 0.01...0.03; cross-hatched: 0.03...0.1; weakly dotted:
0.1...0.3; strongly dotted: 0.3...1; black > 1.

5. RESULTS

    The accuracy with which the theory correctly simulates the dynami-
cal coupling to surface phonons is best tested by comparisons with measured time of flight spectra. Fig.3 shows a particularly interesting
time of flight spectrum, which has been transformed to parallel momentum transfer in place of the flight time [13]. This spectrum illustrates
several interesting resonance effects which have been studied previously
with a more approximate theory [16]. The peak "E" in Fig.3a is attributed
to elastic scattering from surface defects and cannot be simulated by
the present program; the two peaks described by "R" are attributed to
single phonon creation processes near the (11) diffraction peak, while
the peak "B" corresponds to bulk phonons of low frequency (∿ 1•10¹³ rad/
sec). Two final state resonance processes (also called Charybdis) are
expected to occur under these conditions as predicted by the kinematic
relation

$$\frac{(\vec{K}_i + \vec{Q})^2}{\sin^2 \Theta_f} - (\vec{K}_i + \vec{Q} - \vec{G}_{mn})^2 = -\frac{2m}{\hbar^2} |\varepsilon_v| \quad . \tag{16}$$

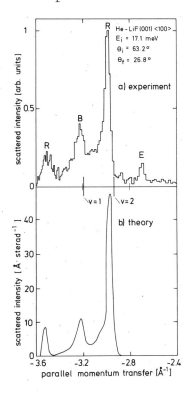

Fig.3: The inelastically scattered intensity as a function of the parallel momentum transfer K_f-K_i of the surface to the atom for He-LiF (001) scattering along the <100> azimuth. The kinematics are specified by the initial energy E_i = 17.1 meV and the initial and final polar angles Θ_i=63.2° and Θ_f=26.8° of the atom. The experimental data (a) derived from a time-of-flight measurement show four peaks attributed to Rayleigh phonons (R), a final state resonance involving bulk phonons (B) and incoherent elastic scattering (E). The conditions for final state resonances with the (11) surface reciprocal lattice vector and the v=1 and v=2 surface bound states are fulfilled for the values of the momentum transfer which are indicated by the two vertical bars. The theoretical results for the same kinematic conditions are shown in (b).

The angular locations of resonances predicted from Eq.(16) are indicated by vertical lines in Fig.3. The two nearest experimental peaks are affected by these resonances. In the case of the bulk phonon the resonance projects out of the broad continuum (see Fig.2) a specific set of states. The Rayleigh phonons always lead to sharp peaks which are further enhanced (or suppressed) by a resonance. Both these effects are well reproduced by the theoretical spectrum shown in Fig.3b. Thus we have here direct evidence that atoms can also be trapped into bound states not only via the creation of a Rayleigh phonon [8] but also via a bulk phonon. The small shifts in position of the experimental and theoretical resonant peaks from the kinematic predictions are not unexpected in view of the approximations made in Eq.(16). The small differences between the experimental and theoretical peak intensities in Fig.3 can be attributed to the finite but small spread in kinetic energy ($\sim$ 0.2 meV) of the incident beam, which has not been accounted for in the calculations. The filling-in of the minima between these peaks is probably due to multiphonon processes not included in the theory. Thus this and other spectra provide convincing evidence that the dynamical coupling is well described by the theory.

To simulate the Scylla angular distributions the differential re-

flection coefficients, Eq.(5), have to be integrated over all final
energies E_f. In Fig.4a we show a comparison of the experimental resonant and non-resonant (background) Scylla final angle distributions for
the <110> direction with the theoretical calculations. Note that the
theoretical ordinate is at the left while the experimental ordinate is
at the right and that the scales have been adjusted to line up the theoretical with the experimental maxima. We recall that the resonant contribution to the measured signal is obtained by subtracting from the
total signal measured at the peak incident angles the average signal
measured at two angles displaced by about 1° to both sides of the "resonant" incident angle. The same procedure was also adopted in the theoretical simulation. The agreement between theory and experiment is very
good both for the resonant and background distributions. There are, however, some discrepancies: in particular, we do not understand why the
theory does not reproduce the experimentally observed minimum at Θ_f=80°.
Although this previously unexplained minimum was considered experimentally significant the theory suggests it may well be an experimental arti-
fact. Fig.4b shows the contributions of different inelastic processes
to the total intensities (resonant plus non-resonant background). It is
gratifying to find that annihilation events can account for almost all
of the intensity in the maximum nicely supporting the model shown in
Fig.1. Moreover we find that the annihilation of Rayleigh phonons con-

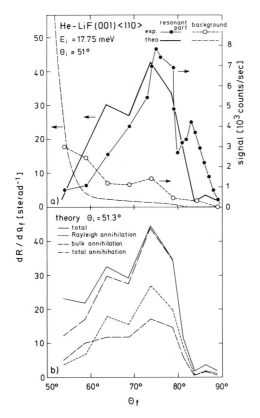

Fig.4: The intensity of
He atoms scattered along
the <110> azimuth of a
LiF(001) surface as a
function of the final
scattering angle Θ_f. The
initial energy and scat-
tering angle are 17.75
meV and Θ_i=51.3° respectively, and correspond to
a resonance with the v=0
surface bound state via
the (10) surface reci-
procal lattice vector.
In (a) the experimental
curves for the resonant
and non-resonant compo-
nents are compared with
the theoretical calcula-
tions. In (b) the dif-
ferent one-phonon contri-
butions to the total
(resonant plus non-reso-
nant) intensity are shown.

tributes 60% to the desorption process, whereas the annihilation of bulk phonons accounts for the 40% remainder.

6. DISCUSSION

In the present paper we have presented the first detailed calculations of single phonon assisted desorption of scattered atoms with specified energy and azimuth from selected surface bound states. The ability of the theory to explain all the detailed elastic and inelastic effects by a simple two-body potential between gas atoms and lattice atoms represents an important step forward in our detailed understanding of the elementary processes leading to accommodation, trapping on and desorption from surfaces.

The calculations have fully confirmed the previously reported model extracted from a large number of angular distributions and a few measured time of flight distributions. From these initial experiments it was not possible to determine the relative contribution of Rayleigh to bulk phonons to the desorption process. The present calculations predict that bulk phonons make a very sizable 40% contribution.

The same theory can be easily adapted to study the Charybdis single phonon adsorption process described in the introduction. Also inelastic processes involving elastic resonances in the initial and final states, which have been observed experimentally [9], can be handled in the same way. Analogous experiments and calculations on metal surfaces are in preparation.

ACKNOWLEDGEMENTS

We are extremely grateful to *V. Celli* for extensive discussions concerning details of the distorted wave Born approximation. We also thank *R. Ahlrichs* for carrying out the first order SCF calculations and *G. Benedek* for providing us with his Green's function program. The calculations were performed on the UNIVAC 1100 computational facility of the Gesellschaft für Wissenschaftliche Datenverarbeitung in Göttingen.

REFERENCES

[1]   D.A. King in Proc. 7th Int.Vac.Congr. and 3rd Int.Conf.Solid Surface, Vienna 1977, p. 769.

[2]   For a modern discussion of precursor states see J.K. Nørskov, A. Huomøller, P.K. Johansson, and B.I. Lundqvist, Phys.Rev.Lett. $\underline{46}$, 257 (1981); J.K. Nørskov, J.Vac.Sci.Technol. $\underline{18}$, 420 (1981).

[3]   W. Brenig; Z.Phys.B - Condensed Matter $\underline{48}$, 127 (1982).

[4]   H.J. Kreuzer, Int.Journal of Mass Spectrometry and Ion Physics $\underline{53}$, 273 (1983); E. Goldys, Z.W. Gortel, and H.J. Kreuzer, Surf.Sci. $\underline{116}$, 33 (1982).

[5]   J. Reyes, I. Romero, and F.O. Goodman; J.Chem.Phys. $\underline{79}$, 5906 (1983).

[6]   G. Brusdeylins, R.B. Doak, and J.P. Toennies; J.Chem.Phys. 75, 1784 (1981).

[7]   These lifetimes are a factor 6 smaller than those reported in Ref. 6. Since the less accurate formula $\tau = h/\Delta E$ was used there instead of $\tau = \hbar/\Delta E$.

[8]   G. Lilienkamp and J.P. Toennies; J.Chem.Phys. 78, 5210 (1983).

[9]   G. Lilienkamp and J.P. Toennies; Phys.Rev. B26, 4752 (1982).

[10] R.B. Doak, U. Harten, and J.P. Toennies, Phys.Rev.Lett. 51, 578 (1983); U. Harten, J.P. Toennies, and Ch. Wöll, in preparation.

[11] G. Boato, P. Cantini, and L. Mattera; Surf.Sci. 55, 141 (1976).

[12] G. Derry, D. Wesner, S.V. Krishnaswamy, and D.R. Frankl; Surf.Sci. 74, 245 (1978).

[13] G. Brusdeylins, R.B. Doak, and J.P. Toennies; Phys.Rev. B27, 3662 (1983).

[14] R. Manson and V. Celli; Surf.Sci. 24, 495 (1971).

[15] G. Benedek, J.P. Toennies, and R.B. Doak; Phys.Rev. B28, 7277 (1983).

[16] D. Evans, V. Celli, G. Benedek, J.P. Toennies, and R.B. Doak; Phys.Rev.Lett. 50, 1854 (1983).

[17] V. Celli, G. Benedek, U. Harten, J.P. Toennies, R.B. Doak, and V. Bortolani; Surf.Sci.Lett., in print.

[18] M. Gell-Mann and M.L. Goldberger; Phys.Rev. 91, 398 (1953).

[19] L. van Hove; Phys.Rev. 95, 249 (1954).

[20] H. Chow and E.D. Thompson; Surf.Sci. 59, 225 (1976).

[21] N. Garcia and J.M. Soler; Surf.Sci. 126, 689 (1983).

[22] For similar approximations specialized to Rayleigh and longitudinal modes see V. Bortolani, A. Franchini, F. Nizzoli, and G. Santoro; Phys.Rev.Lett. 52, 429 (1984).

[23] V. Celli, D. Eichenauer, A. Kaufhold, and J.P. Toennies, in preparation.

[24] K. Mann; Dissertation, Göttingen 1984.

[25] K.T. Tang and J.P. Toennies; J.Chem.Phys., in print (1984).

[26] See also H.J. Böhm and R. Ahlrichs; J.Chem.Phys. 77, 2028 (1982).

[27] A. Dalgarno and W.D. Davison; Mol.Phys. 13, 479 (1967).

[28] L. Miglio, F. Quasso, and G. Benedek; Surf.Sci. 136, L9 (1984).

[29] J.M. Hutson and C. Schwartz; J.Chem.Phys. 79, 5179 (1983).

[30] G. Vidali and M. Cole; Surf.Sci. 110, 10 (1981).

[31] G. Benedek, G.D. Brivio, L. Miglio, and V.R. Velasco; Phys.Rev. B26, 497 (1982).

[32] U. Schröder and V. Nüsslein; Phys. Status Solidi 21, 309 (1967).

[33] G. Benedek, in Dynamic Aspects of Surface Physics, Proceedings of the Int. School of Physics "Enrico Fermi", Course LVIII, edited by F.O. Goodman (1974), p. 605.

[34] R.E. Allen, G.P. Alldredge, and F. de Wette; Phys.Rev. B4, 1648 (1971).

TEMPERATURE EFFECTS IN DIFFRACTIVE ATOM-SURFACE SCATTERING:
PERTURBATION THEORY FORMULAE

V. Celli
Department of Physics
University of Virginia
Charlottesville, VA 22901, USA

A.A. Maradudin
Department of Physics
University of California
Irvine, CA 92717, USA

ABSTRACT

The thermally-induced change in diffracted intensity is expressed in terms of the general effective (optical) potential for atom-surface diffraction. When the surface corrugation is negligible, the decrease of specular intensity is given explicitly in terms of the imaginary part of this potential; a similar formula applies in general for the overall decrease in total diffracted intensity. The thermal effects on the intensities of the individual diffracted beams are also written down explicitly, to leading order of perturbation theory. It is shown how the results reduce to the eikonal and hard-wall formulas in the appropriate limits; the main correction comes from the fact that, if the gas atom-solid atom potential is of the form $\exp(-\beta r)$, the effect of phonon processes involving parallel momentum transfer q is decreased, approximately, by $\exp(-q^2 z_t/\beta)$, where z_t is the average turning point.

1. INTRODUCTION

In a recent paper,[1] to be referred to as I, we wrote down explicit expressions for the effective (optical) potential in atom-surface diffraction, $\langle M \rangle = M_R + i M_I$. The imaginary part of this potential, M_I, arises from the emission and absorption of phonons in the atom-surface collision. These processes cause a decrease in the total elastic intensity, given by

$$\sum_{\vec{F}} |D(\vec{F}, \vec{0})|^2 \quad , \tag{1.1}$$

which can be interpreted as an overall Debye-Waller factor. $D(\vec{F}, \vec{0})$ is given by

15

B. Pullman et al. (eds.), Dynamics on Surfaces, 15–22.
© *1984 by D. Reidel Publishing Company.*

$$D(\vec{F},\vec{0}) = \delta_{\vec{F},\vec{0}} + \frac{m}{\hbar^2\alpha(|\vec{k}+\vec{F}|\omega)} \sum_{\vec{G}\vec{G}'} \int dx_3 \int dx_3' \times \qquad (1.2)$$

$$\times \psi_R^+(\vec{G},\vec{F}|x_3)^* M_I(\vec{G},\vec{G}'|x_3x_3')\psi^+(\vec{G}',\vec{0}|x_3'),$$

where ψ_R^+ is an eigenfunction of M_R and ψ^+ is an eigenfunction of $\langle M\rangle$, both with outgoing-wave boundary conditions. We use the same conventions and notations as in I, except that, for brevity, we omit the subscript $\|$ that was used to denote vectors in the plane of the surface $(x_3=0)$. For example, the outgoing diffracted beams have wave vectors $(\vec{k}+\vec{F},\alpha(|\vec{k}+\vec{F}|\omega))$, while the incoming beam has wave vector $(\vec{k},-\alpha(k\omega))$, with

$$(\vec{k}+\vec{F})^2 + \alpha(|\vec{k}+\vec{F}|\omega)^2 = \vec{k}^2 + \alpha(k\omega)^2 = 2m\omega/h. \qquad (1.3)$$

As usual, $\vec{F}$ is a reciprocal lattice vector ($\vec{G}$-vector) corresponding to diffraction into an open channel, for which $\alpha^2(\vec{F}) > 0$, and the sum in (1.1) is restricted accordingly.

The leading contribution to M_I (see next section) is of order $\langle u_3^2\rangle$, where $\langle u_3^2\rangle$ is the mean square vertical displacement of a surface atom at temperature T; therefore, to this order, we can evaluate D by replacing ψ^+ with ψ_R^+ in Eq. (1.2), and we can further compute ψ_R^+ by using the thermally averaged potential, $\langle V\rangle$, or even the static potential, instead of the full M_R.

It is more difficult to obtain explicit expressions for the temperature effect on the intensity of the individual diffracted beams, because the thermal motion causes a redistribution of diffracted intensity, in addition to the overall decrease given by Eq. (1.1). One can, of course, proceed numerically, by solving the diffraction problem for $\langle M\rangle$ at several temperatures.[2] It is also useful, however, to have approximate formulae in the case of weak diffraction. These can be obtained directly by computing the wave function in perturbation theory: for the non-specular beams one must go to second order in the phonon displacements and to first order in the $\vec{G}\neq0$ Fourier components of $\langle V\rangle$, denoted by $\hat{V}(\vec{G}|x_3)$; for the specular, one should consistently go to order $|\hat{V}(\vec{G}|x_3)|^2$, and thus to fourth order overall. With these wave functions, all the thermally averaged diffracted intensities are correctly obtained to order $\langle u_3^2\rangle|\hat{V}(\vec{G}|x_3)|^2$.

The same results must be obtained by evaluating $\langle M\rangle$ to order $\langle u_3^2\rangle$, and then applying perturbation theory to leading order. The specular amplitude, normalized to the specular amplitude for the thermally and laterally averaged potential, $V(\vec{0}|x_3)$, is (see I, Eq. (4.32))

$$1 + \frac{m}{i\hbar^2\alpha(k\omega)} \int dx_3 \int dx_3' |\chi^+(\vec{0}|x_3)|^2 M_2(\vec{0},\vec{0}|x_3x_3') \qquad (1.4)$$

plus terms of order $|\hat{V}(\vec{G}|x_3)|^2\langle u_3^2\rangle$; the normalized amplitude of the other diffracted beams ($F\neq0$) is written as follows, where χ^+ and Γ^+ denote, respectively, a wave function of the potential $V(\vec{0}|x_3)$ and its Green function:

$$\frac{m}{ih^2\alpha(|\vec{k}+\vec{F}|\omega)} \int dx_3 \int dx_3'' \chi^+(\vec{F}|x_3)^* \chi^+(\vec{0}|x_3'') \times$$

$$\times [\hat{V}(\vec{F}|x_3)\delta(x_3-x_3'') + M_2(\vec{F},\vec{0}|x_3 x_3'') +$$

$$+ \int dx_3' M_2(\vec{F},\vec{F}|x_3 x_3')\Gamma^+(\vec{F}|x_3' x_3'')\hat{V}(\vec{F}|x_3'') +$$

$$+ \int dx_3' \hat{V}(\vec{F}|x_3)\Gamma^+(\vec{0}|x_3 x_3')M_2(\vec{0},\vec{0}|x_3' x_3'')]. \tag{1.5}$$

Here M_2 is the second-order contribution to M (see I, Eq. (3.17b)):

$$M_2(\vec{G},\vec{G}'|x_3 x_3') = \sum_{H\,H'} \int_{BZ} \frac{d^2p}{(2\pi)^2} \int \frac{d\omega'}{2\pi} \{ \sum_{\kappa\alpha} \sum_{\kappa'\beta} W_{\kappa\alpha}(\vec{k}+\vec{G}-\vec{p}-\vec{H}|x_3) \times$$

$$\times C_{\alpha\beta}(\vec{p}-\vec{k},\omega-\omega'|\kappa\kappa') W^*_{\kappa'\beta}(\vec{k}+\vec{G}'-\vec{p}-\vec{H}'|x_3') \} \times$$

$$\times \hat{G}_{\omega'}(\vec{p}+\vec{H},\ \vec{p}+\vec{H}'|x_3 x_3'), \tag{1.6}$$

where $\hat{G}_{\omega'}$ is the Green function of $<V>$ for energy $h\omega'$ (see I, Eq. (2.42)), $C_{\alpha\beta}$ is the Fourier transform of the displacement correlation function, and $W_{\kappa\alpha}$ is essentially the derivative in the α direction ($\alpha = 1,2,3$) of the thermal average of the potential due to the atom layer labeled by κ (see I, Eq. (3.16b)). The topmost surface atoms are labeled $\kappa=0$ and usually give the dominant contribution to the κ, κ' sums in Eq. (1.6).

It will be noted that only the imaginary part of M_2 contributes to the specular intensity, which is the square modulus of the expression (1.4). Thus Eqs. (1.2) and (1.4) are mutually consistent to lowest order, when only F=0 is kept in Eq. (1.1). A detailed discussion of Eq. (1.4) was given in I when the atom-surface interaction is a sum of pairwise Yukawa potentials, with parallel Fourier transform

$$\hat{U}_\kappa(q|x_3) = U_0 e^{-\beta(q)x_3}/2\beta(q) \tag{1.7a}$$

$$\beta(q) = (\beta^2+q^2)^{1/2}. \tag{1.7b}$$

The final result is expressed, approximately, as the previously derived result for the vibrating hard wall,[3] Eq. (2.1), times several correction factors, the most important of which can be identified with the so-called Armand correction.[4] This manifests itself in momentum space as a cutoff in the contribution to $<u_3^2>$ of the phonons with large parallel momentum, according to the approximate cutoff function

$$e^{-2(\beta(q)-\beta)z_t} \simeq e^{-q^2 z_t/\beta}, \tag{1.8}$$

where z_t is the average turning point of $\hat{V}(\vec{0}|x_3)$, measured from the equilibrium position of the surface atoms (x=0). The factor (1.8) is precisely the same that cuts off the cross section for one-phonon exchange in the calculations of Bortolani et al,[5] and gives good agree-

ment with the measured inelastic cross sections for He/Ag(111).[6]
    We report here on the analogous evaluation of the thermal effects
on the individual diffracted intensities, as given by the modulus
square of Eq. (1.5). A fuller derivation of the results will be published separately.
    In Section II we obtain the vibrating hard-wall results by the
methods of perturbation theory. This is quite useful as a guide of
how to proceed in the more general case of the pairwise exponential
potential (1.7), which is discussed in Section III.

2.   THE VIBRATING HARD WALL MODEL

We recall that, for a vibrating hard wall without a static corrugation,
the specular reflection coefficient is[3]

$$1-4\alpha(k\omega) \int \frac{d^2p}{(2\pi)^2} \int \frac{d\omega'}{2\pi} \, C_{uu}(\vec{p}-\vec{k},\omega-\omega') \, \text{Re} \, \alpha(p\omega'), \qquad (2.1)$$

where C_{uu} is the fourier transform of the displacement correlation
function $\langle u(\vec{x};t) \, u(\vec{x}',t')\rangle$, which depends only on $|\vec{x}-\vec{x}'|$ and $t-t'$.
Here $u(\vec{x};t)$ is properly the thermally-induced displacement of the vi-
brating hard wall; within the assumptions of this model, $u(\vec{x};t)$ is
identified with the displacement $u_3(\vec{x}(\ell\kappa);t)$ of the top surface atoms
$(\kappa=0)$ and the $\vec{p}$ integration is restricted so that $\vec{p}-\vec{k}$ is in the first
Brillouin zone of the surface.
    If the displacement u is time-independent, Eq. (2.1) reduces to

$$1-4\alpha(k\omega) \int \frac{d^2p}{(2\pi)^2} \, \langle|\hat{u}(\vec{p}-\vec{k})|^2\rangle \, \text{Re} \, \alpha(p\omega). \qquad (2.2)$$

    This formula can be obtained very simply[7] by solving the
Rayleigh equations for hard-wall scattering to second order, and then
averaging.[8] This suggests the following quick derivation of the
more general result for a vibrating corrugated hard wall. Assume that
the corrugation is $\zeta(\vec{x}) + u(\vec{x})$, where $\zeta(\vec{x})$ is the average corrugation
and $u(\vec{x})$ is a stochastic function. Compute the diffraction amplitudes
for the surface reciprocal vector $\vec{F}$ to first order in the fourier component $\hat{\zeta}(\vec{F})$ and to second order in $\hat{u}(\vec{q})$, using the method of Lopez,
Yndurain and Garcia[7] (the Rayleigh-equation version of the method is
the most convenient), and average over the stochastic distribution of
$u(\vec{x})$. The result is

$$2i\alpha(k)\hat{\zeta}(\vec{F})\{1 -\frac{1}{2} \int \frac{d^2p}{(2\pi)^2} \, \langle|\hat{u}(\vec{p}-\vec{k})|^2\rangle \times$$

$$\times [\alpha^2(k) - 2\alpha^2(p) - \alpha^2(|\vec{k}+\vec{F}|) + 2\alpha(k)\alpha(p) +$$

$$+ 2\alpha(|\vec{p}+\vec{F}|)\alpha(p) + 2\alpha(|\vec{k}+\vec{F}|)\alpha(|\vec{p}+\vec{F}|)]\} \qquad (2.3)$$

where all the α's are evaluated at energy $\hbar\omega$. This can be rewritten in
a more symmetrical form, that satisfies the condition of reciprocity,
by noting that the replacement

$$\alpha^2(k) - \alpha^2(|\vec{k}+\vec{F}|) \rightarrow \alpha^2(p) - \alpha^2(|\vec{p}+\vec{F}|) \tag{2.4}$$

does not change the value of the integral in Eq. (2.3).[9]

To generalize the result to a vibrating wall, replace $<|u(\vec{p}-\vec{k})|^2>$ by $C(\vec{p}-\vec{k}, \omega-\omega')$, evaluate $\alpha(k)$ and $\alpha(|\vec{k}+\vec{F}|)$ at frequency ω, evaluate $\alpha(p)$ and $\alpha(|\vec{p}+\vec{F}|)$ at frequency ω', and integrate over ω' as in Eq. (2.1). To compute the intensity, take the modulus square of Eq. (2.3), and multiply by $\alpha(|\vec{k}+\vec{F}|\omega)/\alpha(k\omega)$ in the usual way. We take the modulus square of the average amplitude, instead of the average of the modulus square of the amplitude, because it is easier to do so, and it gives the same result for the diffracted intensities (this point is discussed in I, for instance).

The thermally-induced change in the intensity of the $\vec{F}$ diffracted beam (usually a decrease) is then given by the factor

$$1 - \text{Re} \int \frac{d^2p}{(2\pi)^2} \int \frac{d\omega'}{2\pi} \, C_{uu}(\vec{p}-\vec{k}, \omega-\omega') \times$$

$$\times [- \alpha^2(p\omega') - \alpha^2(|\vec{p}+\vec{F}|\omega') + 2\alpha(k\omega)\alpha(p\omega') +$$

$$+ 2\alpha(|\vec{p}+\vec{F}|\omega')\alpha(p\omega') + 2\alpha(|\vec{k}+\vec{F}|\omega)\alpha(|\vec{p}+\vec{F}|\omega')] \tag{2.5}$$

This result agrees to order $<u^2>$ with the usual Debye-Waller factor, as given by the eikonal approximation,[10]

$$\exp[-(\alpha(k\omega) + \alpha(|\vec{k}+\vec{F}|\omega))^2 <u^2>] \tag{2.6}$$

in the limit where the momentum and energy exchange in the inelastic collisions are small; more precisely $|\vec{p}-\vec{k}| \ll k$ and $|\omega-\omega'| \ll \omega$, so that $\alpha(p\omega')$ can be replaced by $\alpha(k\omega)$ and $\alpha(|\vec{p}+\vec{F}|\omega')$ can be replaced by $\alpha(|\vec{k}+\vec{F}|\omega)$. For F=0, Eq. (2.5) reduces to Eq. (2.1), which in turn agrees with the usual Debye-Waller factor (2.6) for F=0, in the limit of small $|\vec{p}-\vec{k}|$ and $|\omega-\omega'|$. As in the case of specular scattering, the difference between the hard-wall result (2.5) and the eikonal result (2.6), to order $<u^2>$, is especially notable when $\alpha(p\omega')$ or $\alpha(|\vec{p}+\vec{F}|\omega')$ are imaginary. For this reason the resulting correction has been called the closed channel effect by Meyer.[10] It can be expected to be of the same magnitude as in the specular case, where the Debye-Waller exponent can be reduced by 50% or more by the closed channel effect.[3]

3. THE REPULSIVE EXPONENTIAL POTENTIAL

A fairly realistic potential for the interaction of a closed-shell atom with a monoatomic surface is the sum of Yukawa potentials of the form (1.7). The laterally averaged $<V>$ is then approximately

$$\hat{V}(\vec{0}|x_3) = \frac{U_0}{2\beta a} e^{-\beta x_3} e^{1/2 \, \beta^2 <u_3^2>}, \tag{3.1}$$

and the Fourier components of $<V>$, for $\vec{F} \neq 0$, are (with $\beta^2(F) = \beta^2 + F^2$)

$$\hat{V}(\vec{F}|x_3) = \frac{U_0}{2\beta(F)a_c} e^{-\beta(F)x_3} e^{1/2(\beta^2+F^2)<u_3^2>-1/2<(\vec{F}\cdot\vec{u})^2>} , \qquad (3.2)$$

where a_c is the area of the unit mesh on the surface. The third component of the phonon-induced displacement potentials $W_{\kappa\alpha}$ is given by

$$iW_{\kappa 3}(\vec{q}|x_3) = \frac{U_0}{2} e^{-\beta(q)x_3} e^{1/2(\beta^2+q^2)<u_3^2>-1/2<(\vec{q}\cdot\vec{u})^2>} . \qquad (3.3)$$

We note that, for small q, these can be expressed simply in terms of the x_3 derivative of $\hat{V}(\vec{0}|x_3)$. The basic approximation leading to the final result is to make the replacement

$$e^{-\beta(q)x_3} \simeq e^{-(\beta(q)-\beta)z_t} e^{-\beta x_3} , \qquad (3.4)$$

where z_t is the location of the average turning point.

The other quantity needed in Eq. (1.6) is the Green function matrix $\hat{G}_\omega$. For the calculation of the diagonal elements of M_2 we need only the diagonal element

$$\hat{G}^+_{\omega'}(\vec{p}+\vec{H},\vec{p}+\vec{H}|x_3x_3') = \Gamma^+_{\omega'}(\vec{p}+\vec{H}|x_3x_3') , \qquad (3.5)$$

while for the off-diagonal element $M_2(\vec{F},\vec{0}|x_3x_3'')$ we need also

$$G^+_{\omega'}(\vec{p}+\vec{H}+\vec{F},\vec{p}+\vec{H}|x_3x_3') =$$

$$\int dx_3'' \hat{\Gamma}^+_{\omega'}(\vec{p}+\vec{H}+\vec{F}|x_3x_3'') \hat{V}(\vec{F}|x_3'') \hat{\Gamma}^+_{\omega'}(\vec{p}+\vec{H}|x_3''x_3') . \qquad (3.6)$$

(We recall that $\Gamma(\vec{H}|x_3x_3')$ in Eq. (1.5) stands for $\hat{\Gamma}_\omega(\vec{k}+\vec{H}|x_3x_3')$).

The evaluation of Eq. (1.6) yields intensities proportional to the matrix elements of $\hat{V}(\vec{F}|x_3)$, which is still temperature dependent. In order to fully determine the temperature effects, we must compare with the intensities at some standard temperature, or equivalently with the intensities given by the static potential, which is obtained by setting $<u_3^2> = 0$ and $<\vec{u}^2> = 0$ in Eqs. (3.1) and (3.2). It should be noted that the factor

$$e^{-1/2 \beta^2 <u_3^2>} \qquad (3.7)$$

corresponds to a rigid shift of $<V>$ with respect to the static potential. Because the same factor appears in Eq. (3.3), this rigid shift is of no consequence in the calculation.

With some drastic assumptions based on Eq. (3.4), we can cast the Debye-Waller factor in the same form as the hard wall result (2.5), times an extra factor

$$e^{-<(\vec{F}\cdot\vec{u})^2>} \qquad (3.8)$$

that arises from the exponential on the r.h.s. of Eq. (3.2). The main result of the calculation is that the effective correlation function of the surface profile is now given explicitly in terms of the displacement correlation function of the top surface atoms, as

$$C_{uu}(\vec{p}-\vec{k},\omega'-\omega) = e^{-(\vec{p}-\vec{k})^2 z_t/\beta} C_{33}(\vec{p}-\vec{k},\omega'-\omega|\kappa\kappa) \qquad (3.9)$$

for $\kappa=0$. This relation has been established in I for the specular intensity (2.1), and the exponential cutoff in parallel momentum has been displayed in Eq. (1.8) and given a physical interpretation. From another point of view, we can say that C_{uu} represents the correlation function of the effective hard wall, which is the locus of the turning points in the potential $<V>$, and the exponential factor (1.8) represents a transfer function between the atomic displacements and the location of the turning point.[11] The integral in Eq. (2.5) is formally extended to all values of $\vec{p}$, with C_{33} a periodic function of $\vec{p}-\vec{k}$ over the reciprocal lattice.

It must be mentioned that, in order to arrive at Eq. (3.9), we have dropped the "softness factor" S^2 that multiplies C_{uu} in a more complete formula. The reason for setting $S^2=1$, apart from a desire to obtain a simpler formula, is that the total atom-surface potential also contains the van der Waals attraction, one effect of which is to stiffen the repulsive part, so that the effective β is larger than the one appropriate to the overlap repulsion alone.[12] Approximately, one can use[6]

$$\beta_{eff} = \beta[1 + 2mD/\hbar^2\alpha^2(k\omega)]^{1/2}, \qquad (3.10)$$

where D is the surface well depth. In addition, one should apply the usual Beeby correction, replacing everywhere α^2 by $\alpha^2 + 2mD/\hbar^2$. This has also the effect of reducing the "closed channel" correction, because all the channels in the surface potential well are now available.

The Beeby correction is valid if the vibrating potential does not extend beyond the bottom of the attractive well. If the well bottom shifts appreciably during the vibration, the inelastic scattering can be substantially increased and the apparent Beeby factor corresponds to unphysically large values of the well depth D. In such cases, a complete calculation seems to be necessary, even at low temperatures.

At higher temperatures, one must go beyond perturbation theory. This is easy to do for the two-body Yukawa interaction, Eq. (1.7a). The exact M_2 is of the same form as Eq. (1.6), but C_{33} must now be interpreted as the Fourier transform of

$$\frac{1}{B} [e^{B<u_3(\ell\kappa;t) u_3(\ell'\kappa';t')>} -1] \qquad (3.11)$$

$$B = \beta(\vec{k}+\vec{G}-\vec{p}-\vec{H}) \beta(\vec{k}+\vec{G}'-\vec{p}-\vec{H}').$$

Solution of the Schrödinger equation with the optical potential $<V> + M_2$ seems to give very good results.[2]

REFERENCES

1. V. Celli and A.A. Maradudin, to be published.
2. G. Armand and J.R. Manson, to be published.
3. N. Garcia, A.A. Maradudin and V. Celli, Phil. Mag. $\underline{A45}$ (1982) 384;
    N. Garcia and A.A. Maradudin, Surface Sci. $\underline{119}$ (1982) 384.
4. H. Hoinkes, H. Nahr and H. Wilsch, Surface Sci. $\underline{33}$ (1972) 516;
    G. Armand, J. Lapujoulade and Y. Lejay, Surface Sci. $\underline{63}$ (1977) 143.
    See also Ref. 10.
5. V. Bortolani, A. Franchini, F. Nizzoli, G. Santoro, G. Benedek, and
    V. Celli, Surface Sci. $\underline{129}$ (1983) 507.
6. V. Celli, G. Benedek, U. Harten, J.P. Toennies, R.B. Doak, and
    V. Bortolani, Surface Sci. $\underline{138}$ (1984). (See also Ref. 12.)
7. C. Lopez, F. Yndurain, and N. Garcia, Phys. Rev. $\underline{B18}$ (1978) 970.
8. N. Garcia, V. Celli and M. Nieto-Vesperinas, Optics Commun. $\underline{30}$
    (1979) 279; J. Shen and A.A. Maradudin, Phys. Rev. $\underline{B22}$ (1980)
    4234.
9. G. Brown, V. Celli, M. Coopersmith and M. Haller, Phys. Letters
    $\underline{90A}$ (1982) 361; Surface Sci. $\underline{129}$ (1983) 507.
10. A.C. Levi and H.G. Suhl, Surface Sci., $\underline{88}$ (1979) 133;
    H.D. Meyer, Surface Sci. $\underline{104}$ (1981) 117.
11. J. Lapujoulade, Surface Sci. $\underline{134}$ (1983) L139.
12. V. Bortolani, A. Franchini, N. Garcia, F. Nizzoli, and G. Santoro,
    Phys. Rev. $\underline{B28}$ (1983) 7358.

THE SOFT POTENTIAL IN RESONANT ATOM-SURFACE SCATTERING

J.R. Manson and J.G. Mantovani
Department of Physics and Astronomy
Clemson University
Clemson, SC 29631
U.S.A.

and

G. Armand
Service de Physique des Atomes et des Surfaces
Centre d'Etudes Nucléaires de Saclay
91191 Gif-sur-Yvette CEDEX
FRANCE

ABSTRACT

We consider several aspects of the theory of resonant scattering in low energy atom-surface systems using soft potentials. A method is presented to account for thermal effects in selective adsorption resonances by taking the thermal average of the t-matrix in a decoupling approximation, and comparison is made with experiment. Further calculations indicate that the specular intensity under resonant conditions has a temperature dependent structure which is quite different from that away from resonance. A consideration of the resonant diffraction of 63 meV He atoms by graphite shows that the effects of a soft potential can be as important as thermal attenuation in the manner in which it is commonly introduced into the corrugated hard wall model. A treatment of threshold resonances shows that an emerging beam causes a resonance in all diffracted beams, but for potentials with a soft repulsive part these are too weak to be readily seen in an experiment.

1. INTRODUCTION

In this paper we would like to present the results of several theoretical investigations of atom-surface scattering at low energies using soft potentials. In particular we wish to discuss three different aspects of resonant scattering. By the adjective soft we mean an interaction potential in which the repulsive wall has a finite slope, as opposed to the corrugated hard wall model (CHW). The CHW is an

23

B. Pullman et al. (eds.), Dynamics on Surfaces, 23–36.
© *1984 by D. Reidel Publishing Company.*

attractive model particularly because of its ease of calculation, but it is clear that even though the real repulsive potential is very strong there is some penetration of the wavefunction into the surface and this has a pronounced effect on the relative diffraction intensities (1). A soft potential is particularly important for scattering by metals where the incoming atoms are reflected several atomic units in front of the surface by the exponential tail of the electron cloud. The need for a soft potential to explain the relative intensities of diffraction peaks from metals has been rather thoroughly discussed elsewhere (2,3). Here, we would like to confine our discussion to resonant processes and how they are affected by the softness of the potential. In Section 2 we consider the effects of thermal attenuation in bound state resonances (selective adsorption) and compare the results with He scattered by a stepped copper surface. Section 3 is a discussion of the selective adsorption resonances of He incident on graphite and we show that a soft repulsive wall can modify the resonance signatures as much or more than thermal effects as commonly introduced into the CHW. In Section 4 we consider the threshold resonance which occurs with the emergence of a previously evanescent diffraction beam. Using the example of H_2 scattered by a Cu(100) surface we show that such resonances will be difficult to observe in realistic situations.

2. THERMAL ATTENUATION AND RESONANT SCATTERING

The selective adsorption resonance occurs when a bound state (or adsorption level) becomes accessible to the incident particle. The opening up of this additional degenerate intermediate state gives rise to a characteristic resonance behavior in which the relative intensities of the backscattered diffracted particles can be strongly affected. Elastic calculations have been able to explain the appearance of minima and maxima in the scattered intensities (4,5). However, the neglect of thermal effects can lead to contradictions with experiment in the line shape of a resonance. Better agreement with experiment has been achieved using complex optical potentials (6), and more recently Hutchison (7) has proposed, using the CHW, a method of applying Debye-Waller factors to the scattering amplitudes to account for thermal effects. We discuss here a method for introducing thermal attenuation by use of Debye-Waller factors in the transition matrix formalism. This differs from the approach of Hutchison in that it is particularly well suited to handle the more realistic soft potentials.

We choose to calculate the diffraction intensities using a transition matrix formalism in which the potential V is divided into a distortion term U and a perturbation v, V = U + v. Then the intensity of a beam labelled by the surface reciprocal lattice vector $\underline{G}$ is given by

$$I_G = |\delta_{G,0} - \text{im } t_{Gi}/(\hbar^2\sqrt{k_{iz}k_{Gz}})|^2 \qquad\qquad 2.1$$

where k_{Gz} is the perpendicular wavevector and

$$t_{Gi} = v_{Gi} + \sum_{\ell} v_{G\ell} (E_i - E_\ell + i\epsilon)^{-1} t_{\ell i} \qquad\qquad 2.2$$

and v_{pq} are the matrix elements of v taken with respect to eigenstates of U. Near conditions for a resonance, singularities appear in the Green function of eq. (2.2), but these can be avoided by using projection methods in which eq. (2.2) is written as a pair of coupled integral equations, only one of which involves the resonant bound states. For the case of a single isolated bound state we have

$$t_{Gi} = h_{Gi} + h_{Gb} h_{bi} / (E_i - E_b - h_{bb}) \qquad\qquad 2.3$$

$$h_{pq} = v_{pq} + \sum_{\ell}' v_{p\ell} (E_i - E_\ell + i\epsilon)^{-1} h_{\ell q} \qquad\qquad 2.4$$

where Σ' means the summation does not include the resonant bound state. The h_{pq} are calculated by iteration and the diffracted beam intensities of (2.1) can be evaluated using (2.3) even under resonant conditions. These intensities may be written in the form

where

$$I_G = I_G^0 |1 - ib/(x-i)|^2 \qquad\qquad 2.5$$

and

$$x = [E_i - E_b - \mathrm{Re}(h_{bb})] / \mathrm{Im}(h_{bb}) \qquad\qquad 2.6$$

$$b = N_G h_{Gb} h_{bi} / [(\delta_{G,0} - iN_G h_{Gi}) \mathrm{Im}(h_{bb})] \qquad\qquad 2.7$$

where $N_G = m/(\hbar^2 \sqrt{k_{iz} k_{Gz}})$ is the density of states for perpendicular motion. In a plot of I_G versus incident angle or energy the half width of the resonance structure is $\Gamma = -\mathrm{Im}(h_{bb})$, but the resonance shape is mainly controlled by the value of b. For example if $|\mathrm{Re}(b)| \gg |\mathrm{Im}(b)|$ the line shape is a single minimum for $-2 < \mathrm{Re}(b) < 0$ and a maximum otherwise. Asymmetry in the form of a mixed maximum-minimum structure appears when $\mathrm{Im}(b)$ is non-negligible.

In an exact approach, the thermal average of t_{Gi} would be used to calculate the diffracted intensities of eq. (2.1). The major approximation in our approach is to decouple all of the scattering amplitudes in eq. (2.3) and thermally average each one separately. Thus we follow Hutchison's method of approximating the thermal average by scaling each amplitude by a Debye-Waller factor. The amplitudes h_{Gi}, h_{Gb} and h_{bi} in (2.3) are thermally attenuated by multiplying them by the factors $\exp(-W_{Gi})$, $\exp(-W_{Gb})$, and $\exp(-W_{bi})$ respectively. The exponent W is given by

$$W_{pq} = \tfrac{1}{2}(\Delta k_z)^2 <u_z^2> \qquad\qquad 2.8$$

where $\Delta k_z = k'_{pz} + k'_{qz}$ is the total change in perpendicular momentum calculated in the "Beeby approximation", i.e.

$$(k'_{pz})^2 = k_{pz}^2 + 2mD/\hbar^2 \qquad\qquad 2.9$$

The quantity $<u_z^2>$ is the effective average square displacement of a surface atom in the normal direction, and may involve averaging the motions of several atoms.

The scaling method must be modified when applied to the amplitude h_{bb}. From the optical theorem, $t-t^+ = t^+(G^+-G^-)t$ (which applies equally well to the h operator) the imaginary part of h_{bb} may be related to the probability of a transition from the bound state to the continuum. Thus we have

$$Im(h_{bb}) = -(\hbar/2)w_b \qquad\qquad 2.10$$

where w_b is the total transition rate. Dividing w_b by the flux of particles in the bound state, $1/\tau_b$, gives the corresponding probability P_b. The time τ_b is a semiclassical quantity which can be estimated in several ways. For the potential described below we find that the approach useful in nuclear resonance theory, i.e. to relate τ_b to the inverse of the level spacing of the bound states, gives values which agree fairly well with the period of a classical particle moving in the well.

The probability that a bound particle will remain bound is $1-P_b$, and thermal attenuation is expected to decrease this value. Thus the expression that we use is

$$<Im(h_{bb})> = -\hbar[1-(1-P_b)\exp(-2W_{bb})]/(2\tau_b) \qquad\qquad 2.11$$

with $P_b = 2\tau_b Im(h_{bb})/\hbar$. The thermal attenuation factor is $\exp(-2W_{bb})$ instead of $\exp(-W_{bb})$ since $(1-P_b)$ is a probability and not a probability amplitude. Thermal effects on $Re(h_{bb})$ are ignored since it appears only in eq. (2.6) as a shift in position of the resonance, and it is usually small.

The method of treating thermal attenuation as presented here can now be contrasted with the approach of Hutchison. The energy half-width Γ of the resonance is given by Hutchison as

$$\Gamma = [1-|S(N,N)|] / [|S(N,N)|d\delta_N/dE]$$

where $S(N,N)$ is essentially a scattering amplitude and is scaled by a single Debye-Waller factor $\exp(-W_{NN})$, and δ_N is the phase gained after one period of oscillation in the potential well. In the present paper, Γ is given by $<Im(h_{bb})>$ calculated from eq. (2.11). Clearly both methods show that Γ will increase with temperature. However, the two methods differ in the manner in which thermal effects are introduced into the calculation of Γ.

We have carried out calculations for He at 21 meV scattering from the (113) face of copper, a system for which good experimental data is available (8). As a potential we use a corrugated Morse function

$$V = D[\exp(-2\kappa(z-\phi))/v_o - 2\exp(-\kappa z)] \qquad\qquad 2.12$$

where $\kappa = 1.05\text{Å}^{-1}$, $D = 6.35$ meV, and v_o is the surface average of $\exp(2\kappa\phi)$. The function $\phi(x)$ is the one dimensional corrugation function appropriate to this surface

$$\phi(x) = da_x \cos(2\pi x/a_x) \qquad\qquad 2.13$$

where d = 0.016 and a_x = 4.227Å is the lattice constant for the steps. The distorted potential U is taken as the Morse function corresponding to the surface average of (2.12). For evaluating the Debye-Waller factors the value of $<u_z^2>$ was taken from the data of Lapujoulade et al. (9).

Figure 1 shows the three bound states which are observed experimentally in the specular beam when there is a resonance with the (10) evanescent diffraction channel. The purely elastic calculations show narrow maxima rising almost to unity, and this is easily understood since for all three resonances Re(b) < -2 and |Im(b)| << |Re(b)|. Introducing the effects of thermal attenuation changes the signature of each resonance into a broader miniumum. Inclusion of effects of angular dispersion of the incident beam (not shown in Fig. 1) makes the calculated curves substantially less deep and slightly broader, although still not as broad as the experimental curves. The calculated resonances for n = 1,2 appear at incident angles slightly different from experiment because the bound states of the Morse potential differ from those of the true potential.

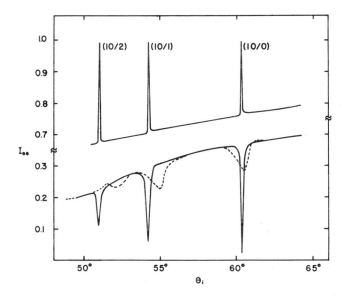

Figure 1. Specular intensity vs. incident angle showing the resonance line shape for He(21meV)-Cu(113). The solid line showing maxima is the elastic calculation, and the solid line with minima is the present theoretical calculation at a temperature of 70 K. Experimental results taken at T = 70 K are shown by the dashed line.

An interesting prediction comes from this work if we examine the temperature dependence of a diffracted beam under exact resonance conditions (i.e. for x=0). In a situation such as that calculated above where $|Re(b)| > |Im(b)|$ the intensity is of the form

$$I_G(T) \cong I_G^0 exp[-2W_{Gi}(T)] [1 + Re\, b(T)]^2 \qquad\qquad 2.14$$

where I_G^0 is independent of temperature T. For $Re(b) < -1$ at $T = 0$ and increasing with T, eq. (2.14) shows that $I_G(T)$ will decrease and pass through a minimum at some temperature T_{min} corresponding to $Re(b) \cong -1$. As $Re\, b(T)$ continues to increase toward zero for $T > T_{min}$ the sum $1 + Re\, (b(T))$ will increase from its minimum value, but this increasing factor is opposed by the term $exp(-2\, W_{Gi})$ leading to the possibility of a relative maximum at a temperature $T_{max} > T_{min}$. Thus for those resonances, which in a purely elastic calculation have $Re(b) < -1$ and $|Re(b)| >> |Im(b)|$, the intensity measured as a function of temperature will show structure containing both a minimum and a maximum.

The curve marked "a" in Fig. 2 shows the temperature dependence of the minimum point of the (10/0) resonance discussed in Fig. 1. A distinct minimum and a broad maximum are observed. Assuming no angular spread in the incident beam and sufficiently high temperatures, the intensity can be cast in the form

$$I = I^\circ e^{-aT} |1+b(t)|^2 \qquad\qquad 2.15$$

where $aT = 2W_{Gi}$ and

$$b(T) = b(0)\, P_b exp(-\alpha T)/[1-(1-P_b)exp(-\gamma T)] \qquad\qquad 2.16$$

with $\alpha T = W_{Gb} + W_{bi} - W_{Gi}$ and $\gamma T = 2W$. For the case at hand α is about an order of magnitude greater than γ. When $|Re(b)| >> |Im(b)|$ the minimum intensity occurs essentially when $b(T) \cong -1$ at the temperature

$$T_{min} \cong (1 + Re[b(0)])/[\gamma(1 - P_b^{-1})] \qquad\qquad 2.17$$

A good approximation for T_{max} is

$$T_{max} = P_b[-Re[b(0)]-2+\sqrt{-8Re[b(0)]\gamma/aP_b}]/(2\gamma) \qquad\qquad 2.18$$

These simple forms for T_{min} and T_{max} indicate that a measurement of the temperature dependent structure of a resonance can give direct information about the assumed behavior of a particle in the bound state.

Fig. 2 also shows the effect of a simple inclusion of incident angular dispersion on the curve. These calculations indicate that the angular dispersion must be less than 0.1° in order for the minimum structure to be clearly observed. However, even if the angular spread is 0.2° or larger there is still a distinct and characteristic difference between the temperature dependence of the intensity at resonance and away from resonance. Fig. 2 shows that for a large angular spread, the intensity, while monotonically decreasing with T, has a positive curvature. The experimental Debye-Waller curves which are measured away from resonance give straight lines or negative curvatures.

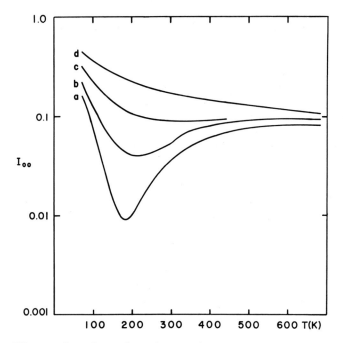

Figure 2. Specular intensity vs. temperature under resonance conditions for the (10/0) resonance of Fig. 1. (a) present theory without angular dispersion, (b) with 0.05° angular dispersion, (c) with 0.1° angular dispersion, and (d) with 0.2° angular dispersion.

3. THE SOFT WALL IN He-GRAPHITE DIFFRACTION

The above section on thermal attenuation in softwall scattering was to a large extent an extension of the earlier work of Hutchison and Celli. Their work is based on the CHW model and was quite successful in explaining the resonance structure for a low energy (21meV) He beam scattered by a graphite surface (7). However, it was later shown that this theory compared rather poorly for the same system when the experiments were done at high energy, 63.8 meV (10). In this section we reconsider the He-Graphite system at the higher energy and show that the effects of using a soft potential, even in a purely elastic calculation, can be of equal or greater importance as the thermal attenuation in the form introduced by Hutchison and Celli.

Again, as in Section 1 we use a transition matrix formalism in which the t-matrix equation of (2.2) is solved by iteration until convergence is obtained. As the potential we again use the corrugated Morse function of eq. (2.12) with parameters $\kappa = 1.31\text{Å}^{-1}$, $D = 15.47$ meV and a two dimensional corrugation function. In the rectangular coordinates x and y it appears as

$$\phi(x,y) = a\beta_{10} \left[\cos\frac{2\pi x}{a} + 2 \cos\frac{\pi x}{a} \cos\frac{2\pi y}{b}\right]$$

$$+ a\beta_{12} \left[\cos\frac{4\pi y}{b} + 2 \cos\frac{3\pi x}{a} \cos\frac{2\pi y}{b}\right]$$

$$+ a\beta_{20} \left[\cos\frac{4\pi x}{a} + 2 \cos\frac{2\pi x}{a} \cos\frac{4\pi y}{b}\right] \qquad (3.1)$$

with $b = 2.465$ Å, $a = b \cos 30°$, $\beta_{10} = -0.013$, $\beta_{12} = 0.01$ and $\beta_{20} = 0.0$.
    The curve marked "A" in Figure 3 shows the experimental data of
Cantini et al. for the specular beam in the case in which the (10)
evanescent beam is in resonance with the lowest four bound states of
the potential. The incident He wavevector is 11.05 Å^{-1} corresponding
to 63.8 meV. Curve "B" of Fig. 3 is the present elastic calculation
using the Morse potential and the corrugation function of (3.1). Curve
"C" is the elastic calculation corrected for the angular spread of the
incident beam.

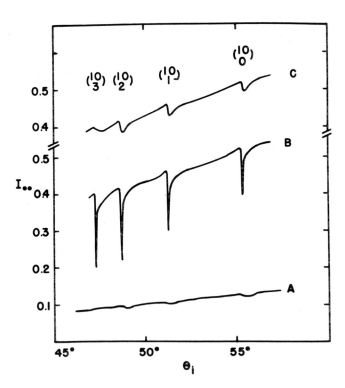

Figure 3. Specular intensity for 63.8 meV He scattered by
graphite. (A) experimental data from ref. (10); (B) present
elastic calculation; (C) present calculation accounting for
angular dispersion of the beam.

Figure 4 gives for comparison the results for the same system reported by Cantini et al. using the CHW, and also the CHW with the Hutchison correction. Clearly, neither the elastic calculation of Fig. 3 nor the Hutchison theory with thermal attenuation agree very well with the very shallow observed resonances. In fact, the best agreement is for the case of the purely elastic softwall calculation corrected for the angular dispersion of the beam. The interesting point is, however, that in comparison with the CHW calculation the present elastic calculation and the Hutchison theory give surprisingly similar results. The Hutchison correction seems to give slightly broader resonance structures, but this is in part an illusion due to the fact that the vertical scale is compressed by the overall Debye–Waller factor which multiplies the curve. This points out that as compared to a CHW calculation, a soft potential can cause changes in resonance intensities which are just as important as the thermal attenuation introduced by the Hutchison method.

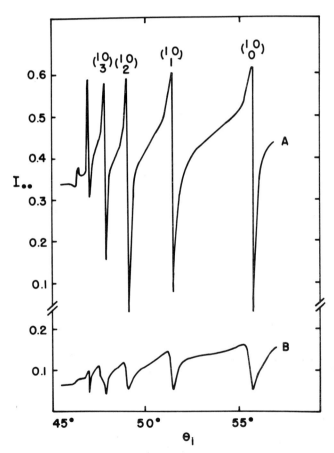

Figure 4. Same system as Fig. 3 (A) CHW calculation, and (B) CHW calculation with thermal attenuation, both from ref. (10).

4.   THRESHOLD RESONANCES

In this section we wish to briefly discuss the effects of a soft
potential on a completely different type of resonance, that which
occurs when an evanescent diffraction beam is just emerging from the
surface.   These surface threshold resonances, which have been discussed
for some time in the context of electron scattering (11,12), have also
been suggested as a possible tool for surface investigation with atom
scattering (13,14).
      The threshold resonance is quite different in character from the
selective adsorption resonances discussed above.   As explained in more
detail below, when a diffraction beam emerges (or disappears) its
intensity as a function of incident beam angle grows from zero with an
infinite slope.   This "sudden" appearance of a new beam causes a
rearrangement of the intensities of the other diffracted beams and they
too will in general have a singularity in slope at the angle of emergence.   A typical example of a calculation is shown in Figure 5 for a
model of H_2 scattered at a Cu(100) surface.   On the basis of CHW

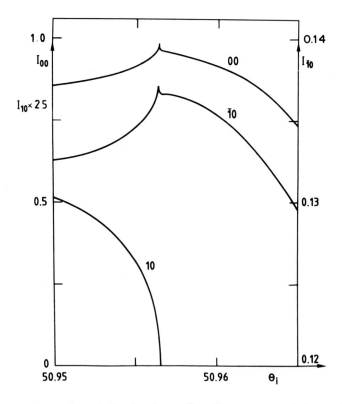

Figure 5.   Calculations for H_2 scattered by a Cu(100) surface
showing the emergence of the (Ī0) diffracted beam.

calculations it has been suggested that these resonances could be useful in obtaining information on surface structure and particularly on the surface corrugation (13,14). Our calculations using more realistic soft potentials, although confirming this point of view, indicate that the threshold resonance is very sensitive to the slope of the repulsive potential and for all but the very stiffest surfaces will be too weak to be seen in an experiment. Indeed, this conclusion is strengthened by the fact that very few threshold resonances have been clearly identified in the experimental atom-surface scattering litera-ture.

The origin of the threshold resonance is most easily demonstrated as follows. If $\underline{N}$ is the surface reciprocal lattice vector of the emergent beam and I_N its intensity, the slope of interest is

$$\partial I_N/\partial\theta_i = (\partial I_N/\partial k_{Nz})(\partial k_{Nz}/\partial\theta_i) + \ldots \qquad 4.1$$

where the other terms in (4.1) are ignored because they are of no consequence to the discussion. Now since $k_{Nz}^2 = k_i^2 - (\underline{K}_o + \underline{N})^2$ with $\underline{K}_o = \hat{e}\, k_i \sin\sigma$ and $\hat{e}$ a unit vector, we obtain immediately

$$\partial k_{Nz}/\partial\theta_i = -\hat{e}\cdot(\hat{e}\, k_i \sin\theta + \underline{G})\, k_i \cos\theta_i/k_{Nz}, \qquad 4.2$$

the important point being that at emergence when $k_{Nz} \to 0$ this deriva-tive blows up as $1/k_{Nz}$. Thus eq. (4.1) shows that $\partial I_N/\partial\theta_i$ will diverge unless I_N goes to zero at least as fast as k_{Nz}^2 for small k_{Nz}. That this is not the case can be seen be looking at the intensity as a function of the transition matrix (2.1) and examining the form of the matrix elements. From eq. (2.1) we have that the intensity of the emerging beam is

$$I_N = |m\, t_{Ni}/\hbar^2|^2/k_{iz}\, k_{Nz} \qquad 4.3$$

and since $0 \leq I_N \leq 1$ we see that the transition matrix must go to zero at least as fast as $\sqrt{k_{Nz}}$ for small k_{Nz}. However, in general we can argue that $t_{pi} \propto p$ for small p. This is because, from eq. (2.2), t_{pi} will have the same limiting behavior as the matrix element v_{pq} except in highly unusual circumstances, and for all potentials to date which admit to an analytical calculation we find $v_{pq} \to p$ as $p \to 0$. (This includes the corrugated Morse and exponential repulsive potentials as well as the CHW and the simple step potential.) Thus $I_N \to k_{Nz}$ for small k_{Nz} and eqs. (3.2) and (3.1) show that the slope of the intensity curve is divergent. This argument would fail for a case in which the matrix element v_{pq} behaved as $p^{3/2}$ for small p but this appears highly unlikely. In fact, even if one considers a potential with an activa-tion barrier in front of the repulsive potential it is rather straight-forward to show that the same linear dependence on p is obtained.

The above arguments justify the statement that the emerging beam appears with an infinite slope in the plot of intensity versus θ_i. Now we need to discuss the other diffracted beams. If we examine $\partial I_G/\partial\theta_i$ where I_G is the intensity of any of the diffracted beams an application of the chain rule for differentiation will produce as one of the terms

$$(\partial I_G/\partial k_{Nz})(\partial k_{Nz}/\partial \theta_i) \qquad\qquad 4.4$$

just as in eq. (4.1). If we look closely at I_G of eq. (2.1) and consider it as an infinite summation of terms coming from the expansion of t_{Gi} in the perturbation series, some of these terms for small k_N will be linear in k_{Nz} which will again lead to divergence. Such terms will come from the δ-function (or pole) contribution in the t-matrix equation (2.2), for example we would have

$$t_{Gi} = v_{Gi} - i\pi m v_{GN} \, v_{Ni}/\hbar^2 k_{Nz} + \dots \qquad\qquad 4.5$$

and the second term on the right will have exactly the behavior in k_{Nz} which will lead to a divergence. Thus it is clear that if there is an infinite slope in the emerging diffraction beam, this divergence will also appear in the other diffracted beams except in the unlikely case of a cancellation of all of the divergent terms generated by the perturbation expansion.

It now remains to consider the effects of a soft potential on these resonances. We illustrate this with a model calculation. Since the intensity of the emerging beam at resonance is small we are justified in using the distorted wave Born approximation, $t_{Ni} \cong v_{Ni}$. Choosing the exponential repulsive potential with a corrugation (i.e. the repulsive part of eq. (2.12)) the relevant matrix elements in the limit of small k_{Nz} are

$$v_{Ni} = \lambda_{NO}\sqrt{\pi} \; p_N \, p_i^2 \, \sqrt{p_i} \, \sinh \pi \, p_i \; / \; (\cosh(\pi p_i)-1) \qquad\qquad 4.6$$

where $p_G = k_G/(2\kappa)$ and λ_{NO} is essentially the Fourier transform of the potential in directions parallel to the surface. Eq. (4.1) for the slope of the emerging intensity becomes

$$\partial I_N/\partial \theta_i = -\pi^3 |\lambda_{No}^2|^2 k_{iz}^4 \, \hat{e}\cdot(\hat{e}k_i\sin\theta_i + \underline{N})k_i\cos\theta_i\exp(-\pi k_{iz}/\kappa)/(4\kappa^5 k_{Nz}) \qquad 4.7$$

where we have made the further simplifying assumption that the arguments of the hyperbolic functions are large. For a soft potential (small κ). The behavior of this expression is dominated by the factor $\kappa^{-5}\exp(-\pi k_{iz}/\kappa)$ which is small. Thus for soft potentials the $1/k_{Nz}$ divergence will be very weak and will appear over a very small angular spread.

This is precisely the sort of behavior exhibited by the typical calculation shown in Figure 5 where we consider H_2 scattered at a Cu(100) face. This is an exact calculation using the method of iteration of the transition matrix equation (2.2) until convergence is obtained. The beam is incident in the [110] direction ($\phi_i = 45°$) with a wave vector $k_i = 8.6$ Å^{-1}. The potential parameters are $D = 21.6$ meV, $2\kappa = 1.94$ Å^{-1} and the lattice spacing is 2.55 Å. The corrugation function is

$$\phi(x,y) = (da/2)[\cos(2\pi x/a) + \cos(2\pi y/a)] \qquad\qquad 4.8$$

with d = 0.036. The (10) diffraction beam is emergent at an incident
angle of slightly less than 50.96°. The steep slope of this curve and
the singularity it causes in the slope of the ($\bar{1}$0) and (00) beams is
evident. However, what is important is the scale. The range of angles
over which the resonance occurs is so small that it could not be
observed with currently available experimental precision. Numerous
other calculations tend to confirm the opinion that with a softwall
repulsive potential the threshold resonances will be too weak for easy
observation (15,16).

5. CONCLUSIONS

We have discussed in this paper three features of resonant scattering
using soft potentials and have made the comparison in each case with
the corresponding calculations using a hard corrugated wall. Section 2
considers the addition of thermal attenuation in selective adsorption
resonances in a manner similar to that introduced by Hutchison and
Celli for the CHW. These effects are able to change narrow elastic
intensity maxima into broader minima, but the signatures are still not
broad enough to agree well with the experimental data for He/Cu(113).
An interesting prediction of this work is that the temperature
dependence of the diffraction intensity at resonance can have a
structure which includes both a maximum and a minimum. This structure
is quite different from the behavior away from resonance and if
observed could give information about the thermal attenuation of
particles trapped in the bound state.

Section 3 is an investigation of selective adsorption resonances
for He scattered by graphite and we find that for higher incident
energies the effects of a soft potential can be as great as those of
thermal attenuation as introduced by Hutchison in the CHW. This
implies that the softness of the repulsion cannot be ignored especially
at energies where the penetration of the particle wave function into
the surface becomes appreciable.

The fourth section is a discussion of threshold resonances generated by soft potentials. We find that the threshold resonance is
characterized by a divergence in the slope of the intensity versus θ_i
curve, not only for the emerging beam, but for the other diffracted
peaks as well. However, we find that the strength of these resonances
is strongly dependent on the slope of the potential and for realistic
soft potentials they will be difficult to observe.

ACKNOWLEDGEMENTS

We would like to thank J. Lapujoulade, B. Salanon and D. Gorse for many
stimulating discussions and C. Manus for his continued support and
encouragement. Two of us (JRM and JGM) would like to thank the Service
de Physique des Atomes et des Surfaces of the C.E.N. Saclay for its
kind hospitality. This work was supported by a NATO Research Grant.

REFERENCES

1. • G. Armand and J.R. Manson: Phys. Rev. Lett. 43 (1979) 1839.

2. J. Perreau and J. Lapujoulade: Surf. Sci. 119 (1982) L292.

3. N. Garcia, J.A. Barker, and I.P. Batra: J. Electron. Spectrosc. 29 (1982) L292.

4. H. Chow and E.D. Thompson: Surf. Sci. 54 (1976) 269; K.L. Wolfe and J.H. Weare: Phys. Rev. Lett. 41 (1978) 1663; N. Garcia, V. Celli and F.O. Goodman: Phys. Rev. B 19 (1979) 634; V. Celli, N. Garcia and J. Hutchison: Surf. Sci. 87 (1979) 112.

5. H. Chow and E.D. Thompson: Surf. Sci. 59 (1976) 225; H. Chow: Surf. Sci. 62 (1977) 487.

6. H. Chow and E.D. Thompson: Surf. Sci. 82 (1979) 1; K.L. Wolfe and J.H. Weare: Surf. Sci. 94 (1980) 581.

7. J. Hutchison: Phys. Rev. B 22 (1980) 5671; J. Hutchison, V. Celli, N.R. Hill and M. Haller: Prog. Astronaut. and Aeronautics 74, S.S. Fisher, ed., (1981) 129; V. Celli and D. Evans: Israel J. Chem. 22 (1982) 289.

8. J. Perreau and J. Lapujoulade: Surf. Sci. 122 (1982) 341.

9. J. Lapujoulade, J. Perreau and A. Kara: Surf. Sci. 129 (1983) 59.

10. P. Cantini, S. Terreni and C. Salvo: Surf. Sci. 109 (1981) L491.

11. E.G. McRae: Surf. Sci. 25 (1971) 491.

12. For a recent review see J.C. LeBossé, J. Lopez, C. Gaubert, Y. Gauthier, and R. Baudoing: J. Phys. C15 (1982) 6087.

13. N. Cabrera and J. Solana, in Proc. Intern. School of Phys. "Enrico Fermi", Ed. F.O. Goodman (Compositori, Bologna, 1974) p. 530.

14. N. Garcia and W.A. Schlup, Surf. Sci. 122 (1982) L567.

15. G. Armand: J. Physique 41 (1980) 1475.

16. G. Armand, J. Lapujoulade and Y. Lejay: in Proc. 4th Intern. Conf. on Solid Surfaces, Cannes, 1980, Eds. D.A. Degras and M. Costa [Suppl. to Le Vide, les Couches Minces (1980), p. 587.

THE DIFFRACTION OF He, Ne AND H_2 FROM COPPER SURFACES

J. Lapujoulade, B. Salanon and D. Gorse

Service de Physique des Atomes et des Surfaces
Centre d'Etudes Nucléaires de Saclay
91191 Gif-sur-Yvette Cedex, France

ABSTRACT

We present an interpretation of diffraction data for He, Ne and H_2 from copper surfaces in term of model potentials. These models are in a good agreement with theoretical calculations. These results prove that atomic (or molecular) beam diffraction can be readily used as a probe for quantitative surface structure determination.

1. INTRODUCTION

Low energy gas beam scattering (especially helium) has been emerging in the last few years as a new and powerful diagnosis in surface physics. Its main advantages are :
 i) no destructive effect on the surface even in the presence of very weakly bonded adsorbed species.
ii) no penetration deeper than the topmost atomic layer of the surface.
    But it remained to prove that the observed diffraction pattern can be unambiguously related to the surface atomic structure with a tractable model. This means that it is possible to build an interaction potential from this structure and then to calculate the diffraction pattern. Many progresses have been made recently and the purpose of this paper is to show it is now possible to get quantitative predictions about diffraction patterns.
    We shall first describe experimental data obtained for the diffraction of He, Ne and H_2 on various crystal faces of copper. Then we shall present the model potential which has been used and discuss how it fits the experimental data.

2. EXPERIMENTAL DATA

The experimental apparatus has been described elsewhere (1,2). The crystal is placed in an ultra-high vacuum chamber (base pressure $\sim 7.10^{-11}$ torr). It can be cooled by liquid nitrogen flow and heated by electron bombardment on its back side. There are facilities for ion

B. Pullman et al. (eds.), Dynamics on Surfaces, 37–45.
© 1984 by D. Reidel Publishing Company.

bombardment cleaning and Auger spectroscopy. The nozzle beam
(Campargue type (3)) delivers atoms (or molecules) of energy adjustable
from 20 to 250 meV. The velocity distribution is about $\Delta V/V \sim 0.06$
and the angular aperture is 0.2°. Scattered atoms (or molecules) are
received into a stagnation ionization detector (2). The incident beam
is chopped at a low frequency (2 Hz) which allows a lock-in amplification of the ion signal. Then the flux sensitivity is about I/I_0
$\sim 10^{-4}$ (I_0 is the incident beam flux $\sim 10^{16}$ atoms/cm².s). Both incidence angle θ_i and in-plane scattering angle θ_r can be continuously
adjusted.
    We chose to study copper surfaces because :
  i) they are quite easily cleaned by ion bombardment and thermal
     annealing,
 ii) they are not very reactive to residual gas contaminants,
iii) they are known not to reconstruct thus their structure is very
     well known from bulk crystallography.
    We select the orientations (100), (110), (112), (115), (117) which
display increasing atomic roughness. The structure of the last four
is indicated in fig. 1. Their atomic corrugation is parallel to the

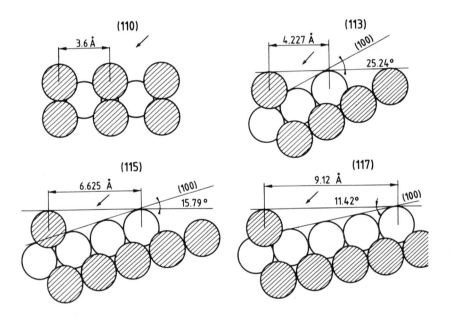

Figure 1. Structure of Cu (110), (113), (115), (117). The figure
presents a section of the crystal by a plane perpendicular to a
|110| direction.

|110| direction. In the |100| direction they are as smooth as the (100)
face. Thus if the incidence plane is chosen to be perpendicular to
the |110| direction, in-plane scattering is expected to be largely
dominant which simplifies greatly both experiments and calculations.

We have studied the diffraction of He on Cu (100)[4], (110)[5], (113)[6], (115)[6], (117)[1] for E$_i$ = 21 and 63 meV, He on Cu (110)[7] for E$_i$ = 77, 125 and 241 meV, Ne on Cu (110)[8], (117)[9] for E$_i$ = 63 meV and H$_2$ on Cu (100)[10], (110)[11], (115)[11] for E$_i$ = 77 meV. In every cases the diffraction pattern has been recorded for various incidence angles. A typical result is shown in fig. 2. Such diffraction patterns are not suitable for direct comparison to calculation. Two kinds of corrections are needed :

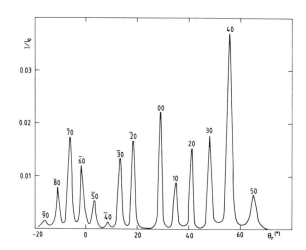

Figure 2. Typical diffraction pattern for He-Cu (115). The ratio of the scattered to incident flux is plotted as function of the in-plane scattering angle θ_r.
Incidence angle θ_i = 28° . Beam energy E$_i$ = 63 meV.

    i) Peak amplitudes must be corrected for instrumental broadening effect : angular apertures, velocity distribution along a procedure previously published (1).
    ii) The elastic peaks are thermally attenuated by a Debye-Waller like factor (4). In order to take this effect into account the thermal dependance of each individual peak has been experimentally determined. This allows an extrapolation to 0 K (12) which gives the peak intensity for an immobile surface lattice.
    Except for helium on the smoothest surfaces Cu(100) and (110) resonances with bound states were readily observed (8, 13). A typical curve is shown in fig. 3. The bound state levels which are deduced are listed on table 1. They are independent of the surface orientation within experimental accuracy.

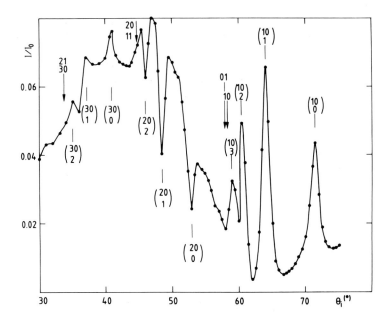

Figure 3. Selective adsorption data for He–Cu (115). The intensity of
the specular peak is plotted as a function of the incidence angle θ_i.
Beam energy E_i = 21 meV.
Resonance label : top label   : resonnant $\vec{G}$ vector index.
                 bottom label : bound state number.

Table 1 – Bound state levels of the potential

		He	Ne	H_2
ε_0	(meV)	4.1	10.8	17.0
ε_1	"	2.2	8.34	10.7
ε_2	"	0.95	6.74	5.7
ε_3	"	0.35	5.36	3.1
ε_4	"		4.16	1.75
ε_5	"		3.17	0.6
ε_6	"		2.47	

3. INTERACTION POTENTIAL

3.1. Theoretical approaches

The gas surface interaction potential is attractive at large distance from the surface due to Van der Waals dispersion forces. Its

asymptotic variation is proportionnal to z^{-3}. This long range interaction smoothens the atomic structure of the surface so that the
attractive part is expected to be only very weakly corrugated. At
short distances the interaction becomes strongly repulsive due to
the electron clouds repulsion. As the electron density of the solid
falls out about exponentially the repulsive part is expected to do
so. This repulsive part is strongly affected by the atomic structure of the surface and its corrugation is responsible for the diffraction phenomena.

There has been recently much effort to calculate this potential
for the helium case. Esbjerg and Norskov (14) using the effective
medium approach have shown that this potential is proportionnal to
the unperturbed local electronic density of the solid $n_o(\vec{r})$:

$$V(\vec{r}) = \bar{\alpha}_{eff} \, n_o(\vec{r})$$

where $\bar{\alpha}_{eff}$ = 255 eV.a$_o^3$ (15).

Thus the isopotential is an electronic isodensity e.g. for E_i = 21 meV
one gets n_o = 8.10^{-5} a.u.

This approximation is only valid at small distances i.e. in the
repulsive region. Then the problem for calculating $V(\vec{r})$ is subject to
the knowledge of $n_o(\vec{r})$. There is presently no self-consistent calculation of $n_o(\vec{r})$ for copper. But it can be approximated, as suggested
first by Haneman and Haydock (16), by a superposition of atomic
densities which can be found in the Clementi and Roetti tables (17).

On the other hand Harris and Liebsch (18) have proposed a calculation of the repulsive part of the potential in which the electronic
properties of the solid are self-consistently taken into account. An
adjustable parameter scales the corrugation. Exchange and correlation
terms are not included in this calculation thus an attractive part
has to be superimposed which is not free from some arbitriness (19).
This calculation is presently only available for He-Cu (110) while
the Esbjerg and Norskov approach can be used in principle for any
structure.

There are no such theoretical developments available for Ne
and H$_2$.

3.2. Empirical form

In any cases it is necessary in order to calculate the diffraction
pattern to assume an analytical from for the potential (with or
without a strong theoretical support). It is useful to expand the
potential in a Fournier series parallel to the surface

$$V(\vec{R},z) = V_o(z) + \sum_{\vec{G} \neq 0} V_{\vec{G}}(z) \; e^{i \, \vec{G}.\vec{R}}$$

One of the most satisfactory form for $V_o(z)$ is the so-called "shifted
Morse hybrid" (SMH) potential (20) :

$$V_o(z) = D_o \ e^{-2\chi_o z} - 2 \ e^{-\chi_o z} - \Delta \qquad z < z_p$$

$$= - \frac{C_3}{(z - z_o)^3} \qquad z > z_p$$

It depends upon four independant parameter D_o, χ_o, C_3, z_o and Δ. z_o and z_p are defined by the matching conditions in z_p. It has the correct dependance in z^{-3} at large distances and behaves exponentially at short distances in agreement with the theoretical predictions. The well depth is $D = D_o (1 + \Delta)$.

For the other Fourier components we have chosen a form which is in agreement with the Harris and Liebsch calculation :

$$V_{\vec{G}}(z) = D \ V_{\vec{G}} \ e^{-2\chi_{\vec{G}} z}$$

The $V_{\vec{G}}$ and $\chi_{\vec{G}}$ have to be adjusted by fitting the peak intensities to the experimental results. The proper variation of the corrugation when the energy increases is particularly sensitive to a suitable choice of the $\chi_{\vec{G}}$'s. The dependance of $\chi_{\vec{G}}$ upon $\vec{G}$ has been explicitly calculated by Harris and Liebsch for the case of He-Cu (110).

4. FIT WITH THIS EXPERIMENTAL DATA

The parameters of the laterally averaged component $V_o(z)$ have been adjusted in order be get the correct values for the bound states which are experimentally determined from the selective adsorption data. The parameters so-adjusted are listed in table 2.

Table 2 - Parameters of the shifted Morse hybrid potential.

	He	Ne	H$_2$
D_o (meV)	5.34	6.34	22.2
Δ	0.13	0.93	- 0.07
D (meV)	6.04	12.4	20.6
χ_o (Å^{-1})	1.0	2.04	0.79
z_p Å	2.65	0.39	1.23
z_o Å	- 3.42	- 4.47	- 3.13
C_3 (meVÅ^3)	320	1200	1000

The other Fourier components are then adjusted to give the correct intensities of diffraction peaks. These intensities are calculated using the iterative T-matrix procedure developped by Armand and Manson (22). When convergent this method gives an exact solution. In order to obtain the matrix elements in an analytic form the SMH potential has been there replaced by a simple Morse potential. This is expected to have a negligible influence on the scattering intensities except in the vicinity of resonances. We found that assuming $\chi_{\vec{G}}$ = 1.5 χ_0 V G we get a good fit while the calculations remains easy. This potential has been called the "Modified corrugated Morse potential" (MCMP) (7).

For the case of He-Cu (110)[7], (113)[6] and Ne (110)[8] we have obtained a very good fit for all the range of incidence parameters (θ_i and E$_i$). The complex values of the $V_{\vec{G}}$'s thus obtained are listed on table 3.

For the more corrugated surfaces Cu (115) and (117) the MCMP potential is no longer convenient since its z < 0 singularity brings about a wrong behaviour at too low an energy (7). Instead we have used the simple corrugated Morse potential (CMP) obtained with $\chi_G = \chi_0$ V $\vec{G}$. Of course its corrugation has not a correct dependancy upon energy so the fit with a set of $V_{\vec{G}}$'s is restricted to a single value of the incidence energy. The $V_{\vec{G}}$'s have to be changed when E$_i$ is changed. Nevertheless a good fit the experimental data is also obtained and the values of $V_{\vec{G}}$ are also listed in table 3 (6).

Table 3 - Fourier components of the potential.

System	Model poten-tial	Energy range (meV)	V$_{10}$	V$_{20}$	V$_{30}$	V$_{40}$
He-Cu (110)	MMCP	20-241	(0.0126 ;0.)	(0.0038;0.)	(0. ; 0.)	(0. ; 0.)
He-Cu (113)	MMCP	20-63	(0.028 ; -0.0004)	(0.001;0.)	"	"
He-Cu (115)	MCP	20	(0.1758 ; -0.0259)	(0.0052 ; 0.0053)	"	"
		63	(0.2141 ; -0.0252)	(0.0131 ; 0.0044)	"	"
He-Cu (117)	MCP	20	(0.1828 ; -0.0836)	(0.0593 ; 0.015)	(0.0116 ; 0.0002)	(0.0017 ; 0.0001)
Ne-Cu (110)	MMCP	63	(0.040 ; 0.)	(0.008 ;0)	(0. ; 0.)	(0. ; 0.)

Note that the absence of out-of-plane components for $V_{\vec{G}}$ is consistent with the diffraction data for He-Cu (100) which show that the diffraction peaks are smaller than 10^{-4} while the specular is equal to 1 within experimental accuracy.

Note that in all these calculations we have also considered χ_0 as an adjustable parameter since its value obtained from resonances data which is valid in the vicinity of the well may not hold for the whole repulsive port. Both values are the same for Helium but for neon there is slight but significant difference. (χ_0 = 19Å^{-1} instead of 2.04Å^{-1} in table 2).

For hydrogen the corrugation is much larger. Diffraction is readily observed on close packed surfaces like Cu (100). Moreover for Cu (110) and (115) resonances with bound states appear also with out of plane G vectors. Thus the out-of-plane corrugation cannot be neglected in this case. So in the absence of measurements of the out-of-plane peaks which are expected to be dominant in some case it was hopeless to try a determination of the potential.

5. DISCUSSION

The potential we have obtained for He-Cu (110) is in a good agreement with that calculated by Liebsch et al. (23). The comparison of all the Helium results with the Esbjerg and Norskoy formula using the best value of α given in ref. 15 i.e. α = 255 eV.a$_0^{-3}$ shows that the calculation gives always a too large corrugation. But we found that a good fit is obtained for every surface orientation by assuming = 600 eV.a^3.

This discrepancy can be explained either by the approximation made by Esbjerg and Norskov or to a deficiency of the superposition method for calculating the electron density of copper. However whatever the origin of this discrepancy the fact that a unique value of is able to fit all the copper results has to be emphasized since this proves that helium beam diffraction can be readily used as a quantitative method in order to analyse surface structures.

We have no theoretical results to be compared to our data for Ne and H_2. We can only qualitatively remark that the corrugation increases in the order He, Ne, H_2 which is also the order of the well depthes. This is obviously due to the increase of the polarizability which increases the attractive forces. Then the classical turning point goes deeper inside the metal towards a higher electron isodensity surface which is more corrugated.

One consequence of these results is that Ne and H_2 would be eventually much more sensitives probes that helium to surface structures.

6. ACKNOWLEDGEMENTS

We wish to thank very much those who have contribute to this work either for the experiment or the theory : G. Armand, F. Fabre, A. Kara, M. Lefort, Y. Le Crüer, Y. Lejay, J.R. Manson, E. Maurel, N. Papanicolaou and J. Perreau.

REFERENCES

1. J. Lapujoulade, Y. Lejay and N. Papanicolaou, Surface Science 90 (1979) 133.
2. J. Lapujoulade, Y. Le Crüer, M. Lefort, Y. Lejay and E. Maurel, Surface Science 118 (1982) 103.
3. R. Campargue, Thesis, Paris (1970), CEA Report R-4213 (1972).
4. J. Lapujoulade, Y. Lejay and G. Armand, Surface Science 95 (1980) 107.

5. J. Perreau and J. Lapujoulade, Surface Science letters 119 (1982) L292.
6. D. Gorse, B. Salanon, F. Fabre, A. Kara, J. Perreau, G. Armand and Lapujoulade, to be published.
7. B. Salanon, G. Armand, J. Perreau and J. Lapujoulade, Surface Science 127 (1983) 135.
8. B. Salanon, Le Journal de Physique (in press).
9. J. Lapujoulade, Y. Le Crüer, M. Lefort, Y. Lejay, E. Maurel and N. Papanicolaou, Le Journal de Physique Lettres 42 (1981) L463.
10. J. Lapujoulade, Y. Le Crüer, M. Lefort, Y. Lejay and E. Maurel, Surface Science letters 103 (1981) L85.
11. J. Lapujoulade and J. Perreau, Physica Scripta T4 (1983) 138.
12. J. Lapujoulade, J. Perreau and A. Kara, Surface Science 129 (1983) 59.
13. J. Perreau and J. Lapujoulade, Surface Science 129 (1983) 59.
14. N. Esbjerg and J.K. Norskov, Phys. Rev. Lett. 45 (1980) 807.
15. M. Marnninen, J.K. Norskov, M.J. Puska and C. Umrigar, Phys. Rev. B 29 (1984) 2314.
16. D. Haneman and R. Haydock, J. Vac. Sci. and Techn. 21 (1982) 330.
17. E. Clementi and C. Roetti, Atomic Data and Nuclear Data Tables 14 (1974) 177.
18. J. Harris and A. Liebsch, J. Phys. C 15 (1982) 2275.
19. P. Nordlander and J. Harris, J. Phys. C 17 (1984) 1141.
20. C. Schwartz, M.W. Cole and J. Pliva, Surface Science 75 (1981) 1.
21. J. Harris and A. Liebsch, Phys. Rev. Letters 49 (1982) 341.
22. G. Armand and J.R. Manson, Le Journal de Physique 44 (1983) 473.
23. A. Liebsch, J. Harris, B. Salanon and J. Lapujoulade, Surface Science 123 (1982) 338.

PROBING SURFACE VIBRATIONS BY MOLECULAR BEAMS: EXPERIMENT AND THEORY

Giorgio Benedek
Gruppo Nazionale di Struttura della Materia del CNR,
Dipartimento di Fisica dell´Universita, Via Celoria 16,
I-20133 Milano, Italy, and
Max-Planck-Institut für Festkörperforschung,
Heisenbergstrasse 1, 7000 Stuttgart 80, FRG

ABSTRACT

The direct measurement of the surface phonon dispersion relations by means of He scattering is opening new possibilities in the physical and chemical characterization of solid surfaces. The recent data so far collected are challenging our present theoretical knowledge of surface vibrations. The intrinsic changes of electronic structure and electron-phonon interaction occuring at the surface appear to have important effects on the surface dynamics. In ionic crystals the enhancement of surface ionic polarizabilities yields a softening of the zone-boundary surface phonons. In noble metals and refractory compounds (TiN(001)) the surface change in quadrupolar deformabilities associated with s-d hybridization gives rise to anomalies in the surface phonon branches. In layered semiconductors such as GaSe(001) the surface optical phonons are unexpectedly found much below the bottom of the bulk optical band.

1. INTRODUCTION

One century ago Lord Rayleigh discovered surface waves in semiinfinite elastic media (1). Since then the theory of surface waves has gradually evolved from the original geophysical realm to the recent impressive applications in signal processing devices, delay lines, etc.(2). These applications, restricted for technical reasons to the continuum limit, were often presented as a good reason for extending the theoretical studies to the dispersive range, where the lattice structure becomes important. However, the spectroscopy of surface phonons, despite their importance, appeared to be much more difficult than that of bulk phonons, since the conventional probes of bulk phonons - cold neutrons -are surface insensitive.

The real breakthrough in surface phonon spectroscopy was achieved less than four years ago. Toennies, Doak and Brusdeylins, by means of

47

B. Pullman et al. (eds.), Dynamics on Surfaces, 47–57.
© *1984 by D. Reidel Publishing Company.*

high-resolution measurements of He beam scattering, were able to obtain for the first time the dispersion relation of Rayleigh waves in different crystal surfaces (3-7). The early theoretical predictions (8-10) could then be verified thanks to the development of intense, highly monochromatic He-beam sources (11).

These measurements are now revealing aspects of surface dynamics that were unexpected of the basis of the underlying bulk dynamics. Many microscopic mechanisms which are forbidden or hindered in the bulk for symmetry reasons, are promoted by a symmetry reduction and manifest themselves at the surface. A typical example is that of surface phonon anomalies induced by electron-phonon coupling, which may eventually occur also in crystals with a regular bulk phonon dispersion. Thus surface dynamics, that paradoxically had its first development in applicative areas, is now rapidly gaining positions among the fundaments of solid state physics.

Most of the existing theories of surface lattice dynamics (12-19) have been conceived for ideal, unrelaxed surfaces. There the dynamical model for a crystal slab is directly derived from that of the bulk with cyclic boundary conditions by cutting the bonds across the boundary. The change of atomic equilibrium positions at the surface (elastic relaxation), of electron density of states (electronic relaxation), and the consequent change in electron-phonon interaction are not taken into account, or, at best, not in a selfconsistent way.

In view of the presently available atom-scattering experiments, we can say that the modern theory of surface dynamics starts from where the ideal-surface models fail to explain the observed dispersion curves. On this basis I shall refrain myself from reviewing much of the excellent work made in the past, and restrict myself to a few calculations that are more or less directly related to the recent atom-scattering data.

2. THE OBSERVATION OF SURFACE PHONONS BY He SCATTERING

The spectroscopy of surface phonons by atom scattering has evident analogies with the neutron spectroscopy of bulk phonons. There is however an important difference. A neutron time-of-flight (TOF) spectrum, as a consequence of wavevector conservation in three dimensions, display a few sharp peaks in correspondence of single bulk phonons. In atom scattering the wavevector is conserved only in two dimensions parallel to the surface and therefore TOF spectra present, in addition to sharp peak for surface phonons, the features of the surface-projected bulk phonon density, possibly including surface resonant modes. There is clearly a larger amount of information in atom scattering TOF spectra, as the surface-phonon dispersion curves and the surface-projected phonon densities are detected at once.

This has posed two difficult problems. On the experimental side, a particularly good resolution is required, say of the order of few wavenumbers. On the theoretical side, a good knowledge of the inelastic scattering cross section as function of energy and parallel momentum transfers is required in order to translate the observed scattering intensities into the corresponding surface phonon densities

This step depends in turn upon our knowledge of the atom–surface potential.

   Although the theoretical foundations date back to the work of Cabrera, Celli, Goodman and Manson (8) only recently the two above requirements have been accomplished and put together in a systematic analysis of TOF spectra (20–24). An example, for He scattering from the (001) surface of LiF along <100 >, is shown in fig. 1. The experimental TOF

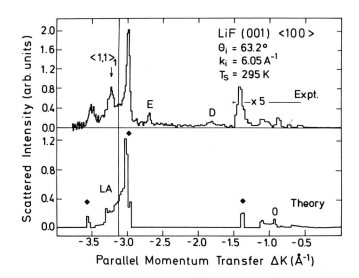

<u>Fig. 1:</u> Inelastic scattering intensity of He from LiF (001) along <100 > for an incident momentum k_i=6.05 $Å^{-1}$, incident angle θ_i=63.2⁰ and 90⁰ scattering geometry (above), and theoretical scattering intensity for a hard–corrugated surface model. E: elastic diffuse peak; D: diffuse diffraction peak.

spectrum for incident momentum k_i=6.05 $Å^{-1}$, incident angle θ_i=63.20⁰, final angle θ_f=90⁰–θ_i, is reproduced in the upper part of the figure as function of the parallel momentum transfer, and is compared with the calculated one–phonon scattering intensity (22). The calculation is based on the breathing shell model for LiF bulk dynamics, the Green´s function (GF) method (25) and a hard–corrugated surface model for atom–surface interaction (20,22). The main experimental peaks are seen to correspond to creation and annihilation processes involving Rayleigh waves (black squares) in different Brillouin zones. The comparison with the theoretical spectrum allows for the interpretation of the minor experimental features in terms of bulk longitudinal acoustic phonons (LA) and optical phonons (O). The edge of LA phonons appears to be amplified in the experiment by an inelastic resonance with a He–surface bound state (26).

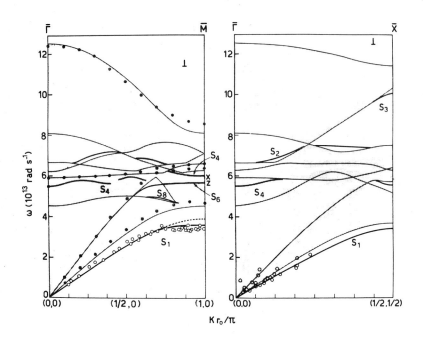

Fig. 2: Surface phonon dispersion curves of LiF(001) calculated by the Green's function method for sagittal polarization. Open circles: He scattering data; black points: neutron scattering data.

    In Fig. 2 the dispersion curves for sagittal polarization in LiF (001) calculated along the symmetry directions by the GF method are shown (27). The heavy lines represent the surface-localized modes and surface resonances. The band edges (thin lines) are compared with neutron data (●) to check the quality of the bulk dynamical model. The experimental points corresponding to Rayleigh wave peaks in TOF spectra (open circles) follow quite well the theoretical dispersion curve (S_1) along the two directions. Slab calculations with either unrelaxed (19) or relaxed (28) surface and ideal-surface GF calculations (25) do not predict correctly the Rayleigh frequency at the $\bar{M}$ point (broken line). An adjustment of bulk fluorine ion polarizability gives a better result (heavy line); still a 17% increase of the surface F^- polarizability is needed to obtain a perfect agreement (27). Among the various mechanisms which can lower the zone-boundary frequency (28), we selected the polarizability change because the misfit is found when the anions move in the $\bar{M}$-point Rayleigh mode, as in LiF, and is not found when the anions are at rest, as in NaF (fig.3, open circles).

    In NaF, fluorine ions move mainly in the optical modes. Since anions have usually a larger interaction with He atoms than cations, owing to their larger size, optical surface phonons could be detected in NaF(001) (fig. 3, black dots) (29). The experimental points are in fair agreement with the theoretical S_2, S_4 and S_6 branches.

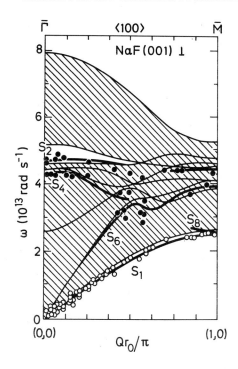

Surface phonon dispersion curves of NaF(001) calculated by the Green function method for sagittal polarization along <100>. Open circles: He scattering data from a low-energy beam; black points: He scattering data from a high-energy beam.

3. SURFACE PHONON ANOMALIES IN METALS AND SEMICONDUCTORS.

The case of LiF(001) gave indication that the change in the electronic structure at the surface may induce important deviations from the ideal-surface behaviour in the surface phonon spectrum. The role of surface electron states and surface electron-phonon interaction becomes apparently dramatic in noble metal surfaces. High-resolution He TOF spectra from Ag(111) (fig. 4) show a double peak, the stronger one corresponding to the Rayleigh mode, the weaker one to a resonance below the LA bulk edge (6,7). The intensity and the downward shift of this resonance with respect to the LA edge are anomalously large, in evident diasgreement with ideal-surface calculations (18,30) (fig. 5: the LA edges are X along <112> and Y along <110>). Such anomalous behaviour has been explained by the Modena group (31), in a slab calculation with a suitable parametrization of bulk dynamics and a softening of in-plane surface force constants. The macroscopic origin of such softening may well be the phonon-induced d-s hybridization of the electronic states, i.e., the large quadrupolar deformability of noble metal ions (32,33). This property of silver ions has been recognized to be responsible for their peculiar behaviour in various dynamical processes (34).

The changes of electronic structure and electron-phonon interaction at the surface are expected to produce large effects on the surface dynamics of semiconductors. The most important one is the instability causing surface reconstruction. Even in layered semiconductors,

where no reconstruction is observed, anomalous surface phonon branches
have been found. Recently Toennies and coworkers (7) have measured the
TOF spectra of He scattering from (0001) surface of GaSe and found a

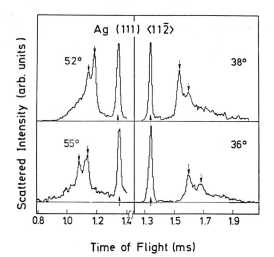

Fig. 4: Time-of-flight spectra of He scattered from Ag(111) along
        <112> at four different incident angles. Downward arrows:
        surface modes; upward arrows: elastic diffuse peak. (From
        ref. 6)

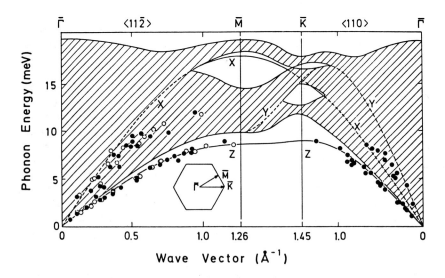

Fig. 5: Dispersion curves of surface phonons in Ag(111) along
        the symmetry directions (from ref. 6). Experimental
        points: He scattering data. Theoretical: GF generating
        coefficients method (Armand and Masri, ref. 27).

fairly large intensity from the optical mode S_2 (fig. 6). Such large
intensity corresponds to a strong localization: the corresponding
dispersion curve falls much below the A_1'-symmetry band from which it
originates. This is seen in fig. 7, where the surface-projected bulk
bands, as derived from Jandl et al. calculation (35) are also plotted
for reference. In a layer compound such a large softening of the surface branch would not be expected. Again a change in the electronic
structure occuring at the surface layer has to be surmised in order to
explain such anomalous behaviour.

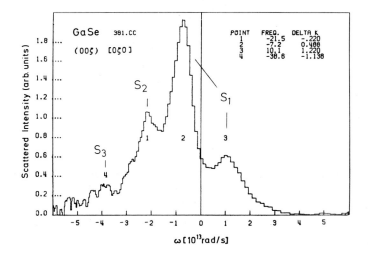

Fig. 6: Time-of-flight spectrum of He scattering from the (0001)
surface of GaSe along <010>. S_1: Rayleigh waves; S_2 and
S_3: optical surface modes.

Schlüter et al. have shown, by means of a pseudopotential calculation of the bulk band structure, that a non-negligible part of the
electronic charge density is contained in the volume between adjacent
layers, due to a certain amount of anion non-bonding atomic-like charge
contained in the various states (36). Moreover the calculated transition function contour plots (36) show that the interlayer electron
density changes appreciably in the principal optical transitions. Consequently, important changes in the electronic charge density and in
the transition function (and therefore in the anion polarizability)
must occur at the surface layer when the next layer is removed.

4. NEW TRENDS

This rapid overview on the recent (some still unpublished) investigations of the surface phonon dispersion in many interesting materials
can hardly be conclusive as these studies are raising many more ques-

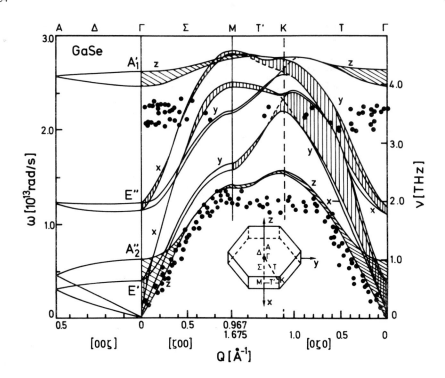

<u>Fig. 7</u>: Dispersion curves of GaSe (from Jandl et al., ref. 59).
Hatched regions represent the surface-projected bulk bands
for sagittal polarization. Black points: He scattering.

tions than those they can answer. Clearly the interpreation of surface
phonon anomalies points directly to the microscopic theory of surface
dynamics, and represents a powerful significant test for the present
models of electron-phonon interaction. An interesting example is offered by the recent cluster-deformability model calculation of TiN(001)
phonon branches.

Superconducting transition metal carbides and nitrides show strong
anomalies in the acoustic phonon branches (37). These anomalies have
been associated with the electron-phonon coupling which determines the
superconducting transition (38). Interesting questions are whether do
anomalies occur in the dispersion of surface phonons; whether surface
anomalies are shifted with respect to the bulk counterpart, as a consequence of the difference in the electron-phonon coupling at the surface; and whether all this has any relevance in surface superconductivity
(39).

In a recent calculation of TiN dispersion curves (40) Miura, Kress
and Bilz have explained the deep anomaly occuring in the LA branch at
2/3 of the zone by a cluster-deformation model. This model is directly
related to the microscopic properties and emphasizes the importance of
excitations of p-d hybridized states near the Fermi surface for the
lattice vibrations in transition metal compounds. A GF calculation of

the surface dynamics of TiN(001) (41) has shown that the dispersion curves of both the Rayleigh wave (S_1) and the quasi-longitudinal resonance S_6 are anomalous in the (100) direction. While the S_6 anomaly remains very close to the bulk anomaly, as a consequence of its prevailing bulk character, the Rayleigh wave anomaly is shifted back to 1/2 of the zone (fig. 8). This indicates that the effective range of the electron-phonon interaction, determining the position of anomalies, is changed at the surface.

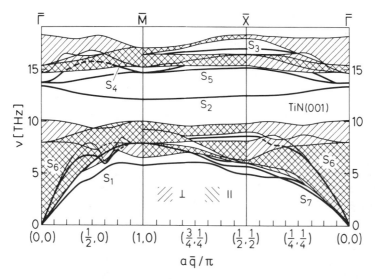

Fig.8: Surface phonon dispersion curves of TiN(001) (from ref. 41)

Like the advent of neutron spectroscopy thirty years ago has stimulated the first microscopic models and theories of lattice dynamics, the present development in surface phonon spectroscopy will trigger more extensive work in the first-principle calculation of surface phonons, particularly in those systems where stability and reconstruction problems (42) have a primary importance. We already have a few but significant examples of these new trends in the microscopic theory of surface vibrations, like the recent investigation by Hanke and Muramatsu (43) on the (2x1) and (7x7) reconstruction of the Si(111) surface, and the remarkable study of surface vibrations in simple metal surfaces accomplished by Beatrice and Calandra (44) by means of a pseudopotential perturbation theory.

Acknowledgments

I am very grateful to Peter Toennies and Guido Brusdeylins (Max-Planck-Institut für Strömungsforschung, Göttingen, FRG) for the continued information on their experimental results and for the many interesting discussions we have in our fruitful collaboration. I am much indebted to Heinz Bilz (Max-Planck-Institut für Festkörperforschung, Stuttgart, FRG) for several illuminating discussions about the role of electron-phonon coupling in phonon anomalies.

REFERENCES

1. Lord Rayleigh, Proc. London Math. Soc. 17 (1887) 4.
2. C.W. Farnell, in Physical Acoustics, Vol. IV, W.P. Mason, ed.
    (Academic Press, 1970), p. 109. An updated extended review has
    been written by A.A. Maradudin, Festkörperprobleme, XXI, J.
    Treusch, ed. (Vieweg, Braunschweig 1981) p. 25.
3. G. Brusdeylins, R.B. Doak and J.P. Toennies, Phys. Rev. Letters
    44 (1980) 1417.
4. G. Brusdeylins, R.B. Doak and J.P. Toennies, Phys. Rev. Letters
    46 (1981) 437.
5. G. Brusdeylins, R.B. Doak and J.P. Toennies, Phys. Rev. B 27
    (1983) 3636.
6. R.B. Doak, U. Harten and J.P. Toennies, Phys. Rev. Letters, 51
    (1983) 578.
7. J.P. Toennies, Proc. Am. Vacuum Soc. Meeting, Boston Nov 1-4,
    1983, to be published in J. Vac. Sci. Technol.
8. N. Cabrera, V. Celli and R. Manson, Phys. Rev. Letters 22 (1969)
    346.
9. R. Manson and V. Celli, Surface Sci. 24 (1971) 495;
    for a review see F.O. Goodman, Crit. Rev. Solid State and Mate-
    rials Sci. 7 (1977) 33.
10. G. Benedek and G. Seriani, Jpn. J. Appl. Phys. Suppl. 2, Pt. 2
    (1974) 545.
11. J.P. Toennies and K. Winkelmann, J. Chem. Phys. 66 (1977) 3965.
12. A.A. Maradudin, E.W. Montroll, G.H. Weiss and I.P. Ipatova,
    Solid State Physics Suppl. 3 (Academic Press, New York 1971).
13. A.A. Maradudin, R.F. Wallis and L. Dobrzinski, Handbook of Sur-
    faces and Interfaces, Vol. 3: ´Surface Phonons and Polaritons´
    (Garland STPM Press, New York, 1980).
14. R.F. Wallis, Rendiconti SIF, Course LII (Academic Press, New York
    and London 1972).
15. T.E. Feuchtwang, Phys. Rev. 155 (1967) 731.
16. V. Bortolani, F. Nizzoli, and G. Santoro in: Lattice Dynamics,
    M. Balkanski ed. (Flammarion, Paris 1978).
17. A.A. Lucas, J. Chem. Phys. 48 (1968) 3156.
18. R.E. Allen, G.P. Alldredge and F.W. de Wette, Phys. Rev. B 4
    (1971) 1648; (1971) 1661; (1971), 1682.
19. T.S. Chen, F.W. de Wette and G.P. Alldredge, Phys. Rev. B 15
    (1977) 1167.
20. G. Benedek and N. Garcia, Surface Sci. L103 (1981) 143.
21. G. Benedek, G. Brusdeylins, R.B. Doak and J.P. Toennies
    J. Phys. (Paris) Colloq. 42 (1981) C6.793.
22. G. Benedek, J.P. Toennies and R.B. Doak, Phys. Rev. B 28 (1983)
    7277.
23. A.C. Levi, Nuovo Cim. 54B (1979) 357.
24. A.C. Levi, G. Benedek, L. Miglio, G. Platero, V.R. Velasco and
    F.G. Moliner, Surf. Sci. (1984).

25. G. Benedek, Surf. Sci. 61 (1976) 603; G. Benedek and L. Miglio, in Ab Initio Calculation of Phonon Spectra, J.T. Devreese ed. (Plenum, New York 1983).

26. D. Evans, V. Celli, G. Benedek, J.P. Toennies and R.B. Doak, Phys. Rev. Letters 50 (1983) 1854.

27. G. Benedek, G.P. Brivio, L. Miglio and V.R. Velasco, Phys. Rev. B 26 (1982) 487.

28. E.R. Cowley and J.A. Barker, Phys. Rev. B 28 (1983) 3124.

29. G. Benedek, G. Brusdeylins, L. Miglio, R. Rechsteiner, J.G. Skofronick and J.P. Toennies, to be published.

30. G. Armand, Solid St. Comm. 48 (1983) 261.

31. V. Bortolani, A. Franchini, F. Nizzoli and G. Santoro, Phys. Rev. Letters 52 (1984) 429.

32. L.D. Marks, V. Heine, D.J. Smith, Phys. Rev. Letters 52 (1984) 656.

33. J.A. Moriarty, Phys. Rev. B 26 1754 (1982).

34. H. Bilz and W. Weber, Proc. Int. Conf. Latent Image in Photography, Trieste 1983 (to be published).

35. S. Jandl, J.L. Brebner and B.M. Powell, Phys. Rev. B 13 (1976) 686.

36. M. Schlüter, J. Camassel, S. Kohn, J.P. Voitchovsky, Y.R. Shen and M. Cohen, Phys. Rev. B 13 (1976) 3534.

37. H. Bilz and W. Kress, Phonon Dispersion Relations in Insulators (Springer V., Berlin, Heidelberg, New York 1979).

38. H. Frölich, Phys. Lett. 35A (1971) 325; W. Weber, H. Bilz and U. Schröder, Phys. Rev. Letters 28 (1972) 600; W. Hanke, J. Hafner and H. Bilz, Phys. Rev. Letters 37 (1965) 1560.

39. R.E. Allen, G.P. Alldredge and F.W. de Wette, Phys. Rev. B 2 (1970) 2570.

40. M. Miura, W. Kress and H. Bilz, Z. Phys. B 54 (1984) 103.

41. G. Benedek, M. Miura, W. Kress and H. Bilz, to be published.

42. A. Fasolino, G. Santoro and E. Tosatti, Phys. Rev. Letters, 44 (1980) 1684.

43. A. Muramatsu and W. Hanke, in: Ab Initio Calculation of Phonon Spectra, J.T. Devreese ed. (Plenum, New York 1983) p. 241; A. Muramatsu and W. Hanke, Phys. Rev. B 27 (1983) 2609. An extended review by these authors is to appear in Physics Reports (North-Holland 1984).

44. C. Beatrice and C. Calandra, Phys. Rev. B 28 (1983) 6130.

AN EXACT THEORY OF INELASTIC ATOM SURFACE SCATTERING
THE "DEBYE-WALLER" FACTOR

G. Armand
Centre d'Etudes Nucléaires de Saclay
Service de Physique des Atomes et des Surfaces
91191 Gif sur Yvette – France
and
J.R. Manson
Department of Physics and Astronomy, Clemson University,
Clemson, SC 29631 – USA

ABSTRACT

The particle surface interaction is decomposed into two parts. The
first one is the static periodic potential and yields the elastic diffraction pattern. The second contains the phonon operators and is responsible for the inelastic effects. In the T matrix equation the elastic
potential is projected out. The total T matrix element is then given by
its elastic component to which is added an expansion containing the
phonon operators and the coupling between elastic and inelastic channels.
A new but simple procedure is presented which allows one to perform the
thermal average to all orders in perturbation. Limiting ourselves to
the calculation of the thermal attenuation of the diffracted beams the
thermal average of T matrix element, which gives directly the diffracted
intensity, is written down explicitely. This general and exact procedure is applied to the scattering of a particle by a flat surface.
With the simple potential model used the influence of the different
parameters is studied. It is demonstrated that the relation between the
reflection coefficient and crystal temperature is not a pure exponential
function. This fact is confirmed by careful examination of the experimental data.

1. INTRODUCTION

Our understanding of low energy atom surface scattering has been considerably improved in the recent years. By using convenient potential
shapes and by solving exactly the scattering equation we are now able to
describe with a great precision the elastic diffraction pattern. Corre-
latively from a given set of diffracted beam intensities an interaction
potential can be constructed.
     The theoretical description of the inelastic scattering is far from
complete. It is of course very much more difficult because it should

59

B. Pullman et al. (eds.), Dynamics on Surfaces, 59–76.
© *1984 by D. Reidel Publishing Company.*

include the interaction of the incident particle with the phonon field.
Some attempts have been made previously with the corrugated hard wall
potential [1] [2] [3]. But the analysis of the elastic scattering data
has shown that this kind of potential is not convenient to represent
the particle surface interaction [4]. It is in fact necessary to model
this interaction by a soft potential like those which are derived from
the so-called Morse corrugated potential [5].

Inelastic theory using soft potentials has been developed either
with the semi-classical approximation or within the framework of quantum
mechanics. The former formalism leads to numerous calculation and suf-
fers limitations as it does not take account of the transition from
continum initial state to all other states particularly the bound states
of the potential [6] [7] [8]. For the later more or less severe approximations are introduced in the calculation [6] or else it is limited
to first order perturbation. The results can be doubtful owing to the fact
that multiple scattering effects are not negligible in this kind of
problem. One approximation we would like to mention in passing is the
neglect of energy exchange of the crystal in the Green function or pro-
pagator of the transition matrix equation [6]. One can show that in
many cases this will lead directly to a divergent perturbation expansion.

In this paper we would like to present a procedure which allows
one to get an exact quantum mechanical solution for the inelastic scattering of atoms by surfaces. It will be applied to the simplest cases:
the scattering by a flat surface. Using a simple soft potential model
which in our opinion is sufficient to yield the salient features of the
inelastic scattering phenomena, we show that the usual Debye-Waller
factor does not give a good representation of thermal attenuation of a
diffracted beam.

2. GENERAL THEORY

Let us consider a particle of mass m scattered by a potential $V(\vec{R}, z, \vec{u})$
where as usual $\vec{R}$ and z label respectively the parallel and normal component of a vector with respect to the surface, and $\vec{u}$ an operator which
takes account of the crystal atom's thermal motion.

The T matrix elements which contain all the information about the
scattering process are solutions of the equation:

$$T = V(\vec{R}, z, \vec{u}) - U(z) + \left[V(\vec{R}, z, \vec{u}) - U(z) \right] G^+ T \tag{1}$$

in which U is the distorted potential and G^+ the Green operator

$$G^+ = \left[E_i - H_o - H^{(c)} + i\varepsilon \right]^{-1}, \quad H_o = -\frac{\hbar^2}{2m} \nabla^2 + U(z). \tag{2}$$

E_i is the total energy of the initial state (crystal + particle) and $H^{(c)}$
the crystal Hamiltonien.

The thermal average of the potential

$$<<V>> = Tr(\exp(-H^{(c)}/kT)V) \left[Tr(\exp(-H^c/kT)) \right]^{-1} \tag{3}$$

has the surface periodicity and yields an elastic diffraction pattern. In order to calculate the scattered intensities it is interesting to write the T matrix element as the sum of the elastic T matrix element and an expansion which contains the effect of the lattice thermal motion. This can be achieved by decomposing the potential into two parts:

$$V-U = v(\vec{u}) + v$$

and carrying out the projection of v. This projection technique is very similar to that used in the calculation of bound state resonances in the purely elastic case [9].

Among the different possibilities of decomposition the most convenient are:

$$v(\vec{u}) = V(\vec{R},z,\vec{u})-<<V(\vec{R},z,\vec{u})>> \quad , \quad v = <<V(\vec{R},z,\vec{u})>> - U(z) \qquad (4)$$

$$v(\vec{u}) = V(\vec{R},z,\vec{u})-V(\vec{R},z,\vec{u}=0) \quad , \quad v = V(\vec{R},z,\vec{u} = 0) - U(z) \qquad (5)$$

In each case the T matrix equations appear as:

$$T = v + v(\vec{u}) + \left[v + v(\vec{u})\right] G^+ T$$

Carrying out the projection one obtains the following set of equations:

$$h = (1-vG^+)^{-1}v \qquad \text{or} \qquad h = v + vG^+h \qquad (6)$$

$$\ell = (1-vG^+)^{-1}v(\vec{u}) \qquad \text{or} \qquad \ell = v(\vec{u}) + vG^+\ell \qquad (7)$$

$$T = h + \ell + \ell G^+T \qquad (8)$$

It can readily be shown that the ℓ equation can be transformed into:

$$\ell = (1+WG^+) \ v(\vec{u}) \qquad \text{with} \qquad W = v + WG^+v$$

and the iteration of equation (8) gives:

$$T = h + (1+WG^+) \ v(\vec{u}) \ (1+G^+h) + \left\{ \sum_{n=1}^{\infty} \left[(1+WG^+)v(\vec{u})G^+ \right]^n \right\}$$

$$(1+WG^+) \ v(\vec{u}) \ (1+G^+h) \qquad (9)$$

The equations giving h and W contain only the periodic potential v, and are independent of the atomic thermal displacement operator $\vec{u}$. They are operators of a purely elastic potential and therefore give the intensities of elastic diffraction peaks. The corresponding equations can be solved exactly (9) (10). Note that in equation (9) we have to calculate matrix element of h and W between different states. Thus we have maintained the distinction between these two operators.

Equation (9) which gives the operator relation for the whole potential is an expansion in power of $v(\vec{u})$ and depicts the multiple scattering which couples elastic and inelastic channels.

The transition rate between initial and final states is proportional to the square modulus of the corresponding T matrix element. After performing the usual Van Hove transformation it has been demonstrated that the intensity of a diffraction peak is given by (6):

$$I_{\vec{G}} = \left| \delta_{GO} - i\pi \frac{2m}{\hbar^2 \, k_{iZ} \, k_{GZ}} \, <<T_{Gi}>> \right|^2 \tag{10}$$

where k_{iZ} and k_{GZ} are respectively the normal components of the specular and diffracted particle momentum.

As this paper is devoted to the calculation of the thermal attenuation of diffracted beams we have only to calculate the thermal average of the corresponding matrix element:

$$<<T>>_{Gi} = \sum_{|n_i>} \frac{\exp - E_i^{(c)}/kT}{Z} <n_i \, \phi_G | T | \phi_i \, n_i>$$

with Z the partition function, $E_i^{(c)}$ the crystal energy. $|n_i>$ and $|\phi_j>$ are an eingenstates of $H^{(c)}$ and H_o respectively.

Putting $<T>_{Gi} = <n_i \, \phi_G | T | \phi_i \, n_i>$ equation (9) gives:

$$<T>_{Gi} = h_{Gi} + <\phi_G | (1 + WG_e^+) <n_i | v(\vec{u}) | n_i> (1 + G_e^+ h) | \phi_i>$$

$$+ <\phi_G | (1 + WG_e^+) <n_i | v(\vec{u}) G^+ (1 + WG^+) v(\vec{u}) | n_i> (1 + G_e^+ h) | \phi_i> + \ldots \tag{11}$$

with $G_e^+ = (E_i^{(p)} - H_o + i\varepsilon)^{-1}$ the Green operator of the purely elastic case.

The evaluation of the thermal average of $<T>_{Gi}$ is the main difficulty due to the presence of the Green operator in the terms of order greater than or equal to three. Therefore it is necessary to transform these terms before performing the average. To this aim we use an integral representation of the Green operator namely:

$$G^+ = \frac{-i}{\hbar} \int_o^\infty \exp[i(c + E_i^{(c)} - H^{(c)} + i\varepsilon)t/\hbar]dt \tag{12}$$

with $c = E_i^{(p)} - H_o$ and $E_i^{(p)}$, $E_i^{(c)}$ respectively the particle and crystal energy in the state i. Thus we have:

$$v(\vec{u})G^+ v(\vec{u}) = \frac{-i}{\hbar} \int_o^\infty \exp(i \, E_i^{(c)} t/\hbar) \, v(\vec{u}) \, \exp(-iH^{(c)}t/\hbar)$$

$$\exp(i(c+i\varepsilon)t/\hbar) \, v(\vec{u}) \, dt$$

Then inserting between $\exp(i \, E_i^{(c)} t/\hbar)$ and $v(\vec{u})$ the factor

$\exp(-i H^{(c)} t/\hbar) \exp(i H^{(c)} t/\hbar) = 1$, one gets:

$$v(\vec{u}) \ G^+ v(\vec{u}) = \frac{-i}{\hbar} \int_0^\infty \exp(i(E_i^{(c)} - H^{(c)}) t/\hbar) \ v(\vec{u}(t))$$
$$\exp(i(c+i\varepsilon) t/\hbar) \ v(\vec{u}) \quad dt$$

$$\langle n_i | v(\vec{u}) G^+ v(\vec{u}) | n_i \rangle = \frac{-i}{\hbar} \int_0^\infty \langle n_i | \ v(\vec{u}(t)) \ \exp(i(c+i\varepsilon) t/\hbar)$$
$$v(\vec{u}) | n_i \rangle \quad dt$$

where $v(\vec{u}(t))$ is the potential in the intereaction picture.

In the same way, one can easily establish that:

$$\langle n_i | v(\vec{u}) G^+ W G^+ \ v(\vec{u}) | n_i \rangle = \left(\frac{-i}{\hbar} \right)^2 \int_0^\infty \int_0^\infty dt_1 \ dt_2 \ \langle n_i | v(\vec{u}(t_1+t_2))$$
$$\exp(i(c+i\varepsilon) t_2/\hbar) \ W \ \exp(i(c+i\varepsilon) t_1/\hbar) \ v(\vec{u}) | n_i \rangle$$

The terms of higher power in $v(\vec{u})$ including W can be transformed in the same way. The number of integrals on different t variables will be equal to the number of Green operators contained in the corresponding term. Then

$$\ll T \gg = h + \langle \phi_G | (1+WG_e^+) \ll v(\vec{u}) \gg (1 + G_e^+ h) | \phi_i \rangle + \langle \phi_G | (1 + WG_e^+)$$
$$\ \ _{Gi} \qquad\quad _{Gi}$$
$$\left\{ \left(\frac{-i}{\hbar} \right) \int_0^\infty \ll v(\vec{u}(t)) \ \exp(i(c+i\varepsilon) t/\hbar) \ v(\vec{u}) \gg \ dt \ + \right.$$
$$\left. + \left(\frac{-i}{\hbar} \right)^2 \int\int_0^\infty \ll v(\vec{u}(t_1+t_2)) \ \exp(i(c+i\varepsilon) t_2/\hbar) \right.$$
$$\left. W \ \exp(i(c+i\varepsilon) t_1/\hbar) v(\vec{u}) \gg \ dt_1 \ dt_2 \right\} (1+G_e^+ h) | \phi_i \rangle + \ldots \quad (13)$$

In the thermal average brackets the remaining exponentials depend only on the particle coordinate (by c). The thermal average can be performed and after that the different integrations. Thus $\ll T \gg$ can be calculated to all orders in the expansion. $\qquad\qquad$ Gi

3. FLAT SURFACE

As an application of the preceeding theory, we consider the scattering of an incident particle by a flat surface. The crystal Hamiltonien is taken in the harmonic approximation. The particle-surface interaction is represented by a soft potential:

$$V(z,\vec{u}) = D \left[\exp(-2\chi(z-u) - 2\chi^2 \ll u^2 \gg) - 2 \exp(-\chi z) \right]$$

Its thermal average is a Morse potential:

$$<<V(z,\vec{u})>> = D\left[\exp(-2\chi z) - 2\exp(-\chi z)\right]$$

The attractive part of this potential is stationary with respect to thermal motion. Its repulsive part is thermally displaced in the direction normal to the surface by the displacement operator u which shall be taken equal to a linear combination of atomic displacements. Therefore the particle can gain or lose energy but the exchange of parallel momentum is precluded. It follows that this description of the physical reality will not be completely correct. However it allows one to see the influence of the softness and of the bound states, particularly the influence of resonances assisted by phonon exchange. Furthermore, it is very convenient to illustrate how the general formula established in paragraph 2 can be managed. The analytical expressions are simple and consequently the numerical calculation is inexpensive allowing one to study easily the influence of the different parameters. It should be considered as a first step toward more realistic model calculations.

First we decompose the potential following equation (4). Chosing for U(z) a Morse potential equal to $<<V(z,u)>>$ it appears that v and $<<v(u)>>$ are equal to zero. Equation (13) reduces to:

$$<<_iT_i>> = \frac{-i}{\hbar} <\phi_i| \int_0^\infty <<v(u(t))\exp(i(c+i\varepsilon)t/\hbar)\ v(u)>>|\phi_i> dt$$

$$+ \left(\frac{-i}{\hbar}\right)^2 <\phi_i| \iint_0^\infty <<v(u(t_1+t_2))\exp(i(c+i\varepsilon)t_2/\hbar)v(u(t_1))$$

$$\exp(i(c+i\varepsilon)t_1/\hbar)\ v(u)>>|\phi_i> dt_1 dt_2$$

$$+ \ldots \tag{14}$$

with $v(u) = D\ e^{-2\chi z}(e^{2\chi u}\ e^{-2\chi^2<<u^2>>}- 1)$.

Here the second and third order terms in Born expansion have been written explicitely.

The thermal average can be readily done leading to:

$$<<_iT_i>> = \frac{-i}{\hbar} <\phi_i|D^2\ \exp(-2\chi z)\int_0^\infty U_1(t)\ \exp(i(c+i\varepsilon)t/\hbar)$$

$$\exp(-2\chi z)|\phi_i> dt$$

$$+ \left(\frac{-i}{h}\right)^2 <\phi_i|\ D^3\exp(-2\chi z)\iint_0^\infty U_2(t_1,t_2)\exp(i(c+i\varepsilon)t_2/\hbar)\exp(-2\chi z)$$

$$\exp(i(c+i\varepsilon)t_1/\hbar)\exp(-2\chi z)|\phi_i> dt_1 dt_2$$

$$+ \ldots \tag{15}$$

with $U_1(t) = \exp(4\chi^2<<u(t)u>>)-1$

$$U_2(t_1,t_2) = \exp\left(4\chi^2\left[<<u(t_2)u>> + <<u(t_1)u>> + <<u(t_1+t_2)u>>\right]\right) - 1$$

$$- (\exp(4\chi^2<<u(t_2)u>>)-1) - (\exp(4\chi^2<<u(t_1)u>> - 1) -$$

$$- (\exp(4\chi^2<<u(t_1+t_2)u>>) - 1)$$

where the correlation function is equal to:

$$<<u(t)u>> = \frac{\hbar}{2M} \int_o^{\omega_{max}} \frac{d\omega}{\omega}\left[(<n_\omega>+1)\exp(-i\omega t)+<n_\omega>\exp(i\omega t)\right]\rho(\omega) \tag{16}$$

M is the mass of a crystal atom, $<n_\omega>$ the mean number of phonons of frequency ω at temperature T and $\rho(\omega)$ the spectral density of u.

In $U_1(t)$, $U_2(t_1,t_2)$... the exponentials are expanded in powers of the correlation function. Taking account of (16) each integral over a t variable is easily calculated and yields a modified Green operator:

$$G^+(\omega_s) = (E_i^{(p)} - H_o + \hbar\omega_s + i\varepsilon)^{-1}$$

ω_s being a sum of n terms ($\pm \omega_1 \pm \omega_2$...) with n the power of the correlation function. There is one $G^+(\omega_s)$ between two factors $\exp(-2\chi z)$ and the calculation of the matrix element $<\phi_i|... |\phi_i>$ is straightforward knowing the set of continum and bound eigenstates of H_o.

Let us illustrate this exact procedure by considering the first term of (15), that is to say the second order term in the Born expansion.

As in a previous paper (9) we define dimensionless quantities with the help of the inverse energy factor $A^2 = 2m/\hbar^2\chi^2$, namely:

$$_iF_i = 4 A^2 {}_iT_i, \quad \Omega = A^2\hbar\omega, \quad \Omega_b = A^2|e_b|, \quad p^2 = A^2 e_p$$

where e_b and e_p are respectively the eingenvalues for bound and continum states of a Morse potential. In this way the reflection coefficient for the specular beam is given by:

$$_iR_i = (1 - i\pi \ _iF_i/4 \ p_i)^2$$

and the first term of (15) is:

$$_iF_i^{(1)} = \frac{m}{M} \int_o^{\Omega_m} \frac{\rho(\Omega)}{\Omega}\left[(<<n_\Omega>>+1) \ Q(-\Omega)+ <<n_\Omega>> \ Q(+\Omega)\right] d\Omega$$

$$+ 2\left(\frac{m}{M}\right)^2 \iint_o^{\Omega_m} \frac{\rho(\Omega_1)\rho(\Omega_2)}{\Omega_1 \Omega_2} \left(\begin{array}{l} (<<n_{\Omega_1}>>+1)(<<n_{\Omega_2}>>+1) \ Q(-\Omega_1-\Omega_2) \\ +(<<n_{\Omega_1}>>+1)<<n_{\Omega_2}>> \quad Q(\Omega_2-\Omega_1) \\ +<<n_{\Omega_1}>>(<<n_{\Omega_2}>>+1) \quad Q(\Omega_1-\Omega_2) \\ +<<n_{\Omega_1}>><<n_{\Omega_2}>> \quad Q(\Omega_1+\Omega_2) \end{array} \right) d\Omega_1 \ d\Omega_2$$

$$+ ... \tag{17}$$

with

$$Q(\lambda) = \int_0^\infty \frac{f(p_i\, p)^2\; dp}{p_i^2 - p^2 + \lambda + i\, \varepsilon} + \sum_b \frac{\ell(p_i\, b)^2}{\pi^2\, (p_i^2 + \Omega_b + \lambda + i\, \varepsilon)}$$

and $p_i^2 = (\vec{k_i}^2 - |\vec{K_i}|^2)/\chi^2$, $\vec{k_i}$ and $\vec{K_i}$ being respectively the incident wave vector and its parallel component to the surface. f and ℓ/π are respectively the continum-continum and continum-bound state matrix elements for the Morse potential [9].

As p_i^2 is the dimensionless normal energy of the incident particle $p_i^2 + \lambda$, in the denominators of Q, is the energy associated with normal motion of the particle which has gained ($\lambda > 0$) or lost ($\lambda < 0$) the phonon energy equal to λA^{-2}. Comparison with the purely elastic case shows that Q is the T matrix element of a particle of normal energy $(p_i^2 + \lambda)A^{-2}$ scattered elastically by the static Morse potential. On the other hand Q can be seen as the matrix element corresponding to the following process: interaction with a transfer of energy λA^{-2} followed by another interaction with the reverse exchange of energy.

We recall that λ stands for $\pm \Omega_1 \pm \Omega_2 \pm \ldots \pm \Omega_n$ where n is the order of the expansion in $_iF_i^{(1)}$. Consequently the nth term involves a process in which n virtual phonons are transfered. The corresponding matrix element is proportional to $(m/M)^n$. It is also proportional to a sum of the Q matrix elements which represent the different ways to achieve the virtual process, weighted by the appropriate bose factor. The sum is again weighted by the product of n spectral densities and integrated n times over all the available frequencies of the crystal. Due to these integrations the resonances with bound states, which are opened by the phonon exchange, are automatically included in the calculation.

We have written down explicitely the second order term in the Born series. We cannot write here the expression for the third and the fourth order term which have been calculated. The analytical expressions show that the total one virtual phonon process is given by the first term of $_iF_i^{(1)}$. Two phonon processes are completely taken into account by expanding up to $<<u(t_1)u>> <<u(t_2)u>>$ in the second, third and fourth order term of the Born series. This is illustrated diagramatically on figure 1. It shows that there is no vertex which does not have a phonon connection. We postulate that this rule can be extended to higher order virtual phonon exchange being unable to present any proof of this fact. This is due certainly to the annulation of the first order term $(<<v(u)>> = 0)$ and perhaps to the particular choice of the potential model.

This rule allows one to predict how many terms should be calculated in order to have the total n phonon process. From the shape of each diagram the corresponding analytical form of the matrix element could be anticipated.

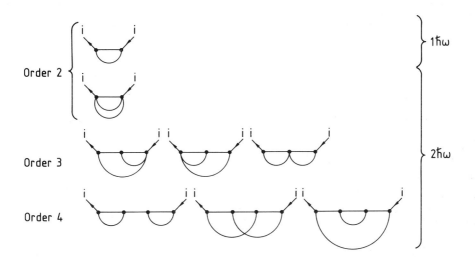

Figure 1. Diagramatic representation of one and two virtual phonon processes. Each phonon connection stands for creation-annihilation and vice et versa. The order in Born expansion is indicated.

4.   RESULTS

The reflection coefficient R for a Helium atom scattered elastically by a flat surface has been numerically calculated taking account only the one virtual phonon process. The matrix element giving the R value is the first term in the expansion of $.F.^{(1)}$. All the ingredients entering in this expression are known except $\rho(\omega)$ which is a quantity depending upon the kind of crystal and its surface.

One knows that the (100) face of Copper behaves practically as a flat surface since the measured intensities of first order diffraction peaks are very small [11]. We take this crystal face as a model of a flat surface.

Due to the finite size of the Helium atom the repulsive part of the potential results from the interaction of the Helium atom with many surface atoms. Therefore its thermal motion u should be directly influenced by the displacement of several surface atoms. It has been proposed to simulate this effect by taking u equal to the average motion of the four atoms belonging to a surface unit cell [11]. In this way:

$$u = \frac{1}{4} (u_1^Z + u_2^Z + u_3^Z + u_4^Z) \qquad (18)$$

with u_j^Z the normal displacement to the surface of the jth atom.

Making the product uu and taking account of the spectral density definition:

$$\rho(\omega) = \sum_{\substack{j\ell \\ jk}} B_{jk}^{6} B_{\ell k}^{\alpha} \delta(\omega^2 - \omega_k^2) \text{ where } B_k^{\alpha} \text{ is the polarisation vector of}$$

a phonon of frequency ω_k for the direction α, it is easily shown that the spectral density of the u operator is given by :

$$\rho(\omega) = \frac{1}{4} \rho_{11}^{ZZ}(\omega) + \frac{1}{2} \rho_{12}^{ZZ}(\omega) + \frac{1}{4} \rho_{13}^{ZZ}$$

where ρ_{11}^{ZZ} is the spectral density of one atom and ρ_{12}^{ZZ}, ρ_{13}^{ZZ} respectively the correlated spectral density between nearest and next nearest neighbours. These different quantities have been calculated previously by the method of generating coefficients of lattice Green function [12]. $\rho(\omega)$ is represented on figure 2.

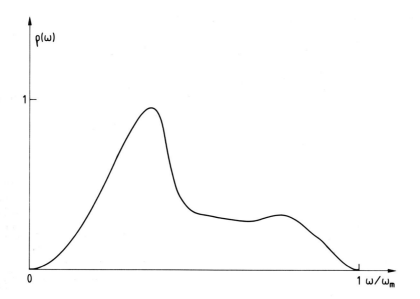

Figure 2. Spectral density of the u operator introduced into the calculation $\omega_m = 7.3 \ 10^{12} \ s^{-1}$.

Central forces between nearest neighbour atoms have been introduced in the crystal model and forces of greater range are assumed to be vanishing. The crystal maximum frequency is $7.3 \ 10^{12} \ s^{-1}$.
Note that the parameter values of the Morse potential for the (100) face of Copper are chosen equal to that of the zero order Fourier component of the potential of the different faces of Copper presently studied [13], namely : $D = 6.35 \ meV \quad 2X = 2.1 \ \mathring{A}^{-1}$

The results presented in the following have been obtained for a parti-
cle incident energy of 21 meV that is to say a wave vector $|k_i| = 6.4 \overset{\circ}{A}^{-1}$.
It is interesting first to consider the effect of potential softness.
In figure 3 the Logarithm of the reflection coefficient versus temperature has been ploted for an exponential potential (D = 0), different
values of the softness parameter 2X, and two incident angles Θ_i. For
comparison, the usual Debye-Waller factor, which can be deduced from a
corrugated hard wall potential calculation (CHW), has been drawn.
It writes :

$$R = \exp - (2|k_i| \cos \Theta_i)^2 <<u^2>> \qquad (19)$$

$<<u^2>>$ being the thermal average of expression (18).
As the softness parameter increases the CHW solution is approached and
at the same time, the elastic specular intensity for a given temperature decreases. In the medium and high temperature range a curvature
can be noticed on the curve relative to exponential potential: the
relation (19) does not hold. This effect does not appear clearly at an
incident angle of 73.5 degrees at the scale of the figure, but it is
noticeable at $\Theta_i = 31.8$ degrees where a decrease of the curvature as 2X
increases could be observed.

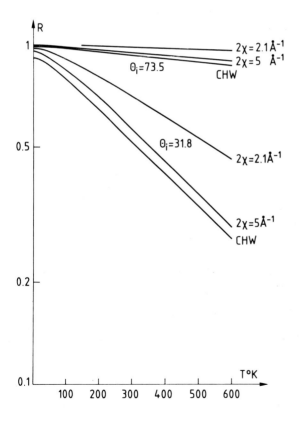

Figure 3. Logarithmic plot
of specular reflection
coefficient versus temperature for an exponential
potential with different
values of the stiffness
parameter 2X. Corrugated
hard wall solution (CHW) –
$k_i = 6.4 \overset{\circ}{A}^{-1}$ –

incident angle $\Theta_i = 73.5$

and 31.8 degrees.

The second important parameter is the well depth D of the Morse potential. The results are shown on figure 4 for $2X = 2.1 \text{ Å}^{-1}$, $\Theta_i = 31.8$
degrees and D varying from 0 to 6.35 meV. Also for comparison the CHW
solution including the so-called Beedy correction for D = 6.35 meV has
been drawn (CHW-B). This correction takes into account the well depth
effect by adding to its normal kinetic energy the energy of the well
depth.

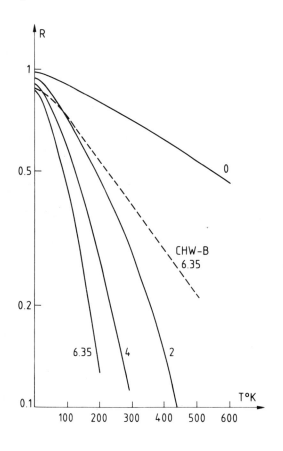

Figure 4. Logarithmic plot
of specular reflection coefficient versus temperature
for different values of the
well depth D – each curve
is labelled by this value in
meV. CHW–B corrugated hard
wall solution with Beedy
correction – incident angle
$\Theta_i = 31.8$ degrees
$k_i = 6.4 \text{ Å}^{-1}$

One notices immediately a strong decrease of the intensity as D
increases. Also in the medium and high temperature range the Log (R)
versus T relation is not a straight line and the curvature is more
pronounced than with the exponential potential. Furthermore the difference between the CHW-B and the corresponding soft potential is very
important.
This difference appears to be also very important at an incidence
angle of 73.5 degrees (figure 5). On this figure can also be seen the
effect of a small variation of the 2χ value around 2.1 Å^{-1}. This yields
an intensity variation which seems to be appreciable on an experimental
point of view only in the high temperature range. The curve labelled C
gives the calculated intensities when bound states are excluded from

the calculation. The comparison with intensity given by the exponential potential shows that the scattering via continum states is, for the greatest part responsible for the strong intensity decrease yielded by the Morse potential.

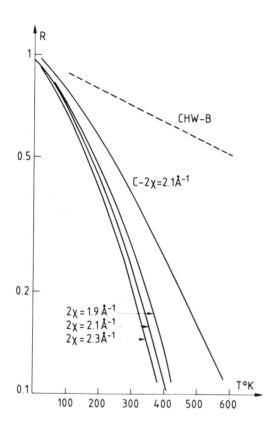

Figure 5. Logarithmic plot of specular reflection coefficient versus temperature for different values of the stiffness parameter. The curve labelled C is obtained by excluding the bound state from the calculation. CHW–B corrugated hard wall solution with Beedy correction –
$D = 6.35$ meV,
$\Theta_i = 73.5$ degrees,
$k_i = 6.4$ Å^{-1}

Note that in the case of figure 4 the calculated results with and without bound states are practically equal.

5. DISCUSSION

It should be keep in mind in this discussion that the t matrix element is given by an integration over variables Ω of a product of three terms:
   − $Q(\lambda)$ which depends upon the potential parameters 2χ and D through the matrix elements f and ℓ/π.
   − $\rho'(\Omega)$ which introduces the vibrational properties of the u operator.
   − $\langle\langle n_\Omega \rangle\rangle$ which depends upon the variable $\hbar\omega/kT$
Of this three factors $\langle\langle n_\Omega \rangle\rangle$ is the only one which is crystal temperature dependent.

Let us keep first this last parameter constant. For low frequency values the spectral density, like the whole crystal spectrum, is proportional to ω^2. Putting $\rho(\omega) = \alpha\omega^2$ the weight factor which multiplies $Q(\lambda)$ is:

$$\frac{\rho(\omega)}{\omega} \, \langle\langle n(\omega) \rangle\rangle = \alpha \, \frac{kT}{\hbar} \, \frac{x}{\exp(x)-1} \quad , \quad x = \frac{\hbar\omega}{kT} \qquad (20)$$

This is a rapidly decreasing function of x. For high ω value $\rho(\omega)$ is much lower than the extrapolation of $\alpha\omega^2$ (see figure 2), and the weight is therefore much smaller than the value given by expression (20). On the other hand the contribution of the continum states to $Q(\lambda)$ is certainly maximum for the low frequency values. In this case the denominator vanishes for p values close to p_i, a region where the matrix element $f(p_i,p)$ is maximum. It can be concluded that the most important contribution to $_iF_i$ comes from the scattering through continum states and virtual phonon exchange of low energy.

In order to complete this analysis it is necessary to look at the effect of bound states. One remembers that they have an appreciable influence at Θ_i = 73.5 degrees whereas their influence is negligible at Θ_i = 31.8 degrees. For these two angles the normal kinetic energy of the incident particle E_i^Z is respectively 1.7 and 15 meV. This different behaviour could be ascribed to a stronger coupling between continum and bound states for the greater incident angle and simultaneously to greater efficiency of resonance induced by the phonon creation in the virtual process. The former effect comes from the increase of the matrix elements ℓ as p_i decreases ($p_i = k_i^Z/\chi$). The latter is a direct consequence of the strong variation of the ponderation factor $\frac{\rho(\omega)}{\omega} \, \langle\langle n_\omega \rangle\rangle$: a resonance can appear only if the involved phonon has an energy greater than E_i^Z. Therefore as E_i^Z increases the ponderation factor decreases rapidly and the resonances are less and less important.

To close this first part of the discussion one can pointed out that the scattering via continum states seems to give the most important contribution to the T matrix element. As an increase of the well depth value yields an increase of the continum-continum matrix element $f(p_i,p)$ the strong influence of this parameter can be understood. On the other hand a variation of χ produces a variation of $f(p_i,p)$ and simultaneously of p_i itself. Thus the influence of this parameter is more difficult to analyze.

Let us now consider the crystal temperature T as a variable, all the other parameters being constant. For $kT/\hbar\omega$ greater than 1, $\langle\langle n_\Omega \rangle\rangle$ is practically a linear function of the temperature and of course increases with this parameter. $_iF_i^{(1)}$ being proportional to $\langle\langle n_\Omega \rangle\rangle$ and $\langle\langle n_\Omega \rangle\rangle+1$ for the one phonon process, this matrix element should be a linear function of the temperature except in the very low temperature region. Thus putting:

$$_iF_i^{(1)} = \beta + \gamma T$$

the reflection coefficient writes :

$$R = \left| 1 - \frac{i\pi}{4P_i} (\beta + \gamma T) \right|^2 = \lambda_0 + \lambda_1 T + \lambda_2 T^2 \tag{21}$$

Obviously the R(T) relation is not an exponential. In order to see the shape of R(T) on a logarithmic plot, we proceed to a cumulant expansion of the relation (21). This leads to:

$$R = \lambda_0 \exp\left[\frac{\lambda_1}{\lambda_0} T + \frac{2\lambda_2\lambda_0 - \lambda_1^2}{\lambda_0^2} T^2 + \ldots \right] \tag{22}$$

Due to terms in T^2, T^3 ... the plot of Log(R) versus T could not be a straight line and this fact is a direct consequence of the shape of the exact expression (17). If we look at the two phonon terms including the contribution of the third and fourth order of the Born expansion it appears that in each term the factors which are temperature dependent

are the products $\left(\begin{array}{c} <<n_{\Omega_1}>> + 0 \\ + 1 \end{array} \right) \left(\begin{array}{c} <<n_{\Omega_2}>> + 0 \\ + 1 \end{array} \right)$. A term proportional to

T^2 will be added in F_i and R will be proportional to a polynomial of degree 4. The cumulant expansion will remain identical to (22) with different coefficients. It does not yield an exponential of the tempe-rature except if all the coefficients of T^n with n greater than 1 vanish. This seems to be improbable except perhaps for a very particular potential shape.

One may argue that this important result comes from the use of a poten-tial for which its thermal average is temperature independent. In order to rule out this objection the calculation has been done with the follow-ing potential:

$$V(z,u) = D \left[\exp(-2\chi(z-u)) - 2 \exp(-\chi z) \right]$$

The equations are equivalent to the preceeding one the well depth value becoming now temperature dependent, namely: $D \rightarrow D \exp - 2\chi^2 <u^2>$.
The calculated reflection coefficients are a little greater than those reported above. The Log(R) versus T plot exhibits the same curvature as expected, since the equations have the same dependence with respect to $<<n_\Omega>>$.

On the other hand the T matrix expression for the one phonon process, using a sum of pairwise potentials extended to all crystal atoms, has been calculated. With this potential parallel momentum exchange is allowed. Although more complicated it depicts the same dependence with respect to $<<n_\Omega>>$ as the preceeding one. The R(T) relation should not be an exponential function. This important result has been confirmed by preliminary numerical calculations.

6.   COMPARISON WITH EXPERIMENTAL RESULTS

Thermal attenuation of the specular beam has been measured for Helium
scattered by the (100) face of Copper [14] which is very nearly a flat
surface.
     A careful examination of the experimental data shows that in the
medium and high temperature regime there are some deviations from the
exponential relation. Qualitatively the shape of the Log(R) versus temperature plot is identical to that given by the calculated results. Thus
it becomes interesting to attempt the comparison even if the potential
used in the present calculation is rather crude. One can at least verify if the curvature exhibited by the experimental and theoretical curves
are the same.
     In order to get the best fit we introduce in the calculation an ad-
justable parameter α which multiplies the spectral density. This quantity becomes $\alpha \rho(\omega)$. It should compensate on the whole the deficiencies
of our potential model, that is to say the lack of parallel momentum exchange between the particle and the phonon field and the limited range
of correlation between crystals atoms. Also there is the fact that the
incident atom is actually scattered by the electrons cloud several atomic units in front of the surface atomic layer, and at this distance the
electron density does not vibrate with as large an amplitude as the atoms.
     For the present calculation the correlation is limited to the four
atoms belonging to a unit cell. Figure 6 and 7 show the best fit obtained with $\alpha = 0.25$.

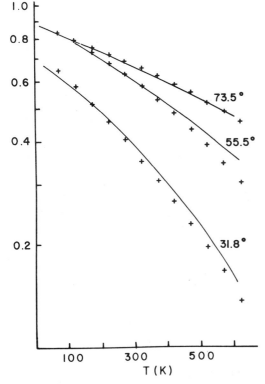

Figure 6. Comparison of
calculated intensities and
experimental data +
$D = 6.35$ meV, $2\chi = 2.1$ $\overset{\circ}{A}^{-1}$,
$k_i = 6.4$ $\overset{\circ}{A}^{-1}$ or $E_i = 21$ meV
adjustable parameter
$\alpha = 0.25$

Note that the calculated curves have been translated vertically in order to compensate the lack of unitarity (at T = 0) of the experimental data.

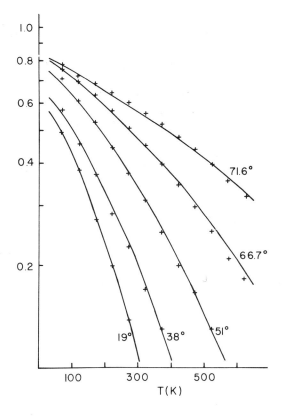

Figure 7. Comparison of calculated intensities and experimental data + D = 6.35 meV, $2\chi = 2.1$ $\overset{\circ}{A}^{-1}$, $k_i = 11$ $\overset{\circ}{A}^{-1}$ or $E_i = 63$ meV–adjustable parameter : $\alpha = 0.25$.

    We emphasize that the fit is obtained with the same value of α whatever the incident angle and energy may be. Some small differences can be observed in the high temperature region (T > 300°K) due probably to double phonons virtual exchange or anharmonic effects which are not included in the calculation.

    However we do not attach to this very good result an absolute meaning. The most important thing which seems to us meaningful is that the general shape, and more precisely the curvature of the experimental and calculated curves are the same in practically all of the temperature interval.

    This confirms our previous statement: the reflection coefficient and crystal temperature are not linked by a simple exponential relation. In this sense one cannot call this effect "a Debye-Waller factor" but rather one should speak of beam thermal attenuation.

7. CONCLUSION

As a contribution to the problem of atom surface inelastic scattering,

we have presented a procedure which allows one to perform the thermal
average and therefore to get an exact expression for the T matrix element to all orders of perturbation. This expansion can be ordered in
number of phonons involved in the studied process.

Analytical expressions and numerical calculations show that the
variation of the reflection coefficient for the specular beam as a
function of the crystal temperature cannot be represented by the usual
Debye-Waller factor. The examination of the experimental data confirm
this statement.

This important conclusion can be extended to the diffracted beams
as it is contained intrinsically in the T matrix equation.

ACKNOWLEDGMENTS

This work was supported in part by the NATO research grant RG 86.81.

REFERENCES

[1]   G. Armand, J.R. Manson, Surf. Science, 80, 532 (1979).

[2]   N. Garcia, G. Benedek, Surf. Science, 80, 543 (1979).

[3]   N. Garcia, A. Maradudin, Surf. Science, 119, 384 (1982).

[4]   J. Perreau, J. Lapujoulade, Surf. Science, 119, L292 (1982).

[5]   B. Salanon, G. Armand, J. Perreau, J. Lapujoulade, Surf. Science,
       127, 135 (1983).

[6]   A.C. Levi, H. Suhl, Surf. Science, 88, 221 (1979).

[7]   H.D. Mayer, Surf. Science, 104, 117 (1981).

[8]   R.B. Gerber, A.T. Yinnon, J. Chem. Phys. 73, 3232 (1980).

[9]   G. Armand, J.R. Manson, Journal de Physique, 44, 473 (1983).

[10]   G. Armand, J.R. Manson, Surf. Science, 119, L299 (1982).

[11]   G. Armand, J. Lapujoulade, Y. Lejay, Surf. Science, 63, 143 (1977).

[12]   G. Armand, Journal de Physique, 38, 989 (1977).

[13]   J. Perreau, J. Lapujoulade, Surf. Science, 122, 59 (1983).

[14]   J. Lapujoulade, Y. Lejay, G. Armand, Surf. Science, 95, 107 (1980).

DYNAMICS AND KINETICS ON SURFACES EXHIBITING DEFECTS

Herschel Rabitz
Department of Chemistry
Princeton University
Princeton, New Jersey 08544

Abstract

This paper considers dynamics and kinetic phenomena on surfaces
with defect structures. Quantum mechanical scattering as well as
kinetics and diffusion processes are treated as examples. In the
case of quantum mechanics, a theoretical formulation is presented
capable of handling atomic disorder on surfaces. The aproach is
based on an approximation which is best in the quantum long wavelength regime. Some simple illustrations considering scattering off
ordered and disordered lattices will be presented. The second part
of the paper considers the steady state non-linear reaction-
diffusion equations used to describe adsorption, desorption, diffusion and reaction on surfaces with macroscopic scale defects.
These latter defects may arise due to inherent faulting of the lattice or foreign material on the surface. Defects within a mean
diffusion length of each other are shown to exhibit cooperativity
in their chemical properties. Finally, in the last portion of the
paper reaction-diffusion models are again considered from a different perspective. In this case, all of the non-linear chemical or
desorptive aspects of the problem are restricted to the edges of the
defect sites and the intervening surface is assumed to be characterized by adsorption, desorption and simple diffusion. These physically realistic models clearly show the capability of multiple
steady states existing on active chemical surfaces. A variety of
non-linear surface phenomena could possibly be important in
appropriate gas-surface systems.

I. Introduction

Gas-surface phenomena has been studied for many years at both
the kinetic level and the microscopic atomic scale. Recently, great
effort has been put forth on experiments and theoretical analysis
involving "clean perfect" surfaces. These studies are largely motivated by the desire to obtain fundamental information under conditions which would be optimum for achieving a clear physical
understanding of the relevant gas-surface processes. Much information has been gained from these studies and vigorous efforts will
surely continue. Nevertheless, it is well-recognized that realistic
chemically reactive surfaces will likely have a high density of
defect structures. Indeed, the presence of defects may be responsible for much of the chemical acitivity of some surfaces. Defects
may arise due to a variety of lattice faulting or atomic disloca-
tions as well as foreign substances residing on the solid surface.

B. Pullman et al. (eds.), Dynamics on Surfaces, 77–88.
© 1984 by D. Reidel Publishing Company.

The study of problems of this type ultimately necessitates a statistical treatment of the defect distributions. Clearly, the
theoretical analysis of these problems is more complex than treatments on well-defined ordered surfaces, and in some cases new techniques may be warranted for this purpose. The present paper
considers this class of problems at both the microscopic atomic
scale and in the macroscopic kinetic regime.

Relatively little effort has thusfar been devoted to atomic
scattering off disordered surfaces.[1] An examination of the
diffraction structure of non-reactive atomic scattering could in
principle be used as a probe of the disordered surface structures.
This technique could be particularly valuable since the scattering
process depends on intermolecular forces of the same general nature
as would arise when utilizing these disordered surfaces for other
chemical processes. In treating quantum scattering off disordered
surfaces, the theoretical treatment must ultimately be practical
enough to admit a statistical average over the defect structures.
In particular, a calculation with a given structure must be sufficiently easy to perform in order for the statistics to be carried
out. This goal becomes a necessary guideline to the development of
appropriate dynamical approximations. In this regard, recently the
sudden approximation was examined as a tool for studying scattering
off disordered surfaces.[2] The simplicity of the sudden approximation
leaves the stochastic aspects of the problem as "bare" as possible.
However, it is desirable to consider approaches applicable in a
regime complimentary to that of the sudden approximation. In particular, low energy collisions on surfaces with high corregation or
giving rise to large momentum transfer would not be amenable to the
sudden technique. The present paper considers an approach which
formally becomes exact in the long wavelength low-energy limit, and
its simplicity should allow for the practical execution of statistical averaging. The technique should also have application to analogous atom-polyatomic gas phase scattering and both types of
applications are being pursued.

The defect structure of concern in the last paragraph are at
the atomic scale and the second portion of this paper is primarily
concerned with larger scale faulting of the surface. Such large
scale faults could arise as mere aggregates of atomic defects or
through inherent crystal flaws. The physical length scales under
consideration here would typically be in the range > 10Å where a
continuum approach is appropriate. Therefore, the relevant processes are described by continuous mass and energy balance
equations. A steady state experiment in this regime might consist of
a catalytically active faulted surface exposed to a steady flux of
reactants with a monitoring of the corresponding product yields. A
series of experiments could be envisioned on surfaces having known
statistically averaged properties. Models and theoretical analyses
of this type essentially bridge the gap between the former quantum
studies and those situations that are believed to exist with

realistic heterogeneous catalysts. The present paper will consider
two models for these phenomena. First, the steady state two dimensional non-linear reaction-diffusion equations will be developed for
processes on an anisotropic (defect structure) surface. The
resultant steady state chemical concentrations will be examined with
the emphasis on the role played by the defects. Special attention
will be given to the potential modification of chemical activity
caused by the defects. In a final related study, the possibility of
multiple steady state solutions on such chemically active surfaces
will be considered. In order to simplify the mathematics in this
case, all the chemistry or other non-linear processes will be confined to the fault structure boundaries. Mathematically this
translates to the non-linearities residing exclusively in the boundary conditions on the differential equations. The origin of the
multiple solutions will be discussed and their consequences considered.

II. An Average Wave Function Approach to Surface Scattering.

The problem under consideration here will consist of atomic
scattering off a rigid surface. Although the treatment can certainly be applied to ordered lattices, the primary goal is the ultimate application to disordered systems. For mathematical simplicity
a plane hard wall reference potential $V_0(z)$ will be assumed to exist

$$V_0(z) \quad = \begin{cases} \infty, \ z > 0 \\ \\ 0, \ z < 0 \end{cases} \tag{1}$$

Therefore the reference hamiltonian $H_0 = -\hbar^2/2m \ \nabla^2 + V_0(z)$ describes
specular scattering of atoms of mass m. The solution to the
corresponding distorted wave Schrodinger equation $H_0\phi(\vec{r}) = \hbar^2 q^2/2m\phi(\vec{r})$ has the form

$$\phi(\vec{r}) = \sin(kz)\exp(i\vec{K}\cdot\vec{R}) \tag{2}$$

where $\vec{q} = k\hat{z} + \vec{K}$ with K being a wave vector parallel to the surface.
By definition, Eq. (2) is only applicable in the region $z < 0$. The
full potential for the problem is the sum of the term in Eq. (1) and
an interaction $V_1(z,R)$ between the scattering beam and atoms
residing on the hard surface. Therefore, the full Schrodinger
equation becomes

$$\left[-\frac{\hbar^2}{2m} \ \nabla^2 + V_0(z) + V_1(\vec{r}) \right] \psi(\vec{r}) = E\psi(\vec{r}) \tag{3}$$

The statistical aspect of the problem resides in the interaction V_1.
The statistics will enter due to the fact that the atoms bound to
the surface contributing to V_1 can be in random locations along or
above the surface. In addition, these atoms could be of a different
nature corresponding to a chemically heterogeneous surface. Since

ultimately no periodicity may exist, a flexible potential form consists of simply a sum of interactions with each of the surface
atoms.

$$V_1(\vec{r}) = \sum_i v_i(\vec{r} - \vec{r}_i) \tag{4}$$

The functional form of the terms in Eq. (4) could be different,
depending on the nature of the surface atoms.

In order to develop a solution to Eq. (3), it is most convenient to rewrite the problem as an integral equation

$$\psi(\vec{r}) = \phi(\vec{r}) + \sum_i \int G(\vec{r},\vec{r}') v_i(\vec{r}-\vec{r}_i) \psi(\vec{r}') d\vec{r}' \tag{5}$$

The Green's function satisfies the equation

$$[E + \frac{\hbar^2}{2m} \nabla^2 - V_0] G(\vec{r},\vec{r}') = \delta(\vec{r}-\vec{r}') \tag{6}$$

with the boundary condition of being zero at the surface. We assume
that each of the terms v_i entering into the potential on the right
hand side of Eq. (5) is a localized function decaying rapidly as a
function of $\vec{r}'$ away from the point $\vec{r}_i$. The Green's function
entering into the integrand of Eq. (5) is a smooth function of $\vec{r}'$,
except for the singularity occuring at $\vec{r}' = \vec{r}$. Therefore, the ker-
nel in Eq. (5) contains a number of strong localized spikes centered at the atomic surface locations. Keeping this in mind, a
natural approximation to consider is the evaluation of the wavefunction in the integrand at just the atomic site locations. Such an
approximation would be the crudest form of a technique described by
Luchka as the method of average functional corrections.[3]
Unfortunately, this zeroth order approach is not adequate since a
more careful examination will show that the resulting implied potential is not hermitian, thereby destroying the basic property of flux
conservation. A more general and physically flexible approximation
based on the same observations can be made having the following structure[4]

$$\psi(\vec{r}) = \phi(\vec{r}) + \sum_i \int d\vec{r}' G(\vec{r},\vec{r}') v_i(\vec{r}'-\vec{r}_i) w_i(\vec{r}')$$
$$\times \int v_i(\vec{r}''-\vec{r}_i) w_i(\vec{r}'') \psi(\vec{r}'') d\vec{r}'' \tag{7}$$

where $w_i(\vec{r})$ can be thought of as a weight function carrying information about the scattering process. The integral

$$\int v_i(\vec{r}''-\vec{r}_i) w_i(\vec{r}'') \psi(\vec{r}'') d\vec{r}'' \tag{8}$$

in Eq. (7) is a localized projection of the wavefunction near the site of the i-th atom. For the special case of $w_i(\vec{r}") = \delta(\vec{r}"-\vec{r}_i)/v_i(0)$ used in Eq. (8) (but not in the Green's function integral of Eq. (7)), the former mentioned pole-like approximation of Eq. (5) will be produced. The essential approximation in Eq. (7) is equivalent to replacing the i-th term of the potential by a non-local interaction

$$v_i(\vec{r}-\vec{r}_i)\psi(\vec{r}) \rightarrow v_i(\vec{r}-\vec{r}_i)w_i(\vec{r}) \int v_i(\vec{r}'-\vec{r}_i)w_i(\vec{r}')\psi(\vec{r}')d\vec{r}' \qquad (9)$$

Therefore, the potential now takes on the form of a projection operator

$$V_1 \approx \sum_i |v_iw_i\rangle\langle v_iw_i| , \qquad (10)$$

with the justification of this structure arising due to the localized nature of the interactions in Eq. (4). In practice, this procedure has a more formal mathematical foundation in the treatment by Luchka and a hierarchy of approximations may be considered based on this concept.[3] Finally, in order to be specific, a choice for the weight $w_i(\vec{r})$ must be made. In order to preserve the norm (potential strength) it is natural to expect that $w_i(\vec{r})$ be proportional to $|\int v_i(\vec{r})d\vec{r}|^{-1/2}$. In addition, due to the role of $w_i(\vec{r})$ in the Green's function integral of Eq. (7), we would also expect the weight to contain at least zeroth order information about the scattering process and this is in accord with expectations from Luchka's technique as extended by Malik and Weare.[3] Therefore, a natural choice to make is

$$w_i(\vec{r}) = \phi(\vec{r}) \; |\int v_i(\vec{r})d\vec{r}|^{-1/2}; \qquad (11)$$

although other choices are possible. The ultimate quality of the answer can be effected by the choice, and variational optimization may be useful here.

Equation (7) may be readily solved by taking moments with respect to the set of functions $w_j(\vec{r})v_j(\vec{r}-\vec{r}_j)$ and truncating the number of surface atoms to a practical size. The resultant operations will produce a set of matrix equations whose dimension is the number of surface atoms regardless of the degree of disorder. Following these procedures, the solution to Eq. (7) may be generated

$$\psi(\vec{r}) = \phi(\vec{r}) + \sum_{ij} \int d\vec{r}'G(\vec{r},\vec{r}')v_i(\vec{r}'-\vec{r}_i)w_i(\vec{r}')$$

$$\times \; (1 - M)_{ij}^{-1} \int v_j(\vec{r}"-\vec{r}_j)w_j(\vec{r}") \; (\vec{r}")d\vec{r}" \qquad (12)$$

where

$$M_{ij} = \int\int d\vec{r}\,d\vec{r}' \; v_i(\vec{r}-\vec{r}_i)w_i(\vec{r})G(\vec{r},\vec{r}')v_j(\vec{r}'-\vec{r}_j)w_j(\vec{r}') \qquad (13)$$

The essentially analytic structure of the solution in Eq. (12) is one of its strong attractions when keeping in mind the desire to perform statistical averages over the potential which is explictly present. At this point, only preliminary applications of this formalism have been made and a few general comments will be presented. First, the basic approximation will become more accurate in the long wavelength limit where the wavefunction entering into the right hand side of Eq. (5) is more slowly varying. Therefore, this technique can be viewed as complementary to the applicability of the sudden approximation. The necessity of a practical evaluation of the integrals in Eq. (13) may put restrictions on acceptable forms for the potential. With the expression in Eq. (11), the integrals required in Eqs. (12) and (13) may be analytically evaluated for Gaussian potentials centered at each atomic site. This form may easily be extended to superpositions of Gaussians, thereby allowing for flexible potential forms. The example shown in Figure 1 is for

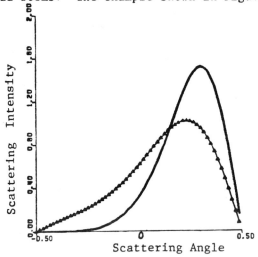

Figure 1. Scattering off a test sample consisting of a 3 x 3 array of atoms on the surface. The scattering is at low energy $qa = 0.5$ where a is the potential range and the incident angle is at -0.2 radians from the normal to the surface. The solid curve is for an ordered array of atoms and the curve denoted by solid triangles comes from a set of 10 statistical samples of the potential range about its nominal value. The disordered surface scattering is more specular and also broader.

scattering off of a 3 x 3 array of atoms on the surface. The effect of potential strength, atomic height from the surface, scattering energy and potential range have been investigated in a number of calculations. A detailed analysis of these factors will be presented in a forthcoming paper.[4]

II. Steady State Reactive Kinetics on Surfaces Exhibiting Defect Structures.

The material in the last section was confined to the quantum mechanical regime of atomic scattering. Ultimately, the processes occuring at that scale would be superimposed and taken to the macroscopic regime of kinetics and diffusion. Length scales in the present case transcend atomic dimensions thereby justifying the continuum picture. The fundamental equations in this limit have the form

$$\frac{\partial Q}{\partial t} = -\vec{\nabla} \cdot \vec{j} + \text{sources} - \text{sinks} \tag{14}$$

where Q is either a species concentration or the energy density, and $\vec{j}$ is correspondingly a mass or energy flux upon the surface. In contrast the sources or sinks correspond to material arriving or leaving the surface associated with the external environment. In the present work thermal diffusion on the surface was neglected, while mass diffusion was treated through Fick's law $\vec{j}_i = -D_i \vec{\nabla} c_i$, where D_i is the i-th diffusion coefficient and c_i is the i-th species concentration. Material sources are assumed to be due to material striking and adhering to the surface at a rate characterized by $s_i \exp[-c_i/c_{io}]$, where s_i is the i-th sticking coefficient and c_{io} is a characteristic saturation concentration coverage. Other forms besides an exponential could be chosen to account for saturation effects if such specific information was available. In addition to material arriving at the surface, it may also desorb with a rate $\Gamma_i c_i$ which is assumed to be proportional to the chemical species concentration through a characteristic constant Γ_i. Finally, the species which have adhered to the surface and may diffuse and also react. In the specific model chosen for illustration, this reaction was assumed to have a bimolecular form $k c_i c_j$. Temperature effects were also included in the form of an energy balance equation connecting the bulk-surface temperature differential to the enthalpy of the reaction, however the present discussion will assume an isothermal system at the mean surface temperature. Finally, to be specific, all the above information may be assembled into a model set of two species reactions in the steady state limit having the form[5]

$$D_i \nabla^2 c_i - k c_1 c_2 - \Gamma_i c_i + s_i \exp(-c_i/c_{io}) = 0, \quad i = 1,2 \tag{15}$$

As indicated above, further generalizations may be readily included as the cases warrant. The problem is finally specified by

prescribing constant flux boundary conditions $\nabla^2 c_i = 0$ on the edge
of a square region having sides of length L. The actual model
calculations performed were done in dimensionless units, and typical
molecular constants give a scale length $L \sim 4000\text{Å}$. The coupled
system of Eqs. (15) was solved by the multigrid technique[6] since it
represents a highly efficient algorithm for this purpose. Although
a full statistical simulation of surface defect structures has not
been performed at this point, the numerical procedures are clearly
efficient enough to carry out such calculations.

The model calculations[5] were performed largely to illustrate the
capability of performing simulations of this type and also to
establish the degree of system sensitivity to chemically active
defect structures. In the domain of area, L^2 four "random" line
defects were located. These defects were assumed to have a domain
of influence falling rapidly in a Gaussian fashion away from their
central locations. The defects were taken to affect the chemical
and transport processes through their distinctive adsorptive or
reactive properties relative to the background surface "plane".
Using the halfwidths of the domains of influence as characteristic
sizes of the defects, the models typically corresponded to defect
coverages of ~10%. An illustration of the results is given in Figure 2

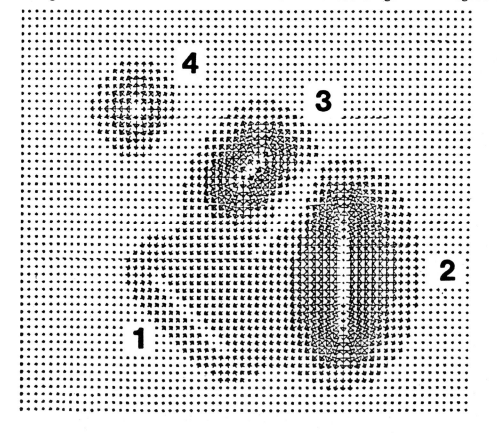

Figure 2. A flux diagram for one chemical species in the vicinity of four absorptive sites. The site labelled 1 is also capable of reactive enhancement of the two chemical species on the surface. It is apparent that sites 1, 2 and 3 are acting in a cooperative fashion with material flowing from the latter two sites to the reactive fault. The size and direction of the arrows is proportional to the species flux vector.

which presents a flux diagram for one of the chemical species. The direction and magnitude of the arrows in the diagram correspond to the concentration flux $\vec{j}_1 = -D_1 \vec{\nabla} c_1$. Site number 1 was chosen to be reactive while all the sites had preferential absorptive characteristics for both chemical species. It is quite evident that a high degree of cooperativity exists between sites 1, 2 and 3 with material flowing from the latter two sites to the chemically active site 1. Site 4 does not exhibit this behavior and in general site-site cooperativity was found to exist when the mean diffusion length was the order of the spacing between defect regions. One of the most important conclusions from this work concerned the net chemical production rate

$$K = \frac{k}{L^2} \iint c_1(x,y) c_2(x,y) \, dx \, dy$$

where the integral is over the surface domain. Dramatic variations in the production rate were found depending on the particular choices of the absorptive and reactive properties of the reactive sites. The study strongly suggests that a thorough examination of such a model on realistic systems would produce interesting insight into the kinetics of catalytic surfaces.

III. The Prospects for Multiple Steady States on Faulted Surfaces.

The differential Eqs. (15) used in the last section to describe chemically active surfaces are of a non-linear nature. In the present case, these non-linearities occured in the reactive and absorptive terms, and in general, other non-linearities could arise. It is well-known that non-linear differential equations of the type in Eq. (15) are capable of having more than one non-trivial solution. A common physical circumstance of this type occurs in flow reactors arising in chemical engineering applications.[7] From this perspective, a chemically active surface may be viewed as simply a two-dimensional flow reactor with materials arriving at the surface, undergoing various physical processes and ultimately leaving unaltered or perhaps as new chemical species. If multiple steady states can exist, they must be taken into account if a proper description of the surface chemical activity is to be realized. Nearby multiple steady states may also "communicate" due to random system fluctuations thereby producing temporal excursions even in steady state systems. Problems of this type can be extremely

serious if undesirable steady states are capable of quenching the
desirable reactions or producing runaway systems. The latter circumstance will likely not occur on solid surfaces due to geometrical
considerations, but a variety of phenomena could occur if multiple
steady states were accessible. For example, otherwise steady
systems might enter into an oscillating mode, and this sort of phenomena has already been seen on a variety of surfaces where temperature effects can couple to chemical processes. In the present
paper, we exclude thermal effects and only consider pure chemistry.

The general non-linear nature of Eq. (15) gives rise to systems
which must be solved numerically as performed above. Therefore to
gain insight into the possibility of multiple steady states, a different model is needed.[8] In this new picture, material arriving at
the surface is assumed to stick with an appropriate probability,
diffuse on the surface, perhaps desorb or perhaps reach boundary
regions where it may react or desorb. The essential point is that
Eq. (14) is now assumed to be entirely linear with all of the chemically or generally non-linear processes occuring at the active sites
where boundary conditions are imposed. The linearity of Eq. (14) is
a reasonable assumption under conditions of low surface coverages
and should be easily realizable in the laboratory. In addition, the
presence of defect sites could readily give rise to the chemistry
exclusively occuring at the edges of these sites. In a similar
vein, even if no reaction exists at these sites, there still could[9]
be non-linear desorptive processes located at the same positions.
Besides the physical clarity of these models, they also have strong
appeal from a mathematical point of view. In particular, the[10]
general solution to linear diffusion equations is well known[10] and
this fact may be utilized to focus on the essential chemical processes occuring in the boundary conditions. Indeed, in many cases,
it may be possible to obtain analytic or near analytic solutions to
reasonable physical problems.

As an illustration of the basic points discussed above, we may
consider a steady-state system in one-dimension (for the sake of
illustration). Furthermore, we will assume only a single chemical
species arriving at the surface and therefore we do not consider
chemical reactivity. The conservation equation on the surface now
takes on the form

$$D \frac{d^2 c}{dx^2} - c/\tau + I = 0 \tag{16}$$

where I is the constant incident flux and $1/\tau$ is the characteristic
rate constant for desorption not in the vicinity of active boundaries. The simplest possible case would consist of a semi-infinite
domain $x > 0$ with the chemically active site located at $x = 0$.
Material is now assumed to desorb on the active site with a rate
given by

$$D \frac{dc}{dx} \Big|_{x=0} = F(c) \Big|_{x=0} \tag{17}$$

where F(c) is a characteristic non-linear function of the concentration at the active surface. In order to mathematically
complete the problem, the concentration $c(\infty)$ is taken as finite.
This latter condition along with Eq. (17) specify the boundary conditions for Eq. (16). The general solution to Eq. (16) has the form

$$c = I\tau + A\exp[-x(D\tau)^{-1/2}] \tag{18}$$

where A is an arbitrary constant. This latter constant is determined by substitution into Eq. (17). Since the latter equation is
non-linear is it quite apparent that one may arrive at multiple
solutions for the constant A and therefore the physical concentration in Eq. (18). As a specific case, we may choose F(c) =
$kc(c_1-c)$ where k is a characteristic desorption rate constant at the
defect and c_1 as a defect saturation concentration. Substitution of
the general solution in Eq. (18) into Eq. (17) will now produce the
determining equation

$$-A(D\tau)^{-1/2} = k(I\tau + A)(c_1 - I\tau - A) . \tag{19}$$

A simple analysis will indicate that two real solutions to Eq. (19)
are possible giving rise to positive definite solutions for the concentration in Eq. (18). These physically acceptable solutions can
occur for a broad class of parameter values. A variety of other
physical choices for F may be taken in Eq. (17) producing similar general conclusions.

The above example simply illustrates the physical possibility
of obtaining multiple steady state solutions to kinetic phenomena on
active faulted surfaces. A host of additional physical questions
must be addressed. For example, the accessibility or stability of
each steady state must be understood. The existence of two or more
isolated and stable steady states would indicate that a system prepared under different conditions might exhibit different reactive or
steady state surface concentrations. Secondly, the statistical
nature of the surface defects must be treated in regard to multiple
solutions. This issue could be quite complicted when the defect density is high enough for the sites to be interactive. However, the
physically realizable case of low surface defect density would be
much simpler and it is certainly the best case for initial study.
The ultimate physical consequences of this phenomena are not clear
at this time but they could have significant implications for heterogeneous chemical reactivity.

Acknowledgement: The author acknowledges support from the Air Force Office of Scientific Research, the Office of Naval Research and the National Science Foundation.

References

1. J. Lapujoulade and Y. Lejay, J. Phys. (Paris) Lett. 38, L303 (1977); J. Horne and D. Miller, J. Vac. Sci. Technol. 13, 355 (1976).

2. J. Gersten, R. Gerber, D. Dacol and H. Rabitz, J. Chem. Phys. 78, 4277 (1983).

3. A. Luchka, The Method of Averaging Functional Corrections (Academic Press, New York, 1965); D. Malik and J. Weare, J. Chem. Phys. 62, 1044 (1975).

4. L. Singh, D. Dacol and H. Rabitz, to be published.

5. F. Grinstein, H. Rabitz and A. Askar, "Steady State Reactive Kinetics on Surfaces Exhibiting Defect Structures", J. Chem. Phys., submitted.

6. W. Hackbusch and U. Trottenberg, Multigrid Methods, in Lecture Notes in Mathematics, No. 960 (Springer-Verlag, New York, 1982).

7. J. Seinfeld and L. Lapidus, Mathematical Methods in Chemical Engineering, (Prentice-Hall, Englewood Cliffs, 1974).

8. A. Askar and H. Rabitz, to be published.

9. J. Tully and M. Cardillo, Science, Vol. 223, p. 445 (1984).

10. J. Crank, The Mathematics of Diffusion, (Clarendon Press, Oxford 1975).

MAGNETIC TRANSITIONS IN HETERONUCLEAR AND HOMONUCLEAR MOLECULE-
CORRUGATED SURFACE SCATTERING

T.R. Proctor and D.J. Kouri
Department of Chemistry and Department of Physics
University of Houston - University Park
Houston, Texas 77004

In this paper we present formal and computational studies of Δm_j -
transitions occuring in heteronuclear and homonuclear molecule - corrugated surface collisions. The model potential is a pairwise additive
one which correctly incorporates the fact that Δm_j - transitions occur
only for corrugated surfaces (provided the quantization axis is chosen
to be the average surface normal). The principal results are; a) Δm_j -
transitions are sensitive to lattice symmetry b) strong selection rules
obtain for specular scattering c) the magnitude of Δm_j - transition
probabilities are strongly sensitive to surface corrugation d) the
ratio of molecular length to lattice dimension (r/a) has a strong influence on the magnitude of Δm_j - transition probabilities (with the
probabilities increasing as (r/a) increases) e) Δm_j - rainbows are
predicted to occur. Computations are reported illustrating the above
for both heteronuclear and homonuclear molecule - corrugated surface
collisions.

1.   INTRODUCTION

Quantum mechanical studies of molecule - surface scattering are now
receiving a great deal of attention.(1-12) An overview has recently
been presented which discussed interesting phenomena in rotationally
inelastic collisions observed experimentally and/or theoretically.(11)
Molecular reorientation by collision with a surface can provide information on surface structure and corrugation when the quantization axis
for the molecular rotational states is taken to be the normal to the
surface.(11-13) For such a choice, $\Delta m_j \neq 0$ transitions occur <u>only</u> due to
the surface corrugation. Recently, Lunz, et.al. (14) report the observation of molecular polarization by surface impact. The experiment
consisted of scattering of NO by Ag (111), such that the NO was initially in the $j \sim =0$ rotor state. The NO molecules were preferentially scattered with their angular momentum parallel to the surface (then $m_{j_f}=0$)
so that $\Delta m_j \sim =0$. This experiment, plus others now being planned, suggest
that quantal calculations of magnetic transitions induced by molecule -
corrugated surface collisions should be carried out. The paper of

89

B. Pullman et al. (eds.), Dynamics on Surfaces, 89–102.
© 1984 by D. Reidel Publishing Company.

Proctor, et.al. (12) reports the first such calculations. In that work, it was found that for homonuclear - square or rectangular corrugated lattice collisions, there occurred very strong effects due to lattice symmetry. Thus, they found that a) selection rules occurred for specular scattering b) Δm_j - transitions are sensitive to surface corrugation c) such transitions depend strongly on diffraction peak d) the ratio of molecular length to lattice dimension is an important parameter and Δm_j - transitions are more probable for longer molecules e) Δm_j - rainbows are expected to accompany Δj - rotational rainbows. Subsequently, Proctor and Kouri (13) have studied computationally the possible existence of magnetic transition rainbows. The present paper is concerned with examining whether similar effects occur when one studies heteronuclear molecule - corrugated surface collisions, as well as the effect of modification of the potential from a purely repulsive one to a potential having both attraction and repulsion. The paper is organized as follows. In Section 2 we briefly discuss the potentials and equations used in our study. Then we present an analysis of selection rules and the m_j - rainbow in Section 3. Finally, in Section 4 we present and discuss our computational results.

2. MODEL POTENTIALS AND BASIC EQUATIONS FOR MOLECULE - SURFACE COLLISIONS

As in the initial study of Δm_j - transitions induced by molecule - corrugated surface collisions, we have adopted the dumbell potential of Gerber, Beard and Kouri.(6) In this model, the potential is assumed to be a sum of two atom - surface interactions. The coordinates are (x,y,z) and the diatom internuclear vector coordinates (r,θ,ϕ), with θ being the polar angle measured from the surface normal and ϕ the azimuthal angle measured relative to the x-axis in the surface. It is easily seen that only when the surface is corrugated will there be a ϕ-dependence to the potential.(11) Thus, the molecule - surface potential is of the form (6)

$$V(xyz\theta\phi) = V_1(x_1y_1z_1) + V_2(x_2y_2z_2) \ , \qquad (1)$$

where V_i is the interaction of surface with atom i, $\underset{\sim}{r}_i = (x_iy_iz_i)$ is the vector from the origin of coordinates (located on the surface) to atom i, and we assume for $V_i(x_iy_iz_i)$ the simple interaction

$$V_i(x_iy_iz_i) = v_i(z_i)[1+\beta_iQ_i(x_iy_i)] \ . \qquad (2)$$

Here, β_i determines the strength of surface corrugation felt by atom i, $Q_i(x_iy_i)$ determines the geometry of the surface as seen by atom i and $v_i(z_i)$ is the portion of the interaction which depends on how high above the surface atom i is. For the homonuclear case, β_1 and β_2 are equal and the forms of the functions v_i and Q_i are independent of the index i. In our calculations, $Q(x_iy_i)$ is taken to be of the form

$$Q(x_i, y_i) = \cos\left(\frac{2\pi x_i}{a}\right) + \cos\left(\frac{2\pi y_i}{b}\right) , \tag{3}$$

where a and b are lattice dimensions in x and y respectively. We shall consider both an exponential repulsion and Morse potential form for $v(z_i)$. In the case of a heteronuclear interacting via a sum of repulsive atom – surface potentials, the dumbell potential becomes

$$V = A_1 e^{-\alpha_1 z} \; e^{-\alpha_1 \frac{dm_2}{M} \cos\theta} \; [1 + \beta_1 \{ \cos\frac{2\pi x}{a} \cos\left(\frac{2\pi dm_2}{Ma}\right) \sin\theta\cos\phi$$

$$-\sin\left(\frac{2\pi x}{a}\right) \sin\left(\frac{2\pi dm_2}{Ma}\right) \sin\theta\cos\phi\right) + \cos\left(\frac{2\pi y}{b}\right) \cos\left(\frac{2\pi dm_2}{Mb}\right) \sin\theta\sin\phi$$

$$-\sin\left(\frac{2\pi y}{b}\right) \sin\left(\frac{2\pi dm_2}{Mb}\right) \sin\theta\sin\phi\}]$$

$$+ A_2 \; e^{-\alpha_2 z} \; e^{+\alpha_2 \frac{dm_1}{M} \cos\theta} \; [1 + \beta_2 \{ \cos\left(\frac{2\pi x}{a}\right) \cos\left(\frac{2\pi dm_1}{Ma}\right) \sin\theta\cos\phi$$

$$+\sin\left(\frac{2\pi x}{a}\right) \sin\left(\frac{2\pi dm_1}{Ma}\right) \sin\theta\cos\phi\right) + \cos\left(\frac{2\pi y}{b}\right) \cos\left(\frac{2\pi dm_1}{Mb}\right) \sin\theta\sin\phi$$

$$+\sin\left(\frac{2\pi y}{b}\right) \sin\left(\frac{2\pi dm_1}{Mb}\right) \sin\theta\sin\phi\}] \; . \tag{4}$$

Here, d is the interatomic distance of the diatom. This reduces to the form used earlier in the homonuclear case (where $m_1/M = m_2/M = 1/2$). When one uses a Morse potential for each of the atom – surface interactions, then the result for a heteronuclear molecule is

$$V = A_1 \; [e^{-2\alpha_1 (z-z_1)} \; e^{-2\alpha_1 \frac{dm_2}{M} \cos\theta} - 2 \; e^{-\alpha_1 (z-z_1)} \; e^{-\alpha_1 \frac{dm_2}{M} \cos\theta}]$$

$$\times [1 + \beta_1 \{ \cos\left(\frac{2\pi x}{a}\right) \cos\left(\frac{2\pi dm_2}{Ma}\right) \sin\theta\cos\phi - \sin\left(\frac{2\pi x}{a}\right) \sin\left(\frac{2\pi dm_2}{Ma}\right) \sin\theta\cos\phi$$

$$+\cos\left(\frac{2\pi y}{b}\right) \cos\left(\frac{2\pi dm_2}{Mb}\right) \sin\theta\sin\phi - \sin\left(\frac{2\pi y}{b}\right) \sin\left(\frac{2\pi dm_2}{Mb}\right) \sin\theta\sin\phi\}]$$

$$+A_2 \; [e^{-2\alpha_2 (z-z_2)} \; e^{2\alpha_2 \frac{dm_1}{M} \cos\theta} - 2 \; e^{-\alpha_2 (z-z_2)} \; e^{\alpha_2 \frac{dm_1}{M} \cos\theta}]$$

$$\times [1 + \beta_2 \{ \cos\left(\frac{2\pi x}{a}\right) \cos\left(\frac{2\pi dm_1}{Ma}\right) \sin\theta\cos\phi + \sin\left(\frac{2\pi x}{a}\right) \sin\left(\frac{2\pi dm_1}{Ma}\right) \sin\theta\cos\phi$$

$$+ \cos\left(\frac{2\pi y}{b}\right) \cos\left(\frac{2\pi dm_1}{Mb}\right) \sin\theta\sin\phi + \sin\left(\frac{2\pi y}{b}\right) \sin\left(\frac{2\pi dm_1}{Mb}\right) \sin\theta\sin\phi\}] \tag{5}$$

where z_i is the minimum of the Morse potential for the i th atom inter-

acting with the surface. In the case of a homonuclear, this reduces
to

$$V = 2A \; e^{-2\alpha(z-z_o)} \{ \cosh(\alpha d \cos\theta)[1+\beta[\cos\left(\frac{2\pi x}{a}\right) \cos\left(\frac{\pi d}{a} \sin\theta\cos\phi\right)$$

$$+ \cos\left(\frac{2\pi y}{b}\right) \cos\left(\frac{\pi d}{b} \sin\theta\sin\phi\right)]] + \beta\sinh(\alpha d \cos) [\sin\left(\frac{2\pi x}{a}\right)$$

$$x \;\; \sin\left(\frac{\pi d}{a} \sin\theta\cos\phi\right) + \sin\left(\frac{2\pi y}{b}\right) \sin\left(\frac{\pi d}{b} \sin\theta\sin\phi\right)]\}$$

$$- 4A \; e^{-\alpha(z-z_o)} \{ \cosh\left(\frac{\alpha d \cos\theta}{2}\right)[1+\beta[\cos\left(\frac{2\pi x}{a}\right) \cos\left(\frac{\pi d}{a} \sin\theta\cos\phi\right)$$

$$+ \cos\left(\frac{2\pi y}{b}\right) \cos\left(\frac{\pi d}{b} \sin\theta\sin\phi\right)]] + \beta\sinh\left(\frac{\alpha d \cos\theta}{2}\right) [\sin\left(\frac{2\pi x}{a}\right)$$

$$x \; \sin\left(\frac{\pi d}{a} \sin\theta\cos\phi\right) + \sin\left(\frac{2\pi y}{b}\right) \sin\left(\frac{\pi d}{b} \sin\theta\sin\phi\right)]\} \qquad (6)$$

It is interesting to note that for the Morse atom-surface interaction,
one does <u>not</u> obtain a potential of the form of a single Morse potential (for the molecular center of mass) times an angular dependent
function except when the molecule is oriented at an angle of $\theta=\pi/2$ with
respect to the surface normal.
        The Schrodinger equation for the collision is

$$[-\frac{\hbar^2}{2m} \{ \frac{\partial^2}{\partial x^2} + \frac{\partial^2}{\partial y^2} + \frac{\partial^2}{\partial z^2}\} + \frac{1}{2I} \; j^2 + V(xyz\theta\phi)]\psi = E\psi \; . \qquad (7)$$

We shall give results for which the diffractive degrees of freedom,
ρ ($\equiv x$ and y), are treated within the sudden approximation.(15)   The
(θ,ϕ) rotor degrees of freedom will also be treated by two methods in
different calculations. These include an exact close coupling treatment of the rotor and also the rotational sudden approximation in which
we do the integrals over (x,y,θ,ϕ) numerically. In the diffractive
sudden – rotational close coupling case (DSCCR) we neglect the (ρ) –
kinetic energy in Eq. (7) to write

$$\{-\frac{\hbar^2}{2m} \frac{\partial^2}{\partial z^2} + \frac{1}{2I} \; j^2 + V(\rho,z,\theta,\phi)\psi = E\psi \; . \qquad (8)$$

We then expand ψ in the rotor basis set $\{Y_{jm_j}(\theta\phi)\}$, so that

$$\psi(j_o m_{jo}|\rho z\theta\phi) = \sum_{j'm'_j} Y_{j'm'_j}(\theta\phi)\psi(j'm'_j|j_o m_{jo}|\rho z) \; , \qquad (9)$$

where the $\psi(jm_j \; j_o m_{jo}|\rho z)$ satisfy

$$\left\{ \frac{\hbar^2}{2m} \frac{d^2}{dz^2} + E - \frac{\hbar^2 j(j+1)}{2I} \right\} \psi(jm_j | j_o m_{jo} | \rho z)$$

$$= \sum_{j'm'j} V(jm_j | j'm'_j | \rho z) \psi(j'm'_j | j_o m_{jo} | \rho z) . \qquad (10)$$

This equation is solved subject to the usual boundary conditions to yield the probability amplitude $S(jm_j \, j_o m_{jo} | \rho)$ for the rotor transition $j_o m_{jo} \rightarrow jm_j$ induced by collision of the molecule with the point $\rho \equiv (x,y)$ in the surface. The physical probability amplitude is then obtained via

$$S(jm_j mn | j_o m_{jo} 00) = \frac{1}{A} \int d\rho \, \exp(i\underset{\sim}{G}_{mn} \cdot \rho) \, S(jm_j | j_o m_{jo} | \rho) , \qquad (11)$$

where $\underset{\sim}{G}_{mn}$ is the reciprocal lattice vector for the (x,y) motion and A is the area of the unit cell. In the case of the full sudden (FS) approximation, the equation becomes

$$\left\{ -\frac{\hbar^2}{2m} \frac{d^2}{dz^2} - E + V(\rho,z,\theta,\phi) \right\} \psi = 0, \qquad (12)$$

and the resulting fixed ρ, θ, ϕ S-matrix $S(\rho,\theta,\phi)$ ($\equiv \exp(2i\eta(\rho,\theta,\phi))$) is related to the physical S-matrix via

$$S(jm_j mn | j_o m_{jo} 00) = \frac{1}{A} \int d\rho \, d(\cos\theta) d\phi \, Y^*_{jm_j} (\theta\phi)$$

$$\exp(i\underset{\sim}{G}_{mn} \cdot \rho) \, \exp(2i\eta(\rho,\theta,\phi)) \, Y_{j_o m_{jo}} (\theta\phi) . \qquad (13)$$

For the potential under consideration, the phase shift for Eq. (12) can not be obtained analytically and is instead obtained by a numerical evaluation of the WKB approximation expression for the phase shift. The terms in the potential V involving $\cos\theta$ are related to forces felt by the molecule due to its tilt with respect to the average surface normal. The terms $(m_i d/Ma)\sin\theta\cos\phi$ and $(m_i d/Mb)\sin\theta\sin\phi$, i=1,2 essentially represent projections of the vector from the center of mass of the molecule to atom i on the molecular axis taken along the x and y axes in the surface (measured in units of the lattice constants a and b). These latter terms are responsible for inducing the magnetic transitions. We now turn to the results of calculations using the above potential forms.

3. FORMAL ANALYSIS

3.1 Selection Rules For Specular Scattering For Heteronuclear Molecule-Corrugated Surface Collisions

In the first study (12) of homonuclear molecule - corrugated surface collisions, it was found that within the specular peak, a selection rule was observed such that only $\Delta m_j = \pm 4t$, $t = 0,1,2\ldots$ was obeyed for the case of a square lattice and $\Delta m_j = \pm 2t$, $t = 0,1,2,\ldots$ for a rectangular lattice. Outside the specular peak, no selection rules were found to hold and all Δm_j transitions were allowed for both lattices. Simple arguments rationalizing this were given based on local symmetry of the potential. We now wish to show that these simple local symmetry arguments also apply when one considers heteronuclear molecule – corrugated surface collisions. The basic assumption again is that the dominant contribution to the quantal amplitude for specular scattering comes from points (x,y) in the lattice close to those which give rise to classical specular scattering. This condition for classical specular scattering is

$$\frac{\partial V}{\partial x} = \frac{\partial V}{\partial y} = 0. \tag{14}$$

For vertical incidence ($\theta = 0$), this results in specular scattering from the points $x = 0$, $\frac{a}{2}$, a and $y = 0$, $\frac{b}{2}$, b. These are exactly the same points as in the homonuclear case. Furthermore, they are the same for the more general Morse potential given in Eq. (5) above. Now if one looks in the neighborhood of these points, the potential for the square lattice has the symmetry

$$V(zxy\theta\phi) = V(zxy\theta, \phi + \frac{\pi}{2}) \quad , \tag{15}$$

and the initial state $m_{j_0} = 0$ has the same symmetry. We then expect the allowed final states to have this symmetry, so that

$$m_j(\phi + \frac{\pi}{2}) = m_j\phi + 2\pi N, \text{ N an integer.} \tag{16}$$

The result is

$$m_j = 4N \quad , \tag{17}$$

which corresponds to a selection rule

$$\Delta m_j = \pm 4t, \; t = 0, \; 1, \; 2, \; \ldots \quad . \tag{18}$$

In the case of a rectangular lattice, the local symmetry is

$$V(zxy\theta\phi) = V(zxy\theta, \; \phi + \pi) \tag{19}$$

so that

$$m_j = 2N, \text{ N an integer.} \tag{20}$$

This implies the selection rule

$$\Delta m_j = \text{even.} \tag{21}$$

We comment that an exact evaluation of the quantum amplitude will
sample lattice points for which the symmetry of Eq. (15) does not hold
Thus, a full close coupled calculation may show a propensity for Δm_j=
+4t, t=0,1,2,... rather than an exact selection rule. In fact, preliminary results of Lill and Kouri (16) show precisely such a propensity.

3.2 Rainbow Scattering In Δm_j - Transitions

The possible occurance of magnetic transition rainbows was first predicted by Proctor, et.al.(11-12) Subsequent calculations by Proctor
and Kouri (13) are strongly suggestive that for a model Cl_2/exponential
repulsive potential, such rainbows have been seen. Under certain circumstances, it is expected that the magnetic rainbow will be manifested
by a significant increase in the transition probabilities for larger
Δm_j. In addition, the results of Proctor and Kouri (13) show oscillations in the transition probability as a function of final m_j. These
oscillations appear to be consistent with interpreting the results as
showing a magnetic rainbow. The first prediction of the magnetic rainbow phenomenon used a stationary phase evaluation of the full sudden ex-
pression for $S(jm_jmn|j_om_{j_0}00)$ given in Eq. (13). It should be noted
that due to the semiclassical nature of a stationary phase rainbow
treatment, there is some question of applying it to systems in which
the Δm_j transitions allowed are small. For instance, a uniformized
semiclassical treatment of rainbows may shift the rainbow maximum to
smaller values of the variable. When the allowed Δm_j is small, this
can have a substantial effect. A second note of caution is that generally such a rainbow analysis is most suitable when the transition
probabilities are strong. In our case, the Δm_j transitions are relatively weak. However, we believe the rainbow interpretation is valid
qualitatively and therefore present it. In the analysis of Proctor,
et.al. (12), both the θ and ϕ integrals were evaluated by stationary
phase and the m_j-rainbow was predicted to accompany rotational-j state
rainbows. Subsequently, the first computations which may possibly show
an m_j-rainbow was for conditions for which no rotational-j rainbow
occurred. Proctor and Kouri (13) analyzed the m_j-rainbow using a
heuristic stationary phase method only for the ϕ-integral in Eq. (13)
so that the question of a j-rainbow never entered. In the present
cases of heteronuclear-repulsive corrugated surface and homonuclear-
Morse corrugated surface collisions, we again use Eq. (13) along with a
heuristic stationary phase method to analyze the conditions for the
m_j-rainbow. In the general case (including both potentials (4) and (5)
and their homonuclear specializations), our potential can be written
in the form

$$V(zyx\theta\phi) = A_1 f_1(z, \theta) \; [1+\beta_1 g_1(xy\theta\phi)]$$

$$+ A_2 f_2(z\theta)[1+\beta_2 g_2(xy\theta\phi)] \qquad (22)$$

$$= V_o(z,\theta) + \beta_1 F_1(xy\theta\phi)A_1 f_1(z\theta)$$

$$+ \beta_2 F_2(xy\theta\phi)A_2 f_2(z,\theta) \; . \qquad (23)$$

We assume that β_1 and β_2 are small so that the potential is dominated by $V_o(z\theta)$. Then the WKB phase shift is approximately

$$\eta(\theta\phi xy) \stackrel{\sim}{=} \eta_o(\theta) - \beta_1\eta_1(\theta\phi xy) - \beta_2\eta_2(\theta\phi xy) , \tag{24}$$

where

$$\eta_o(\theta) = \lim_{z\to\infty} \left(\int_{z_t}^{z} \{ [k_z^2 - \frac{2\mu}{\hbar^2} V_o(z'\theta)]^{1/2} - k_z\} \, dz' \right)$$

$$-k_z z_t \tag{25}$$

and

$$\eta_i(xy\theta\phi) = A \frac{i\mu}{\hbar^2} \quad F_i(xy\theta\phi) \int_{z_t}^{\infty} dz \quad \frac{f_i(z\theta)}{\sqrt{k_z^2 - \frac{2\mu}{\hbar^2} V_o(z\theta)}} \tag{26}$$

We note that the $(xy\theta)$ dependence of $\eta_i(xy\theta\phi)$ does not occur inside the integral in Eq. (26). Furthermore, the $(xy\theta)$ dependence of $\eta_i(xy\theta\phi)$ is identical to that of the potential. Thus, defining

$$\lambda_i(\theta) = A_i \frac{\mu}{\hbar^2} \int_{z_t}^{\infty} dz' \quad \frac{f_i(z\theta)}{\sqrt{k_z^2 - \frac{2\mu}{\hbar^2} V_o(z\theta)}} , \tag{27}$$

we have that

$$\eta(\theta\phi xy) = \eta_o(\theta) - \beta_1\lambda_1(\theta) F_1(xy\theta\phi) - \beta_2\lambda_2(\theta)F_2(xy\theta\phi) . \tag{28}$$

Now the stationary phase evaluation (with respect to ϕ) of Eq. (13) becomes

$$S(jm_j mn|0000)$$
$$= \frac{j_1}{\sqrt{2} \ ab} \int d\rho \int d(\cos\theta) P_{jm_j}(\cos\theta) \exp(i\underset{\sim}{G}_{mn}\cdot\rho) \ F_{\Delta m_j}(\rho\theta) \tag{29}$$

where

$$F_{\Delta m_j}(\rho\theta) = \sqrt{\frac{i}{2\pi}} \quad \frac{1}{\sqrt{\frac{\partial^2}{\partial\phi^2} 2\eta(\rho\theta\phi)}}$$

$$x \ \exp[i(-m_j\phi + 2\eta(\rho\theta\phi)]\}_{\phi=\phi_\nu} , \tag{30}$$

with ϕ_ν being determined by the condition

$$\Delta m_j = \frac{2\partial \eta}{\partial \phi} (\rho \theta \phi) \quad . \tag{31}$$

In light of Eq. (28), this is approximately

$$\Delta m_j = -2 \ [\beta_1 \lambda_1(\theta) \frac{\partial F_1}{\partial \phi} + \beta_2 \lambda_2(\theta) \frac{\partial F_2}{\partial \phi} \ ] \quad . \tag{32}$$

The condition for rainbow scattering is that not only do the ϕ_ν satisfy Eq. (32), but also that

$$\frac{\partial^2 \eta}{\partial \phi^2} \bigg|_{\phi=\phi_\nu} = 0 \tag{33}$$

or that

$$\beta_1 \lambda_1(\theta) \frac{\partial^2 F_1}{\partial \phi^2} \bigg|_{\phi=\phi_\nu} -\beta_2 \lambda_2(\theta) \frac{\partial^2 F_2}{\partial \phi^2} \bigg|_{\phi=\phi_\nu} = 0 \quad . \tag{34}$$

The (approximate) stationary phase condition Eq. (32) implies that points of maximum ϕ-force result in the largest final-m_j values (as is physically reasonable). The (approximate) Δm_j-rainbow condition Eq. (34) says that the rainbow occurs due to the points in the surface which give rise to the extrema in the ϕ-force. This suggests that the occurrance of a large probability in the larger Δm_j-transitions is a signature of the m_j-rainbow. Of course, the above corresponds to a "primitive" semiclassical treatment. The exact quantum results may show the rainbow maximum shifted with respect to the above result. We now turn to consider the results of computations for several model systems.

4. COMPUTATIONAL RESULTS AND DISCUSSION

We have carried out extensive calculations of molecule - corrugated surface collisions using both the exponential repulsion atom-surface and the Morse atom - surface interactions. In the former case, the present calculations are for a heteronuclear diatom while for the latter, we consider again mass and rotor parameters appropriate to the H_2 molecule. Our intent in this case is to compare with the scattering of H_2 by the repulsive exponential interaction reported earlier by Proctor, et.al. (12) in order to examine the effect of the potential well.

In Table 1 we give results for H_2 scattered by a corrugated surface in which each of the H-atoms interacts with the corrugated surface via a Morse potential.

$j\,m_{j_0}$	jm_j	$n=m=0$	$n=1,m=0$	$n=m=0^*$	$n=1,m=0^*$
00	00	0.4391(0)	0.1040(0)	0.3747(0)	0.1105(0)
00	2,-2	0	0.1101(-4)	0	0.5383(-5)
00	2,-1	0	0.1900(-4)	0	0.4074(-5)
00	2,0	0.3413(-3)	0.5226(-4)	0.1436(-2)	0.3058(-3)
2,-2	00	0	0.1101(-4)	0	0.5383(-5)
2,-2	2,-2	0.8185(0)	0.4223(-1)	0.6014(0)	0.8369(-1)
2,-2	2,-1	0	0.1168(-3)	0	0.2893(-3)
2,-2	2,0	0	0.1556(-4)	0	0.2814(-4)
2,-2	2,1	0	0.1265(-7)	0	0.3365(-7)
2,-2	2,2	0.1055(-7)	0.1501(-8)	0.2641(-7)	0.1450(-8)

Table 1. Collision of H_2 with Corrugated Surface Having
Morse Potentials for the Atom - Surface Interaction. The
parameters are E_{tot} =0.1 eV A=0.3 eV α=1.9395Å^{-1} a=b=4Å β=0.1
Results for H_2 with corrugated surface having exponential repulsion for the atom - surface interaction are denoted with
*. The results are independent of the value of A and have
E_{tot}=0.1 eV, the range parameter α=1.9395A^{-1}, β=0.1 and
a=b=4Å.

Also shown are our earlier results for H_2 scattered by the sum of exponential repulsion atom - surface potentials for each of the H - atoms.
It is clear that qualitatively, the results are very similar. In particular, as expected, the specular peak selection rules in the two cases
are identical. However, outside the specular peak, the Morse potential
does appear to result in somewhat larger probabilities for the $\Delta m_j \neq 0$
magnetic transitions. In spite of this, these probabilities remain
very small (still about 2-5 times smaller than the Δm_j=0 one for j =
m_{j_0} =0). For this case, the unit cell is much larger than the diatom
length and as noted earlier (12), this leads to weaker effective corrugation and low probabilities for magnetic transitions. In Table 2
we give the diffraction summed - rotationally resolved transition
probabilities for this same situation. These results show once more
that overall, the Morse potential interaction leads to more magnetic
transitions. Next, we reduced the dimension of the lattice cell to a
value comparable to the diatom length, but left all other system parameters unchanged. Sample results for this case are shown in Table 3
below. Of course, the specular peak selection rules are unchanged.
However, now the probabilities for $\Delta m_j \neq 0$ transitions are even more
enhanced and most significantly, one even sees the suggestion of a
magnetic transition rainbow! A hint of similar behavior is also present in the H_2-exponential repulsion corrugated surface result for the
first diffraction spot n=1, m=0. However, no such behavior appears in
the diffraction summed results for this system. In the case of the
H_2-Morse corrugated surface results, a m_j-rainbow is suggested not only
in the n=1, m=0 diffraction spot but also in the diffraction summed results. Thus, it appears that reducing the ratio of the lattice size
to molecular length, combined with introducing a potential well in the
interaction, leads to an enhancement of the m_j-transitions and may
yield magnetic rainbow scattering. The effect of the well is to accell-

$j_o m_{jo}$	jm_j	$P(jm_j \ j_o m_{jo})$	$P(jm_j \ j_o m_{jo})*$
00	00	0.9975(0)	0.9955(0)
	2,-2	0.4888(-4)	0.2577(-4)
	2,-1	0.1086(-3)	0.2911(-4)
	2,0	0.8409(-3)	0.3014(-2)
2,-2	00	0.4888(-4)	0.2577(-4)
	2,-2	0.9993(0)	0.9979(0)
	2,-1	0.5508(-3)	0.1712(-2)
	2,0	0.6545(-4)	0.1262(-3)
	2,1	0.7119(-7)	0.2125(-6)
	2,2	0.2290(-7)	0.4505(-7)

Table 2. Diffraction Summed – Rotational State Resolved Transition Probabilities for the H_2 – Morse Corrugated Surface Collision. The parameters are E_{tot}=0.1 eV A=0.3 eV α=1.9395Å^{-1} a=b=4Å β=0.1. Results for H_2 – exponential repulsion corrugated surface collision with E_{tot}=0.1 eV, A=0.3 eV, α=1.9395Å^{-1}, a=b=4Å, β=0.1 are denoted with *.

$j_o m_{jo}$	jm_j	n=m=0	n=1,m=0	n=m=0*	n=1,m=0*
00	00	0.9117(0)	0.1748(-1)	0.9136(0)	0.1872(-1)
	2,-2	0	0.1347(-2)	0	0.6356(-3)
	2,-1	0	0.1802(-3)	0	0.3819(-4)
	2,0	0.3315(-3)	0.6766(-3)	0.2406(-2)	0.6857(-4)

Diffraction Summed – Rotationally Resolved Probabilities

$j_o m_{jo}$	jm_j	Morse Interaction	Exponential Interaction*
00	00	0.9838(0)	0.9915(0)
	2,-2	0.5432(-2)	0.2585(-2)
	2,-1	0.8194(-3)	0.2274(-3)
	2,0	0.3656(-2)	0.2908(-2)

Table 3. Collision of H_2 with Corrugated Surface Having Morse Potentials for the Atom – Surface Interaction. The parameters are E_{tot}=0.1 eV A=0.3 eV α=1.9395Å^{-1} a=b=1Å β=0.1. Results for H_2 with corrugated surface having exponential repulsion for the atom – surface interaction are denoted by *. These results are independent of A and are for E_{tot}=0.1eV, α=1.9395Å^{-1}, a=b=1Å, β=0.1.

erate the molecule so that the impact with the surface effectively occurs at higher energy. Indeed, it is very possible that higher energy collisions of the H_2 with the exponential repulsion – corrugated surface (with small enough lattice length) may show m_j – rainbow scattering.

The next series of calculations were done for a heteronuclear diatom colliding with the exponential repulsion corrugated surface. We took the H_2 molecule parameters and simply doubled them for one of the "H" atoms. In Table 4 we give results for a square lattice having a cell length of 4A. (This is large so we do not expect to see any m_j - rainbow scattering). The first point to note is that within the

$j_o m_{jo}$ jm_j		n=m=0	n=1,m=0	n=m=0*	n=1,m=0*
00	00	0.1344(0)	0.7236(-1)	0.3747(0)	0.1105(0)
	1,-1	0	0.1095(-3)	0	0
	1,0	0.7221(-1)	0.3421(-1)	0	0
	2,-2	0	0.1417(-4)	0	0.5383(-5)
	2,-1	0	0.1551(-4)	0	0.4074(-5)
	2,0	0.4542(-2)	0.1907(-2)	0.1436(-2)	0.3058(-3)

Diffraction Summed Probabilities

$j_o m_{jo}$ jm_j		Heteronuclear	Homonuclear
00	00	0.6684(0)	0.9955(0)
	1,-1	0.8123(-3)	0
	1,0	0.3070(0)	0
	2,-2	0.7741(-4)	0.2577(-4)
	2,-1	0.1511(-3)	0.2911(-4)
	2,0	0.1687(-1)	0.3014(-2)

Table 4. Collision of Heteronuclear Diatom with Corrugated Surface Having Exponential Repulsions for the Atom - Surface Interactions. The parameters are E_{tot}=0.1 eV m_2=2m_1=2m_H A_2=2A_1=0.6eV α_2=2 α_1=3.8790 a=b=4Å β_2=2 β_1=0.2 r_e=0.7406Å. Results for homonuclear with A=0.3 eV, α=1.9395, a=b=4Å, E_{tot}=0.1 eV, β=0.1 are denoted by *.

specular peak, the same selection rules are observed as the homonuclear case. Thus, one finds that Δm_j=+4t, t=0,1,2...   Outside the specular peak, one sees all transitions being allowed. The size of the magnetic transition probabilities appear to be larger for the heteronuclear system, but not a great deal larger. The main effect of going to a heteronuclear is that the j=1 state steals a significant amount of probability from the rotationally elastic transition.

Finally, in Table 5 we give results for the heteronuclear system with a rectangular lattice having a=4A, b=2A. Again, all other parameters of one of the atoms are half those for the other. We see that other than eliminating the Δj=even selection rule, going to the heteronuclear system does not change the selection rule on m_j, namely Δm_j= even within the specular peak. One still does not see sufficiently large $\Delta m_j \neq 0$ transition probabilities to merit the suggestion of an m_j - rainbow (this in spite of the fact that the 00→2,-2 diffraction summed probability is larger than that for 00→2,-1; it is however, _much_ smaller than that for 00→2,0). Again, we see that the j=0→j=1 transition steals a large amount of probability from the j=0→j=0 transition. We expect that a heteronuclear molecule - corrugated surface system with smaller lattice parameters (a,b) should show substantial $\Delta m_j \neq 0$ probabilities, including an m_j - rainbow.

An important aspect of the present results (both formally and

$j_o m_{jo}$ $j m_j$		$n=m=0$	$n=1,m=0$	$n=m=0*$	$n=1,m=0*$
00	00	0.1635(0)	0.6295(-1)	0.4408(0)	0.8136(-1)
	1,-1	0	0.3030(-3)	0	0
	1,0	0.8793(-1)	0.2613(-1)	0	0
	2,-2	0.2043(-4)	0.1046(-3)	0.1029(-4)	0.5070(-4)
	2,-1	0	0.4764(-4)	0	0.1466(-4)
	2,0	0.4769(-2)	0.1635(-2)	0.1572(-2)	0.2331(-3)

Diffraction Summed Probabilities

$j_o m_{jo}$ $j m_j$		Heteronuclear	Homonuclear
00	00	0.6753(0)	0.9956(0)
	1,-1	0.1191(-2)	0
	1,0	0.3003(0)	0
	2,-2	0.5455(-3)	0.1939(-3)
	2,-1	0.2679(-3)	0.6607(-4)
	2,0	0.1610(-1)	0.2970(-2)

Table 5. Collision of a Heternuclear Diatom with Corrugated Surface Having Exponential Repulsions for the Atom – Surface Interactions. The parameters are E_{tot} =0.1eV m_2=2m_1=2m_H A_2=2A_1=0.6eV α_2=2α_1=3.8790 a=4Å b=2Å β_2=2β_1=0.2 r_e= 0.7406Å. Results for homonuclear with A=0.3eV, α=1.9295, a=4Å, b=2Å, E_{tot}=0.1eV, β=0.1 are denoted by *.

computationally) is the observation that, unlike gas phase scattering, going from a homonuclear to heteronuclear molecule appears to have no effect on the m_j – selection rules (which however, only occur in the specular peak). Of course, one does see the breaking of the Δj=even selection rule. A second interesting result is the observation of Δm_j- scattering for the H_2 molecule which is suggestive of a rainbow, albeit when the lattice dimension is made small. The earlier calculations which strongly suggested that Δm_j – rainbows had been seen were for the Cl_2 molecule scattering off an exponential repulsion – corrugated surface. In general, we believe that magnetic rainbow scattering will occur either as the energy is raised (so that eventually Δj – rainbow scattering occurs and is accompanied by a m_j – rainbow) or the lattice length is decreased relative to the molecular length. This latter effect appears to increase the effect a given corrugation has on the molecule. Thus, we suggest that if one wants to use molecular scattering off surfaces to study surface structure, the maximum effect will be obtained if one uses long rod – like molecules rather than short ones. Third, we are continuing to study the question of how magnetic rainbows are manifested, selection rules and propensities in molecule- surface scattering. In particular, Lill and Kouri (16) now have the first preliminary (numerically) exact calculations (i.e., close coupling on both rotation and diffraction) as well as formal analysis establishing the occurrence of such selection rules or propensities. Fourth, and finally, it appears that the relative strength of the $\Delta m_j \neq 0$ transitions (compared to the corresponding Δm_j=0 transition) is greater in the nonspecular peaks than for the specular one. The quantity sensitive to this effect is the relative strength ratio

$$\frac{P_{j'm'_j, mn \leftarrow jm_j 00}}{P_{jm_j, mn \leftarrow jm_j 00}},$$

where $m'_j \neq m_j$. Evidence for this comes from the $j, m_j=0,0 \rightarrow j'=2, m'_j=-2$ transition in Table 5 and the $2,-2 \rightarrow 2,2$ transition in Table 1. The interpretation appears to be that specular transitions are dominated by high local symmetry and low effective corrugation ($x=0$, $a/2$ in the square lattice case). Diffraction probes regions of higher local corrugation and lower symmetry which enhances the relative strength of the $\Delta m_j \neq 0$ transitions and also leads to the absence of selection rules in the nonspecular Δm_j - transitions.

5. ACKNOWLEDGEMENTS

Support of this research under National Science Foundation Grant CHE-8215317 is gratefully acknowledged. We also gratefully acknowledge helpful suggestions regarding this work by Prof. R.B. Gerber.

6. REFERENCES

1. G. Wolken, Jr., J. Chem. Phys. 59, 1159 (1973); Chem. Phys. Lett. 22, 373 (1973).
2. R.B. Gerber, A.T. Yinnon, Y. Shimoni and D.J. Kouri, J. Chem. Phys. 73, 4397 (1980).
3. D.E. Fitz, L.H. Beard and D.J. Kouri, Chem. Phys. 59, 257 (1981).
4. J.E. Adams, Surf. Sci. 97, 43 (1980).
5. D.E. Fitz, A.O. Bawagan, L.H. Beard, D.J. Kouri and R.B. Gerber, Chem. Phys. Lett. 80, 537 (1981).
6. R.B. Gerber, L.H. Beard and D.J. Kouri, J. Chem. Phys. 74, 4709 (1981).
7. A.O. Bawagan, L.H. Beard, R.B. Gerber and D.J. Kouri, Chem. Phys. Lett. 84, 339 (1981).
8. R. Schinke, Chem. Phys. Lett. 87, 438 (1982).
9. K.B. Whaley, J.C. Light, J.P. Cowin and S.J. Sibener, Chem. Phys. Lett. 89, 89 (1982).
10. C.F. Yu, C.S. Hogg, J.P. Cowin, K.B. Whaley, J.C. Light, and S.J. Sibener, Israel J. Chem. 22, 305 (1982).
11. D.J. Kouri and R.B. Gerber, Israel J. Chem. 22, 321 (1982).
12. T.R. Proctor, D.J. Kouri and R.B. Gerber, J. Chem. Phys. (in press).
13. T.R. Proctor and D.J. Kouri, unpublished.
14. A.C. Lunz, A.W. Kelyn and D.J. Auerbach, Phys. Rev. B25, 4273 (1982).
15. R.B. Gerber, A.T. Yinnon and J.N. Murrell, Chem. Phys. 31, 1 (1978).
16. J.V. Lill and D.J. Kouri, unpublished results.

RAINBOWS AND RESONANCES IN MOLECULE-SURFACE SCATTERING

Reinhard Schinke
Max-Planck-Institut für Strömungsforschung
3400 Göttingen, Federal Republic of Germany

The effect of rotational rainbows in molecule-surface scattering is investigated for NO-Ag(111). It is demonstrated that recent experimental state distributions can be qualitatively repruduced with one potential energy surface without inclusion of the thermal motion of the surface atoms. We conclude that the interaction potential should be quite asymmetric, i.e., the interaction of N and O with the metal surface should be different. The mean rotational energy transfer increases linearly with the collision energy and the surface temperature which is in accordance with experimental findings. The second topic deals with selective adsorption resonances in H_2 (and isotope variations)-metal surface scattering. The effect of rotationally mediated selective adsorption for HD as impinging molecule and its dependence on the potential energy surface is discussed. A quantitative prediction of the potential anisotropy is made for H_2-Ag(110) using the experimentally observed magnetic sublevel splitting of selective adsorption resonances. It agrees very well with a recent estimation based on ab initio methods.

1. INTRODUCTION

Molecule-surface scattering is currently studied by many research groups both experimentally and theoretically. A rather long list of references (although by far from complete) can be found in Ref.1. Several examples of molecule-surface scattering will be discussed at this meeting. Two goals of these studies are to understand the dynamics of gas -surface collisions on a microscopical level and to extract information about the interaction potential from experimental data. It is hoped that such information might eventually lead us to an understanding of macroscopical processes like de- or adsorption, surface reactions, catalysis, etc.

The determination of potential energy surfaces by fitting of experimental data is usually simplified if the observables, i.e., the final state distributions exhibit distinct and predictable structures which can be directly correlated with the potential or at least parts of it.

B. Pullman et al. (eds.), Dynamics on Surfaces, 103–116.
© *1984 by D. Reidel Publishing Company.*

Such structures indeed exist in molecule-surface scattering and can be classified as *rotational rainbows* [2,3] or *rotationally mediated select-ive adsorption resonances* [4-6]. Rotational rainbows are classical feat-ures and exist whenever the wavenumber of the translational motion is large (high energies, heavy molecules) and many rotational states are energetically open. Their observation yields primarily information about the anisotropy of the repulsive branch of the interaction potential. Re-sonances are quantal features and exist whenever the wavenumber is small (low energies, light molecules) and only few quantum states are energe-tically open. They yield primarily information about the attractive part of the potential. Both effects will be discussed for two examples in connection with recent experiments.

2. ROTATIONAL RAINBOWS IN NO-Ag(111) SCATTERING

Rotational rainbows have been observed for many systems in gas phase collisions [2]. In molecule-surface collisions they have been de-tected only for NO-Ag(111) at normal collision energies E_n ranging between 0.2 eV and 1 eV [7]. A typical experimental distribution in form of a Boltzmann plot is shown in Fig.1(c). It consists of a nearly linear part at low energy transfer and a broad shoulder at large energy transfer. While the latter part was immediately interpreted as a rota-tional rainbow [7] it was speculated [8] that the linear part originates from molecules which are temporarily trapped at the surface thus leading to a partly equilibration with the surface temperature. However, the corresponding Boltzmann temperature is almost independent on the surface temperature but dependent on the collision energy which is in contra-diction with the trapping-desorption mechanism. These arguments together with other experimental evidences [7] convinced several researchers [9-11] that both parts of the distribution are caused by a direct scat-tering mechanism.

In Ref.11 we presented a possible explanation of the experimental distributions in terms of rotational rainbows. We assumed a potential energy surface of the form

$$V(z,\gamma) = A(z-\alpha \cos \gamma - \beta \cos^2\gamma)^{-8} - B(z-\delta \cos \gamma)^{-3} , \qquad (1)$$

where z is the distance of the molecular center-of-mass from the sur-face and γ the orientation angle with respect to the surface normal. The scattering calculations are performed within the exact close coupl-ing (CC) method and the energy sudden (ES) approximation. The surface is assumed to be rigid and flat. The parameters in Eq.(1) are determined to give best agreement with the experiment for all energies and are given in Ref.11.

In Fig.1(a) we show ES $0 \rightarrow j$ rotational excitation probabilities for $E_n=0.65eV$. The two distinct maxima at $j \simeq 10$ and 28 are rotational rainbows as confirmed by the corresponding classical excitation function $J(\gamma)$ shown in Fig.2 of Ref.11. $J(\gamma)$ is usually [2,3] a sin-type function with a maximum and a minimum at around $\gamma \sim 45°$ and $135°$, re-spectively. The classical transition probability is proportional to

$|dJ/d\gamma|^{-1}$ and thus the two extrema lead to two singularities. In a quantal calculation they are changed into broad Airy-type maxima which are shifted into the classically forbidden region. The oscillations are due to quantal interference of either four or two complex contributions to the scattering amplitude. Fig.1(b) shows the same data as in (a) but plotted in the Boltzmann representation. The two rainbows, especially that at low energies, are less clearly visible, but the linear fall off at small E_{rot} is already indicated. If we take into account the average over the initial rotational state distribution which roughly corresponds to $T_{rot}=50°$ [12] we obtain the results in Fig.1(c). Now, all oscillations are completely smeared out and good qualitative agreement with the experiment is achieved. The agreement is of course not perfect. First, the amplitude of the theoretical distribution at large energy transfers is too large by roughly a factor of 1.5-2 and second, the fall off to larger energies is too steep. While the rainbow maxima are approximately determined by the first order derivative of V with respect to γ, the intensities are roughly determined by the second order derivative [2]. Therefore, fitting of rainbow intensities is always more difficult than fitting the rainbow locations.

The same potential fits also the distributions for other experimental collision energies. In Fig.2 we show ES and CC results for those energies which are also probed experimentally [7]. The arrows mark the classical rainbow singularities as obtained from the corresponding excitation functions [11]. Each curve shows the linear part at low energy transfers and a pronounced shoulder at large transitions. The more inelastic rainbow shows the usual increase with the collision energy [2]. Figure 2 has to be compared with Fig.1 of Ref.7. The parameter variations had been performed within the ES approximation which is computationally very convenient. The time-consuming CC calculations were finally made to assure that no additional error is introduced due to the ES approximation. The main difference between both calculations is a shift of the more inelastic rotational rainbow due to violation of energy conservation. Besides that the overall behaviour is the same in both cases.

Our conclusions from these simple calculations are the following: (a) It is possible to describe the experimental distributions by a direct scattering mechanism without the assumption of a trapping-desorption channel. Within the calculation both the linear part at low energies and the broad shoulder at high energies are due to rotational rainbows, very pronounced dynamical features. (b) If this explanation is -at least qualitatively- correct the NO-Ag interaction potential should be quite asymmetric (in the present case $\alpha=\beta$ in Eq.(1) [11]) to yield two rainbows which are widely separated, i.e., the interaction of N and O with the metal surface should be quite different. Only one rainbow would be obtained for homonuclear molecules ($\alpha = \delta = 0$ in Eq.(1)). (c) One should be careful in presenting rotational state distributions as Boltzmann-plots because distinct dynamical structures can easily get lost. A linear presentation of the theoretical $P(j)$ versus j would exhibit two distinct maxima, one at low and one at high transitions. Such a behaviour is indeed observed if the original experimental data are plotted in this way [13]. (d) We cannot claim that the potential determined in this study is the true NO-Ag potential. It is very likely that other

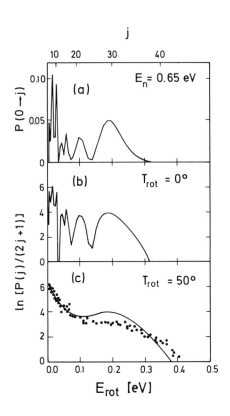

Fig.1(a): Theoretical energy sudden 0→j transition
probabilities versus rotational state j (upper scale)
or rotational energy transfer E_{rot} (lower scale).
(b) The same data as in (a),
however, plotted as a Boltzmann distribution. (c) Comparison of experimental [10]
and theoretical distribu-
tions. The latter are averaged over the initial state
distribution corresponding
to a temperature of
T_{rot}=50°.

potentials give even better
agreement with the experiment. The rainbows primarily determine the anisotropy
of the repulsive branch of
the potential. Conclusions
about the well region are
much less direct.

Very recently rotation-
al energy transfer in NO-Ag
scattering was experimentally reconsidered but at energies below $E_n \approx$
0.2 eV [14]. It was observed that the mean rotational energy transfer
increases linearly with both the collision energy and the surface temperature. Theoretically we define

$$\langle E_{rot} \rangle = \sum_j B j(j+1) \, P(0 \to j) \tag{2}$$

with B the rotational constant of NO. The results for the rigid surface model as obtained from the CC and the ES calculations with the potential of Ref.11 are plotted in Fig.3. One clearly sees a linear dependence over the entire range of normal energies which is easily explained in terms of the rainbow picture. The main contribution in Eq.(2)
stems from the region around the more inelastic rotational rainbow. How-
ever, it has been shown [2] that $j_R \sim E_n^{1/2}$ such that $\langle E_{rot} \rangle \sim E_n$.
The proportionality constant is a measure of the rotational coupling for
the particular system. Also shown in Fig.3 are the experimental results
[14] extrapolated to zero surface temperature. The agreement is only
fair. However, the potential was originally deviced to reproduce the
high energy data. With decreasing energy the well region plays a gradually more important role. Changing this part of the potential could very
likely yield a better agreement with the experimental energy transfer.
This is not attempted here because the relation between $\langle E_{rot} \rangle$ and the
potential is not direct and it would be easily possible to arrive at

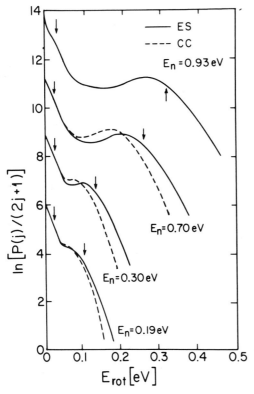

Fig.2: Theoretical CC and ES rotational state distributions of NO scattered from Ag(111).

wrong conclusions. We only wanted to demonstrate that the overall variation with E_n is correctly described.

The surface temperature dependence has recently been studied in Ref.1 using a sudden treatment of the surface phonons. We extended the potential in Eq.(1) to include the motion of the surface atoms, i.e.,

$$V(z,\gamma,u) = V_{rep}(z,\gamma) \cdot (1+\varepsilon u) + V_{att}(z,\gamma) , \quad (3)$$

where u is the normal displacement of a surface atom from equilibrium. ε

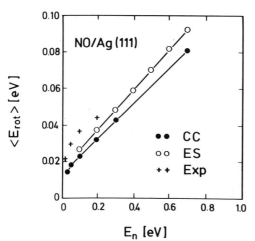

Fig.3: Mean rotational energy transfer for NO-Ag scattering versus normal collision energy. Comparison between theoretical CC and ES calculations and experimental data [14].

is the vibrational-translational coupling strength and V_{rep} and V_{att} are the repulsive and attractive parts of Eq.(1), respectively. Results for $E_n=0.65$eV and two values of ε are shown in Fig.4 and we note that the mean rotational energy transfer increases linearly with the surface temperature T, i.e., $\langle E_{rot}\rangle \sim T$ and the proportionality constant depends on the coupling strength. The argumentation goes as

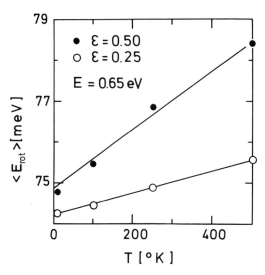

Fig.4: Mean rotational energy transfer for NO-Ag scattering versus surface temperature. ε is the phonon coupling strength of Eq.(3).

follows: The Debye-Waller factor is an increasing function of the energy transfer and consequently higher transitions are less attenuated than lower transitions. This leads to a shift of the rainbow to larger energy transfers such that $<E_{rot}>$ increases with T . The most crucial point in this pure model study is the coupling with the phonons. Neither the form nor its strength is even qualitatively known and therefore we do not attempt to make a quantitative comparison with experiment. Again, it was our goal to demonstrate that the experimental findings are in qualitative accord with the rainbow picture. Incidentally we note that increasing the surface temperature has only little influence on the overall state distributions shown in Figs. 1 and 2 which has been also observed experimentally [10].

3. THE H_2-Ag INTERACTION POTENTIAL

Scattering of H_2, D_2 and HD molecules from metal surfaces has been the topic of several recent experimental studies [4,15-18] aimed to obtain the isotropic as well as the anisotropic parts of the interaction potential. Selective adsorption resonances mediated through temporarily excited closed diffractional [16-18] or rotational states [4,15] are the main features in these experiments and yield information about the bound states of the molecule-surface system. The rotational coupling for the homonuclear molecules H_2 and D_2 is naturally very weak such that rotationally inelastic processes are extremely unlikely. In these cases scattering from faces with moderate corrugation yields the bound states of the spatially averaged isotropic potential via diffractionally mediated selective adsorption. Because of the shift between the center-of-mass and the center-of-symmetry the rotational coupling is strong for HD and rotationally inelastic transition probabilities are large. In that case rotationally mediated selective adsorption resonances can be easily resolved [4,15]. Because of the strong coupling between translation and rotation zeroth order perturbation theory is not applicable and the isotropic and the anisotropic parts of the interaction potential must be fitted simultaneously.

In addition to the experimental studies several theoretical investigations have been performed either dealing with the potential energy surface [19,20] or dynamical calculations [5,6,21]. In the following we will present some new calculations for HD-Ag and comment on the possibility to extract the anisotropic potential from selective adsorption measurements.

According to the work of *Harris et al.* [19,20] we write the H_2-metal potential as

$$V(z,\gamma) = V_0^{rep}(z) \; [1+\alpha P_2 (\cos \gamma)] + V_0^{att}(z) \; [1+\beta P_2 (\cos \gamma)] \; , \quad (4)$$

where P_2 is the second order Legendre polynomial. The proportionality constant for the attractive part is known as $\beta \approx 0.05$ for H_2 [19]. If we assume that the isotropic potentials V_0^{rep} and V_0^{att} are known from other experimental sources (for example, diffractionally mediated selective adsorption resonances) the proportionality constant for the repulsive branch is the only open parameter. For HD the potential in Eq.(4) has to be properly transformed to take the mass asymmetry into account.

In Fig.5 we show $j=0 \to 0$ rotationally elastic transition probabilities versus normal incident kinetic energy E_n. The thresholds for the various transitions are at 11.02 meV, 33.11 meV and 65.98 meV for $j=1$, 2 and 3, respectively. The repulsive and attractive potentials are

$$V_0^{rep}(z) \quad = \quad D_0 (1 + \lambda/p)^{-2p} \; , \quad (5)$$

$$V_0^{att}(z) \quad = \quad -2D_0 (1 + \lambda/p)^{-p} \quad (6)$$

with the parameters $D_0 = 31.7$ meV, $\lambda = 1.11$ $\mathring{A}^{-1}$ and $p = 4.06$ determined to fit selective adsorption resonances in H_2-Ag(110) scattering [17]. They are used without any modifications in the subsequent calculations. The distinct structures in Fig.5 are due to rotationally mediated selective adsorption including $j=2$ and 3 as closed rotational states. The energies of these resonances are given in terms of first order perturbation theory by

$$E_R(jmn) \quad = \quad \varepsilon_n^{(o)} + \varepsilon_j + \Delta(jmn) \quad , \quad (7)$$

$$\Delta(jmn) \quad = \quad (-1)^m (2j+1) \sum_{\lambda \neq 0} \begin{pmatrix} j & \lambda & j \\ 0 & 0 & 0 \end{pmatrix} \begin{pmatrix} j & \lambda & j \\ m & 0 & -m \end{pmatrix}$$
$$\int_0^\infty dz \; \phi_n(z) \; V_\lambda(z) \; \phi_n(z) \quad , \quad (8)$$

where j,m and n are the rotational, magnetic and vibrational quantum numbers, respectively. $\varepsilon_n^{(o)}$ and ε_j are the free vibrational and rotational energies and Δ is the first order perturbation correction. V_λ are the Legendre expansion coefficients and ϕ_n are the free vibrational wavefunctions. For flat surfaces we have the selection rule $\Delta m=0$ and thus $m=0$ for initially not-rotating molecules.

In Table 1 we list the experimentally determined resonance energies [15] together with the zeroth order predictions (Eq.(7) without the

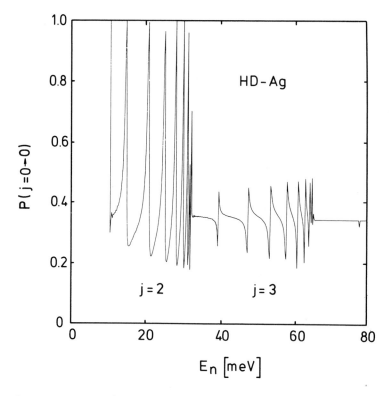

Fig.5: Theoretical j=0→1 transition probability versus normal kinetic energy E_n . The potential is that of Eqs.(5-7)
with α=0.2 and β=0.05 .

third term) and calculated resonance energies using the potential described above with α=0.1 , 0.2 and 0.3 , respectively. The theoretical
values are given as deviations from the experimental results. The assignment is given in the first column of the table. The resonances below the 0→1 threshold are obtained from a delay time analysis of the
0→0 S-matrix [5]. The calculations exhibit a few more resonances which
are not observed experimentally probably because of the limited resolution power.
    The strong dependence of $E_R^{ex}-E_R^{(o)}$ on the rotational state j
indicates the strong coupling between the rotational and the translational, i.e., vibrational degree of freedom. In order to demonstrate
the influence of the rotational coupling we analyzed exact close coupling calculations for three values of the strength parameter α in Eq.
(4). Although α is varied over a wide range a systematic behaviour is
difficult to extract from Table 1. The j=1 resonances are best described by α=0.1 while the j=3 resonances are best reproduced with α=0.3.
The value of α=0.2 seems to give the best overall agreement. However,

Table 1: Experimental (E_R^{ex}), zeroth order ($E_R^{(o)}$) and calculated (E_R^{cal}) resonance energies for HD-Ag in meV

(j,n)	E_R^{ex}	$E_R^{ex}-E_n^{(o)}$	$E_R^{ex}-E_R^{cal.}$		
(1,2)	2.5	1.1	$-0.7^{(a)}$	$-1.1^{(b)}$	$-1.5^{(c)}$
(1,3)	6.2	0.4	-0.1	-0.4	-0.5
(1,4)	8.7	0.1	-0.1	-0.1	-0.1
(1,5)	10.2	0.1	0	-0.1	-0.2
(2,1)	14.4	-2.4	-0.4	-1.0	-1.7
(2,2)	20.9	-2.6	-0.1	-0.5	-0.9
(2,3)	25.9	-2.0	-0.4	0.1	-0.3
(2,4)	29.7	-1.0	-1.1	0.9	0.7
(3,0)	40.2	-0.1	0.2	0	-0.8
(3,1)	48.6	-1.1	1.0	0.4	-0.2
(3,2)	55.2	-1.2	1.4	1.0	0.4
(3,3)	60.0	-0.8	1.8	1.4	1.0
(3,4)	64.1	0.6	0.3	0.5	0.3

(a) $\alpha=0.1$; (b) $\alpha=0.2$; (c) $\alpha=0.3$ in Eq.(4).

a clear decision is difficult to make although α is increased by a factor of *three* in Table 1. This observation might have, of course, several reasons: First, variations of the isotropic part could give a better agreement. Second, the potential model is not sufficiently flexible, and third, more parameters are needed. *To summarize our experience we state that the rotationally mediated selective adsorption resonances are not suitable to directly determine the anisotropy of the H_2 -surface interaction potential.*

A very direct way to determine α in Eq.(4) has become possible through a nice experiment due to *Chiesa et al.* [17]. According to Eqs. (7,8) the energy of a molecule adsorbed on a surface depends on both the rotational state j and the magnetic quantum number m. Taking into account the degeneracy of $+m$ and $-m$ states each bound state is split into $j+1$ levels. The +/- degeneracy is strictly valid only for φ-independent potentials (flat surface). *Chiesa et al.* [17] resolved ordinary diffractionally mediated selective adsorption resonances in scattering of H_2 molecules from a moderately corrugated Ag(110) surface. They used natural H_2 with a mixture of $j=0$ and $j=1$ states and probed scattering energies below the first threshold such that only rotationally elastic transitions are involved. The main observation in Ref.17 was that some of the observed resonances are split into two levels and this splitting was interpreted as being due to $j=1,m=0$ and $j=1,m=\pm1$ H_2 rotational states.

The amount of this magnetic sub-level splitting depends directly on the potential anisotropy. Using Eqs.(4) and (8) the first order correction term is

$$\Delta(jmn) = (-1)^m (2j+1) \begin{pmatrix} j & 2 & j \\ 0 & 0 & 0 \end{pmatrix} \begin{pmatrix} j & 2 & j \\ m & 0 & -m \end{pmatrix}$$
$$\int_0^\infty dz \, \phi_n(z) \, [\alpha \, V_o^{rep}(z) + \beta \, V_o^{att}(z)] \, \phi_n(z) . \tag{9}$$

With V_o^{rep}, V_o^{att} and β fixed $\Delta(jmn)$ is a linear function of α and thus the measurement of the Δm-splitting offers a very direct way to determine the potential anisotropy.

We performed exact close coupling scattering calculations aimed to investigate the magnetic sub-level splitting and to extract a realistic value for α from the measurements [21]. We extended the potential in Eq.(5) to qualitatively describe the surface corrugation, i.e.,

$$V_o^{rep}(z,x) = D_o [1+\lambda(z-h \cos\frac{2\pi}{a} x) / p]^{-2p} \tag{10}$$

with $h=0.05$ Å and $a=4.084$ Å and all the other parameters are the same as above. As usual, the initial angle distribution is analysed to give the bound state energies of the spatially averaged potential, ε_n. Calculations were performed for $j=m=0$, $j=1$, $m=0,\pm1$ and $j=2$, $m=0,\pm1,\pm2$ yielding (jm)-dependent bound levels $\varepsilon_n(jm)$. In Fig.6 we show for $n=2$ $\varepsilon_n(jm)$ normalized to the free energies $\varepsilon_n^{(o)}$ of the spatially and orientationally averaged potential as a function of the coupling parameter αD_o with $\beta=0$ in Eq.(4). As expected from Eq.(9) the shift away from the zeroth order energies and consequently the Δm-splitting varies linearly with the coupling parameter.

The sub-level splitting depends on the vibrational state n as shown in Fig.7 where we plot $\Delta\varepsilon_n=\varepsilon_n(j=1,m=0) - \varepsilon_n(j=1,m=\pm1)$ for $\alpha D_o=2$ meV. The coupling strength was chosen to give best agreement with the experimental values [17] which are also presented in Fig.7. In a second set of calculations we took the realistic value $\beta=0.05$ and re-adjusted α to give best agreement with the experiment and obtained finally $\alpha=0.2$. This value disagrees significantly from the values $\alpha=-0.05$ as obtained from fitting rotationally mediated selective adsorption resonances [15] and $\alpha=0.08$ as obtained from earlier theoretical considerations [20]. Incidentally we note that the latest value derived from refined *ab initio* calculations is $\alpha=0.18$ which is in excellent accord with our prediction extracted from fitting experimental results [22]. Since α enters Eq.(9) linearly its determination via magnetic sub-level splitting is very direct. A change of $\pm20\%$ would lead to not-acceptable Δm-splittings.

Finally we note that the potential used in the first part of this section to fit the rotationally mediated selective adsorption resonances in HD-Ag(111) scattering is equivalent to the potential originally derived for H_2-Ag(110) [17] with $\alpha=0.2$. The agreement with the experimental resonance energies in Table 1 is not perfect but acceptable if the reader keeps in mind the model character of these calculations.

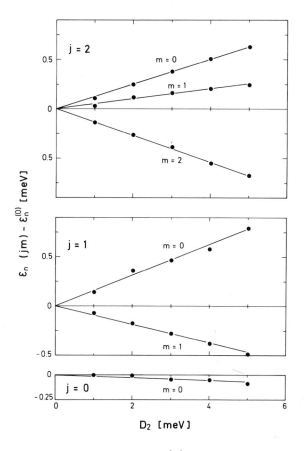

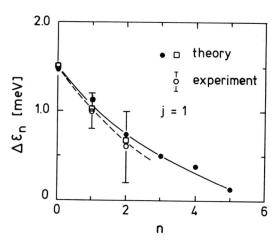

Fig.6: The shift $\varepsilon_n(jm) - \varepsilon_n^{(o)}$ for $n=2$ versus coupling
parameter αD_o and $\beta=0$ in Eq.(5).

Fig.7: Magnetic sub-level splitting $\Delta\varepsilon_n = \varepsilon_n(j=1, m=0) - \varepsilon_n(j=1, m=\pm1)$ versus vibrational quantum number. Comparison between experiment [17] and two theoretical calculations: $\bullet$, $\alpha D_o = 2$ meV, $\beta=0$ and $\square$, $\alpha=0.2$, $\beta=0.05$.

4. CONCLUSIONS

We presented calculations for two molecule-surface scattering experiments: NO-Ag and H_2-Ag. Because of the short wavelength and the large number of internal states the first example is governed by classical mechanics. The model calculations which we presented describe the overall state distributions observed experimentally reasonably well. Within the framework of this study both the linear shape of the Boltzmann representation as well as the broad shoulder at large energy transfer are interpreted as rotational rainbows, i.e., dynamical structures which stem primarily from the anisotropy of the repulsive potential branch. The linear part at low transition energies is the result of an accidental interplay between the high intensity rainbow maximum at low energies, the Boltzmann representation of the data and substantial averaging over the initial rotational state distribution. Essential for this interpretation is a large asymmetry of the NO-metal interaction. This is, however, the only conclusion which we can derive from our model study. More precise information about the potential, especially the well region would require more experimental results as well as refined theoretical investigations. It is definitely possible that other potentials are also consistent with the experiment. Incidentally we note that *Tanaka* and *Sugano* [23] independently come to conclusions very similar to ours. The main drawback of our calculations is the neglect of energy transfer from the molecule to the surface phonons. Future dynamical studies should take this energy transfer channel into account in order to prove or disprove the possibility of a trapping-desorption mechanism.

Because of the relatively long wavelength and the small number of rotational states involved the second example, H_2-Ag, is governed by quantum mechanics and resonances are the features of interest. In this case we are quite free from speculations and can make quantitative conclusions about the interaction. The potential surface is naturally similar to the well studied He-metal case extended to include the anisotropy, i.e., the rotational degree of freedom. The parameters of the spatially and orientationally averaged potential can be conveniently obtained from selective adsorption resonances mediated through diffractionally trapped molecules. Once this part is determined, the potential model contains only one extra parameter, the proportionality constant of the repulsive branch anisotropy. In principle, it can be determined by an analysis of rotationally mediated selective adsorption. However, we find such a procedure not very unique, i.e., the acceptable range of possible values is too large. A much more direct way to determine the anisotropy parameter has become possible due to the observation of magnetic-sub-level splitting of adsorbed H_2 molecules in rotational states $j \geqq 1$. We extracted a value $\alpha=0.2$ from fitting experimental data for H_2-Ag which is in good agreement with $\alpha=0.18$ as recently derived from *ab initio* calculations.

ACKNOWLEDGEMENTS

The author gratefully acknowledges the contributions of his co -workers V. Engel and H. Voges (Göttingen). Many fruitful discussions with A. Luntz (IBM, San José) have deepened his understanding of the experimental NO-Ag results. Finally, he is grateful to L. Mattera (Genova) for the permission to use the experimental H_2-Ag(110) magnetic sub-level results prior to publication.

REFERENCES

[1] R. Schinke and R.B. Gerber; 'Phonon sudden theory of Debye-Waller attenuation: Temperature dependence of rotational energy transfer in molecule-surface scattering', submitted for publication.

[2] R. Schinke and J.M. Bowman; 'Rotational rainbows in atom-diatom scattering' in *Molecular Collision Theory*, ed. J.M. Bowman (Springer, Berlin, 1983).

[3] R. Schinke; 'Rotational rainbows in diatom(solid) surface scattering', J.Chem.Phys. 76, 2352 (1982).

[4] J.P. Cowin, C.F. Yu, S.J. Siberer and J.E. Hurst; 'Bound level resonances in rotationally inelastic HD/Pt(111) surface scattering', J.Chem.Phys. 75, 1033 (1981).

[5] R. Schinke; 'Bound-level resonances in diatom-(rigid) surface scattering', Chem.Phys.Lett. 87, 438 (1982); 'Rotationally mediated selective adsorption in rigid rotor/rigid surface scattering', Surface Sci. 127, 283 (1983).

[6] K.B. Whaley, J.C. Light, J.P. Cowin and S.J. Sibener; 'Calculation of rotationally mediated selective adsorption in molecule surface scattering: HD on Pt(111)', Chem.Phys.Lett. 89, 89 (1982).

[7] A.W. Kleyn, A.C. Luntz and D.J. Auerbach; 'Rotational energy transfer in direct inelastic surface scattering: NO on Ag(111)', Phys.Rev.Lett. 47, 1169 (1981).

[8] E. Zamir and R.D. Levine; 'Energy transfer in the scattering of molecules from surfaces with applications to NO', Chem.Phys.Lett. 104, 143 (1984).

[9] J.E. Hurst, G.D. Kubiak and R.N. Zare; 'Statistical averaging in rotationally inelastic gas-surface scattering: The role of surface atom motion', Chem.Phys.Lett. 93, 235 (1982).

[10] J.A. Barker, A.W. Kleyn and D.J. Auerback; 'Rotational energy distribution in molecule surface scattering: Model calculations for NO/Ag(111)', Chem.Phys.Lett. 97, 9 (1983).

[11] H. Voges and R. Schinke; 'A double rainbow interpretation of rotational energy transfer in energetic NO/Ag(111) collisions', Chem.Phys.Lett. 100, 245 (1983).

[12] A.W. Kleyn; private communication.

[13] A.C. Luntz; private communication.

[14] G.D. Kubiak, J.E. Hurst, H.G. Rennagel, G.M. McClelland and R.N. Zare; 'Direct inelastic scattering of nitric oxide from clean Ag(111): Rotational and fine structure distributions', J.Chem. Phys. 79, 5163 (1983)

[15] C.F. Yu, C.S. Hogg, J.P. Cowin, K.B. Whaley, J.C. Light and S.J. Sibener; 'Rotationally mediated selective adsorption as a probe of isotropic and anisotropic molecule-surface interaction potentials: HD(J)/Ag(111)', Israel J. of Chem. $\underline{\underline{22}}$, 305 (1982).

[16] J. Lapujoulade and J. Perreau; 'The diffraction of molecular hydrogen from upper surfaces', Physica Scripta $\underline{\underline{T4}}$, 138 (1983).

[17] N. Chiesa, L. Mattera, R. Musenich and S. Salvo; to be published.

[18] C.F. Yu, K.B. Whaley, C.S. Hogg and S.J. Sibener; 'Selective adsorption resonances in the scattering of n-H_2, p-H_2, n-D_2, and o-D_2 from Ag(111)', to be published.

[19] J. Harris and P.J. Feibelman; 'Asymmetry of the van der Waals interaction between a molecule and a surface', Surface Sci. $\underline{\underline{115}}$, L133 (1982).

[20] J. Harris and A. Liebsch; 'On the physisorption interaction of H_2 with Cu-metal', Physica Scripta, to be published.

[21] R. Schinke, V. Engel and H. Voges; 'Magnetic sub-level splitting of selective adsorption resonances in molecule/surface collisions and its relation to the interaction anisotropy', Chem.Phys.Lett. $\underline{\underline{104}}$, 279 (1983).

[22] J. Harris; private communication.

[23] S. Tanaka and S. Sugano; 'A theoretical approach to rotationally inelastic scattering of a heteropolar rigid rotor by rigid and flat surfaces', Surface Sci. $\underline{\underline{136}}$, 488 (1984); 'Effect of initial rotational distribution and softness of surface on inelastic scattering of heteropolar rotor by surfaces', to be published.

ENERGY REDISTRIBUTION IN DIATOMIC MOLECULES ON SURFACES

M. Asscher[+] and G. A. Somorjai

Materials and Molecular Research Division, Lawrence
Berkeley Laboratory and Department of Chemistry,
University of California, Berkeley, CA 94720, U.S.A.

1. ABSTRACT

Translational and internal degrees of freedom of a scattered beam of NO
molecules from a Pt(111) single crystal surface were measured as a
function of scattering angle and crystal temperature in the range
450-1250K. None of the three degrees of freedom were found to fully
accommodate to the crystal temperature, the translational degree being
the most accommodated and the rotational degree of freedom the least.
A precursor state model is suggested to account for the incomplete
accommodation of translational and vibrational degrees of freedom as a
function of crystal temperature and incident beam energy. The vibrational accommodation is further discussed in terms of a competition
between desorption and vibrational excitation processes, thus providing
valuable information on the interaction between vibrationally excited
molecules and surfaces. Energy transfer into rotational degrees of
freedom is qualitatively discussed.

2. INTRODUCTION

Energy transfer processes at gas-solid interfaces are of great importance for the understanding of the dynamics of interactions between gas
phase atoms or molecules and solid surfaces. For atomic scattering
studies, the momentum changes as measured by angular and translational
energy distributions of the scattered particles provide most of the
necessary information to describe the collision process(1). For molecules, however, changes in the internal energy states (rotation and
vibration) can also occur in addition to changes in the translational
energy of the incident species. Determination of the internal state
distributions of molecules scattered from surfaces received an in-

+ Present address: Department of Physical Chemistry,
                   The Hebrew University of Jerusalem,
                   Jerusalem 91904, ISRAEL.

B. Pullman et al. (eds.), Dynamics on Surfaces, 117–134.
© *1984 by D. Reidel Publishing Company.*

creased attention in recent years. They were monitored by employing
laser induced fluorescence or two photon ionization techniques that
provide information on stronlgy as well as weakly interacting gas
molecule-surface systems(2-10). In most of these studies the rotational distribution of the scattered molecules were measured, and were
found to be at or close to a Boltzmann distribution with rotational
temperatures lower than that of the crystal temperature. When higher
incident energies were used in the NO/Ag(111) system deviation from the
Boltzmann distribution occurs at the high rotational states, due to
a rotational rainbow effect(4). Vibrational excitations as a result
of the interaction of molecules with surfaces were observed only in a
few strongly interacting molecular-surface systems(7,11,12). In these
cases only partial accommodation of the molecules with the surface was
reported(7). The only system for which complete energy transfer and
accommodation information exists, e.g. for translation rotation and
vibration, is the scattering of NO molecules from the Pt(111) single
crystal surface(2,7,8).

In this paper we shall present the experimental data obtained for
the redistribution of the translational, rotational and vibrational
states of supersonic NO molecules upon scattering from a Pt(111) surface. We found that the accommodation on the surface is best, but not
complete, for translation, then for vibration and the poorest for the
rotational degrees of freedom. A precursor state model will be presented with an approximate one dimensional potential for the NO/Pt(111)
system, which accounts well for the translational and vibrational data.
A more detailed analysis of the vibrational excitation process of NO,
while it is in the chemisorption state, will be discussed and be shown
to provide a unique insight into the interaction between vibrationally
excited molecules and metal surfaces. Finally the poor rotational
accommodation will be rationalized in terms of models that were recently suggested for rotational excitation of molecules on surfaces.

3. EXPERIMENTAL

The experiments to be described involve a supersonic molecular NO beam
scattered off a Pt(111) single crystal surface under UHV conditions.
Different detectors were employed for monitoring the translational
energy and the internal energy distribution of the scattered molecules.
For the translational energy distributions measurements, a time of
flight detector, equipped with a two stages differentially pumped
quadrupole mass spectrometer, rotatable around the crystal for angular
resolution was utilized. The details of this apparatus and the time of
flight results were given previously(13,14). Briefly, a supersonic NO
beam is generated by expanding through a 75μm nozzle with stagnation pressure of 200 torr. The beam scattered from a Pt(111) disc
located at the center of a UHV chamber at a base pressure of low 10^{-10}
range. A pseudorandom slit chopper is mounted on the detector, thus
chopping the scattered beam instead of the incident beam. This way the
time of flight measurements of the scattered molecules are not affected
by the unknown surface residence time. The incident beams' mean kinet-

ic energy was 265K, 615K and 1390K, by antiseeding with xe, pure NO and seeding with He respectively. The crystal temperature range was 475-1200K. The measurements were done at two scattering angles, e.g. 7° from the normal to the surface and at specular angle which is 51° from the normal(14).

Internal states e.g. vibrational and rotational states distributions determination of the scattered NO molecules required an optical detector(15). The time of flight detector was removed, therefore, and a set of quartz deflecting prisms and a focusing lens allowed a tunable UV laser beam at 225 and 236 nm ranges to be focused 2.5 cm above the crystal, intersecting the NO molecules at the scattering plane. At the focus of the laser, a two photon ionization process occurs via a bound electronic excited state of the NO molecules ($A^2\Sigma^+(v'=0)$). An electron multiplier was attached to the detector, at a fixed distance from the laser focus, and by applying bias voltages of 2000-3500V, the NO^+ ions were collected. Tuning the laser around 225 nm, the rotational states distribution of the ground vibrational state of NO was measured, while at 236 nm the rotational states distribution of molecules that were two photon ionized from the first vibrationally excited state could be detected. Thus relative ion currents at these two wavelengths provided vibrational and rotational state distributions. Details of the relevant spectroscopy of NO and the laser system are described elsewhere(15). Both the supersonic beam source and the Pt(111) sample were identical in the time of flight and the internal state distribution experiments.

The sample cleaning and surface composition determination were carried out with typical UHV techniques e.g. Ar^+ ion sputtering followed by oxygen treatment and annealing and Auger electron spectroscopy (AES) respectively. If scattering experiments were longer than 20 minutes at crystal temperature above 900K, oxygen coverages were built up to 0.1 of a monolayer were detected by AES, presumably due to slow NO decomposition at surface defects. To prevent further oxygen accumulation, experiments aimed at the determination of rotational spectra were interrupted to regulate the clean surface.

4. RESULTS AND DISCUSSION

The interaction between NO molecules and the Pt(111) single crystal surface is the first and yet the only molecule-surface interaction example for which a complete characterization of energy transfer and excitation of the three degrees of freedom of the scattered molecule, e.g. translation, rotation and vibration were obtained experimentally (7,14,15). The purpose of this paper is, therefore, to summarize the experimental results and to present new models which will account for the observed energy accommodation. As described in previous studies of the angular and velocity distributions of the NO/Pt(111) system, scattering occur by at least two different mechanisms, usually referred to as direct inelastic and trapping desorption processes(8,14,16). No attempt was made, however, to use this data for better understanding of the interaction potential between the NO molecules and the Pt(111)

surface. In the first section the experimental results will be presented, in the second section we shall present a model, based on the
assumption of a precursor state prior to chemisorption, which combines
the data from both translational and internal (vibrational in this
case) energy accommodation studies . In the third section a kinetic
model for the vibrational excitation process at the chemisorption state
will be outlined and be shown to provide an important insight into the
interaction between vibrationally excited molecules and metal surfaces.
Finally the poorly accommodated rotational degree of freedom will be
discussed in view of recent interpretations of such observations that
appeared in the literature.

4.1 Experimental Results

The experimental data for translational and vibrational energy exchange
of NO molecules in a supersonic beam, upon scattering from a Pt(111)
crystal surface are briefly outlined in this section. Time of flight
distribution measurements were taken at three incident beam energies
$<E>/2K=265K$, 615K and 1390K and were scattered from crystal at temperature range of 475-1200K. In Fig. 1 a typical time of flight distributions are presented for an incident beam of 615K with the detection at
7° from the surface normal and at the specular angle. Note that the
fit to a Boltzmann distribution (solid line in Fig. 1) is better near
the surface normal than at the specular angle. The accumulation of
scattered molecules' average kinetic energy (temperature) as a function
of crystal temperature for three different incident kinetic energies
are shown in Fig. 2. The detector is at 7° from the normal to the
surface. Note that even at large scattering angle where contribution
from direct inelastic scattering is negligible, there is some memory of
the incident beam energy. The full line is the calculated scattered
molecules' temperature which will be discussed in the next section.
Similar results were obtained at the same incident beam energies with
the detector at the specular angle of 51° from the normal to the surface. These results are summarized elsewhere(14) and discussed within
the framework of the precursor model (next section) elsewhere(19). A
typical angular distributions of the scattered NO molecules are shown
in Fig. 3. Note the increased flux at the specular angle by increasing
the crystal temperature. Similar effect is seen at a given crystal
temperature by increasing the incident beams energy(14).
    The vibrational temperature of scattered NO molecules as a function
of crystal temperature is presented in Fig. 4. Such a presentation
assumes a Boltzmann vibrational distribution and is derived from the
measurements of the ratio of scattered $NO(v"=0)$ molecules (7,15). Note
that here there is no complete vibrational accommodation even at the
lowest crystal temperature. The incident beam's vibrational energy was
not measured but is estimated to be 200K. The results on the rotational excitation will be presented and discussed in sec. 4.4.

4.2   A Precursor State Model for Vibrational and Translational
      Energy Accommodation.

Fig. 5 describes the main features of the proposed precursor state. The
shallow physisorption state has a well of E_∞ kcal/mole while a barrier
of E_s kcal/mole separates the precursor and the deep chemisorption well.
We assume that the incident molecules are trapped in the precursor state
with a unity probability. It is further assumed that the molecules are
equilibrated with the surface in their vertical momentum component
parallel to the surface is conserved. The distinction between the vertical

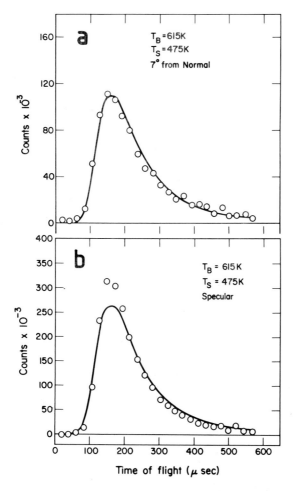

XBL 818-11064A

Fig. 1:   Time of flight distributions for a) T_B=615K, T_S=475K
          detector at 7° from the surface normal b) T_B=615K,
          T_S=475K, specular detection. The solid curves are
          the Maxwellian distributions at the corresponding
          surface temperature.

and horizontal components is based on the "hard cube" model which is
believed to be applicable for physisorbed states such as our precursor
state where the lateral motion of the molecules is practically free.
Since the diffusion parallel to the surface should not affect the
probability of chemisorption, one can assign the crystal temperature,
T_s to the molecules. Then the probability, P, for chemisorption is given
by:

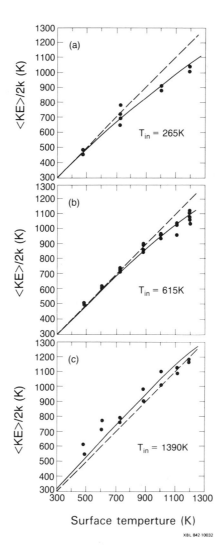

Fig. 2   Translational temperature of scattered NO molecules from PT(111)
surface. Solid circles denote the experimental measurements at
7° from the normal to the surface, the full line is the calculated temperature (sec. 4.2 eq 3) and the dashed line represents
a complete accommodation.

$$P = \frac{k_1}{k_1 + k_2} = \frac{1}{1 + \frac{k_2}{k_1}} \qquad (1)$$

where k_1 and k_2 are the rate constants for chemisorption and desorption from the precursor state, respectively. If first order processes are assumed, described by an Arrhenius rate constant, eq. 1 becomes:

$$P = [1 + \exp(-\Delta E/kT_s)]^{-1} \qquad (2)$$

where $\Delta E = E_\infty - E_s$. Note that the preexponential factors for these processes are assumed to be identical. Being a shallow well, the residence time of the molecule in the precursor state should be very short, thus vibrational accommodation is not expected. Moreover the coupling between translational and vibrational degrees of freedom is predicted to be weak(18). Once chemisorbed, however, both momentum components as well as the vibrational degree of freedom are accommodated to the surface temperature.

If we denote θ_{in} as the incident beam angle with respect to the surface normal, T_{in}^t as the incident beam translational temperature, defined as $\langle E_k \rangle/2k$, where $\langle E \rangle$ is the average incident kinetic energy, and T_f^t is the final translational temperature, then the expression for the final translational temperature is given by:

$$T_f^t = P \cdot T_s + (1-P)\left\{T_s \cdot \cos^2\theta_{in} + T_{in}^t \cdot \sin^2\theta_{in}\right\} \qquad (3)$$

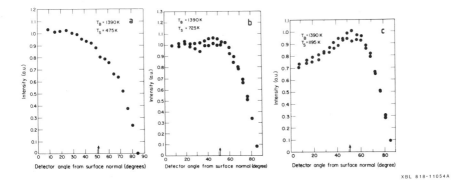

XBL 818-11054A

Fig. 3: Angular distributions for T_B = 1390K
a)T_s=475K; b)T_s = 725K; c)T_s = 1195K.

The first term in the right hand side of eq. 3 denotes a complete
accommodation from the chemisorbed state, while the second term partial
accommodation due to the precursor state. The final vibrational
temperature is calculated similarly. If the molecules desorb from the
precursor state they will retain their initial vibrational temperature,
which was not measured in this study but may be estimated at a maximum
of $T_{in}^V=200K(15)$. The molecules that desorb from the chemisorbed state
were completely accommodated with the surface. The resulting final
vibrational temperature is:

$$T_f^V = P \cdot T_s + (1-P) \cdot T_{in}^V \qquad (4)$$

Note that in this model, only one adjustable paramter e.g. ΔE, is
used for evaluating both the vibrational and the translational temperatures of the scattered molecule.

The vibrational accommodation coefficient γ_V is usually defined
as:

$$\gamma_V = \frac{T_f^V - T_{in}^V}{T_s - T_{in}^V} \qquad (5)$$

By using eq. 2 and 4, eq. 5 becomes:

$$\gamma_V = P = [1 + \exp(-\Delta E/kT_s]^{-1}) \qquad (6)$$

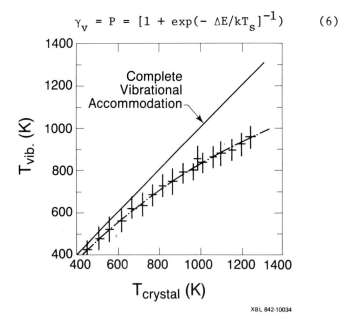

XBL 842-10034

Fig. 4: Vibrational temperature as a function of crystal temperature.
         Crosses are experimental data (derived from the measure-
         ments of the ratio of NO(v"=1)/NO(v"=0) of scattered
         molecules). The dashed-dotted line are the calculated
         values (see section 4.2, eq. 4) and the full line represents
         a complete accommodation.

Thus the vibrational accommodation coefficient can be interpreted as the probability of being chemisorbed from the precursor state. The translational accommodation coefficient is similarly transformed to:

$$\gamma_t = P \cdot \sin^2\theta_{in} + \cos^2\theta_{in} \qquad (7)$$

We found that with $\Delta E = 2.5 \pm 0.2$ kcal/mole the agreement with the the experimental data for NO scattering from the Pt(111) crystal surfaces is excellent.

In Fig. 4 a fit of eq. 4 to the measured final vibrational temperature as a function of crystal temperature is shown. The dash-dotted line is the calculated T_f^v from the precursor model, while the crosses are the experimental vibrational temperatures, measured as the intensity ratio of scattered NO(v"=1) to NO(v"=0)(15).

In Fig. 2, the calculated final translational temperatures (dashed lines) are shown together with the experimentally measured points (filled dots) at 7° from the normal to the surface and at three different incident translational temperatures(14). The agreement is again remarkably good.

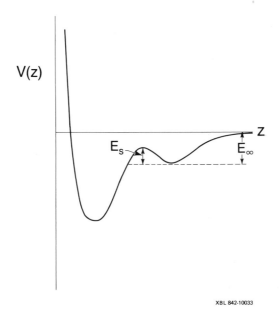

XBL 842-10033

Fig. 5:   A one dimensional potential describing the interaction between a diatomic molecule and a metal surface.

It should be noted that in this model we assumed a unity sticking coefficient into the precursor state. This is not accurate since, as mentioned earlier, there is a noticeable contribution to the scattering process from molecules which undergo direct inelastic scattering. This fraction enhances the flux of molecules scattered into the specular direction (see Fig. 3). The contribution of these molecules near the normal to the surface, however, is minimal, since the angular distribution of direct inelastically scattered molecules is centered near the specular angle(8,14,16) therefore is assumed to be negligible. In this case, the non unity sticking probability into the precursor state should not affect the measurements of either the translational or the vibrational temperatures which are presented in Figs. 2 and 4 since those were measured at the normal to the surface.

When measured at the specular angle(14), however, one has to consider this portion of the molecules and to take their contribution to the scattered translational temperature into account. A more detailed analysis of this kind is presented elsewhere(19).

The precursor model described here is not new, it has been utilized before to describe the adsorption process. It was suggested by Lennard Jones(20) in the early days of adsorption studies on metal surfaces, that the interaction potential should have two minima, a shallow one due to a physisorption state and a deeper one, closer to the metal which represents the chemisorption state. Later two different precursor mechanisms were suggested by Kisliuk(21) and Ehrlich(22). Kisliuk's mechanism attributes the shallow minima to molecules physisorbed on top of already occupied chemisorption sites prior to chemisorption. this mechanism was recently invoked to explain the chemisorption kinetics of NO on Pt(111) surfaces at high coverages(8,23). This model, however, cannot be aplied to the current study where most of the measurements were carried out at crystal temperatures where the NO coverage is very small(14,15). The precursor mechanism suggested by Ehrlich (22), describes the two adsorption wells situation with different rates for removal of molecules from the precursor state either to the chemisorption well or to the gas phae. The model presented here, therefore, is very similar to the one suggested by Ehrlich. Similar precursor state was recently suggested also as a possible explanation for the small sticking probability of N_2 molecules on Fe(110)(24). In this case, the activation energy for chemisorption from the precursor state exceeds the initial kinetic energy of the incident N_2 molecules, making this system a case of activated adsorption. Note that in the present model ΔE is positive, therefore there is no energy barrier higher than the initial kinetic energy, thus high sticking probability is possible, as reported for the NO/Pt(111) system(8,16,23). Note also that the experimental data available from refs. 14 and 15 are not sufficient to obtain information about the well depth of the precursor (E_∞) or the barrier for chemisorption (E_s).

4.3 Vibrational Excitation and Deexcitation of the Chemisorbed
      Molecules

The precursor model presented in the previous section does not consider the source for the vibrational excitation of the NO molecules on the Pt(111) surface nevertheless the prediction of the extent of vibrational excitation is very good (see Fig. 4). If one extends this model to higher crystal temperatures than what was employed in the experiments, however, one obtains the following expression for the vibrational temperature (see eq. 4) of the scattered molecules:

$$T_f^V = \quad 1/2(T_{in}^V + T_s) \tag{8}$$
$$(T_s \to \infty)$$

This expression is obtained since $p \to 1/2$ at $T_s \to \infty$. In fact, for the NO/Pt(111) system with $\Delta E = 2.5$ kcal/mole, already at $T_s \geqslant 1200$, T_f^V can be obtained from eq. 8. The question is whether this is a physically sound prediction.

The residence time of NO molecules on the Pt(111) terrace, using the kinetic parameters suggested by Serri et al(16) ($\nu = 10^{16}$ sec^{-1}; $E_d = 25$ kcal/mole) is calculated to be $3.7 \cdot 10^{-12}$ sec at 1200K and drops to $5.5 \cdot 10^{-14}$ sec at 2000K. Unless an unreasonably high vibrational excitation mechanism is assumed for the adsorbed NO molecules, one must consider the possibility that desorption of ground state molecules may compete with the vibrational excitation. In the following, we shall examine this question by developing a simple kinetic model, that takes into account the various kinetic processes which take place during the residence time of a chemisorbed molecule on the surface. The results of this model and the informtion that may be obtained from it as to the interaction between vibrationally excited molecules and surfaces, will be discussed and compared to the previous model.

Consider a population of adsorbed molecules on a metal surface established during a molecular beam-surface scattering experiment as in (15). The flux of molecules into the surface is $F(T_s) = F_o \cdot S(T_s)$ where $S(T_s)$ is the sticking coefficient. The molecules which stick to the surface may be vibrationally excited by their interaction with the metal or desorb as ground state molecules. The rate equations describing the concentration of ground (N_a^o) and vibrationally excited (N_a^1) molecules on the surface are given by:

$$\frac{d\ N_a^o(T_s)}{dt} = F(T_s) - k_{ex}(T_s) \cdot N_a^o(T_s) + k_{dex} \cdot N_a^1(T_s) - k_d^o(T_s) \cdot N_a^o(T_s) \tag{9a}$$

$$\frac{d\ N_a^1(T_s)}{dt} = k_{ex}(T_s) \cdot N_a^o(T_s) - k_{dex} \cdot N_a^1(T_s) - k_d^1(T_s) \cdot N_a^1(T_s) \tag{9b}$$

In eqs. 9, $k_d^o(T_s)$ and $k_d^1(T_s)$ are the desorption rate constants of the ground and vibrationally excited molecules respectively, while k_{ex} and k_{dex} are the vibrational excitation and deexcitation rates respectively. In the experiment of ref. 15 the desorbing flux of NO_g^1 was compared with that of NO_g^o. Once the concentrations on the surface are known, these values can be obtained as follows:

$$N_g^o = N_a^o \cdot k_d^o(T_s) \tag{10a}$$

$$N_g^1 = N_a^1 \cdot k_d^1(T_s) \tag{10b}$$

Applying steady state conditions on eqs. 9, (which is typical to the experiment in ref. 15), one gets the concentraiton of N_a^o and N_a^1 at steady state:

$$N_a^o(T_s) = \frac{F(T_s)[k_{dex}(T_s) + k_d^1(T_s)]}{K_{dex}(T_s) \cdot k_d^o(T_s) + k_d^1(T_s)[k_{ex}(T_s) + k_d^o(T_s)]} \tag{11a}$$

$$N_a^1(T_s) = \frac{k_{ex}(T_s) \cdot N_a^o(T_s)}{k_{dex}(T_s) + k_d^1(T_s)} \tag{11b}$$

Four rate constants determine the concentrations on the surface:
k_d^o, k_d^1, k_{ex} and k_{dex}. The value of k_d^o can be determined from
thermal desorption or modulated molecular beam experiments. These
experiments measure the total flux of ground and vibrationally excited
molecules, but at the temperature range at which these techniques apply, the fraction of excited molecules is less than 2% and is therefore
negligible. We assume that due to a weak coupling between the molecule-
surface bond and the internal stretch of the adsorbed molecule, which
originates from a large difference in these frequencies, the desorption
rates for ground and vibrationally excited molecules are identical –
$k_d^o = k_d^1 = k_d$. We also assume that k_d may be described as a first
order process. For the case of NO/Pt(111), k_d was carefully measured
(16). The values of k_{dex} and k_{ex} however are unknown, and so are their
possible crystal temperature dependencies. Recent theoretical(25–28)
as well as experimental(29) studies on the nature of the deexcitation
process, predicted very weak if any crystal temperature dependence, as
long as an electronic excitation/deexitation mechanism in the metal
is involved in quenching the vibrational energy of the adsorbed molecules. If equilibrium between the adsorbed molecules and the surface
exists the ratio k_{ex}/k_{dex} may be approximated by:

$$\frac{k_{ex}}{k_{dex}} = \exp[-Ev/kT_s] \tag{12}$$

where E_v is the vibrational energy spacing of the adsorbed molecule and
k is the Boltzmann constant. We assume that at higher crystal
temperatures, where the concentrations of N_a^o and N_a^1 may deviate
from the Boltzmann expression, the ratio between the rate constants is
still given by eq. 12. With the above model, we attempt to fit the
data for the NO/Pt(111) system(15). It should be mentioned that due to
the large temperature range of the experiment, the preexponential factor
used was not the fixed number given by Serri et al(16) sec^{-1}, but rather
a crystal temperature dependent parameter suggested by Redondo et
al(30). These authors developed a theory where this preexponential
factor is related to microscopic parameters of the system such as the
bending and stretching frequencies of the molecule–surface bond(30).
For the NO/Pt(111) system the resulting preexponential is
$v = 1.15 \cdot 10^{18}/T_s$ sec^{-1}, which is somewhat smaller than the value suggested
by Serri et al(16). In Fig. 6 the resulting fit is shown. These are
calculated for three values of the vibrational deexcitation rate con-
stant $k_{dex} = 1 \times 10^{13}$, 5×10^{11} and 5×10^{10} sec^{-1} as indicated in Fig. 6. Two
facts emerge from this fit: a. This model does not predict the ob-

served vibrational temperature. b. The high temperature trend is very
different from the precursor model, and predicts an increased deviation
of the vibrational temperature $T_f^V(T_s)$, from the crystal temperature.
In this model $T_f^V(T_s)$ has a maximum which corresponds to the value of
k_{dex} and thus also of k_{ex} (see eq. 12): $T_f^V(T_s)$ shifts to higher
values at higher T_s as k_{dex} increases. One may improve the quality of
the fit by using the assumption that $k_d^1=0.67k_d^0$ which was suggested
(however not well understood) in ref. 15. If such an assumption is
made the predicion of the kinetic model will look as in Fig. 7. Here
the ratio NO(1)/NO(0) is calculated and compared with the experimental
values. While such an approach provides a rather accurate prediction
of the unknown rate constants $k_{dex}=4\times10^{11}sec^{-1}$, it is still unclear
whether the assumption made for the values of k_d^1 and k_d^0 is justified.

For the NO/Pt(111), however, one may combine the two models presented
here and evaluate a quite accurate lower limit for the rate constant
k_{dex}, without the need to assume different desorption rates for the
ground and vibrationally excited molecules. It is suggested that
instead of assuming a full accommodation for those molecules which
reside in the chemisorption well (eq. 4), the vibratinal temperature
predicted by this kinetic model should be used. For the NO/Pt(111)
case(15), the highest crystal temperature was 1245K. It is clear from
Fig. 6, that in order to obtain a more accurate fit and therefore better
values for k_{dex}, experimental values at higher crystal temperatures of
1600 to 1800K are necessary. One can, however, argue using the
existing data, that the fit shown in Fig. 4 could not be as accurate

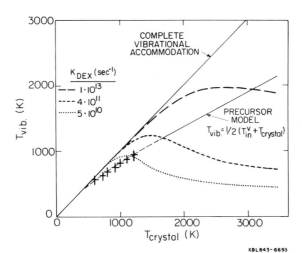

XBL843-6693

Fig. 6: Vibrational temperature of Scattered NO molecules from
         a Pt(111) surface for three values of the deexcitation
         rate constant $k_{dex}= --1\times10^{13},---5\times10^{11}$ and $....5\times10^{10}sec^{-1}$.
         The crosses are some of the experimental vibrational
         temperature values. The light solid line is the precursor
         model prediction at high temperatures. The heavy solid line
         presents a complete vibrational accommodation.

unless $k_{dex} > 1 \times 10^{12} sec^{-1}$.

In conclusion, it appears that for the NO/Pt(111) system the best
way to explain the data given in ref. 15, is to assume a precursor
state adsorption mechanism. The vibrational excitation occurs only
inside the chemisorption well, where at high crystal temperatures, the
competition between the desorption of ground state molecules and the
vibrational excitation process should be considered. It is believed
that the present model could be best utilized for somewhat less strongly interacting systems than the NO/Pt(111). In such systems deviations
from either full accommodation or from the precursor model predictions
should be observed at lower crystal temperatures, therefore easier to
measure experimentally. In such cases one may obtain better values for
the deexcitation and excitation rates. Finally, the lower limit for
$k_{dex}(10^{12} sec^{-1})$ obtained from the model described here, is well within
the range predicted by recent theories (25–28) to reflect an interaction between vibrationally excited molecules and electronic states in
the metal. This supports the assumption we made in the model as to the
temperature independence of k_{dex}. The value of at least $10^{12} sec^{-1}$ for
k_{dex} seem to agree with the electron–hole pair excitation as the most
probable mechanism for vibrational deexcitation in the NO/Pt(111)
system, since this mechanism predicts values in this order of magnitude
(25,31). Multiphonon excitation can be ruled out for this system,
since rates that are 5 orders of magnitude slower were predicted for
this mechanism(26).

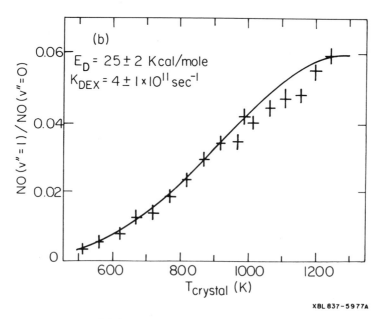

XBL 837-5977A

Fig. 7: The ratio NO(1)/NO(0) as a function of crystal temperature.
         The solid line represents the calculated values using the
         kinetic model with $K_{dex}=4 \cdot 10^{11} sec^{-1}$ and $E_d=25kcal/mole$,
         while the crosses are from the experiment (15).

4.4 Rotational Energy Accommodation of the Scattered NO Molecules.

The rotational temperature of NO scattered from Pt(111) surface was
measured in this study and found to be lower than the crystal temperature. In Fig. 8, the rotational temperature is shown as a function of
crystal temperature. The details of the experiment are given
elsewhere(15). Very similar results were recently reported for the
same system by Segner et al(8). It is evident that the accommodation
is rather poor, and the rotational temperature seem to level off at
around 450K, regardless of the crystal temperature. None of the models presented here treated the rotational excitation process. It is
not clear whether a different rotational temperature is expected from
molecules scattered from the precursor-physisorption state as compared
with those which undergo the trapping desorption process. On the
contrary, it seems that molecules which experience only weak interaction with the surface, have somewhat higher rotational temperature,
even though their incident rotational temperature is very cold. This
is shown by the measurements carried out by Segner et al.(8), who
scattered NO from an oxygen covered Pt(111) surface and observed
slightly higher rotational temperature as compared with the clean metal
surface. Similar observation was made in the present study, where at
specular angle (see Fig. 8) the contribution from direct inelastically
scattered molecules increase the overall rotational temperature(15).
Similar results were obtained also for the weak interaction between NO

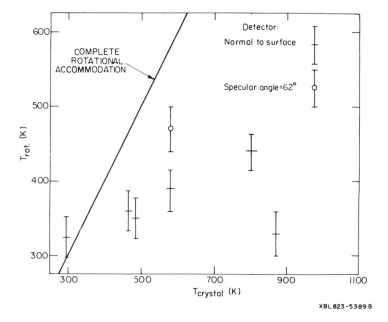

Fig. 8:   Rotational temperature of scattered NO molecules from
          a Pt(111) surface.

and Ag(111)(4). In such a case the precursor model is not expected to account correctly for the processes that are behind the rotational excitation mechanism. If exit channel effects are important in the dynamics of desorption(5) and direct inelastic scattering of NO molecules, then possibly there is only little difference in the outcome rotational distribution of the scattered molecules by the two mechanisms. At high incident translational energies, however, an efficient surface mediated translational to rotational energy transfer must be considered for weakly interacting gas-surface systems(4). It is believed, therefore, that more detailed dynamical treatments, such as those by Gadzuk et al(32), Zamier et al(33) and Tanaka et al(34) are necessary to understand and predict the crystal temperature dependence of the observed rotational temperature.

5. CONCLUSION

The energy transfer and accommodation data involving the translational and vibrational energy distributions of NO molecules scattered from a Pt(111) surface were analyzed in terms of a precursor model. With a single parameter, e.g. the difference in activation energy (ΔE) for desorption (E_∞) and chemisortpion (E_s) from the precursor state, we could account quite accurately for the data taken at different scattering angles, crystal temperatures and various incident kinetic energies. The vibrational excitation data was further treated by a kinetic model from which a lower limit for the deexcitation rate constant of vibrationally excited NO molecules on the Pt(111) surface of $1 \times 10^{12} sec^{-1}$ was estimated. Using this model, a crystal temperature dependence for the excitation process, possibly via an electron hole pair annihilation mechanism is suggested. The rotational excitation in this NO/Pt(111) system is poorly accommodated to the surface. Further study of the detailed dynamical effects during the desorption process is necessary in order to understand this process.

ACKNOWLEDGMENTS

The contributions of Dr. E. Pollak and Dr. Y. Zeiri to the development of the models are greatly acknowledged. This work was partially supported by the Director, Office of Energy Research, Office of Basic Energy Sciences, Chemical Sciences Division of the U.S. D.O.E. under contract No. DE-AC03-76F00098.

REFERENCES

1. a. F.O. Goodman and H.Y. Wachman, Dynamics of Gas-Surface Interactions", Academic Press (1976). b. M.J. Cardillo; Ann. Rev. Phys. Chem., 32, 331(1981) and references therin.

2. F. Frankel, J. Hager, W. Krieger, H. Walther, C.T. Campbell, G. Ertl, H. Kuipers and J. Segner, Phys. Rev. Lett., 46,152 (1981).

3. G.M. McClelland, G.D. Kubiak, J.G. Rennagel and R.N. Zare, Phys. Rev. Lett., 46, 831 (1981).

4. A.W. Kleyn, A.C. Luntz and D.J. Auerbach, Phys. Rev. Lett., 47, 1169 (1981).

5. R.R. Cavanagh and D.S. King,Phys. Rev. Lett., 47, 1829 (1981).

6. J.W. Hepburn, E.J. Northrup, G.L. Ogram, J.C. Polanyi and J.M. Williamson, Chem. Phys. Lett., 85, 127 (1982).

7. M Asscher, W.L. Guthrie, T.-H. Lin and G.A. Somorjai, Phys. Rev. Lett., 49, 76 (1982).

8. J. Segner, H. Robota, W. Vielhaber, G. Ertl, F. Frenkel, J. Hager, W. Krieger and H. Walther; Surf. Sci., 131, 273 (1983).

9. J.S. Hayden and G.J. Diebold, J. Chem. Phys., 77, 4767 (1982).

10. H. Zacharias, M.M.T.Loy and P.A. Roland; Phys. Rev. Lett., 49, 1790 (1982).

11. D.A. Mantel, Y.-F. Maa, S.B. Ryali, G.L. Haller and J.B. Fenn, J. Chem. Phys., 78, 6338 (1983).

12. J.B. Cross and J.J. Valentini, 3rd International Conference on Vibrations at Surfaces, 1982.

13. S.T. Ceyer, W.J. Siekhaus and G.A. Somorjai, J. Vac. Sci. Technol., 19, 726 (1981).

14. W.L. Guthrie, T.-H. Lin, S.T. Ceyer and G.A. Somorjai, J. Chem. Phys., 76, 6398 (1982).

15. M. Asscher, W.L. Guthrie, T.-H. Lin and G. A. Somorjai, J. Chem. Phys., 78, 6992 (1983).

16. J.A. Serri, M.J. Cardillo and G.E. Becker, J. Chem. Phys. 77, 2175 (1982).

17. E.K. Grimmelmann, J.C. Tully and M.J. Cardillo, J. Chem.Phys. 72, 1039 (1980) and references therin.

18. A.O. Bawagan, L.H. Beard, R.B. Gerber and D.J. Kouri, Chem. Phys. Lett., 84, 339 (1981).

19. M. Asscher, E. Pollak and G.A. Somorjai, to be published.

20. J.E. Lennard Jones, Trans. Farad. Soc., 28, 333 (1932).

21. P. Kisliuk, J. Phys. Chem. Solids, 3, 95 (1957).

22. G. Ehrlich, J. Phys. Chem. Solids, 1, 3 (1956).

23. C.T. Campbell, G. Ertl and J. Segner, Surf. Sci., 115, 309
     (1982).

24. J. Boheim, W. Brenig, T. Engel and V. Leuthausser,
     Surf. Sci., 131, 258 (1983).

25. B.N.J. Persson, J. Phys. Chem., 11, 4251 (1978).

26. V.P. Zhdanov and K.I. Zamarev, Catal. Rev. Sci. Eng. 24, 373
     (1982).

27. M.A. Kozhushner, V.G. Kustarev and B.R. Shub, Surf. Sci., 81,
     261 (1979).

28. L.E. Brus, J. Chem.Phys., 73, 940 (1980).

29. J. Misewick, C.N. Plum, G.Blyholder, P.L. Houston and R.P. Merril,
     J. Chem. Phys., 78, 4245 (1983).

30. a. A. Redondo, Y. Zeire and W.A. Goddard III, Phys. Rev. Lett.,
        49, 1847 (1982). b. Y. Zeiri, A. Redondo and W.A. Goddard III,
           Surf. Sci. 136, 41 (1984).

31. a. B.N.J. Persson and R. Ryberg, Phys. Rev. Lett., 48, 549 (1982).

     b. Ph. Avouris and B.N.J. Persson, J. Phys. Chem. 88, 837 (1984).

32. J. W. Gadzuk, U. Landman, E.J. Kuster, C.L. Cleveland and
     R.N. Bornett, Phys. Rev. Lett., 49, 426 (1982).

33. E. Zamir and R.D. Levine, Chem. Phys. Lett., in press.

34. S. Tanaka and S. Sugano, Surf. Sci., 136, 488 (1984).

LASER STUDIES OF VIBRATIONAL ENERGY EXCHANGE IN GAS–SOLID COLLISIONS

V. A. Apkarian, R. Hamers, P. L. Houston, and J. Misewich
Department of Chemistry,
Cornell University,
Ithaca, New York 14853, U.S.A.
   and
R. P. Merrill
School of Chemical Engineering
Cornell University
Ithaca, New York 14853, U.S.A.

ABSTRACT

Recent measurements on the probability of vibrational deactivation in gas–solid collisions are summarized. The experiments were performed by using a tunable infrared laser source to excite gas–phase molecules vibrationally before their collision with the surface and by measuring the population of vibrationally excited molecules through their time-resolved infrared fluorescence. On polycrystalline silver surfaces, the deactivation probability for $CO(v=2)$ was found to decrease from 0.33 at 300 K to 0.20 at 440 K, while the deactivation probability for $CO_2(101)$ was found to decrease from 0.72 at 300 K to 0.37 at 440 K. In contrast, the probability for dectivation of $CO_2(001)$ was found to be 0.16 and to vary only slightly with temperature. These observations are consistent with a mechanism involving trapping followed by electron-hole pair formation. In separate experiments, state-resolved scattering of rotationally cold NO from an Ir(111) surface held at 80K has demonstrated the exchange between translational and rotational energy. A similar technique employing vibrational excitation of the incoming molecular beam should provide a very detailed picture of vibrational deactivation at surfaces.

1. INTRODUCTION

An understanding of how the vibrational modes of gas-phase molecules are coupled to a solid surface during gas-solid collisions is important both to such practical areas as catalysis, corrosion, and heat transfer and to more basic problems concerning the nature of the gas-surface potentials. Despite the importance of this area, there is still little information on the probability of vibrational deactivation during a gas-surface encounter. Nor has there emerged a clear mechanism for such energy exchange. In this paper we summarize the findings of previous experiments in our laboratory[1] and introduce preliminary results which suggest that much more detailed measurements are possible in the future.

To date, most applications of laser techniques to the study of gas-surface dynamics have involved probing the internal degrees of freedom of molecules on their departure from a collisional encounter.[2] Rotational and occasionally vibrational distributions have been measur-

135

B. Pullman et al. (eds.), Dynamics on Surfaces, 135–148.
© *1984 by D. Reidel Publishing Company.*

ed for NO scattered from a variety of metals,[3-7] for HF and CO scattered from LiF(001),[8,9] and for NO desorbing from Ru(001)[10] or OH desorbing from polycrystalline silver.[11]

More recently, a few reports have appeared in which the laser was used to prepare the initial state of a molecule prior to its collision with a surface. Loy et al.[12] have employed vibrational excitation in conjunction with multiphoton ionization to study the internal energy of NO scattered from LiF, CaF_2, and Ag(110) surfaces. For the LiF and CaF_2 surfaces a survival probability of less than 2.3×10^{-4} was reported and an upper limit of 2×10^{-3} was placed on the probability for energy transfer into high rotational levels of v=0. For the Ag(110) surface the survival probability was considerably higher (ca. 10%), even though the translational velocity of the scattered products was completely accommodated to the surface temperature.

The role of vibration in the interaction of CO and CO_2 with polycrystalline platinum surfaces has recently been studied by Mantell et al.[13] A beam of vibrationally cold molecules was scattered from a hot platinum surface and the vibrational distribution of the scattered molecules was examined by Fourier transform emission spectroscopy. The vibrational accommodation coefficient for the CO_2(001) mode was found to drop from 0.22 at a surface temperature of 700 K to 0.16 at 1500 K. A vibrational accommodation coefficient of 0.7 for CO was also reported.

In our method, CO or CO_2 is vibrationally excited by its absorption of a tunable, pulsed laser. The population of the vibrationally excited molecules is then monitored by detecting their infrared radiation, and the surface deactivation probability is extracted from the decay rate of the fluorescence.

2. EXPERIMENTAL

A specially designed cylindrical cell for our studies is illustrated in Fig. 1. It is attached to an ultra high vacuum system pumped by an ion

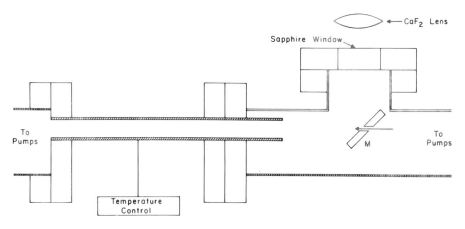

Figure 1. Schematic Diagram of Experimental Apparatus.

pump and a titanium sublimation pump with a base pressure of 8×10^{-11} torr. Laser radiation, either from a pulsed CO_2 laser operating at 10.6 μm or from tunable laser operating at the difference frequency between a dye laser and the fundamental of a YAG laser, is focused through an aperture in a gold coated mirror located inside the cell and passed down the center line of the cylinder. Depending on the laser frequency, CO_2 molecules are excited to the (001) or (101) level, or CO molecules are excited to their v=2 level. Population in these levels is monitored by a time-resolved infrared fluorescence technique. Parallel rays are reflected by the gold coated mirror, collected with an f/1 CaF_2 lens, and detected by a 4-°K mercury doped germanium semi-conductor element. The fluorescence signal is then digitized using a Biomation 805 transient recorder and fed to an LSI-11 computer for signal averaging to enhance the signal-to-noise ratio. The data are fit to a single exponential decay using an iterative least squares fitting technique.

A tungsten filament wrapped with silver is mounted in the apparatus so that it is slightly off-center but parallel to the walls of the cylinder. By heating the filament resistively, silver can be rapidly evaporated onto the walls of the cylinder. During evaporation the maximum pressure recorded in the UHV system is 7×10^{-8} torr. The deactivation measurements were completed immediately after deposition of the film.

3. RESULTS

A typical decay signal for $CO_2(100) \leftarrow (101)$ fluorescence is shown in Fig. 2. For this experiment the pressure of CO_2 was 0.5 mtorr, the

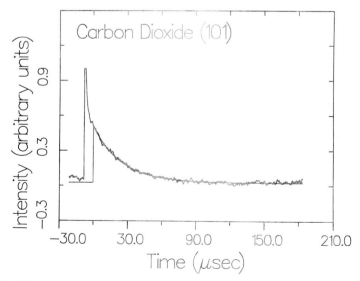

Figure 2. Fluorescence Signal from $CO_2(101)$. The smooth line is a fit to the signal of a single exponential decay.

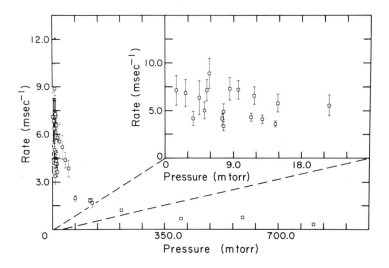

Figure 3. Pressure dependence of $CO_2(001)$ Deactivation Rate.

surface was silver, and the temperature was 300 K. The signal is the
result of averaging 5000 laser shots. A least squares single exponential fit to the data is given by the smooth line in the figure. Because the spike at early times is residual laser scatter, the starting
point for analysis is delayed until the detector has completely recovered from the laser scatter, as determined by signals obtained with
the cell evacuated to 10^{-9} torr. Similar signals were obtained for the
$CO_2(001)$ and CO $(v=1) \leftarrow (v=2)$ fluorescence. In the case of CO however,
greater signal averaging was required since the transition strength is
weaker than that for CO_2.

   For the experiments involving $CO(v=2)$ or $CO_2(101)$, a filter cell
containing CO or CO_2 was used to block any CO $(v=0) \leftarrow (v=1)$ fluorescence
or $CO_2(000) \leftarrow (001)$ fluorescence. No difference in signal was observed
when comparing the signals obtained with and without gas in the filter
cell.

   The dependence of the decay rate of the $CO_2(001)$ fluorescence on
pressure is illustrated in Fig. 3. Although this particular data is
for the relaxation of $CO_2(001)$ on stainless steel surfaces, similar
data was obtained for all three excited molecules on clean silver surfaces. The probability for deactivation was determined from the low
pressure fluorescence decay rates as discussed in Section 4, below.

   In the case of $CO_2(101)$ and $CO(v=2)$, the vibrational deactivation
probabilities were found to decrease with increasing temperature. On
freshly deposited silver the deactivation probability for the $CO_2(101)$
vibrational level fell from 0.72 at 300 K to 0.37 at 440 K. This temperature dependence is illustrated in Fig. 4, which displays the logarithm of the deactivation probability as a function of 1000/T. For the
$CO(v=2)$ vibration, the deactivation probability decreased from 0.33 at

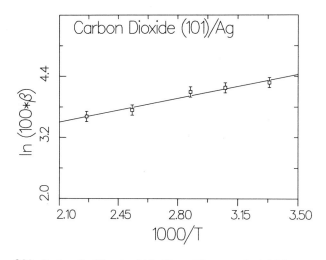

Figure 4. Temperature Dependence of $CO_2(101)$ Deactivation Probability on Silver.

300 K to 0.20 at 440 K. The probability for deactivation of $CO_2(001)$ was found to be insensitive to temperature within the experimental uncertainty.

Increasing the surface exposure also caused the deactivation probability to fall for both $CO_2(101)$ and $CO(v=2)$. The effect of allowing the silver to sit in the vacuum system for roughly one day was to decrease the probability of deactivation by roughly 30-40%.

4. DISCUSSION

4.1 Signal Analysis

There are four contributions to the decay of the infrared fluorescence signals: radiative decay of the excited vibrational level, effusion of excited molecules out of the cylindrical cell, gas-phase intermolecular energy transfer from the initially produced vibrational level, and relaxation as a result of gas-solid collision.

In the case of $CO_2(001)$ it has been shown that the radiative decay and the gas-phase intermolecular energy transfer are negligible under the conditions of our experiments.[1a] Even the effusion out of the cell amounts to only a very minor correction. In the case of $CO_2(101)$ and $CO(v=2)$, however, while effusion and radiative decay make only small contributions to the fluorescence decay, the gas-phase energy transfer is not negligible even at the lowest pressures employed (0.5-1.0 mtorr). Correction for this decay channel, which amounted to 10% in the worst case, was made using either gas-phase collisional deactivation rates reported in the literature[14] or rates measured by separate experiments in our laboratory.[1b]

The surface deactivation probability was calculated after correcting the observed fluorescence decay rate for effusion, radiative decay, and self deactivation. At the lower pressures used in these experi-

ments, the mean free path was large enough compared with the cell diameter so that the deactivation probability is related simply to the
fluorescence decay:[1]

$$\beta = 1 - \exp(-kd/c), \hspace{3cm} (1)$$

where k is the corrected decay rate, d is the cell diameter and c is
the two-dimensional average velocity. The resulting surface deactivation probabilities extrapolated to zero pressure are listed in Table
I.

TABLE I: Deactivation Probabilities on Silver

Molecule	Temperature ($^\circ$K)	β ($\pm$ 10%)
$CO(v=2)$	300	0.33
	348	0.29
	395	0.28
	440	0.20
$CO_2(101)$	300	0.72
	325	0.65
	348	0.60
	395	0.42
	440	0.37
$CO_2(001)$	298	0.16

A cold gas filter cell filled with CO or CO_2 was used in the experiments probing the relaxation of $CO(v=2)$ and $CO_2(101)$; however, no
effect on the amplitude or rate of decay was observable when compared
with experiments performed without the filter cell. The CO fluorescence was thus due entirely to the $(v=1)\leftarrow(v=2)$ transition, while the
CO_2 fluorescence was all due to the $(100)\leftarrow(101)$ transition. It thus
appears that $CO(v=2)$ and $CO_2(101)$ are deactivated at the silver surface
without going through the $CO(v=1)$ or $CO_2(001)$ intermediate levels.

4.2 The Nature of the Surface

The geometry of our experimental cell does not facilitate in situ analysis of surface composition and structure, so we can only speculate
about the cleanliness of our surface. As in our previously reported
study of the surface deactivation of $CO_2(001)$,[1] polycrystalline silver
surfaces were prepared by evaporation of silver from a tungsten filament wrapped with silver and mounted parallel to the center axis of the
cylindrical cell. Also as in our previous work, the gases used were
the purest commercially available: 99.999% CO_2 and 99.99% CO, both
gases having ppm quantities of $CO(CO_2)$, nitrogen, oxygen, hydrogen and

hydrocarbons. Each gas was further purified by passing it over traps with molecular sieves and activated charcoal. We will briefly review the conjecture on the nature of the silver surface presented in Ref. 1.

The major components in the residual gas of the vacuum system are $CO(CO_2)$ and hydrogen. From helium scattering data, the adsorption probabilities of CO and hydrogen on silver are vanishingly small.[15] Also, CO_2 is known not to be adsorbed on silver.[16] The component of the residual gas with the highest sticking probability (3×10^{-3}) is oxygen;[17] however, our experiments typically required an oxygen exposure of only about 10^{-5} torr seconds, which would produce only about 0.01 monolayer of oxygen. Fresh silver was evaporated between experimental runs to ensure the cleanest surface possible with our apparatus.

4.3 Comparison to Other Experiments

The major finding of the current experiments is that the probability for vibrational deactivation of $CO_2(001)$, $CO_2(101)$, and $CO(v=2)$ in surface collisions with polycrystalline silver is very high. Recent reports from other laboratories are consistent with this conclusion and suggest that high deactivation probabilities, at least for CO_2 and CO, may be the rule rather than the exception.

The accommodation of internal energy has been measured previously by Rosenblatt and his coworkers using an oscillating surface technique.[18] For CO_2 on Cu an internal energy accommodation coefficient of 0.45 was found, whereas our results at room temperature for $CO_2(001)$ and $CO_2(101)$ on silver give deactivaton probabilities of 0.16 and 0.72, respectively. Mantell et al.[13] have used Fourier transform emission spectroscopy to study the distribution of internal energy in CO and CO_2 vibrationally excited by their collision with a hot platinum surface. An increase in the surface temperature from 700 K to 1600 K caused the accommodation coefficient for the asymmetric stretching mode of CO_2 to drop from 0.22 to 0.16. An accommodation coefficient of 0.7 was obtained for the vibrational excitation of CO. The results obtained by Loy and his co-workers[12] on the vibrational relaxation of NO at CaF_2, LiF, and Ag(111) surfaces also suggests that the probability for vibrational exchange is high, although it is hard to understand why it should be nearly unity on the former two crystals and substantially less on Ag(111). While it is difficult to make quantitative comparisons between these results and ours, the qualitative features are the same: high probabilities are found for vibrational energy transfer between the surface and the incoming molecules, and the accommodation decreases with increasing temperature.

4.4 The Mechanism of Vibrational Relaxation

The mechanism for interchange of energy between a excited moleule and a surface is still unclear. One mechanism that can probably be discounted is relaxation via exclusive excitation of surface phonons, since the molecular vibrational frequencies are much higher than the surface Debye frequencies. For example, in $CO_2(101)$ the molecular vibration frequencies. For example, in $CO_2(101)$ the vibrational frequency is 4260

cm^{-1} but the surface Debye frequency for silver is only 107 cm^{-1}.[19]
Thus, the transfer of a large fraction of the molecular vibrational
energy would require a high-order multiphonon process that is not likely to be efficient.
     Gerber et al.[20,21] have considered the dynamics of a vibration-
ally excited diatomic molecule colliding with a rigid, smooth surface.
They conclude that the dominant mechanism of vibrational deactivation
is through energy transfer from vibration to rotation; i. e., that the
scattered molecules will be vibrationally relaxed but rotationally excited.   Tully, on the other hand, has found that when the surface is
allowed to move in response to the incoming diatomic, very little vi-
bration-to-rotation energy transfer occurs; in fact, even the total
probability for vibrational deactivation is small.[22]   Both of these
theoretical treatments neglect the direct interaction of the incoming
transition dipole with the conduction electrons of the metallic surface.
     In contrast, Persson and Persson maintain that for short distances
energy transfer from the dipole to form electron-hole pairs is the dominant deactivation mechanism.[23]   They calculate a lifetime of 10^{-11} sec
for a CO molecule adsorbed on a copper surface.   Thus, a vibrationally
excited gas-phase molecule colliding with a metal surface might have a
significant probability for losing its energy to conduction electrons,
especially if the molecule is trapped at the surface for a few vibrational periods.   The rate of transfer is predicted to vary inversly as
the fourth power of the molecule-surface distance.   Thus, the decrease
in deactivaton probability with exposure observed in this work may be
related to the fact that increased exposures lead to overlayers which
prevent the excited molecule from coming as close to the surface.   The
probability is also predicted to vary strongly with the square of the
dipole matrix element connecting the initial and final vibrational
states.   Thus, deactivation of $CO_2(101)$ to $CO_2(001)$, which is dipole
forbidden, should not be observed, in agreement with the fact that we
see no fluorescence from the $CO_2(001)$ level following surface deactivation of $CO_2(101)$.   Deactivation of CO(v=2) to CO(v=1) would be allowed
by the theory, but it may be that the CO molecule interacts long enough
with the surface to undergo two sequential deactivations.

4.5 Trapping and Temperature Dependence

An estimate of the trapping probability of CO_2 on silver is 0.15.   This
value was calculated from the angular scattering distributions reported
by Asada[24] by computing the ratio of diffusely emmitted to specularly
scattered molecules.[25]   Since it is unlikely that our evaporatively deposited surfaces are atomically smooth, the trapping probability is
likely to be higher in our experiments.   Further evidence that the
trapping probability can be significant comes from a comparison of our
systems with rare gases.   The fraction trapped depends on the ratio of
the well depth of the gas-surface interaction potential to the incident
kinetic energy.   From the experiments of Asada,[24] the well depth of CO_2
on Ag is estimated to be 6200 cal/mol, a value which is comparable to
the well depth of 5800 cal/mole estimated by correlation with gas-phase
Lennard-Jones parameters.[26]   This well depth makes CO_2 analagous to

Xe/Ag(111), which has a well depth of 6100 cal/mol.[25] The trapping probability for Xe/Ag(111) was found to be 0.42, larger than the estimate for CO_2. Kr/Ag(111) has a well depth of 1400 cal/mol and a trapping probability of 0.20.

In a scheme where the deactivation is due to trapping followed by deexcitation and subsequent reemission, the temperature dependence of the deactivation probability should be given by that which governs the residence time. The residence time is given by:

$$\tau = [(n+1)\nu]^{-1} \exp(E_d/kT_s), \qquad (2)$$

where ν is the fundamental vibrational frequency of the complex formed between the trapped gas molecule and the surface atom, E_d is the activation energy for reemission, which is measured as the positive difference in energy between the vacuum level and the occupied level in the potential well.[26] If we assume that the deactivation probability is proportional to the residence time, the temperature dependence of the deactivation probability would be:

$$\ln \beta \simeq constant + E_d/kT_s \qquad (3)$$

Thus, a plot of $\ln \beta$ versus $1/T_s$ should give a straight line whose slope is determined by the activation energy for reemission. Figure 4 displays the data for $CO_2(101)$ and shows the least squares linear fit. The activation energies determined in this manner are given in Table II along with a comparison to the relevant rare gas data. While the

TABLE II: Values of E_d

System	E_d (cal/mol)
$CO_2(101)$	1320†
$CO(v=2)$	730†
Xe	2362*
Kr	359*

†This work
*Reference 26

activation energies obtained from our results are somewhat smaller than might be expected by comparison to Xe, they still suggest strongly that the deactivation is intimately connected with trapping at the surface. In this view, the decrease in accommodation with increasing temperature observed both in this work and in the results of Mantell et al.[13] is due primarily to the decrease in surface residence time.

5. STATE-RESOLVED SCATTERING

In order to understand completely the dynamics of gas–surface interactions one would like ideally to select the internal and translational
energies of the molecular reactant, the crystal face and temperature of
the surface, and the angle of incidence for the gas–surface collision.
One would then like to measure the internal and translational energies
of the products as well as their angular distribution with respect to
the surface. In conjunction with the experiments described in the preceding four sections, we have recently assembled an apparatus which attempts to achieve these ideals by coupling molecular beams, UHV surface
analysis, and lasers. Our state-to-state scattering system consists b-
asically of a supersonic molecular beam, a tunable infrared laser for
state preparation, a single crystal whose surface can be analyzed under
UHV conditions, and a mass spectrometer equipped for detection by both
electron impact and, for state-specific detection, laser multiphoton
ionization. A schematic diagram is shown in Fig. 5.

   Initial state selection is achieved by optical excitation with a
tunable F-center laser (Burleigh) operating in the infrared and pumped
by a krypton ion laser (Coherent, CR-3000K). It provides cw output
from 2.3–3.2 μm with powers on the order of 1–10 mW and a linewidth of
roughly 1 MHz. The infrared laser is coupled to the molecular beam in
the second-stage pumping region via a multipass reflector.[27]

   Single crystal surfaces are held on a precision manipulator which
is designed to allow the crystal to be cooled by liquid nitrogen or
heated by electron bombardment. The crystals are analyzed by LEED and
Auger spectroscopy after having been cleaned by an argon ion source.

   Detection of the scattered products in our system is accomplished

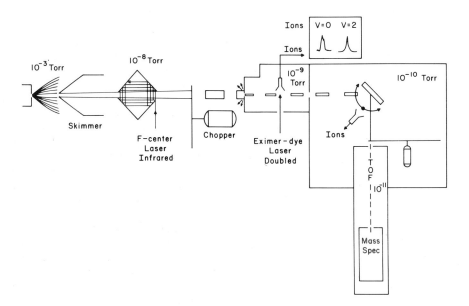

Figure 5. Schematic Diagram of the Molecular Beam Scattering Apparatus

either by a quadrupole mass spectrometer or by laser-induced multiphoton ionization (MPI). Both the mass spectrometer and the MPI detection system are mounted entirely within the vacuum system on a turntable which can position the ionization region to within 0.1° around the axis parallel to the crystal face and to within 0.001" in and out of the principal scattering plane. These manipulations are accomplished through bellows-isolated feed-throughs with stepping motors interfaced to a lab-dedicated microcomputer. Time-of-flight measurements are made with a second, fixed-angle mass spectrometer which is differentially pumped.

Detection by multiphoton ionization employs an excimer-pumped dye laser (Lambda Physik 2002, EMG 101). The laser beam enters the ultra-high vacuum chamber through a sapphire window, is reflected by a prism along the axis of rotation for the turntable described in the previous paragraph, and is finally reflected by two more prisms out a radial arm from the rotation axis and up through a lens to the ionization area located immediately in front of a channeltron detector. By doubling the visible output of the laser in nonlinear crystals, it is possible to obtain wavelengths as low as 217 nm. The multiphoton ionization technique uses two or more of these laser photons to cause a transition from a selected rovibronic level of the probed molecule to the ionization continuum; most often the transition is enhanced by a resonant intermediate state. The resulting ions are multiplied by a channeltron and detected as current pulses. The technique of multiphoton ionization has been used previously[6] and has been found to be sensitive to fewer than 10^5 molecules/cm^3 of NO, for example.

Preliminary experiments in this new apparatus have demonstrated intriguing dynamical effects even for molecules which start in their ground vibrational state. A multiphoton ionization spectrum of the NO(v=0) in the incoming beam of our apparatus is shown in Fig. 6a. Our calculations show that we are able to detect a single NO molecule in the focal volume of the laser. A spectrum of the NO(v=0) scattered from an Ir(111) surface held at 80°K is shown in Fig. 6b. Analysis of the spectrum shows that the rotational distribution of the scattered NO, while non-Boltzmann, is quite hot; rotational levels with

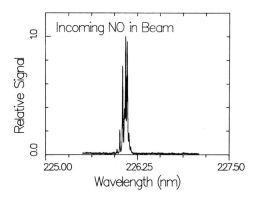

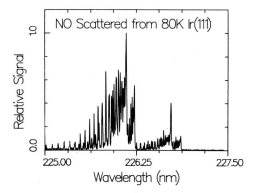

Figure 6. MPI Spectra of NO

J<31/2 are fit by temperatures of 147K and 121K for the $^2\Pi_{1/2}$ and $^2\Pi_{3/2}$ states respectively, while those with' J>31/2 in the $^2\Pi_{1/2}$ state are more strongly populated than a temperature of 147K would indicate. Since the temperature of the scattered molecules is higher than the rotational temperature of the incoming beam and also higher than the surface temperature, translation-to-rotation energy transfer must have occurred. The fact that upper rotational levels are more strongly populated may indicate the presence of a rotational rainbow.

The apparatus described above will provide a detailed picture of the interaction of state-selected molecules with surfaces. The translational velocity of the beam can be varied by seeding in different carrier gases. Its vibrational energy can be changed by selective excitation with the tunable laser. The crystal face and temperature of the surface as well as the orientation with respect to the molecular beam are easily varied. Finally, state-selective detection provides the identity, direction, and internal energy distribution of product molecules.

6. CONCLUSION

The vibrational relaxation of CO(v=2), CO_2(001), and CO_2(101) on collision with polycrystalline silver surfaces has been found to be extremely efficient. It is likely that the deactivation is due largely to trapping followed by transfer of vibrational energy to form electron-hole pairs. This mechanism would account (1) for the approximate magnitude of the deactivation probabilities, (2) for the decrease in accommodation with increasing temperature, observed both in our results on CO(v=2) and CO_2(101) and in others,[13] (3) for the observed absence in population of the CO_2(001) intermediate level, and (4) for the decrease in deactivation probability with increasing surface exposure.

While the proposed mechanism might also account for the results of Rosenblatt[18] and of Mantell et al.,[13] some controversy still remains in light of the observations made by Loy and his coworkers.[12] The trapping/electron-hole formation mechanism cannot account for the very high probability of deactivation observed for NO(v=1) on surfaces such as LiF which have no conduction electrons, especially in view of the fact that smaller deactivation probabilities are observed for Ag(111). Continued work in our laboratory will be aimed at examining the deactivation of CO(v=2) and CO_2(001,101) on LiF surfaces and in measuring the state-to-state reaction probabilies for deactivation in beam-surface experiments.

ACKNOWLEDGEMENTS

This work was supported by the Army Research Office under grants DAAG-29-80-K-0020 and DAAG29-84-K-0016.

REFERENCES

1.   a. J. Misewich, C. N. Plum, G. Blyholder, P. L. Houston and R. P. Merrill, J. Chem. Phys. **78,** 4245 (1983).
     b. J. Misewich, P. L. Houston, and R. P. Merrill, 'Vibrational Relaxation of Carbon Dixoide(101) and Carbon Monoxide(v=2) During Gas-Surface Collisions,' submitted.
2.   G. S. Selwyn and M. C. Lin, 'Laser Studies of Surface Chemistry,' to be published.
3.   a. G. M. McClelland, G. D. Kubiak, H. G. Renegal and R. N. Zare, Phys. Rev. Lett. **46,** 831 (1981).
     b. G. D. Kubiak, J. E. Hurst, Jr., H. G. Rennagel, G. M. McClelland, and R. N. Zare, J. Chem. Phys. **79,** 5163 (1983).
4.   a. A. W. Kleyn, A. C. Luntz and D. J. Auerbach, Phys. Rev. Lett. **47,** 1169 (1981).
     b. A. C. Luntz, A. W. Kleyn, and D. J. Auerbach, J. Chem. Phys. **76,** 737 (1982).
     c. A. W. Kleyn, A. C. Luntz, and D. J. Auerbach, Phys. Rev. **B25,** 4273 (1982).
     d. A. W. Kleyn, A. C. Luntz, and D. J. Auerbach, Surf. Sci. **117,** 33 (1982).
5.   a. F. Frenkel, J. Häger, W. Krieger, H. Walther, C. T. Campbell, G. Ertl, H. Kuipers, and J. Segner, Phys. Rev. Lett. **46,** 152 (1981).
     b. F. Frenkel, J. Häger, W. Krieger, H. Walther, G. Ertl, J. Segner, and W. Bielhaber, Chem. Phys. Lett. 90, 225 (1982).
     c. J. Segner, H. Robota, W. Vielhaber, G. Ertl, F. Frenkel, J. Haeger, W. Krieger and H. Walther, Surf. Sci., to be published.
6.   a. M. Asscher, W. L. Guthrie, T.-H. Lin, and G. A. Somorjai, Phys. Rev. Lett. **49,** 76 (1982).
     b. M. Asscher, W. L. Guthrie, T.-H. Lin, and G. A. Somorjai, J. Chem. Phys. **78,** 6992 (1983).
7.   J. S. Hayden and G. J. Diebold, J. Chem. Phys. **77,** 4767 (1972).
8.   D. Ettinger, K. Honma, M. Keil and J. C. Polanyi, Chem. Phys. Lett. **87,** 413 (1982).
9.   J. W. Hepburn, F. J. Northrup, G. L. Olgram, J. C. Polanyi and J. M. Williamson, Chem. Phys. Lett. **85,** 127 (1982).
10. a. R. R. Cavanagh and D. S. King, Phys. Rev. Lett. **47,** 1829 (1981).
     b. D. S. King and R. R. Cavanagh, J. Chem. Phys. **76,** 5634 (1982).
11. a. L. D. Talley, D. E. Tevault, and M. C. Lin, J. Chem. Phys. **72,** 3314 (1980).
     b. L. D. Talley and M. C. Lin, Chem. Phys. **61,** 249 (1981).
12. a. H. Zacharias, M. M. T. Loy and P. A. Roland, Phys. Rev. Lett. **49,** 1790, (1982).
     b. M. M. T. Loy, H. Zacharias, and P. A. Polland, 'Time-Resolved, State-Selective Study of Vibrationally Excited NO Scattering From Ag(110),' private communication.
13. D. A. Mantell, S. B. Ryali, G. L. Haller and J. B. Fenn, J. Chem. Phys. **78,** Part II, 4250 (1983).
14. J. Finzi and C. B. Moore, J. Chem. Phys. **63,** 2285 (1975).

15. R. Sau and R. P. Merrill, Surf. Sci. 34, 268 (1973).

16. A. W. Czanderna, J. Colloid and Interface Sci. 22, 482 (1966).

17. M. Bowker, M. A. Barteau and R. J. Madix, Surf. Sci. 92, 528 (1980).

18. a. G. Rosenblatt, Acc. Chem. Res. 14, 42 (1981).
    b. G. Rosenblatt, R. Lemons and C. Draper, J. Chem Phys. 67, 1099 (1977).

19. E. R. Jones, J. T. McKinney and M. B. Webb, Phys. Rev. 151, 476 (1966).

20. R. B. Gerber, L. H. Beard and D. J. Kouri, J. Chem. Phys. 74, 4709 (1981).

21. A. O. Bawagan, L. H. Beard, R. B. Gerber and D. J. Kouri, Chem. Phys. Lett. 84, 359 (1981).

22. a. J. C. Tully, Acc. Chem. Res. 14, 188 (1981).
    b. J. C. Tully and M. J. Cardillo, Science 223, 445 (1984).

23. B. N. J. Persson and M. Persson, Surf.Sci. 97, 609 (1980).

24. H. Asada, Jpn. J. Appl. Phys. 20, 527 (1981).

25. a. A. C. Stoll, D. L. Smith and R. P. Merrill, J. Chem. Phys. 54, 163 (1971).
    b. W. H. Weinberg and R. P. Merrill, J. Chem. Phys. 56, 2881 (1972).
    c. R. Sau and R. P. Merrill, Surf. Sci. 32, 317 (1972).

26. W. H. Weinberg and R. P. Merrill, J. Vac. Sci. Technol. 10, 411 (1973).

27. T. E. Gough, D. Gravel, and R. E. Miller, Rev. Sci. Instrum. 52, 802 (1981).

ADSORPTION OF ATOMIC HYDROGEN ON LiF SURFACES

V. Levy-Duval, M. Shapiro, Y. Zeiri
Department of Chemical Physics
The Weizmann Institute of Science
Rehovot 76100, Israel

and

R. Tenne
Department of Plastics Research
The Weizmann Institute of Science
Rehovot 76100, Israel

ABSTRACT

A theory based on semi-empirical valence bond formalism describing interactions and charge transfer between atoms and ionic surfaces is developed. The interaction of a hydrogen atom with a LiF crystal is computed explicitly. The contribution of the charge transfer structures in stabilizing the adsorbed atom is shown to be substantial. The sites at which energetic barriers for diffusion on the surface and penetration into the bulk are most favourable are also computed. The stability of interstitial hydrogen is discussed.

I. INTRODUCTION

The adsorption of hydrogen on ionic surfaces is a question of fundamental importance to a wide variety of fields including the photoelectrolysis of water, the performance of solar cells and catalysis [1,2]. Besides quantitatively understanding the adsorption energy the preferred adsorption sites and the probability of diffusion of hydrogen on the surface, one would like to understand to what extent the electronic properties of the adsorbate are being affected by the nearby presence of an ionic surface. If significant changes are found one could go a step further to explain the greater reactivity of adsorbed species to such reaction involving for example electrophylic or nucleophylic attacks on double bonds.

In a previous paper [3] it was shown that the electronic structure of an atom or an ion may be strongly modified when it is adsorbed on an ionic surface. It was found that at close enough atom-surface separations near degeneratices (intersections) between ionic and atomic states

149

B. Pullman et al. (eds.), Dynamics on Surfaces, 149–167.
© 1984 by D. Reidel Publishing Company.

exist, in contrast to the behaviour of the atom far from the surface. As a result efficient electron transfer (tunneling) between the adsorbate and the surface can take place [3].

In the present paper we probe this question further by considering the adsorption of a hydrogen atom on an ionic surface, such as that of a LiF crystal. In particular, we consider the role of the various ionic channels, arising from a charge transfer between H, F$^-$ and Li$^+$, in promoting the adsorption of hydrogen at various surface sites and in stabilizing the interstitial structures.

In order to calculate the relevant potential-surfaces we make use of the Semi-empirical Valence Bond (SVB) methodology [3-5]. This approach, developed originally to treating electron transfer reactions between isolated atoms and molecules [4], was actually applied to a molecular system analogous to the present one, namely the H+LiF$\rightleftharpoons$Li+HF exchange reaction [4]. Contrary to the molecular case, we can essentially exclude the exchange channel, in which hydrogen replaces the surface Li$^+$ ion at most energies of interest. However, as we show below, the three basic structures, schematically written as H(Li$^+$F$^-$), H$^+$(LiF$^-$) and H$^-$(Li$^+$F), considered in the molecular case, also determine the H-LiF crystal interaction.

II. SEMI-EMPIRICAL VALENCE BOND

We consider three Valence Bond structures, one "covalent",

$$\psi_1 = \hat{A}\{ [\prod_{k=1}^{M} L_k^+ F_k^-] H(1) \} \tag{1}$$

and two "ionic",

$$\psi_2 = \hat{A}\{ [\prod_{k=1}^{M} L_k^+ F_k^-] L_M(1) H^+ \} \tag{2}$$

and

$$\psi_3 = \hat{A}\{ [\prod_{k=1}^{M-1} L_k^+ F_k^-] L_M^+ F_M^+ F_M(1) H(2) \bar{H}(3)] \} \tag{3}$$

In the above, L_k^+ denotes the wavefunction of all electrons belonging to the k'th Li$^+$ ion, F_k^- the wavefunction of the k'th F$^-$ ion, H(1) (H(2), H(3)), a Slater-type 1S orbital centered about the hydrogen atom, $L_M(1)$ a Slater-type 2S orbital of the Li closest to the adsorbed hydrogen and $F_M(1)$ a Slater-type 2p orbital of the fluorine closest to the adsorbed hydrogen. In Eqs. (1-3) $\hat{A}$ stands for an antisymmetrizer of all electrons and 1,2,3 are the valence electrons indices. A bar above the orbital designates that the electron is allocated a β spin function.

The first structure (ψ_1) describes an atom adsorbed on a perfect ionic crystal surface. The second structure (ψ_2) describes the situation after an electron has been transferred from the hydrogen atom to the Li atom (L_M) closest to it, and the third structure (ψ_3) depicts the opposite case in which an electron has been transferred to the hydrogen from the fluorine atom (F_M) closest to it.

The Born-Oppenheimer (adiabatic) wavefunction is expanded in terms of the three VB structures as,

$$\Psi = C_1(R)\psi_1(q,R) + C_2(R)\psi_2(q,R) + C_3(R)\psi_3(q,R) \tag{4}$$

where q is a collective electronic coordinate and R is the nuclear geometry, in particular – the H–surface separation. In order to find the $C_i(R)$ expansion coefficients we solve a (3x3) Configuration Interaction (CI) problem, expressed as a set of secular equations

$$\det(\underline{\underline{H}} - E\underline{\underline{S}}) = 0 \tag{5}$$

where $\underline{\underline{H}}$ and $\underline{\underline{S}}$ are the Hamiltonian and overlap matrices in the $(\psi_1\psi_2\psi_3)$ basis set.

In order to calculate the Hamiltonian matrix we write the electronic Hamiltonian for M nuclei and N electrons

$$H = -\sum_{i=1}^{N}\frac{1}{2}\nabla_i^2 - \sum_{i=1}^{N}\sum_{p=1}^{M}\frac{Z_p}{|r_i - R_p|} + \sum_{i=1}^{N}\sum_{j<i}\frac{1}{|r_i - r_j|}$$
$$+ \sum_{p=1}^{M}\sum_{q<p}Z_pZ_q/|R_p - R_q| \tag{6}$$

where i,j are electronic indices p,q are nuclear indices and Z_p, Z_q are nuclear charges, as

$$H = \sum_{p=1}^{M}H_p + \sum_{p=1}^{M}\sum_{q<p}V_{pq} \tag{7}$$

where

$$H_p = \sum_{i=1}^{N(p)}\{-\frac{1}{2}\nabla_i^2 - \frac{Z_p}{|r_i - R_p|}\} + \sum_{i=1}^{N(p)}\sum_{j<i}1/|r_i - r_j| \ , \tag{8}$$

$$V_{pq} = -\sum_{i=1}^{N(p)}Z_q/|r_i - R_q| - \sum_{j=1}^{N(q)}Z_p/|r_j - R_p| + \sum_{i=1}^{N(p)}\sum_{j=1}^{N(q)}1/|r_i - r_j|$$
$$+ Z_pZ_q/|R_p - R_q| \ . \tag{9}$$

In the above $N(p)$ is the number of electrons to be associated with the p nucleus.

As in the Diatomics In Molecules (DIM) method [6] we rewrite Eq.(7) using diatomic Hamiltonians, defined as

$$H_{pq} = H_p + H_q + V_{pq} \ . \tag{10}$$

Each diatomic Hamiltonian describes the motion of $N(p)+N(q)$ electrons in the field of the p and q nuclei and is therefore the true Hamiltonian for a diatomic molecule defined in this manner. Substituting Eq. (10) in Eq. (7) results in the following form

$$H = \sum_{p=1}^{M} H_p + \sum_{p=1}^{M} \sum_{q<p} (H_{pq} - H_p - H_q)$$

$$= \sum_{p=1}^{M} \sum_{q<p} H_{pq} - (M-2) \sum_{p=1}^{M} H_p$$

(11)

The introduction of empirical information is done via the separation of the electrons in each diatomic Hamiltonian into <u>valence</u> electrons and <u>core</u> electrons. According to this we now write H_{pq} as

$$H_{pq} = H_{pq}^V + H_{pq}^C + V_{pq}^{cc}$$

(12)

where the diatomic Hamiltonian for the valence electrons is

$$H_{pq}^V = H_p^V + H_q^V + V_{pq}^V$$

(13a)

$$H_{pq}^C = (H_p - H_p^V) + (H_q - H_q^V)$$

(13b)

in which

$$H_p^V = \sum_{i=1}^{N_V(p)} \{-\frac{1}{2}\nabla_i^2 - \frac{[Z_p - N_c(p)]}{|r_i - R_p|}\} + \sum_{i=1}^{N_V(p)} \sum_{j<i} 1/|r_i - r_j|$$

(14)

and similarly for H_q^V and

$$V_{qp}^V = - \sum_{i=1}^{N_V(p)} [Z_q - N_c(q)]/|r_i - R_q| - \sum_{j=1}^{N_V(q)} [Z_p - N_c(p)]/|r_j - R_p|$$

$$+ \sum_{i=1}^{N_V(p)} \sum_{j=1}^{N_V(q)} 1/|r_i - r_j| + [Z_p - N_c(p)][Z_q - N_c(q)]/|R_p - R_q|$$

(15)

where

$$N_c(p) + N_v(p) = N(p) .$$

(16)

In all the applications below $N_V(p) < 2$.
The valence diatomic Hamiltonian describes the motion of the $N_V(p) + N_V(q)$ valence electrons in the field of the nuclei screened by the $N_c(p)$ and $N_c(q)$ core electrons. The remaining "core-core" potential V_{pq}^{cc} is composed of two terms: V_{qp}^C – the interaction between the core electrons and the other nucleus as well as the interaction between core electrons associated with different nuclei, and V_{qp}^{vc} – the valence-core interaction Explicitly

$$V_{pq}^{cc} = V_{pq}^c + V_{pq}^{vc}, \text{ where}$$

(17a)

$$V_{pq}^c = - \sum_{i}^{N_c(p)} Z_p/|r_i - R_q| - \sum_{j}^{N_c(q)} Z_p/|r_j - R_p| + \sum_{i}^{N_c(p)} \sum_{j}^{N_c(q)} 1/|r_i - r_j|$$

$$+ \frac{N_c(p)Z_q + N_c(q)Z_p - N_c(q)N_c(p)}{|R_p - R_q|}$$

(17b)

and

$$V_{pq}^{vc} = - \sum_i^{N_V(p)+N_v(q)} \{N_c(p)/|r_i-R_p|+N_c(q)/|r_i-R_q|\} \tag{17c}$$

$$+ \sum_i^{N_v(p)+N_V(q)} \sum_j^{N_c(p)+N_c(q)} 1/|r_i-r_j|$$

As an example to the use of the above formulae we look at the three structure adsorption models discussed above. In order to compute the H_{11} matrix element we consider the following factorization of the Hamiltonian,

$$
\begin{aligned}
H &= \sum_{p=1}^{2M+1} \sum_{q<p} H_{pq} - (2M-1) \sum_{p=1}^{2M+1} H_p \\
&= \sum_{p=1}^{2M} \sum_{q<p} H_{pq} - (2M-2) \sum_{p=1}^{2M} H_p + \sum_{k=1}^{M} (H_{HL_k^+} + H_{HF_k^-}) \\
&\quad - \sum_{k=1}^{M} (H_{L_k^+} + H_{F_k^-}) - 2MH_H + H_H
\end{aligned}
\tag{18}
$$

This form allows us to identify H as being composed of

$$H = H_{LiF} + H_{INT} + H_H \tag{19}$$

where the Hamiltonian of the pure crystal is,

$$H_{LiF} = \sum_{p=1}^{2M} \{ \sum_{q>p}^{2M} H_{pq} - (2M-2)H_p \} \tag{20a}$$

and the interaction Hamiltonian is,

$$H_{INT} = \sum_{k=1}^{M} \{ H_{HL_k^+} + H_{HF_k^-} - H_{L_k^+} - H_{F_k^-} - 2H_H \} \tag{20b}$$

Given the form of the first structure (Eq. (1)) and Eqs. (19,20) we obtain that

$$H_{11} = \langle \psi_1|H|\psi_1 \rangle = \langle [\prod_{k=1}^{M} L_k^+F_k^-]H(1)|H_{LiF}+H_{INT}+H_H|\sum_{\hat{P}_i} (-1)^{P_i} \hat{P}_i[\prod_{k=1}^{M} L_k^+F_k^-]H(1) \rangle \tag{21}$$

where $\hat{P}_i$ are permutation operators and i goes over all the electrons. Because of the poor overlap between the valence and core electrons we neglect the exchange between them. Under this assumption we can write the H_{11} matrix element as

$$H_{11} = E_{LiF} + E_H + V_H^{POL} + \sum_{k=1}^{M} \{V_{HL_k^+} + V_{HF_k^-}\} \tag{22}$$

where E_{LiF} is the energy of the LiF crystal, E_H is the energy of the hydrogen atom, V_H^{POL} is the hydrogenic polarization energy, and the diatomic interactions are defined as,

$$V_{HL_k^+} = <H(1)|H_{L_kH}^+|H(1)> - E_{Li^+} - E_H \tag{23a}$$

$$V_{HF_k^-} = <H(1)|H_{F_k^-H}|H(1)> - E_{F_k^-} - E_H \tag{23b}$$

Setting $N_v(H)=1$ $N_v(L_k^+)=0$ in Eq. (15) leads the following explicit expressions,

$$V_{HL_k^+} = <H(1)|-1/|r_1-R_{L_k}||H(1)> + V_{HL_k^+}^{cc} + 1/|R_H-R_{LK}| \tag{24a}$$

Substituting $N_v(F_k^-)=10$, $Z_{F_k^-}=9$ in Eq. (15) results in,

$$V_{HF_k^-} = <H(1)|1/|r_1-R_{F_k}|||H(1)> + V_{HF_k^-}^{cc} - 1/|R_H-R_{F_k}| \tag{24b}$$

The integral over valence electron can, at times, be computed exactly
[7]. Alternatively point charge approximation [4,8] can be used,
to yield

$$<H(1)|- 1/|r_1-R_{L_k}||H(1)> = -1/R_{HL_k} - \alpha(L_i^+)/2R_{HL_k}^4 \tag{25}$$

$$<H(1)|1/|r_1-R_{F_k}|H(1)> = 1/R_{HF_k} - \alpha(F^-)/2R_{HF_k}^4 \tag{26}$$

where $R_{HL_k}=|R_{11}-R_{L_k}|$, $R_{HF_k}=|R_H-R_{F_k}|$.

The last term in Eqs. (25,26) accoutns for the polarization of the Li$^+$
or F$^-$ core, where $\alpha(Li^+)$ $(\alpha(F^-))$ denote the polarizability of the Li$^+$(F$^-$)
ions.

    The core-core potentials are obtained from empirical information
about stable diatomics such as LiH or HF. The exact procedure used for
extracting the core-core potentials has been described elsewhere [4].
The actual form used here was

$$V^{cc} = B \exp(-bR) \tag{27}$$

where the values of B and b taken from the work of Finzel [9] are
given in Table 1.
    The polarization of the hydrogen, V_H^{POL} - Eq. (22), due to the
combined effect of all charges on the crystal surface, cannot be
included in the $V_{HL_k^+}$ and $V_{HF_k^-}$ terms, because the field felt by the
hydrogen is a vector sum of contributions from all crystal ions. We
therefore write

$$V_H^{POL} = \alpha(H)\vec{E}^2/2 \tag{28}$$

where

$$\vec{E} = \sum_{k=1}^{M} \left\{ \frac{(\vec{R}_H-\vec{R}_{L_k^+})}{|\vec{R}_H-\vec{R}_{L_k^+}|^3} - \frac{(\vec{R}_H-\vec{R}_{F_k^-})}{|\vec{R}_H-\vec{R}_{F_k^-}|^3} \right\} \tag{29}$$

The value used for the polarizabilities of H, Li^+ and F^- are also given in Table 1.

Table 1.   Atomic and Molecular Parameters[1]

Diatom	$B^{[2]}$			$b^{[2]}$	
Li^+H	232.51			2.01	
F^-H	1194.70			2.06	
Atom	I.P.	E.A.		α	ρ
F	0.6402	0.1257		4.04917	2.6
F^-				7.0862	
H	0.49982	0.02773		4.5	1.0
H^-				93.22	
Li	0.1981	0.02206		164.14	0.65
Li^+				0.1930	

[1]All values in atomic units
[2]Defined in Eq. (27)

The calculations of other matrix elements proceed in a similar fashion. For H_{22} we obtain

$$H_{22} = E_{LiF} + \sum_{k=1}^{M-1} (V^v_{L_ML_k^+} + V^v_{L_MF_k^-} + V_{H^+L_k^+} + V_{H^+F_k^-})$$

$$+ V_{H^+F_M^-} + V_{H^+L_M} + V^v_{L_MF_M^-} + H^v_{L_M}$$

(30)

where as before E_{LiF} is the energy of a perfect LiF crystal, $V^v_{L_ML_K^+}$, $V^v_{L_MF_K^-}$ are the interactions of the L_M valence electron with the L_k^+ and F_k^- cores

$$V^v_{L_ML_k^+} = \langle L_M(1)|-1/|r_1-R_{L_k}||L_M(1)\rangle \cong -1/R_{L_ML_k} - \alpha(L_i)/2R^4_{L_ML_k}$$ (31)

$$V^v_{L_MF_k^-} = \langle L_M(1)|1/|r_1-R_{F_k}||L_M(1)\rangle \cong 1/R_{L_MF_k} - \alpha(Li)/2R^4_{L_MF_k}$$ (32)

The other integrals are evaluated as

$$V_{H^+L_k^+} = 1/R_{HL_k} + V^{cc}_{H^+L_k^+}$$ (33)

$$V_{H^+F_k^-} = -1/R_{HF_k} + V^{cc}_{H^+F_k^-}$$ (34)

$$V^v_{L_MF_M^-} = 1/R_{L_MF_M} + V^{cc}_{L_MF_M^-} - \alpha(Li)/2R^4_{L_MF_M}$$ (35)

$$V_{H^+L_M} = V^{cc}_{H^+L_M^+} - \alpha(Li)/2R^4_{HL_M} \tag{36}$$

$$H^v_{L_M} = -IP(Li) \tag{37}$$

where IP is an ionization potential (see Table 1). By convention the energy of H^+ is taken to be zero.

Similarly for the H_{33} matrix element, we have, using Eq. (3),

$$H_{33} = E_{LiF} + E_{H^-} + V^{POL}_{H^-} + \sum_{K=1}^{M-1} \{-V^v_{F^-_M L^+_K} - V^v_{F^-_M F^-_k} + V_{H^- L^+_K} + V_{H^- F^-_K}\}$$
$$+ V_{H^- F_M} + V_{H^- L^+_M} - V^v_{F^-_M L^-_M} - H^v_{F^-_M} \tag{38}$$

This expression is similar to Eq. (30) save for the existence of a hole in the F^-_M core, which results in <u>substraction</u> of such terms as $V^v_{F^-_M L^+_K}$ and $V^v_{F^-_M F^-_K}$. The diatomic integrals of Eq. (38) are similar to those appearing in Eqs. (31-37). Explicitly,

$$V^v_{F^-_M L^+_K} = -1/R_{F_M L_K} - [\alpha(F^-) + \alpha(Li^+)]/2R^4_{F_M L_K} \tag{39}$$

$$V^v_{F^-_M F^-_K} = 1/R_{F_M F_K} - \alpha(F^-)/R^4_{F_M F_K} \tag{40}$$

$$V_{H^- L^+_K} = V^{cc}_{HL^+_K} - 1/R_{HL_K} - [\alpha(H^-) + \alpha(Li^+)]/2R^4_{HL_K} \tag{41}$$

$$V_{H^- F^-_K} = V^{cc}_{HF^-_K} + 1/R_{HF_K} - [\alpha(H^-) + \alpha(F^-)]/2R^4_{HF_K} \tag{42}$$

$$V_{H^- F_M} = V^{cc}_{HF^-_M} - \alpha(F)/2R^4_{HF_M} \tag{43}$$

$$H^v_{F^-_M} = EA(F) \tag{44}$$

where EA(A) denotes electron affinity of an (A) atom. Values used here are given in Table 1.

We now turn our attention to the computation of off-diagonal elements. These can be computed using all three factorizations of H (e.g. Eq. (20)). We thus need to symmetrize the resulting off-diagonal matrix element. Thus when we compute $\tilde{H}_{21}$ we use the factorization of Eqs. (18-20). We obtain,

$$\tilde{H}_{21} \equiv \langle \psi_2 | H | \psi_1 \rangle = \tag{45}$$

$$\langle [\prod_{K=1}^{M} L^+_K F^-_K] L_M(1) H^+ | H_{L_i F} + H_{H^+} + \sum_{K=1}^{M} \{H_{HL^+_K} + H_{HF^-_K}$$

$$- H_{L^+_K} - H_{F^-_K} - 2H_H\} | \sum_i (-1)^{i\hat{P}_i} \hat{P}_i [\prod_{K=1}^{M} L^+_K F^-_K] H(1) \rangle$$

$$\stackrel{\sim}{=} (E_{L_i F} + E_H) S_{12} + \sum_{K=1}^{M} \langle L_M(1) | H_{HL^+_K} + H_{HF^-_K} - H_{L^+_K} - H_{F^-_K} - 2E_H | H(1) \rangle$$

$$\overset{\sim}{=} (E_{L_iF} + E_H)S_{12} - <L_M(1)|1/|r_1-R_{L_M}||H(1)> \tag{45) cont.}$$

where we have neglected all three (and higher) center integrals. The assumption of orthogonality between the core and valence electrons reduces the S_{12} matrix element $(S_{12} \equiv <\psi_1|\psi_2>)$, to the following

$$S_{23} \overset{\sim}{=} S_{L_MH} = <2S_{L_M}|1S_H> . \tag{46}$$

Analytic formulae for evaluation of the overlap integrals were given by Mulliken [10]. The $-<L_M(1)|1/|r_1-R_{LM}||H(1)>$ integral of Eq. (45) is evaluated as, [10]

$$-<L_M(1) 1/|r_1-R_{L_M}||H(1)> \overset{\sim}{=} \frac{S_{L_MH}}{2} [-\frac{1}{R_{L_MH}} -\rho(L_i)] \tag{47a}$$

where $\rho(A)$ is the screening constant of the outer electron on atom A. The values used here for the various atoms are given in Table 1.

When we approximate $\tilde{H}_{12} \equiv <\psi_1|H|\psi_2>$ we use factorization applicable to the proton adsorbed onto LiF with excess electron (see Eq. (30)). We obtain

$$\tilde{H}_{12} \overset{\sim}{=} [E_{LiF}(e^-)-IP(Li)]S_{12}-<H(1)|1/|r_1-R_H||L_M(1)> \tag{48}$$

where $E_{LiF}(e^-)$ is the energy of a LiF crystal with one extra electron (assumed to reside one one of the Li atoms). This energy is given, according to Eq. (30), as

$$E_{LiF}(e^-) = E_{LiF} + \sum_{K=1}^{M-1} (V^v_{L_ML_K^+} + V^v_{L_MF_K^-}) \tag{49}$$

The $V_{H^+L_K^+}$, $V_{H^+F_K^-}$ terms were not included in Eq. (48) since they involve three-center integrals. As in Eq. (47a) we evaluate the $-<H_1(1)|1/|r_1-R_{LM}|L_M(1)>$ integral as,

$$-<H(1)|1/|r_1-R_{L_M}|L_M(1)> \overset{\sim}{=} \frac{S_{L_MH}}{2} [-\frac{1}{R_{L_MH}} -\rho(H)] . \tag{47b}$$

Finally we symmetrize the results to obtain,

$$H_{12} = H_{21} = (\tilde{H}_{12} + \tilde{H}_{21})/2 . \tag{50}$$

The coupling between the first and third configuration is slightly more complicated since it involves more valence electrons. Explicitly,

$$\tilde{H}_{31} = <[\prod_{K=1}^{M-1} L_K^+F_K^-]L_M^+F_M^+F_M(1)\overline{H}(2)H(3)|H_{LiF}+H_H$$

$$+ \sum_{K=1}^{M} \{H_{HL_K^+} + H_{HF_K^-} - H_{L_K^+} - H_{F_K^-} - 2H_H\}|\sum_{P_i} (-1)^{i}\hat{P}_i[\prod_{K=1}^{M-1} L_K^+F_K^-] \tag{51}$$

$$L_M^+F_M^+F_M(1)\overline{F}_M(2)H(3)> .$$

However neglecting all three-center integrals and terms containing S^2, S^3

etc. leads to the simplified expression,

$$\tilde{H}_{31} = (E_{LiF} + E_H)S_{13} + V_{HF_M^-}S_{F_MH} \tag{52}$$

where $V_{HF_M^-}$ is calculated using Eqs. (24,25) and S_{13} can again be approximated as S_{F_MH} by neglecting all higher powers of the overlap integrals.

Using the same approximations the $\tilde{H}_{13}$ integral is given as,

$$\tilde{H}_{13} = (E_{LiF}(h^+) + E_{H^-})S_{31} + V_{H^-F_M}S_{F_MH} \tag{53}$$

with $V_{H^-F_M}$ given by Eq. (43), and $E_{LiF}(h^+)$ the energy of the LiF crystal with one extra hole (identified here with the presence of one neutral F atom) is given by

$$E_{LiF}(h^+) = E_{LiF} - \sum_{K=1}^{M-1} \{V_{F_M^-L_K^+}^v + V_{F_MF_K^-}^v\} - V_{F_M^-L_M^+}^v \tag{54}$$

As above we symmetrize the $\tilde{H}_{13}$ and $\tilde{H}_{31}$ matrix elements.

The interaction between the second and third configuration involves changing the assignment of two electrons, essentially by moving one electron from the neutral Li atom and one electron from F^- ion to the H^+ core - to form H^-. The H_{23}, H_{32} matrix elements thus involve terms proportional to $S_{F_MH} \cdot S_{L_MH}$ and higher orders. Since these terms are consistently neglected here, we set,

$$H_{23} = H_{32} = 0 . \tag{55}$$

The above development completes the derivation of our 3 state model. We have expressed all the Hamiltonian matrix elements in terms of simple diatomic integrals. If, as mentioned, above one identifies the L_M or F_M sites as those closest to the H atom, the rest of the calculation amounts to solving a 3x3 set of secular equations (Eq. (5)). In this way we concentrate on a set of 3 localized surface states, whose localization is induced by the adsorbate. In order to generalize this picture to include the conduction band we need to consider other $(\psi_2'\psi_3')$ charge transfer configurations. These configurations are identical to ψ_2 and ψ_3 except that the F_M or L_M atoms are not necessarily those closest to H. The present model allows for the direct calculation of the coupling between ψ_2' and ψ_2, ψ_3' and ψ_3 as well as the coupling between ψ_1 and the ψ_2', ψ_3' configurations. This would result in a much larger matrix which would be difficult to diagonalize. However this is only necessary if we are after the continuum of states belonging to the band. For the sake of calculating the set of discrete surface states below the band the matrix can be truncated to include only sites in the proximity of the adsorbate. This procedure will be followed in a future application [11]. In the present paper we only use the more limited three-state model outlined in these sections. The resulting potential surfaces for a hydrogen on a LiF surface and the hydrogen embedded in the LiF crystal bulk are given in the following section.

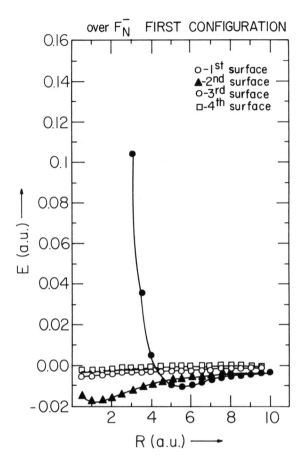

Figure 1. Interaction of neutral hydrogen with the first four planes of the LiF crystal above the F^- site.

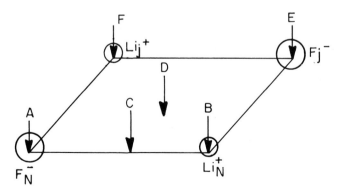

Figure 2. Location of the four adsorption sites (A,B,C,D) studied in this paper.

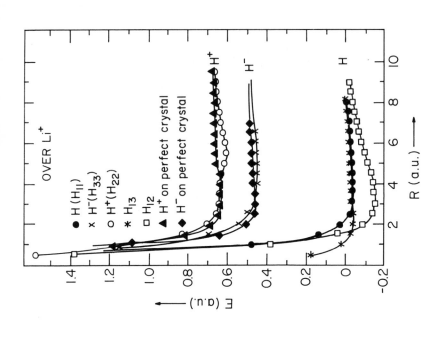

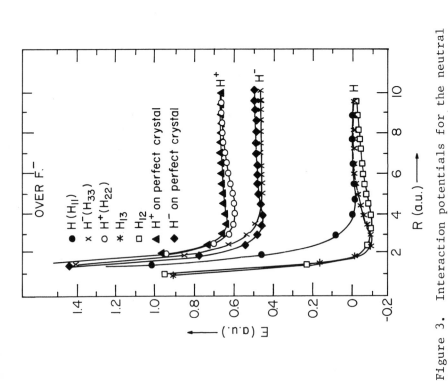

Figure 3. Interaction potentials for the neutral (H_{11}) and chargetransfer (H_{22} and H_{33}) H+LiF system above the F$^-$ site. Also shown are the interactions of H$^+$ and H$^-$ with a perfect LiF crystal and the H_{12}, H_{13} coupling terms.

Figure 4. Interaction potentials above the Li$^+$ site. The notation follows that of Fig. 3.

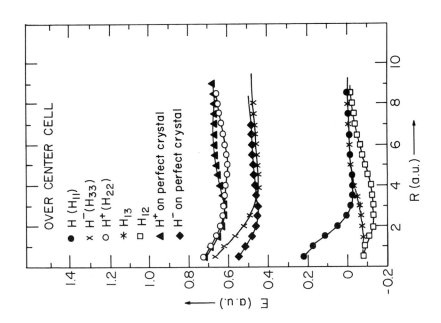

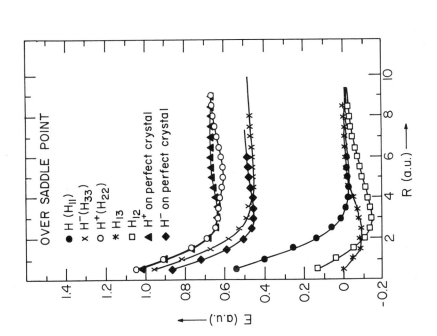

Figure 5. The same as Fig. 3 for the saddle point
(point C of Fig. 2) site.

Figure 6. The same as Fig. 3 for the cell-center
(point D of Fig. 2) site.

III. RESULTS AND DISCUSSION

We first consider the adsorption of neutral hydrogen uncoupled to the charge-transfer configurations. This amounts to calculating the H_{11} matrix element of Eq. (22). In Fig. 1 we plot the computed interaction energies of a hydrogen adsorbed on an F^- crystal-ion site as a function of z-the H-LiF vertical separation. This is done with respect to the first four planar crystal layers. As shown in Fig. 1 the interaction is, as expected, dominated by the outermost layer of atoms. However, the second planar layer is quite important in helping reduce the repulsive energy at very close separations. The third and fourth layers contribute very little at the equilibrium distance of 5 a.u. (see Table II). We may conclude that the summation of the interaction energy over crystal planes converges well by the time we are four layers deep.

Since hydrogen is a small probe we expect it to feel a large degree of surface roughness. Our calculations reveal that the different character of the V_{Li^+H} and V_{F^-H} interactions (see Eqs. (23,24)) tends to enhance this effect. As an illustration of this effect we probe the interaction energies at four different crystal sites, A, B, C, D, where, as shown in Fig. 2, A refers to the F^- site, B to the Li^+ site, C to the midpoint between F^- and Li^+ ("saddle point") site and D to the cell-center site.

The results are summarized in Table II. Quite clearly, the preferred adsorption site for neutral hydrogen is that of the Li^+ ion, where the adsorption energy (=D_e - the well depth of Table II) is largest. Clearly the hydrogen prefers to be as far away from the F^- ion as possible. However as the hydrogen gets closer to the surface (this would necessitate energies of a few eV), the effective surface roughness changes, since the interaction with Li^+ becomes repulsive too. Under these circumstances, the hydrogen is funnelled to the saddle point (site C) or the cell-center (site D) sites. This is of importance when we consider the possibility of penetration of the hydrogen into the bulk. As shown below the center-cell channel is the one most likely to be the one occupied by interstitial hydrogen.

We next turn our attention to the ionic charge-transfer configurations. The interaction potential of H^+ on LiF with excess surface electron (H_{22}) and that of H^- on LiF with excess surface hole (H_{33}) are plotted together with H_{11} in Figs. 3-6. Also plotted are the off-diagonal coupling potentials (H_{12} and H_{13}) as well as the interactions of H^+ and H^- with a perfect LiF crystals. (The interactions of ions with a perfect crystal are identical in shape, but not in absolute energies, to the interactions of these ions with a charged LiF crystal in which the extra charge is located at a remote site with respect to the adsorbed ion).

As shown in Table II, even when the surface is charged, H^+ prefers to be adsorbed on the (neutral) F site and H^- on the (neutral) Li site. Quite clearly (see Figs. 5,6) the short-range repulsive part of the potential is milder at the saddle point and center-cell configurations. This demonstrates again the change in the nature of surface roughness at different vertical adsorbate-crystal separations.

The effect of the charge transfer configurations is best felt when

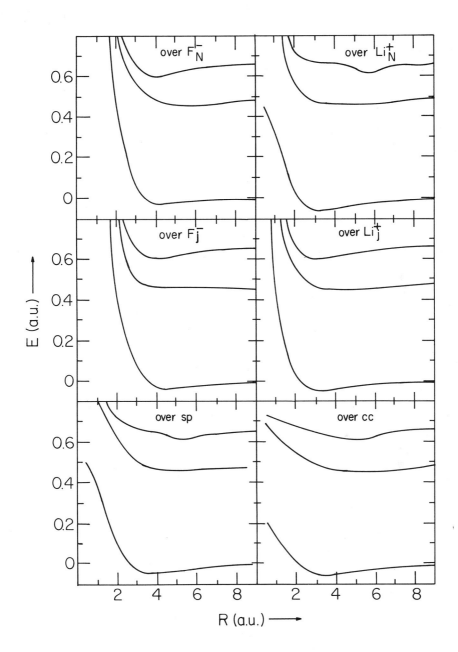

Figure 7.   Three lowest adiabatic (Born-Oppenheimer) H-LiF interaction potentials at six different sites. F_N^- is the location of the excess positive surface charge ($F^- + h^+$) and L_N^+ is the location of the excess negative surface charge ($Li^+ + e^-$). F_j^- and L_j^+ are the two other ions in the unit surface cell (see Fig. 2). sp is the saddle point site (point C in Fig. 2) and cc is the cell-center (point D in Fig. 2).

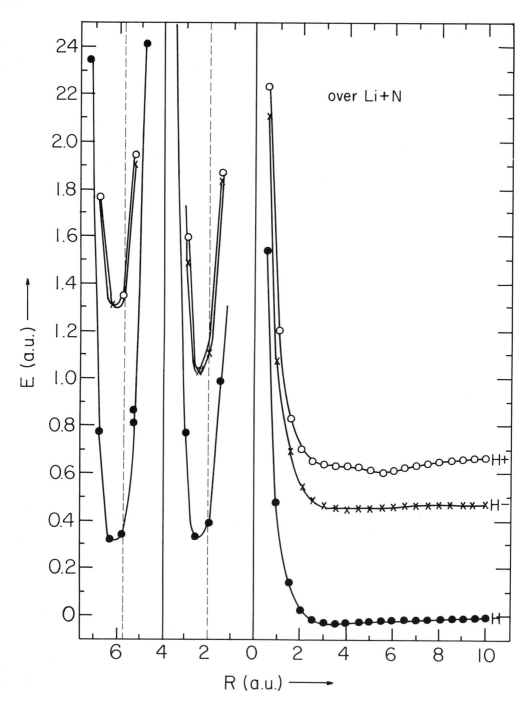

Figure 8. Interaction potentials inside and outside the crystal
surface at site B of Fig. 2. The two solid vertical lines indicate the
centers of the two uppermost LiF crystal layers.

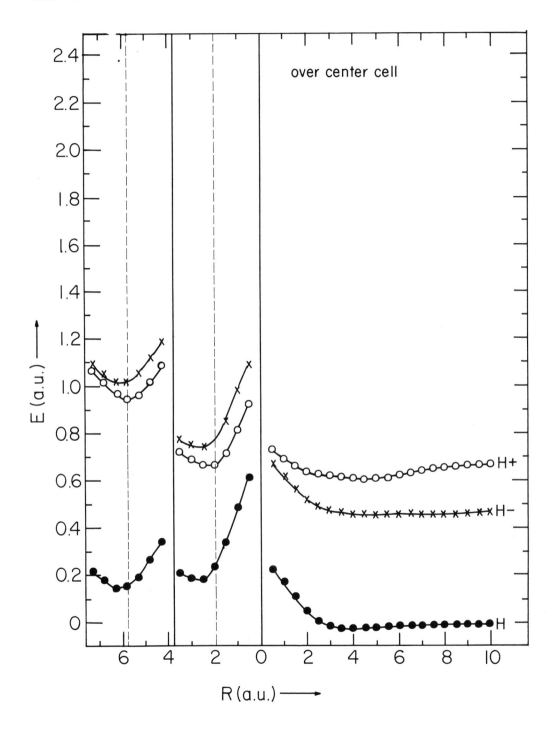

Figure 9.   The same as Fig. 8 at site D of Fig. 2.

we solve the three-configuration model to obtain the adiabatic (Born-Oppenheimer) potentials. These are plotted at various crystal sites in Fig. 7. The adsorption energy and equilibrium surface-adsorbate separations are summarized in Table III. Most noticeable is the marked increase in adsorption energy over sites B, C and D, due to the presence of the charge-transfer channels.

Contrary to the previously studied case [3] of Ag^+ and Br^- adsorbed on silver halides, the adsorption energies of the charge-transfer states is not larger than that of the neutral species. As a result the vertical separation between neutral hydrogen and H^+ or H^- is not substantially decreased when hydrogen is adsorbed to the surface. Thus no new spectroscopy or new surface induce resonances are expected to be observed. This is contrary to the situation noted in the case of halide ions adsorbed onto a silver-halide crystal [3].

Finally we examine the potential felt by interstitial hydrogen and the barriers for penetration into the bulk. These are shown in Figs. 8 and 9 at the Li^+ and the center cell sites. It is evident that interstitial hydrogen can be quite stable especially at the center-cell position.   At the center-cell position its energy is only 3eV higher than that of free hydrogen. If in addition we allow the crystal ions to be distorted from their natural configuration, we expect that LiF crystal can accomodate hydrogen with a marginal cost of energy. Such optimization will be the subject of a forthcoming work [11].

It is interesting to note that interstitial H^+ is more stable than interstitial H^-. Thus as H^- enters the bulk it tends to exchange one or two electrons with the LiF crystal.

IV.   CONCLUSION

The simple VB model for adsorption of hydrogen on LiF was shown to yield interesting predictions regarding the role of charge transfer configurations in the adsorption and penetration of hydrogen into the bulk. Surface roughness was shown to be a sensitive function of adsorbate-surface separation. The accomodation of interstitial hydrogen was shown to be a likely event. Of extreme interest would be a study, based on the present model in which crystal geometry would be optimized to show to what extent hydrogen distorts the crystal and to what extent such distortions, which bring about strong coupling to surface phonons, can effectively increase the adsorption energies. Such effects can be studied within the framework of the present model in a straightforward manner.

Table II. Adsorption Energies and Equilibrium Distances (in a.u.)

	$H+LiF$		$H^{+}+LiF(e^{-})$		$H^{-}+LiF(h^{+})$		$H^{-}+LiF$		$H^{+}+LiF$	
SITE[a]	De[b]	Re[c]	De	Re	De	Re	De	Re	De	Re
A	0.011	5.0	0.073	4.0	0.010	5.0	0.014	4.5	0.028	3.5
B	0.024	3.5	0.058	5.25	0.018	4.5	0.019	4.5	0.029	3.0
C	0.015	4.5	0.059	5.25	0.013	5.0	0.016	4.5	0.032	3.0
D	0.019	4.0	0.064	5.0	0.014	5.0	0.016	5.0	0.037	2.5

[a] Site notation according to Fig. 2
[b] Adsorption energy (depth of attractive well)
[c] Equilibrium vertical distance from surface

Table III. Adiabatic Adsorption Energies and Equilibrium Separation
(in a.u.)

State	E1		E2		E3	
Site	De	Re	De	Re	De	Re
A	0.02	4.5	0.02	6.0	0.05	4.2
B	0.05	3.2	0.02	4.5	0.05	3.5
C	0.04	3.2	0.01	5.5	0.05	5.5
D	0.04	3.5	0.01	5.5	0.05	5.0

REFERENCES

1.   R.W. Newns, *Phys.Rev.* 178, 1123 (1969).
2.   T.R. Knowles, *Surf.Sci.* 101, 224 (1980); T.R. Knowles, *J.Electroanal. Chem.* 150, 365 (1983).
3.   Y. Zeiri, M. Shapiro and R. Tenne, *Chem.Phys.Lett.* 99, 11 (1983).
4.   Y. Zeiri and M. Shapiro, *Chem.Phys.* 31, 217 (1978); M. Shapiro and Y. Zeiri, *J.Chem.Phys.* 70, 5264 (1979).
5.   Y. Zeiri, M. Shapiro and R. Tenne, *J.Chem.Phys.*, in press.
6.   F.O. Ellison, *J.Am.Chem.Soc.* 85, 3540 (1963).
7.   R.M. Stevens,*"Molecular Integral Package for Slater Type Orbitals"*, privately distributed computer program; V.R. Saunders, *"Atmol 3, Molecular Integrals Package"*, RL-76-104 Atlas Computing Division Ruthrford Laboratory, Chilton, England, 1976.
8.   H.A. Pohl, R. Rein and K. Appel, *J.Chem.Phys.* 41, 3385 (1964).
9.   H.U. Finzel, *F.Sci.* 49, 577 (1975).
10. R.S. Mulliken, *J.Chim.Phys.* 46, 497 (1949); 46, 675 (1949).
11. Y. Zeiri, M. Shapiro and R. Tenne, to be published.

Thermodynamic Implications of Desorption from Crystal Surfaces

M. J. Cardillo & J. C. Tully
AT&T Bell Laboratories
Murray Hill, New Jersey 07974

ABSTRACT

Thermal desorption of molecules from surfaces provides a unique opportunity to relate kinetic measurements to equilibrium properties. The transition state can frequently be quantitatively characterized, and dynamical effects can be independently assessed via the sticking probability. Thus desorption rates reflect directly thermodynamic properties of the adsorbate-substrate system. Competition among product channels, energy and angular requirements for adsorption, effects of steps or defects, mobility, and orientational constraints on adsorbates can all be understood with this framework. Significant deviations from Arrhenius behavior may result from incomplete energy equilibration, coverage dependent configurational entropy, or most strikingly, from phase transitions involving the adsorbate layer.

I. Introduction

The use of quasi-equilibrium theories, particularly transition state theory (TST), to describe rate processes often appears to be limited to pedagogical purposes, that is, to illustrate at a molecular level the effects of potential energy barriers, vibrational frequencies, torsional motion, isotopic substitution, etc., on reaction rates. One difficulty in applying transition state theory to observable systems in a quantitative and predictive way is associated with defining and/or observing the transition state species. Another problem is associated with the validity of the TST approximation itself, i.e., with determining the *net* rather than total flow of reactants to products through some suitable boundary in configuration space.[1] Trajectories may loop, jitter, or turn back, thereby complicating the counting of the true flux. However, rates of desorption of molecules into the vacuum provide a nearly ideal situation in which in many cases these difficulties are absent or can be greatly reduced. The configuration space boundary between reactants and products may be taken as a plane parallel to the surface and sufficiently removed from the surface to be out of the influence of the surface potential. The free molecule and isolated substrate, with known partition functions, becomes the transition state species. Desorption into the vacuum is inherently free of return trajectories. The only error introduced by the TST assumption arises from the contribution to the flux through the boundary from trajectories which approach the surface and then are reflected back, without "sticking". It can be shown under quite general conditions that the dynamical correction to TST in this case is the thermally averaged sticking probability.[2]

B. Pullman et al. (eds.), Dynamics on Surfaces, 169–180.
© 1984 by D. Reidel Publishing Company.

For first order desorption into the vacuum the rate constant can thus be written as

$$k_d = \frac{kT}{h}\sigma\, e^{-\Delta G/kT} = \frac{kT}{h}\sigma\, e^{\Delta S/k}e^{-\Delta H/kT} \tag{1}$$

where k and h are the Boltzmann and Planck constants, T is temperature, σ is the sticking probability, and ΔG, ΔS, and ΔH are the free energy, entropy, and enthalpy differences between the gaseous species and the adsorbed state, respectively. Note this is essentially a classical mechanical theory despite the appearance of the h in front which serves to cancel an extra h in the partition function for the adsorbed state.[1] The form of Eq. 1 allows a close identification of the enthalpy and entropy of desorption with the activation energy and pre-exponential term in an experimental Arrhenius description of the rate constant,

$$k_d = Ae^{-E_a/kT}\,. \tag{2}$$

In this paper we informally discuss a selection of recent results and examples of desorption from crystal surfaces which may be interpreted within this equilibrium rate theory context and which provide interesting conclusions regarding the pre-exponential of the Arrhenius rate constant and the entropy of the adsorbed state. We discuss some generalizations concerning equilibration and detailed balancing, adsorbate entropies, and the relation between isotherms and the desorption rate.

II. The Equilibrium Hypothesis and Detailed Balancing: Elementary Examples

An illustration of a simple application of an expression like Eq (1) is the quasi-equilibrium model of gas-surface interactions developed by Batty and Stickney.[3] A reactant is assumed to "equilibrate" with a surface with a probability $\epsilon(P,T)$ and products then desorb in equilibrium proportions as prescribed by their thermodynamical properties as listed, for example, in the JANAF tables.[4] The parameter $\epsilon(P,T)$ is analogous to the sticking probability in Eq. 1 but empirically fit using mass balance. Results of the quasi-equilibrium model[5] for the oxidation of tungsten at high temperatures and low pressure are shown in Fig. 1, where it is seen that quite good fits to a variety of product desorption rates are obtained with only a one parameter (ϵ) variation. There are, of course, dynamical details which must enter into the function $\epsilon(P,T)$. However, the basic postulate of the achievement of a near equilibrium distribution of products prior to desorption in these experiments is essentially confirmed.

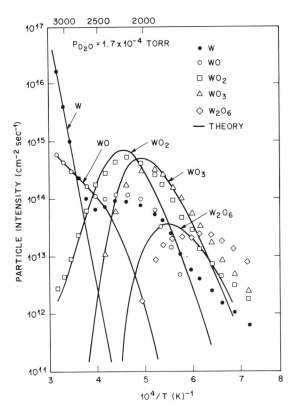

Fig. 1   Volatization rates of tungsten products of oxidation by D_2O. Solid lines are the rates calculated by the quasi-equilibrium model using standard free energy values.

A second example of the utilization of the quasi-equilibrium hypothesis has been given in the molecular beam measurements of the activated dissociative adsorption of H_2 on a series of Cu single crystals.[6] As illustrated in Fig. 2, the incident energy and angle dependence of the dissociation probability of H_2 on Cu(100), Cu(110), and the stepped Cu[3(100)x(100)] surfaces at elevated temperatures is indicative of a kinetic process impeded by a potential barrier. These velocity and angle dependent probabilities may be assumed to determine the fraction of an incident equilibrium distribution striking a surface which goes on to dissociate, as a function of angle. In accordance with detailed balancing, folding these angle dependent probabilities into Maxwell-Boltzmann distributions at the crystal temperature yields the sharp angular distributions of desorption plotted in Fig. 2. To a large extent, therefore, the results of these adsorption and desorption experiments, both far from equilibrium conditions, represent equivalent dynamical processes and can be interpreted using the principal of detailed balancing.

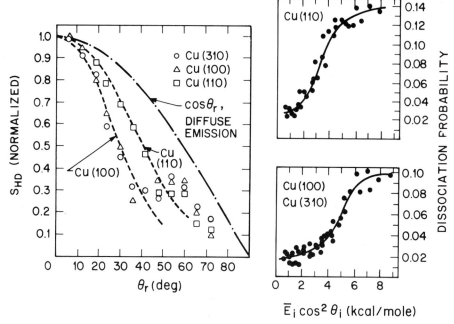

Fig. 2   Kinetic energy and incident angle dependent H_2 dissociative adsorption
probabilities for Cu single crystals.   Experimental angular distributions of H_2
desorption. Smooth lines result from an equilibrium construction using the
probability data .

III. Entropy of Constraint: Site Distributions and Orientational Anisotropy

The distribution of sites available to an adsorbate enters into the entropy of the
adsorbate and thus can strongly affect the prefactor in the Arrhenius form of a desorption
rate constant. This distribution will naturally be sensitive to coverage and a simple
example with the Langmuir isotherm shows this. However, even at very low coverages the
ever-present low density of defect sites may contribute strongly to the entropy of desorption
if the defects are effectively sampled.

Under equilibrium conditions at temperature T and coverage Θ the rate of desorption R_d
can be written

$$R_d = F(\Theta,T)\sigma(\Theta,T) \qquad (3)$$

where $F(\Theta,T)$ is the flux per unit area striking the surface and σ is the sticking probability
as defined in Eq. 1. At the equilibrium pressure P

$$F(\Theta,T) = (2\pi mkT)^{-1/2}P(\Theta,T). \qquad (4)$$

In analogy to experiment we define an activation energy to be

$$E_a = \frac{-\partial \ln R_d}{\partial \beta} \tag{5}$$

where $\beta = (kT)^{-1}$. We then define a prefactor

$$\nu = R_d e^{\beta E_a} \tag{6a}$$

or

$$\nu_n = (\Theta N_a)^{-n} \nu \tag{6b}$$

where ν is in molecules (sec. unit area)$^{-1}$, independent of the order of desorption and ν_n is the usual prefactor for an assumed n^{th} order desorption process. N_a is the site density.

We can write for the desorption rate assuming $\sigma \sim 1$,

$$R_d = F(\Theta,T) = Q_1 Q_2 \beta^{-1} e^{-\beta E_1}, \tag{7}$$

where $-E_1$ is the zero of energy (binding energy). We have factored the partition functions in the following manner

$$Q_1 = N_a Q_{gas}/hQ_{ads} \tag{8}$$

with Q_{gas} representing the free molecule, excluding translation perpendicular to the surface, and Q_{ads} representing an isolated adsorbed molecule. We assume Q_1 is approximately temperature independent unless the nature of the substrate changes (e.g. substrate phase transitions).

Q_2 is a configurational factor. For the example of a simple Langmuir isotherm

$$Q_2 = \frac{\Theta}{1-\Theta} \tag{9}$$

The activation energy and prefactor for an Arrhenius rate constant are simply obtained as

$$E_{act} = E_1 + kT \quad ; \quad \nu = Q_1 Q_2 ekT \tag{10}$$

where $Q_1 ekT/N_a$ is of the order of 10^{13} sec^{-1} for an atom at room temperature. This simple example ($\sigma = 1$, Langmuir isotherm) illustrates the coverage dependent nature of the prefactor, that is a configurational entropy, which increases as $(1-\Theta)^{-1}$ as the available site density decreases due to increasing coverage. Note that the activation energy remains constant with coverage in this model. In general for more complex or realistic isotherms, there still remains a significant dependence on coverage in the prefactor due to the availability of sites.[7]

At low coverage an interesting constráint occurs due to the inevitable presence of a low density of defect or step sites on single crystals. A well documented example of this is the desorption of NO from Pt(111) single crystals as shown in Fig. 3. These differences in rates of desorption are associated with variations in step densities and coverage for each set of experimental conditions. The GSG rate[8] (extrapolated from lower temperatures) results from thermal desorption, a "high" coverage measurement with respect to step density, and therefore represents desorption from ideal terrace sites. It is much faster than the others due to the smaller binding energy of NO at terraces (25 kcal/mole) compared to steps (34 kcal/mole).[9] The other five rates are from low coverage modulated molecular beam measurements.[10-12] For the four (111) surfaces, the rates converge at high temperatures suggesting reasonable relative accuracy in these results. The deviations at lower temperatures have been demonstrated to be the result of saturating the strongly binding step sites. Of particular interest are the differences between the LS(557) (16% steps)[10] and the low step density surfaces, which in the linear regime are nearly parallel and represent the difference due to the different number of step sites. This difference in prefactors is the configuration entropy of assigning the NO adsorbate ground state to steps of different density or availability at low coverage. For equal activation energies and $\sigma \sim 1$, this configurational entropy term $S \sim k \ln \Omega$ put into Eq. 1, yields the differences between these rates within experimental uncertainty. A microscopic rate model confirms these results.[9] This interpretation rests on the assumption that populations at the steps and terraces achieve local equilibrium. Rapid surface diffusion rates, compared to desorption rates, are required for this and are concluded from this study.

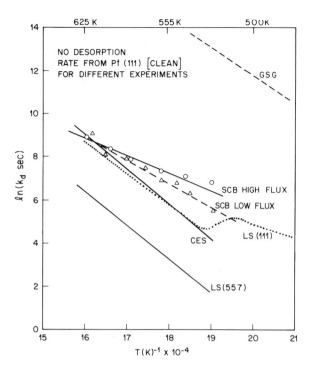

Fig. 3   Thermal desorption rates of NO from clean Pt(111) and vicinal Pt(111) surfaces measured at different laboratories; GSG-Ref. 8 (Thermal Desorption); SCB-Ref. 12 (Molecular beam (MB)); LS-Ref. 10 (MB); CES-Ref. 11 (MB); LS 557-Ref. 10 (MB-stepped Pt(111).

Another interesting application of transition state theory is to describe the NO desorption rate from Pt(111) terraces as measured by GSG[8] as shown in Fig. 3 using thermal desorption. The thermal desorption rate constant is

$$k_d = 10^{16} \, e^{\frac{-25 \text{kcal/mole}}{kT}}.$$

E_a = 25 kcal/mole is approximately the binding energy of NO to Pt(111) as expected. However 10^{16} s^{-1} is three to four orders of magnitude higher than both the molecule-surface vibrational frequency and typical prefactors observed in atomic desorption. An evaluation of the partition functions, however, illustrates how the strong orientational anisotropy of the NO molecule, that is the entropy of orientational constraint, accounts for this value.

Evaluating Q_{gas} from the molecular constants of NO and taking for Q_{ads} the following frequencies, *i)* 306 cm^{-1} NO-Pt stretch[14], *ii)* 465 cm^{-1} frustrated rotation[14], and 60 cm^{-1} frustrated lateral translation (estimated), at 300 K the prefactor is

$$A = \frac{kT}{h} \frac{Q_{gas}}{Q_{ads}} = 2 \times 10^{16} sec^{-1} .$$

This estimate assumes separable harmonic motions. Trajectory simulations of NO desorption, using a complete multidimensional force field developed to describe NO scattering and which does not require separability into harmonic motions, generates approximately the same value for the prefactor. If the molecule-surface potential is taken to be isotropic with respect to orientation of the NO bond, but with the same well depth, the values $E_a = 25$ kcal/mole, $A = 5 \times 10^{13}$ sec^{-1} are obtained from the simulations. As has been pointed out by others,[7] the factor of 400 is thus a measure of the reduced entropy due to constraining the orientation of the adsorbed molecule.

In the last two examples, we have used experimental desorption rates, analyzed by TST, to extract information about the adsorbate; i.e., about the reactant. This is in contrast to the usual strategy in gas-phase or solution phase kinetics, where it is hoped that the reactants are well understood and rate data is employed to obtain hints about the transition state. As mentioned above, for desorption to the vacuum the transition state is well characterized, and kinetic data can provide information about the binding of the adsorbate to the surface that may be difficult to obtain by spectroscopic or other static techniques.

IV. Temperature Dependence of Sticking Probability

The thermodynamic implications of Arrhenius parameters may be complicated by the deviation of the sticking probability from unity. Let us consider the following form

$$\sigma(T) = e^{-\alpha(T)} \tag{11}$$

where α is a non-negative function of T. We define an activation energy and prefactor for unit sticking probability as

$$E_a^o = -\frac{\partial}{\partial \beta} \ln F , \tag{12}$$

$$\nu^o = F \, e^{\beta E_a^o} . \tag{13}$$

Then the observed activation energy and prefactor are

$$E_a = E_a^o + \frac{\partial \alpha(T)}{\partial \beta} \; , \tag{14}$$

$$\nu = \nu^o \, e^{-\alpha(T) + \beta \frac{\partial \alpha}{\partial \beta}} \; . \tag{15}$$

For adsorption without an activation barrier $\sigma(T)$ can be assumed to decrease monotonically. Then

$$\frac{\partial \alpha(T)}{\partial \beta} < 0 \; . \tag{16}$$

Thus E_a and ν decrease with increasing T, partially compensating each other in the total rate.[15] Computer simulations of Ar and Pt desorption from Pt(111) show this effect clearly (Table I).[16]

Table I. Effect of temperature dependent
sticking probability on Arrhenius parameters.

	T(K)	E_a(kJ/mole)	ν (s^{-1})
Ar	60	8.8	5×10^{11}
	143	7.8	1.6×10^{11}
	300	5.0	0.4×10^{11}
Xe	125	28.8	6.8×10^{11}
	333	21.6	0.9×10^{11}

For the case of activated adsorption the sticking probability typically rises with incident energy, then drops off. For a barrier height E_2 we can approximate the function α of Eq. (11) by

$$\alpha(T) = \beta E_2 + \alpha'(T) \; . \tag{17}$$

Then

$$\frac{\partial}{\partial \beta} \alpha'(T) < 0 \text{ for all T} \; , \tag{18}$$

and

$$E_a = E_a^0 + E_2 + \frac{\partial}{\partial \beta} \alpha'(T) ,$$

(19)

$$\nu = \nu^0 \, e^{-[\alpha'(T) + \beta \frac{\partial}{\partial \beta} \alpha(T)]} .$$

(20)

Thus the barrier height simply adds to the activation energy and both E_a and ν decrease with temperature as in the case of unactivated adsorption.

V. Desorption and the Equilibrium Isotherm

In this last example we illustrate the variation in desorption kinetics of an atom or molecule for which there exists a hypothetical isotherm with attractive adsorbate interactions so that at equilibrium a transition into an ordered phase occurs above certain coverages. The isotherm is sketched in Fig. 4. For simplicity we assume $\sigma(\Theta,T)=1$. We have indicated the regimes a-d for which the desorption kinetics behave qualitatively differently. Recall from Eqs. 3 and 4 that the desorption rate is proportional to the equilibrium pressure required to maintain the coverage so that the abscissa could be replaced by R_d. At lower pressures and coverages than the region marked (a), desorption kinetics are not linear due to the presence of surface defects at which the adsorption energies are greater than at terraces (Fig. 3). Through region (a) the coverage is linear in pressure (as in the Langmuir isotherm) and desorption is first order. In region (b) there is a transition, which may be sharp, as the nucleation regime is approached and the pressure (desorption rate) is a variable function of coverage $(0<n<1)$. In region (c) the pressure remains constant as Θ is increased corresponding to the phase transition. This region is associated with zero order desorption kinetics, a thermodynamic result independent of kinetic modeling. Region (d) corresponds to increased coverage in an unspecified manner above the ordered region which we have continued arbitrarily. Isothermal desorption experiments would show this sequence in reverse, i.e. a region of zero order kinetics followed by a transition to first order kinetics. A clear recent example of this was published by Opila and Gomer for Xe desorption from W(110).[17] It is of interest in this study that the activation energies of desorption for the first, second, and third layers of Xe were shown to be less than the enthalpies of adsorption. This may be due in large part to the temperature dependence of the sticking probabilities as discussed in Sec. IV.

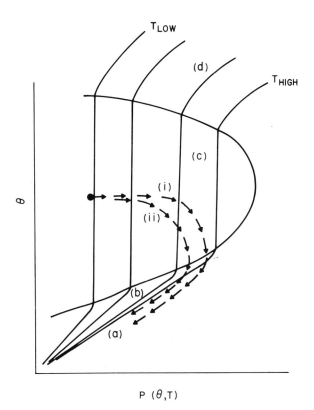

Fig. 4   Hypothetical isotherm for an adsorbate with attractive interactions. Regions (a)-(d) correspond to different equilibrium desorption kinetics as indicated in the text. Two thermal desorption paths for different heating rates are sketched.

It is useful to consider temperature programmed thermal desorption spectroscopy (TDS) from a system exhibiting this isotherm. We have sketched in the trajectories followed by TDS for two heating rates $(\frac{\partial T}{\partial t})_i > (\frac{\partial T}{\partial t})_{ii}$. After establishing a coverage the temperature is increased until a rate of desorption is reached which starts to noticeably affect the coverage and the onset of a desorption signal is observed experimentally. The initial kinetics of the experiment thus involve traversing the zero order region. As the coverage decreases and the temperature increases, the desorption rate peaks $(P_{max}(\Theta,T))$. At this point the trajectory may or may not have crossed from zero order, through variable order, to first order. Eventually first order kinetics are achieved and finally at low coverages $(\Theta \sim$ defect density) non-linear kinetics enter again. Different heating rates reach different peak desorption rates and cross into the linear kinetics at different points. It is clear that the interpretation of the TDS in terms of Arrhenius parameters by assuming first order desorption is not meaningful for this isotherm. Indeed for most realistic isotherms a meaningful interpretation of A and E_a is restricted to a small region where the pressure is linearly related to the coverage Θ. Forcing of the Arrhenius form leads to extreme variations of A and E_a with Θ such as reported in [7] and discussed by Leuthausser[18] and Johansson.[19]

CONCLUSION

In this paper we have discussed the applicability of quasi-equilibrium concepts to thermal desorption from crystal surfaces. In particular we have illustrated the utility of detailed balancing and the applicability of transition state theory. Detailed balancing provides the correspondence between adsorption and desorption and allows the identification of equivalent dynamical processes which can then be modeled in the direction more easily described. Transition state theory provides a useful basis for the discussion of desorption rates and, with appropriate consideration of the sticking probability, allows the description of these rates to be interpreted in terms of the thermodynamic state of the adsorbates. The importance of measuring the sticking probabilities becomes clear in this discussion, which when combined with equilibrium isotherms can clarify the origin of what sometimes appears as puzzling behavior in the temperature and coverage dependence of desorption rates.

References

[1]   P. Pechukas in *Dynamics of Molecular Collisions,* Part B, ed. W. H. Miller (Plenum, NY, 1976), p.269.

[2]   E. K. Grimmelmann, J. C. Tully and E. Helfand, J. Chem. Phys. **74**, 5300 (1981).

[3]   J. C. Batty and R. E. Stickney, J. Chem. Phys. **51**, 4475 (1969); J. C. Batty and R. E. Stickney, Oxidation of Metals **3**, 311 (1971).

[4]   JANAF Thermochemical Tables, Ed. D. R. Stull (Dow Chemical Co., Midland, Mich., 1965).

[5]   M. J. Cardillo and Y. Look, Surface Sci. **66**, 272 (1977).

[6]   M. J. Cardillo, M. Balooch, R. E. Stickney, Surface Sci., **50**, 263 (1975).

[7]   See, for example, D. Menzel, in *Chemistry and Physics of Solid Surfaces IV*, ed R. Vanselow and R. Rowe (Springer-Verlag, Berlin, 1982), p. 389.

[8]   R. J. Gorte, L. D. Schmidt and J. L. Gland, Surf. Sci. **109**, 367 (1981).

[9]   J. A. Serri, J. C. Tully and M. J. Cardillo, J. Chem. Phys. **79**, 1530 (1983).

[10] T. H. Lin and G. A. Somorjai, Surf. Sci. **107**, 573 (1981).

[11] C. T. Campbell, G. Ertl and J. Segner, Surf. Sci. **115**, 309 (1982).

[12] J. A. Serri, M. J. Cardillo and J. E. Becker, J. Chem. Phys. **77**, 2175 (1982).

[13] J. C. Tully and M. J. Cardillo, Science **223**, 445 (1984).

[14] H. Ibach and S. Lehwald, Surface Sci. **76**, (1978); G. Pirug, H. P. Bonzel, H. Hopster, H. Ibach, J. Chem. Phys. **71**, 593 (1979).

[15] E. F. Greene, P. J. Estrup, M. J. Cardillo, and J. C. Tully, unpublished.

[16] J. C. Tully, Surface Sci. **111**, 461 (1981).

[17] R. Opila and R. Gomer, Surface Sci. **112**, 1 (1981).

[18] U. Leuthausser, Z. Physik **B37**, 65 (1980).

[19] P. K. Johansson, Chem. Phys. Lett. **65**, 366 (1979).

ACCOMMODATION AND ENERGY TRANSFER IN MOLECULE-SURFACE SCATTERING

R.D. Levine
The Fritz Haber Molecular Dynamics Research Center
The Hebrew University
Jerusalem 91904, Israel

The implications of detailed balance are explored for collisions with a surface at a finite temperature. In particular it is shown that for a process taking place on the surface, the surface can shift the equilibrium constant provided that the surface temperature is kept different from the ambient gas temperature. The magnitude of the effect is governed by the energy accommodation of the surface. Another way of stating the result is that measuring the rates of both a forward and a reverse process does provide dynamical information. This is not the case for collisions in the gas phase.

1. INTRODUCTION

Changes in the internal energy of molecules upon collision with a surface can be primarily due to energy exchange with the translation. The role of the surface is then largely that of a catalyst. It is essentially unchanged (not only chemically but also in terms of its physical state) and it acts as a rigid wall. Computing the energy exchange can then be done as for a gas phase collision and the role of the internal motion of the solid is only somewhat to smear out the results computed for a rigid surface [1]. There is however an increasing body of evidence that the thermal motion of the solid can soak up or provide significant amounts of energy in a molecule-surface collision. Such deeply inelastic processes where many phonons can be excited or deexcited are seen directly in atom-surface scattering [2-4] and there is no reason to exclude their role in molecule-surface collisions. Indeed there is also experimental evidence for their role, [5].

Needless to say, the distinction between 'energy transfer' and accommodating collisions is not sharp. For convenience we can however regard energy transfer collisions strictly as those involving no net change in the energy of the phonons. This will regard the one or few phonon processes as part of the accommodating collisions. When multiphonon processes are probable the contribution from few phonon processes is small. There will however be situations when that is not the case [6]. It is then reasonable to consider separately the few- and the

181

B. Pullman et al. (eds.), Dynamics on Surfaces, 181–189.
© *1984 by D. Reidel Publishing Company.*

multi-phonon processes.

Elsewhere we have discussed multiphonon energy transfer in atom surface scattering [7] and the phenomenological determination of the relative contribution of the possible mechanisms in terms of the observed final state distribution [8]. It was also pointed out in [8] that an effective accommodation coefficient can be defined even when several mechanisms contribute. Such an effective accommodation coefficient will depend (in the simplest case, linearly) on both the incident kinetic energy and the surface temperature. Such behaviour is indeed observed [9].

Here we discuss the formulation of detailed balance for scattering off surfaces at a finite temperature, T_s. The two limiting types of behaviour (pure energy transfer and pure accommodation) are readily recovered as special cases. In general however, because the temperature characteristic of the incident translation can be quite different from the temperature of the solid, the statement of detailed balance is more complex than that of a pure gas phase collision [10,11]. The surface temperature provides therefore an additional parameter, not available in the gas phase, for affecting the equilibrium constant of a process.

2. PRELIMINARIES

The differential cross section for scattering of a collimated molecular beam off a surface (at a temperature T_s) into a final solid angle element $d\Omega'$ and translational energy in the range E_T', $E_T' + dE_T'$ is given by [7,12]

$$d^3\sigma/dE_T'd\Omega' = p'^2\cos\theta' P(\omega,q) \tag{1}$$

Here p' is the final wavevector $E_T' = \hbar^2 p'^2/2\mu'$. (It is usually denoted by k but there are two other k's in our problem, the rate constant and Boltzmann's constant, so to avoid confusion we are taking the liberty of using p for the wavevector. In atomic units where $\hbar=1$, the momentum is even numerically equal to the wavevector). q is the momentum transfer parallel to the surface and ω is the energy transfer to the surface

$$\omega = E_i - E_f . \tag{2}$$

E_i and E_f are the initial and final energies of the molecule. Using j to designate the internal energy levels

$$E_i = E_T + E_j$$
$$E_f = E_T' + E_j' \tag{3}$$

$$\omega = -\Delta E_T - \Delta E_j . \tag{4}$$

Note that for a particular $j \to j'$ transition ω and ΔE_T do not necessarily have to have opposite signs. If $\Delta E_j < 0$ then $\omega > 0$ whenever $\Delta E_T < 0$ (but $\Delta E_T > 0$ does not necessarily imply $\omega < 0$) and if $\Delta E_j > 0$ then $\omega < 0$

whenever $\Delta E_T > 0$ (but $\Delta E_T < 0$ does not imply $\omega > 0$). For use in section 6 we already note here the modification of equations (2)-(4) for a chemical reaction, where ε_0 and ε_0' are the ground state energies of the reagents and products

$$E_i = E_T + E_j + \varepsilon_0$$

$$E_f = E_T' + E_j' + \varepsilon_0' \tag{5}$$

$$\omega = -\Delta E_T - \Delta E_j - \Delta \varepsilon_0 . \tag{6}$$

Here $\Delta \varepsilon_0 = \varepsilon_0' - \varepsilon_0$ and is positive when the products lie higher in energy than the reagents.

The probability P in (1) is obtained by averaging the probability of a completely state selected (state of molecule and state of solid) event over the distribution of the initial states of the solid (and, of course, summing over final states. The conservation of energy condition (2) restricts the range of the possible final states of the solid. The complete details will be found elsewhere [7,12]). P will, in general, depend not only on the net energy and momentum transfer to the solid but also on the initial and final state of the molecule and on the temperature.

3. DETAILED BALANCE

At complete equilibrium the rate of any process, detailed as it may be, must be balanced by the rate of the reversed process. We specify the initial state by the internal energy level of the molecule, its kinetic energy and the direction of incidence. Similarly for the final state. Hence a state is specified by the triplet j, E_T, Ω and the detailed equilibrium we consider is

$$(j, E_T, \Omega) \rightleftharpoons (j', E_T', \Omega') .$$

The rate of the forward reaction is given by a product of the following:
(i) the concentration of molecules in the energy level j

$$[A(j)] = \{g_j \exp(-E_j/kT_s)/Q_{in}\}[A] . \tag{7}$$

[A] is the total concentration of molecules, g_j is the degeneracy of energy level j and Q_{in} (the internal 'partition function') insures that $[A] = \Sigma_j [A(j)]$. (ii) the flux of molecules incident on the surface

$$F(E_T, \Omega | T_s) = v \cos\theta \rho(E_T, \Omega) \exp(-E_T/kT_s)/Q_T . \tag{8}$$

Here $\rho(E_T, \Omega) dE d\Omega$ is the translational density of states [12]

$$\rho(E_T,\Omega)\,dE_T d\Omega = (2\pi)^{-3}d\underline{p}$$

$$= (2\pi)^{-3}p^2 d p d\Omega \tag{9}$$

$$= (2\pi h)^{-3}\mu \hbar p d E_T d\Omega.$$

Recall that $E_T = \hbar^2 p^2/2\mu = \mu v^2/2$ and so

$$v\rho(E_T,\Omega) = h^{-1}(p^2/4\pi^2)\,, \tag{10}$$

or, integrating over all incident directions

$$v\rho(E_T) = h^{-1}(p^2/\pi)\,. \tag{11}$$

h is Planck's constant, $\hbar = h/2\pi$. Q_T in (8) is the translational partition function, $v\cos\theta$ is the velocity normal to the surface.

The final factor in the forward rate is (iii) the differential cross section $d^3\sigma/dE_T'd\Omega'$. Note that we could have used the (differential) rate constant $v\cos\theta\, d^3\sigma/dE_T'd\Omega'$ but then the second factor should have been taken as the number density

$$n(E_T,\Omega|T_s) = \rho(E_T,\Omega)\exp(-E_T/kT_s)/Q_T\,. \tag{12}$$

The forward rate is

$$R\{(jE_T\Omega) \to (j'E_T'\Omega')\} =$$

$$[A(j)]F(E_T,\Omega|T_s)d^3\sigma/dE_T'd\Omega' = \tag{13}$$

$$([A]/Q)\,g_j\exp(-E_i/kT_s)(4\pi^2 h)^{-1}p^2\cos\theta p'^2\cos\theta'\vec{P}$$

$Q = Q_{in}Q_T$ is the total partition function of the molecule.

The rate of the reversed detailed process is given via a similar reasoning by

$$R\{(j'E_T'\Omega') \to (j,E_T,\Omega)\}$$

$$([A]/Q)\,g_j'\exp(-E_f/kT_s)(4\pi^2 h)^{-1}p^2\cos\theta p'^2\cos\theta'\overleftarrow{P}. \tag{14}$$

Equating the two rates we obtained the detailed balance condition:

$$g_j\exp(-E_i/kT_s)P(i \to f) = g_j'\exp(-E_f/kT_s)P(f \to i)\,. \tag{15}$$

As a check we derive from (15) the detailed balance condition for the (less) detailed equilibrium

$$(E_T,\Omega) \rightleftarrows (E_T',\Omega')\,.$$

Now, according to the canon [10]

$$P\{(E_T,\Omega) \rightarrow (E_T',\Omega')\} =$$

$$\sum_j([A(j)]/[A]) \sum_{j'} P\{(j,E_T,\Omega) \rightarrow (j',E_T',\Omega')\}. \tag{16}$$

Using (7) and (15)

$$\exp(-E_T/kT_s)P\{(E_T,\Omega) \rightarrow (E_T',\Omega')\} =$$

$$\exp(-E_T'/kT_s)P\{(E_T',\Omega') \rightarrow (E_T,\Omega)\} \tag{17}$$

which is the result previously derived for the special case of atom-surface scattering [7].

The derivation of the detailed balance condition (17) was simple since the incident velocity is sharply defined. Such is not the case for the complementary problem - that of changes in the internal state irrespective of changes in translational energy or direction of motion. The relevant details will become clear from the derivation in the next section. Here we only point out that the final result is the same as when we simply use detailed balance for the process j ⇄ j' in a gas at equilibrium with the surface. Introducing $k(j \rightarrow j')$ as the rate constant

$$[A(j)]k(j \rightarrow j') = [A(j')]k(j' \rightarrow j) \tag{18}$$

or

$$k(j \rightarrow j') = (g_{j'}/g_j)\exp[-(E_{j'}-E_j)/kT_s]k(j' \rightarrow j) . \tag{19}$$

4. DISEQUILIBRIUM

Thus far we have assumed that the gas and surface are in thermal equilibrium. At this point consider the more general case where the translational motion of the gas is in thermal equilibrium but at a temperature T not necessarily equal to the surface temperature T_s. Due to collisions with the surface there will be changes in the internal states of the molecules. Our purpose is to relate the forward, $k(j \rightarrow j')$ and reverse, $k(j' \rightarrow j)$, rate constants under these, non-equilibrium, conditions.

The rate constant $k(j \rightarrow j')$ is obtained by averaging the differential rate constant

$$k(j,E_T,\Omega \rightarrow j',E_T',\Omega') = vd^3\sigma/dE_T'd\Omega' \qquad (\cos\theta\ ?) \tag{20}$$

over a thermal (at a temperature T, T not necessarily equal to T_s) distribution of kinetic energies and directions of incidence and summing over all final kinetic energies and directions,

$$k(j \rightarrow j') = \int d\Omega \int dE_T F(E_T,\Omega|T) \int d\Omega' \int dE_T'(d^3\sigma/dE_T'd\Omega') . \tag{21}$$

By detailed balance, we have for a surface at the temperature T_s:

$$[A(j)]F(E_T,\Omega|T_S)d^3\sigma/dE'_T d\Omega' =$$

$$[A(j')]F(E'_T,\Omega'|T_S)d^3\sigma/dE_T d\Omega. \tag{22}$$

Also, the reversed rate is given by

$$k(j'\to j) = \int d\Omega' \int dE'_T F(E'_T\Omega'|T) \int d\Omega \int dE_T d^3\sigma/dE_T d\Omega . \tag{23}$$

Now both (21) and (23) are to be evaluated at the temperature T. But T appears only in the flux term. Explicitly

$$F(E_T,\Omega|T) = [Q_T(T_S)/Q_T(T)]\exp[-\frac{E_T}{k}(\frac{1}{T} - \frac{1}{T_S})]F(E_T,\Omega|T_S). \tag{24}$$

Using (24) in (21) and then detailed balance implies that

$$k(j\to j') = ([A(j')]/[A(j)])_{T_S}<\exp[\frac{\Delta E_T}{k}(\frac{1}{T} - \frac{1}{T_S})]>k(j'\to j). \tag{25}$$

In (25) we have introduced the notation

$$<\exp[\frac{\Delta E_T}{k}(\frac{1}{T} - \frac{1}{T_S})]> =$$

$$\frac{\int d\Omega' \int dE'_T \int d\Omega \int dE_T \exp[\frac{\Delta E_T}{k}(\frac{1}{T} - \frac{1}{T_S})]F(E'_T\Omega'|T_S)(d^3\sigma/dE_T d\Omega)}{\int d\Omega' \int dE'_T \int d\Omega \int dE_T F(E'_T\Omega'|T_S)(d^3\sigma/dE_T d\Omega)} . \tag{26}$$

The denominator in (26) is just $k(j\to j')$ and the entire term is the averaged change in the exponential factor in $j'\to j$ collisions. When T is very near T_S so that the exponential can be expanded up to first order only

$$<\exp[\frac{\Delta E_T}{k}(\frac{1}{T} - \frac{1}{T_S})]> \simeq 1 + \frac{<\Delta E_T>}{k}(\frac{1}{T} - \frac{1}{T_S}) \tag{27}$$

$<\Delta E_T>$ is similar but not identical to the quantity considered in the definition of an accommodation coefficient [13]. The difference is that in (27) $<\Delta E_T>$ is the mean change in the translational energy for $j \to j'$ collisions only. If $T=T_S$ then the exponential term is unity and we recover (19).

There is an alternative way of writing (25). In (25) the ratio of concentrations is to be evaluated at the temperature T_S of the surface. Using (7) and (4) we see that an equally valid form is

$$k(j\to j') = ([A(j')]/[A(j)])_T<\exp[-\frac{\omega}{k}(\frac{1}{T} - \frac{1}{T_S})]>k(j'\to j) . \tag{28}$$

5. ACCOMMODATION vs. ENERGY TRANSFER

The equilibrium constant for the surface mediated process

$$A(j) \rightleftarrows A(j')$$

is

$$K = k(j \rightarrow j')/k(j' \rightarrow j). \tag{29}$$

When the translational and surface temperatures are equal K is given by the usual ratio of concentrations (cf. (19)). If they are not, we have two alternative (equivalent) forms

$$K = \frac{(\frac{[A(j')]}{[A(j)]})_{T_s} <\exp[\frac{\Delta E_T}{k}(\frac{1}{T} - \frac{1}{T_s})]>}{(\frac{[A(j')]}{[A(j)]})_{T} <\exp[-\frac{\omega}{k}(\frac{1}{T} - \frac{1}{T_s})]>} \tag{30}$$

If the surface is strictly rigid, $\omega = 0$ and

$$K = K(T) \equiv ([A(j')]/[A(j)])_T = (g_j'/g_j)\exp(-\Delta E_j/kT). \tag{31}$$

The equilibrium constant is the concentration ratio computed at the translational temperature. This is the norm for bimolecular gas phase collisions where, by necessity, $\omega = 0$. On the other hand, if the change in the internal energy is completely at the expense of the surface, $\omega = -\Delta E_j$ and

$$K = K(T_s) \equiv ([A(j')]/[A(j)])_{T_s}. \tag{32}$$

The equilibrium constant is again the concentration ratio but now, at the surface temperature.

In general, we will not have either of these limiting forms and only the general conclusion (30) will apply. In terms of the 'gas phase' equilibrium constant

$$K = \frac{K(T_s) <\exp[\frac{\Delta E_T}{k}(\frac{1}{T} - \frac{1}{T_s})]>}{K(T) <\exp[-\frac{\omega}{k}(\frac{1}{T} - \frac{1}{T_s})]>} \tag{33}$$

As an example, say the translational motion is colder than the surface, $T < T_s$. Then such transitions which predominantly cool the surface ($\omega < 0$) are going to be favoured (since the entire exponent will be positive so that $K > K(T)$). It is thereby possible to achieve an excess (over thermal at T) population in those final states which require $\omega < 0$. If the transition not only cools the surface but also warms the translation (ΔE_T predominantly positive) then one can even have $K > K(T_s)$.

It is perhaps worthwhile to stress that the result is definitely state-dependent and that the direction of the effect depends on the internal states j and j'. The best demonstration is to calculate the overall forward and reverse rates

$$k(T) = \sum_{j}([A(j)]/[A])_T \sum_{j'} k(j \to j') . \tag{34}$$

If the distribution of the internal states is characterized by the same temperature as the translation the forward and reverse rates must be the same and K=1. Indeed one can verify by an explicit computation that the (state dependent) averages which appear in (25) or (28) do average out (over all initial states) to unity.

6. REACTIONS

The discussion in sections 3-5 was limited to inelastic collisions. There is nothing in the formalism which requires this limitation and the results can indeed be generalized. We quote here only the final results for the overall equilibrium constant for a chemical reaction which takes place on a surface at the temperature T_s. The temperature in the gas (both translational and internal) is T. Then

$$K = K(T) <\exp[- \frac{\omega}{k}(\frac{1}{T} - \frac{1}{T_s})]> . \tag{35}$$

Here K(T) is the thermodynamic equilibrium constant for the reaction at the temperature T. If there is no energy accommodation by the surface $\omega = 0$ and K=K(T). Similarly, if the gas is in thermal equilibrium with the surface $T=T_s$ so that $K=K(T_s)$. This is as it should be since at thermal equilibrium the presence of a catalyst cannot shift the equilibrium. The point is however that one need not work under conditions of thermal equilibrium.

If the gas is colder than the surface $T<T_s$ then the equilibrium will be shifted in the direction which cools the surface ($\omega <0$). If the reaction is very endoergic ($\Delta\varepsilon_0 >> kT_s$) then $-\Delta\varepsilon_0$ will be the major contribution to ω (cf. equation (6)) and the reaction will be shifted towards the higher energy product.

The shift in the equilibrium constant predicted by (35) is definitely dependent on the surface. Different surfaces will accommodate the reagents and products to different extents and hence will differ in the average value of $<\omega>$. Thus while the sign of the effect is in the direction which is intuitively reasonable, its magnitude does depend on the interaction dynamics.

7. CONCLUDING REMARKS

The active role of the surface in providing (or accommodating) energy in chemical processes which takes place on the surface has been discussed. It is important to emphasize that the surface does not act merely as a heat bath. The relevant parameter is not the exoergicity (ΔE_j or $\Delta E_j + \Delta \varepsilon_0$) but that amount of the exoergicity which is taken up by the surface [i.e. ω, cf. (4) or (6)]. How much of the exoergicity is disposed of by the surface (rather than by the translation) is very much dependent on the detailed dynamics of the process. Different surfaces will very much differ in this respect. A simple example is vibrational excitation of NO. NO has a high vibrational frequency and is unlikely to be excited in a gas phase collision since the probability of the translation to undergo the large change in momentum required for vibrational excitation is very low. (This is the exponential gap behaviour, [14]). NO is vibrationally excited on collision with a Pt surface [15] (but not, say, on Ag). Recently it has been fashionable to speculate on the ability to bridge the exponential gap with laser photons [14]. Surfaces provide us with the much simpler possibility of bridging the gap with phonons.

REFERENCES

1. See, for example, Hurst, J.E., Jr., Kubiak, G.D. and Zare, R.N.: 1982, Chem. Phys. Lett. 93, p. 235; Barker, J.A., Kleyn, A.W. and Auerbach, D.J.: 1983, Chem. Phys. 97, p. 9.
2. Janda, K.C., Hurst, J.E., Becker, C.A., Cowin, J.P., Auerbach, D.J. and Wharton, L.: 1980, J. Chem. Phys. 72, p. 2403.
3. Hurst, J.E., Wharton, L., Janda, K.C. and Auerbach, D.J.: 1983, J. Chem. Phys. 78, p. 1559.
4. Guthrie, W.L., Lin, T.-H., Ceyer, S.T. and Somorjai, G.A.: 1982, J. Chem. Phys. 76, p. 6398.
5. See, for example, Hepburn, J.W., Northrup, F.J., Ogram, G.L., Polanyi, J.C. and Williamson, J.M.: 1982, Chem. Phys. Lett. 85, p. 127.
6. See, for example the presentations by J.P. Toennies and by S.J. Sibener in this volume.
7. Meyer, H.-D. and Levine, R.D.: 1984, Chem. Phys. 85, p. 189.
8. Zamir, E. and Levine, R.D.: 1984, Chem. Phys. Lett. 104, p. 143.
9. Kubiak, G.D., Hurst, J.E., Jr., Rennagel, H.G., McClelland, G.M. and Zare, R.N.: 1983, J. Chem. Phys. 79, p. 5163.
10. Kaplan, H., Levine, R.D. and Manz, J.: 1976, Chem. Phys. 12, p. 447.
11. Levine, R.D.: 1976, *The New World of Quantum Chemistry*, Pullman, B. and Parr, R., eds. (Reidel, Dordrecht) p. 103.
12. Meyer, H.-D.: 1981, Surf. Sci. 104, p. 117.
13. See, for example, Somorjai, G.A.: 1981, *Chemistry in Two Dimensions* (Cornell University Press).
14. Ben-Shaul, A., Haas, Y., Kompa, K.L. and Levine, R.D.: 1981, *Lasers and Chemical Change* (Springer, Berlin).
15. Asscher, M., Guthrie, W.L., Lin, T.-H. and Somorjai, G.A.: 1983, J. Chem. Phys. 78, p. 5250.

ALKALI ATOMS ON SEMICONDUCTOR SURFACES: THE DYNAMICS OF DESORPTION AND OF SURFACE PHASE TRANSITIONS

E. F. Greene, J. T. Keeley, and D. K. Stewart
Chemistry Department
Brown University
Providence, Rhode Island 02912
U.S.A.

ABSTRACT

Beams of alkali atoms incident on heated surfaces of semiconductors produce ions by losing electrons to the surfaces. Potassium atoms on surfaces of Si(100) and graphite serve as examples of how surface ionization can provide information about the surfaces and their interactions with the atoms. The Si surface appears to undergo a phase transition at T_c = 980 ± 20K. At higher temperatures the atoms move freely over what we presume is a relatively smooth surface. At lower temperatures the atoms collect in a small number of special sites where they are bound more tightly and concentrated so they interact with each other but remain essentially in equilibrium with a small fraction that move over the remaining surface. On highly oriented pyrolytic graphite K ions are bound on at least three different kinds of sites between which the ions move at an appreciable rate but not rapidly enough to reach equilibrium.

1. INTRODUCTION

Progress in our understanding of the complex process of chemisorption has come from many different kinds of experiments. Included in this group is surface ionization, a process in which a neutral atom strikes a surface, exchanges an electron with the surface, and then leaves as an ion, either positive or negative.

In this paper our aim is to show for two special cases, potassium atoms striking silicon and graphite, how studies of surface ionization can help to describe the nature of the surface and its interactions with the ionizing atoms.

Although surface ionization has been much studied since its discovery[1] in 1922 and is the subject of some excellent reviews,[2] relatively little attention has been paid to ionization on semiconductors. Two recent papers[3,4] from this laboratory describe

191

B. Pullman et al. (eds.), Dynamics on Surfaces, 191–202.
© *1984 by D. Reidel Publishing Company.*

experiments with four alkali atoms and thallium on Si(100) and (111) and refer to earlier work by others. Several workers have studied the ionization of alkali metals on graphite[5] or on carbon covered metal surfaces.[6,7] For some of these systems surface ionization has the notable advantage of high sensitivity. Furthermore, the background is low, and single ions can be counted. This often means that chemisorption can be studied at coverages as low as 10^{-6} of a monolayer where one may hope that the atoms not only have a negligible effect on the nature of the surface but instead serve as useful probes of its characteristics.

2. APPARATUS AND EXPERIMENTAL PROCEDURE

The apparatus is shown schematically in Figure 1. Three separately

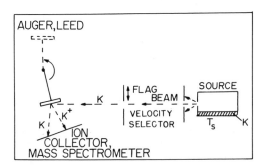

Figure 1. Schematic drawing of apparatus. The source containing potassium, K, is maintained at a temperature T.

pumped chambers contain respectively the source for the atoms, a velocity selector and a beam flag, and the surface to be studied. The latter is normally placed to intercept the beam from the source but may be moved into a position for analysis by Auger spectroscopy or low energy electron diffraction. The surface is heated electrically by conduction, and its temperature T is determined by resistivity measurements calibrated with an optical pyrometer.

The principal measurement made is of the current of positive ions $I^+(t)$ from the surface at constant T as this current varies with time t after the beam of atoms is successively turned on and off. At steady state this is $I_{ss}^+ = n_o e \sigma_o \beta_+$ where n_o is the flux of neutral atoms arriving from the source, σ_o their sticking probability, e the charge on the proton, and β_+ the fraction of atoms that leave as ions after sticking on the surface. Thus the efficiency of ionization is I_{ss}^+/I_o^+ where I_o^+ is $n_o e$. A useful expression for β_+ for a given kind of surface and atom in the beam is named for Saha and Langmuir,

$$\beta_+ = \alpha_+/(1+\alpha_+), \quad \alpha_+ = (q_+/q_o)\exp[(\Phi-IP)e/RT]. \tag{1}$$

Here α_+ is the ratio of the fluxes of positive ions to neutral atoms desorbing from the surface, q_+/q_o the corresponding ratio of their partition functions, Φ the work function of the surface, IP the

ionization potential of the atoms, and R the gas constant. The
equation may by derived by assuming that the rate of desorption to a
vacuum is equal to that at equilibrium and then finding the latter.
This is done by using kinetic theory and σ_0 for the incoming particles
to find the rate of adsorption and thus also the rate of desorption to
which it is equal at equilibrium. An analogous kind of equilibrium
may be assumed in deriving the Richardson-Dushman equation for the
thermionic emission of electrons from a hot surface. Results reported
in the literature show that the Saha-Langmuir equation is valid for
many systems.[2] This appears to be true for alkali metals on Si
also.[3,4]

3. RESULTS

We obtain the values of I_o^+ needed to test the Saha-Langmuir equation
by keeping the source at such a low temperature that the mean free
path of atoms in it is much greater than the diameter of the aperture
from which the atoms emerge. Thus they effuse, and we find I_o^+ from
the temperature of the source T_s, the vapor pressure of the atoms, and
the geometry of the apparatus. We estimate that this can give values
of I_o^+ with an uncertainty of ca. 10%. As an example of measuring the
ionization efficiency, we consider K on Si. Because the work function
for Si (Φ=4.80eV) is substantially greater than the ionization
potential of K (IP=4.34eV) the Saha-Langmuir equation predicts β_+=0.99
in our temperature range while we measure values of β_+=0.92 $\pm$ 0.1.
This agreement shows that at most a small fraction, less than about
10% of the incident atoms, reflect without equilibrating with the
surface (i.e. σ_0=1). Furthermore, insertion of the velocity selector
in the path of the beam permits measurement of the dependence of β_+ on
the velocity of the incoming atoms. That this is found to be
Maxwellian down to the lowest kinetic energy used (E_k≃0.01eV) not only
confirms that the atoms are effusing but also indicates that there is
no barrier to adsorption as high as 0.01eV.[4]

The rate of approach of I^+ to I_{ss}^+, its steady state value, is
of particular interest. Some examples are given in Figure 2 which
shows how I^+ varies as the surface is successively exposed to, and
shielded from, the atoms in the beam. A quantitative fit to these
profiles is obtained from

$$d[M]/dt = n_o A^{-1} - k_1[M] - k_2[M]^2 \qquad (2)$$

where [M] is the number of atoms adsorbed per unit area, A is the area
of the surface exposed to the beam, and k_1 and k_2 are respectively
first and second-order rate coefficients for desorption. In some
cases the term in k_2 makes a negligible contribution to d[M]/dt though
this is not so in the middle panel of Figure 2 where the slope of the
profile first increases after the beam is turned on. This shows that
the term is significant and thus that there is appreciable interaction
between the adsorbed atoms even though, as is shown below, the

coverage $\theta = [M]/[M]_0$ is only 10^{-4} of a monolayer in this example. Here $[M]_0 = 10^{15}$ cm^{-2} is the concentration in a monolayer. The two equal areas A_1+A_1' and A_2 represent the integrals of the difference between the instantaneous and steady state currents of ions over the time since the beam is turned on (t_1) or off (t_2). At the lower temperatures desorption is slow, so the temperature of the crystal must be raised (t_3) to give complete desorption in a reasonable time.

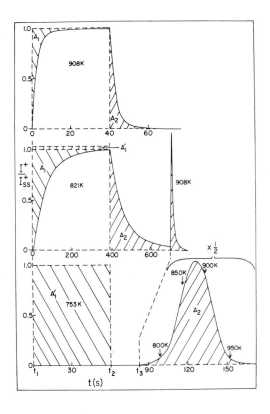

Figure 2. Positive ion current I^+ divided by the steady state current I_{ss}^+ vs. time for K on Si(100)-s, the stable form of the surface. Beam on at t_1, off at t_2, temperature raised at t_3. Areas A_1 measure amount adsorbed at t_2.

That A_1 nearly equals A_2 at 821 and 908K indicates that there is negligible diffusion of K into the bulk of the crystal. For these runs the areas $A_2 \simeq A_1$ also are measures of $[M]_{ss}$ the concentration of adsorbed atoms on the surface at steady state (at 753K a steady state is not reached by t_2). However, there is a complication. In our temperature range of 800-950K the Si surface can be prepared in two forms. From the one we call m the desorption is relatively fast while from the other, which we call s, it is slower. In Figures 3 and 4 the points show the temperature dependence of k_1 and k_2 (k_2 is appreciable for the s surface only) while the lines give the best fits to the Arrhenius expression for the rate coefficient $k_i = A_i \exp(-E_{ai}/RT)$. In Figure 5 the points give the measured $[M]_{ss}$ while the lines are obtained from the values of k_1 and k_2 and Equation (2) with d[M]/dt set equal to zero for the steady state. The agreement of the lines and points shows that the measurements of k_1, k_2, and $[M]_{ss}$ are consistent. Another check for the s surface is that, as Figure 3 shows for $I_o^+ < 10^{-9}$ A, the k_1 are independent of I_o^+ for three runs in

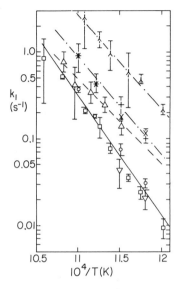

Figure 3. Rate coefficient k_1 for first order desorption of K from Si(100) vs. reciprocal temperature. $\square$, $\bigcirc$, $\triangledown$ (solid line), $\triangle$ (dashed line) stable surface s; λ, $\times$, $+$ (dot-dashed line) metastable surface m. Incident beam I_o^+(A): λ, $\square$: 2.0×10^{-11}; $\times$, $\bigcirc$: 2.0×10^{-10}; $+$, $\triangledown$: 8.6×10^{-10}; $\triangle$ 2.3×10^{-9}.

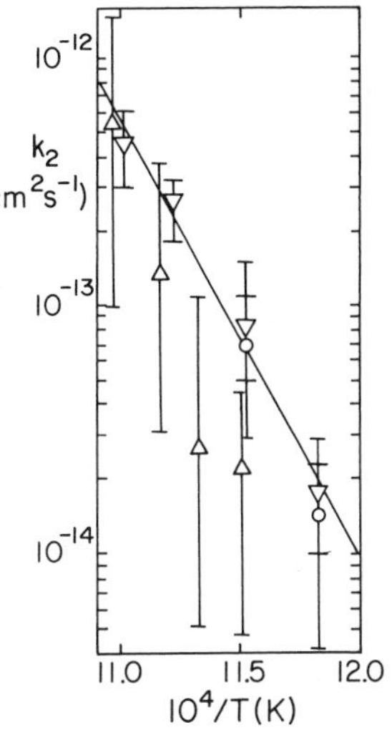

Figure 4. Rate coefficient k_2 for second order desorption of K from Si(100)-s vs. reciprocal temperature. I_o^+ as in Figure 3.

which I_o^+ differs by a factor of 43. For the m surface two of the runs agree but the points for another ($I_o^+ = 2 \times 10^{-11}$ A) are somewhat higher although the Arrhenius parameters agree within our experimental uncertainty. This may mean that the m surface differs slightly from one preparation to another.

To prepare the m surface we heat the s one to 1200K for 30s and then decrease the temperature abruptly to T where we measure the k_i. Figure 6 shows how k_1 changes with time after the heating. We consider the s surface stable and the m one metastable in this temperature range. Figure 7 shows schematically how the surface can be cycled through its two forms. The transition temperature T_c is located by heating the s surface to T' and then returning it to T.

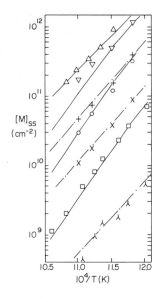

Figure 5. Concentration of K atoms adsorbed on Si(100) at steady state vs. reciprocal temperature. I_o^+ and s, m surfaces as in Figure 3. Points measured;
lines calculated from k_1 and k_2.

When T' becomes greater than T_c, the surface converts to the m form,
and this is detected by the increase in k_1. In this way we find
$T_c = 980 \pm 20K$.

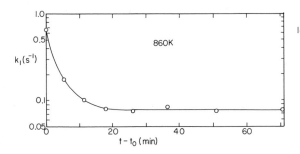

Figure 6. Variation of the first order
rate coefficient k_1 at 860 K for desorption of K from Si(100) with time after
t_o when surface is cooled from 1200K.

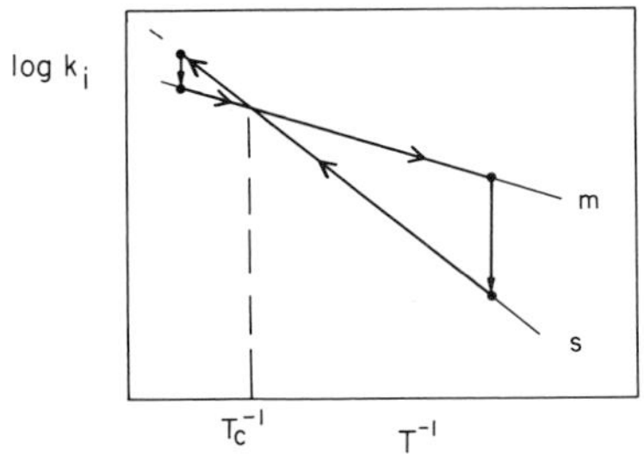

Figure 7. Schematic plot
illustrating preparation
of metastable surface m
from stable one s. T_c
temperature of transition.

When we use a sample of highly oriented pyrolytic graphite
(HOPG)[8] instead of Si, we also get desorption profiles, but they
differ from those in Figure 2 in that there is no second-order
process. Furthermore the first-order part becomes more complex,
separating into three processes of very different rates. Three
examples are shown in Figure 8. At 1418K only one process has a
desorption rate slow enough for us to measure (we find $A_1=A_2$). At

1046K there is only a negligible response from this process while the next faster one is measurable (we find $A_3=A_4$, but A_1' is defined to be equal to A_2 that is produced by raising T at t_3). Finally at 985K the fastest process alone determines the profile while the two slower ones only contribute to the desorption after the temperature is raised first at t_3 and then again at t_4. The temperature dependences of the three rate coefficients k_1', k_1'', and k_1''' are shown in the lower part of Figure 9 while the upper part gives the fractions B_i of the incident flux I_o that desorbs at each of the three rates. Note that the B_i vary with T.

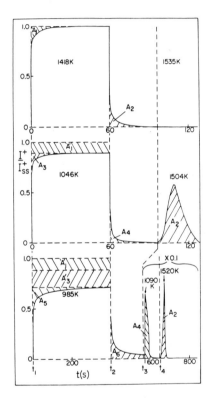

Figure 8. Positive ion current I_+^+ divided by steady state current I_{ss}^+ vs. time for K on highly oriented pyrolytic graphite (HOPG). Beam on at t_1, off at t_2, temperature raised at t_3 and t_4. Areas A_1 measure the amount adsorbed at t_2.

Table 1 gives the values for the parameters A_i and E_{ai} for K on Si(100) and on graphite (HOPG).

4. INTERPRETATION

We separate the results for Si(100) into two classes. In the first we put those that are determined directly, i.e. with little dependence on models. These include:
1) the sticking probability of K is near unity;

2) any barrier to adsorption is less than ca. 0.01eV;
3) the ionization probability is what would be expected at equilibrium, Equation (1);
4) a change occurs in the nature of the surface near 980K;
5) on the m surface at low coverages the atoms do not interact while on the s one they do;
6) the fraction of K penetrating into the bulk is less than about 10%;
7) the atoms are highly mobile on the s surface.

The last point follows because at equilibrium if the incident atoms strike the surface randomly and almost always stick, they must also desorb from all parts of the surface. Since they also interact on

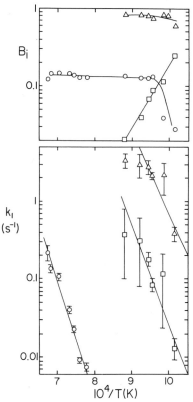

Figure 9. Dependence on reciprocal temperature of three individual first-order rate coefficients k_{1i} that together describe the desorption of K from graphite (HOPG) and of the fractional contributions B_i made by each k_{1i}. In order of decreasing rate they are: $k_1{}'$ (O), $k_1{}''$ (□), $k_1{}'''$ (△).

the s surface, even though the average coverage is low ($\bar{\theta} \ll 1$), they must move freely on it until they find and occupy a small fraction of special sites where they are bound more strongly than on the major part of the surface. The term in k_2 in Equation (2) appears because the concentration of atoms on these sites is greater than $\bar{\theta}$ by a factor that is the ratio of the total area to that occupied by the special sites.

In the second class we put those results that we try to understand by comparison with the predictions of plausible models.

1) Activation energy. Because the potential barrier to adsorption

from the gas phase must be small, if present at all, the potential
energy decreases monotonically as an atom approaches the surface until
it reaches its equilibrium position. This suggests that the parameter
E_a is approximately equal to the energy of desorption ΔE_o. If, as is
likely, the adsorbed state of K is essentially ionic, the main
contribution to the binding may be the ion's attraction to an image
charge in the surface. Equating this electrostatic energy to E_a gives

Table 1. Arrhenius parameters for the desorption of K from Si(100)
and graphite.

	Si(100)					Graphite		
	k_1-m	k_1-m^a	k_1-s	k_2 -s	k_1'-s^c	k_1'	k_1"	k_1'''
E_a(eV)	2.22	2.16	2.75	3.16	1.96	2.12	2.54	2.89
$\log_{10}A(s^{-1})$	12.2	12.4	14.4	5.2^b	10.6	10.5	11.2	9.4

^{a}least intense beam, I_o^+=2.0x10^{-11}A
^{b}units (cm^{2}s^{-1})
^{c}most intense beam, I_o^+=2.3x10^{-9}A

reasonable values[4] of 2.7 and 3.1 Å for the distance of the adsorbed
ion from the effective surface for the s and m surfaces respectively.

2) The preexponential factor, A. In transition state theory A is
determined principally by the activation entropy for desorption.
Again, in the absence of a barrier to adsorption the activated complex
may be expected to resemble the products of the reaction, in this case
the desorbed ion. Thus the experimental value of A is a measure of
the entropy of desorption, or, since the entropy of the gas is known,
the entropy of the surface plus the adsorbed ions. If the surface is
not much changed by the presence of a small number of these ions
($\bar{\theta}$<<1), A measures the entropy of the adsorbed ions. For the m
surface the experimental value ($\log_{10}A_1$ (s^{-1})=12.6 ± 1) agrees well
with the one calculated for particles K_{ads} moving freely as a 2-D gas
at 900K ($\log_{10}$ A(s^{-1})=12.7). For the s surface the larger value of A
(Table 1) implies a smaller entropy of K_{ads} or necessarily a more
nearly localized adsorption if the surface is uniform. This is
unrealistic because the adsorbed particles must interact to produce a
second-order desorption. Consistency is possible if the K_{ads} have
their entropy reduced by being restricted to a small number of special
sites. These might by similar to the ridges seen on the reconstructed
7x7 surface of Si(111) using reflection electron microscopy[9] or the
deep holes observed on the same surface by scanning tunnelling
microscopy.[10]

Our picture of the surface is that it undergoes a phase
change as the temperature increases through T_c from the reconstructed
one having special sites with a higher binding energy to a relatively

smoother and disordered or even melted one where the binding energy is lower. Such a transition has not been reported for Si(100), although it is well-known[11,12] for Si(111) where in our experiments[4] we see corresponding behavior at T_c = 1120 ±40K.

A model having a small number of special sites can be tested by exposing the crystal to a sufficiently large I_o so the special sites tend to become saturated. In our case this should make the rate faster and more nearly first order with a smaller value of E_a because the desorbing atoms only have access to the majority of sites on which they are less strongly bound than on the special ones. This is just what we find for the most intense beam we have used (I_o^+=2.3x10^{-9}A). The results for k_1 indicated by the points (△) and dashed line in Figure 3 have a larger magnitude and smaller temperature dependence than those for the less intense beams. In addition the corresponding points for k_2 (△) in Figure 4 have a smaller magnitude and larger uncertainties because they represent a smaller fraction of the measured rate. Table 1 gives the values for the Arrhenius parameters. We find that the model remains a useful one.

As Figure 8 shows, the profiles for the desorption of K from graphite differ from those on Si. Again there are deductions we can make that do not depend strongly on models:
1) The existence of processes with three different rates shows that there are at least three kinds of sites (a, b, and c) at which K can be bound,

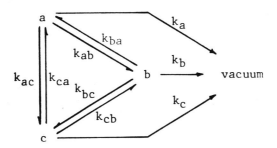

2) The variation with temperature of the relative amounts of K_{ads} desorbing in the three processes shows that K_{ads} must by able to move from one kind of site to another.
3) However, this motion cannot be so rapid as to keep the populations on the various sites in equilibrium with each other, for then solving the kinetic equations shows that the whole population desorbs with a single rate coefficient that is a composite of the k's for the individual steps.

We would like to be able to make a model in which we could

use the Arrhenius parameters we find for the three experimental rate coefficients k_1', k_1'', and k_1''' to deduce the parameters for each of the elementary rate coefficients (e.g. k_a, and k_{ca} in the scheme above). Unfortunately, as is well-known, this inversion requires accurate data, so even small uncertainties may leave it ambiguous. For example, even when we see only two first-order rates, as we do for K desorbing from a carbon layer on SiC, we cannot get reliable values for the A_i and E_{ai} of the four elementary steps involved. However, one can expect that a combination of measurements of this kind with others on the same systems may provide the extra information needed to yield the Arrhenius parameters and an improved understanding of this surface.

6. CONCLUSION

Detailed measurements of the rates of desorption of atoms, particularly at low coverages, can be interpreted to give information not only about the interaction of atoms with the surfaces of semiconductors but also changes in the nature of the surfaces themselves.

ACKNOWLEDGEMENTS

We thank the U.S. Department of Energy, Division of Chemical Sciences for its support of this work and the National Science Foundation for providing a research assistantship to D.K.S. through the Materials Research Laboratory of Brown University.

REFERENCES

1. K. H. Kingdon and I. Langmuir, Phys. Rev. **21**, 380 (1923); I. Langmuir and K. H. Kingdon, Science **21**, 58 (1923); H. E. Ives, Phys. Rev. **21**, 385 (1923).
2. E. Ya. Zandberg and N. I. Ionov, Surface Ionization (Keter, Jerusalem, 1971); N. I. Ionov, Progr. Surf. Sci. **1**, 237 (1972).
3. E. F. Greene, J. T. Keeley, and M. A. Pickering, Surf. Sci. **120**, 103 (1982).
4. E. F. Greene, J. T. Keeley, M. A. Pickering, and D. K. Stewart, Surf. Sci. (in press).
5. E. Ya. Zandberg and V. I. Paleev, Sov. Phys.-Tech. Phys. **9**, 1575 (1965).
6. J. O. Olsson and L. Holmlid, Mater. Sci. Eng. **42**, 121 (1980).
7. E. V. Rut'kov and A. Ya. Tontegode, Sov. Phys.-Tech. Phys. **27**, 589 (1982).
8. Union Carbide Corporation, Carbon Products Division.
9. N. Osakabe, Y. Tanishiro, K. Yagi, and G. Honjo, Surf. Sci. **109**, 353 (1981).
10. G. Binnig, H. Rohrer, Ch. Gerber, and E. Weibel, Phys. Rev. Lett.

    50, 120 (1983).

11. J. J. Lander, Surf. Sci. 1, 125 (1964).

12. P. A. Bennett and M. B. Webb, Surf. Sci. 104, 74 (1981); M. B. Webb and P. A. Bennett, J. Vac. Sci. Technol. 18, 847 (1981).

SOME RECENT RESULTS IN SURFACE DIFFUSION STUDIES BY THE FLUCTUATION
METHOD

Robert Gomer
The James Franck Institute and Department of Chemistry
The University of Chicago
Chicago, Illinois 60637, U.S.A.

ABSTRACT

This paper presents a brief summary of the fluctuation method for
measuring the surface diffusion of adsorbates on single crystal planes
of field emitters. The extension of the method to measuring diffusion
anisotropy by using a rotatable rectangular slit is also presented. The
results of anisotropy measurements for oxygen and for hydrogen and deuterium on the (110) plane of tungsten will be presented. Oxygen shows
the theoretically expected anisotropy, hydrogen and deuterium do not,
and reasons for this difference are discussed. Anisotropies are also
seen in the mean square fluctuations and the origin of this effect is
explained in terms of correlation lengths. Finally, results of tritium
diffusion on W(110) are given and contrasted with those for hydrogen and
deuterium. It is shown that the magnitudes of the tunneling diffusion
coefficients for the three isotopes can be explained by assuming mass
renormalization, resulting from substrate-adsorbate interaction. The
variation of prefactors with isotope and coverage in the thermally activated regime turns out to be quite unexpected and cannot be explained at
this time.

1. THE FLUCTUATION METHOD

This method (1, 2) relies on the fact that the field emission microscope makes it possible to examine emission from a small region $\leqslant 100$ Å
in linear dimension, of a field emitter. This probed region can therefore be embedded in one of the substantially larger, well developed low
index single crystal planes of the emitter. The presence of adsorbed
atoms on the substrate surface leads to work function and hence emission
changes. When a field emitter uniformly covered with adsorbate is heated,
adsorbed atoms begin to diffuse and thus there are small but measurable
current changes within the probed region, resulting from changes in
adsorbate concentration within it. It is possible to measure the time
autocorrelation function $f_i(t)$ of these fluctuations defined as

B. Pullman et al. (eds.), Dynamics on Surfaces, 203–213.
© *1984 by D. Reidel Publishing Company.*

$$f_i(t) = <\delta i(0) \; \delta i(t)>/<i^2> \tag{1}$$

where $\delta i = i - \bar{i}$ and to show that this is simply related (1, 2) to the adsorbate number autocorrelation function $f_n(t)$. The latter can be expressed as

$$f_n(t) = <\delta n(0) \; \delta n(t)> = \frac{<(\delta n)^2>}{A} \int_A d^2r \int_A d^2r' \frac{e^{-|r-r'|^2/4Dt}}{4\pi Dt} \tag{2}$$

where A is the area of the probed region and D the (chemical) diffusion coefficient. $\delta n = n - \bar{n}$, with n the number of adatoms in A. For a circular probed region of radius r_o the time τ_o

$$\tau_o = r_o^2/4D \tag{3}$$

occurs quite naturally in Eq. 2 which can be written as a function of t/τ. τ_o has the physical significance of the relaxation time, in which a fluctuation builds up or decays. It is the time required for a mean square displacement $<r^2>$ of a diffusing atom, or the time required for a delta-function concentration fluctuation to evolve into a Gaussian with second moment r_o^2. $f_n(0) \equiv <(\delta n)^2>$ corresponds (2, 3) to the mean square number fluctuation, which is proportional to the ad-layer compressibility:

$$f_n(0)/\bar{n} = K \, k_B Tc = [(\partial/k_B T)/\partial \ln c]^{-1} \tag{4}$$

where K is the 2-dimensional compressibility and μ the chemical potential of the ad-layer. c is the mean coverage in atoms/cm^2. It is possible to compare experimentally obtained plots of $f_i(t)/f_i(0)$ vs. $\log t$ with similar theoretical curves plotted against $\log(t/\tau_o)$ and thus to obtain τ_o and D. The experiment can be carried out at virtually constant adsorbate concentration, but this can be varied widely, as can be temperature.

It can be shown (2) that the quantity obtained corresponds to the chemical diffusion coefficient, i.e., that appearing in Fick's law, rather than to a tracer diffusion coefficient which determines the mean square displacement of a single atom. It can also be shown (2) that Eq. 2 is value for interacting systems and even in a two phase coexistence region.

It has recently proved possible to extend the method to a determination of diffusion anisotropy, by using a probed region in the form of a long and narrow rectangular slit (4). When such a slit is lined up along one of the principal axes of the diffusion tensor, i.e. when the latter is diagonal, the correlation function decomposes into a product of two one-dimensional functions identical in form and differing only in the numerical values of their relaxation times (5):

$$f_n = f_n(t/\tau_x) \cdot f_n(t/\tau_y) \tag{5}$$

Here $f_n(t/\tau_x)$ is given by

$$f_n(t/\tau_x) = erf(\tau_x/t)^{\frac{1}{2}} + (t/\pi\tau_x)^{\frac{1}{2}}(e^{-\tau_x/t}-1) \tag{6}$$

with a corresponding expression for $f_n(t/\tau_y)$. The relaxation times are

$$\tau_x = a^2/D_{xx} \tag{7a}$$

$$\tau_y = b^2/D_{yy} \tag{7b}$$

where the slit dimensions parallel to the y and x axes are respectively 2b and 2a. In the experimental setup used b/a = 10; consequently the decay of the correlation function in the y direction would be 100-fold slower than in the x direction if $D_{yy} = D_{xx}$. Even if D_{yy} exceeds D_{xx} considerably it is still possible to measure only D_{xx}. If the slit is then rotated 90° so that the short dimension is parallel to the y axis D_{yy} can be determined.

2. DIFFUSION ANISOTROPY

Let us turn now to a consideration of some recent results, starting with anisotropy measurements for oxygen and hydrogen diffusion on the tungsten (110) plane by Tringides and Gomer (4). If adsorption causes no breaking of symmetry, it can be shown (5) that the principal axes of a given plane are determined by that plane's symmetry only. In general, inspection reveals along which directions a concentration gradient does not lead to a diffusion current in any other direction, i.e., which are principal axes. More formally it is possible to deduce these by requiring that a symmetry operation leave the diffusion tensor and hence its matrix elements invariant (5). In this way one can show that any set of orthogonal axes are principal axes for (100) and (111) planes, i.e. that these planes are totally degenerate, while planes with rectangular unit cells have axes parallel to the sides of the rectangle. This is illustrated in Fig. 1 for a bcc (110) plane. If the directions in which actual jumps occur coincide with the principal axes the ratio of D_{xx}/D_{yy} cannot be predicted a priori. However, if the jump directions differ from the principal ones, for instance if they correspond to the directions labelled x_1 and x_2 in Fig. 1 the ratio D_{xx}/D_{yy} is fixed. For the case shown in Fig. 1 we have

$$D_{yy}^{*}/D_{xx}^{*} = \langle y^2 \rangle / \langle x^2 \rangle = \tan^2\theta \frac{\langle x_1^2 \rangle + \langle x_2^2 \rangle + 2\langle x_1 x_2 \rangle}{\langle x_1^2 \rangle + \langle x_2^2 \rangle - 2\langle x_1 x_2 \rangle} \tag{8}$$

for the ratio of tracer diffusion coefficients D^{*} which are proportional to mean square displacements. If the displacement along x_1 is uncorrelated with that along x_2 the term $\langle x_1 x_2 \rangle = 0$ and

$$D_{yy}^{*}/D_{xx}^{*} = \tan^2\theta = 2 \tag{9}$$

with the numerical value 2 corresponding to $\theta = 54.7°$.

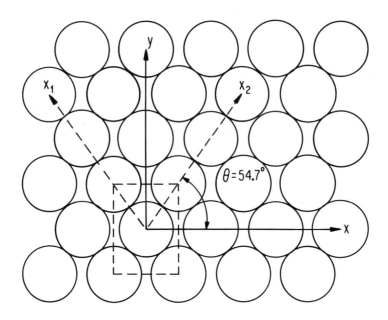

Fig. 1   Drawing of a bcc (110) plane. x and y are the principal axes for diffusion, x_1 and x_2 the postulated actual jump
directions.   The dashed rectangle is the non-primitive unit
cell.

The quantity measured experimentally, however, is not D^*_{yy}/D^*_{xx} but
the ratio of chemical diffusion coefficients. It is possible to give a
simple plausibility argument why this ratio should be equal to that of
the tracer diffusion coefficients when the term $\langle x_1 x_2 \rangle = 0$. It is easy
to show from general considerations (2) that

$$\frac{D_{yy}}{D_{xx}} = \frac{D^*_{yy} + \frac{1}{2N} \int_o^\infty dt \sum_{i \neq j} \langle v^y_i(t) v^y_j(0) \rangle}{D^*_{xx} + \frac{1}{2N} \int_o^\infty dt \sum_{i \neq j} \langle v^x_i(t) v^x_j(0) \rangle} \tag{10}$$

where N is the total number of particles in the system and v^x and v^y the
components of the velocity along the x and y directions. If it is assumed that correlation exists only between neighboring ad-atoms and that
the time over which such correlations exist is limited to $\frac{1}{2} v^{-1}$, where
v is the effective jump frequency, it can be shown that the sum of velo-
city-velocity terms cancels, so that

$$\frac{D_{yy}}{D_{xx}} = \frac{D^*_{yy}}{D^*_{xx}} = \tan^2 \theta \tag{11}$$

The results for O adsorption (4) are in accord with the predictions
of Eq. 11. Within experimental error a ratio of 2 ± 0.2 is observed for
all coverages $\theta < 0.55$; for $\theta = 0.55$ the highest coverage studied,

values closer to 1 were found except at high temperature, T = 721 K. It was verified that other orientations of the slit, i.e., along orthogonal but otherwise arbitrary axes x', y', yielded smaller ratios of the apparent diffusion coefficients $D_{y'y'}/D_{x'x'}$, than along the principal axes, a result expected theoretically from simple considerations of matrix algebra (5). This result suggests that the diffusion of O, although known to depend strongly on O-O interactions (from the fact that using appropriate interactions in a Monte Carlo simulation (6) reproduces the experimentally observed (7) increase in activation energy with coverage) behaves as if O atoms diffused by jumping from one "hourglass" well in Fig. 1 to the next over the barrier consisting of 2 adjacent W atoms with no bias as to jumps left or right, up or down. The exception to this is the coverage where the p(2x1) structure is essentially complete, or even overcomplete. It is not surprising that this could lead to some deviations, although the detailed reasons are not understood at present.

In view of the oxygen results it was somewhat of a shock to find that hydrogen and deuterium (4) showed virtually no anisotropy at any temperature. For all coverages except $\theta = 0.25$, the lowest coverage easily studied, $D_{yy}/D_{yy} = 1$ within experimental error and this ratio was seen also for all other slit orientations. At $\theta = 0.25$ $D_{yy}/D_{xx} = 1.3$ for both hydrogen and deuterium. Formally a ratio of 1 can be explained by assuming that $\langle x_1 x_2 \rangle \neq 0$. This can be interpreted by saying that a jump along the positive x_2 direction is more likely to be followed by one along the negative x_2 direction than along either the positive x_1 or x_2 direction, with analogous probabilities for an initial jump along $\bar{x}_1$. A physical explanation might be sought along the following lines: There is some evidence that the actual binding sites for H are the ends of the hourglass regions shown in Fig. 1. If there were a potential barrier in the middle between these, comparable in height to those separating hourglass regions from each other, such an effect would obviously occur, but would have to be temperature dependent, since the degree of reflection by the barrier in the middle would be governed by a Boltzmann factor. Such a dependence is not observed, however. If both ends of the hourglass, i.e., both sites within an hourglass could be simultaneously occupied an analogous effect would result. If all hourglass sites could be doubly occupied the saturation coverage would be H/W = 2. There is no real evidence for or against this; to the best of my knowledge no really reliable absolute coverage measurements are available. Perhaps the strongest, if indirect evidence against such a high coverage is the fact that only p(2x1), p(2x2) and p(1x1) Leed pattern are observed (8). If the p(1x1) pattern at saturation corresponded to 2 rather than 1 H atom per hourglass one might expect more structures at intermediate coverages than have been seen. It is quite possible of course that double occupancy can occur to a very limited extent during diffusion. Since $D_{yy}/D_{xx} = 1$ even at temperatures where diffusion is thought to occur by tunneling (9) this would be necessary. Arguments analogous to those leading to Eq. 11 show that

$$\frac{D_{yy}}{D_{xx}} \cong \frac{1-q-p/6}{1+q+p/6} \tan^2\theta \qquad (12)$$

if

$$\frac{\langle x_1 x_2 \rangle}{\langle x_1^2 \rangle} = -q \qquad (13)$$

for a single particle and if the relative probability of simultaneous jumps along x_1 and $-x_2$ or $-x_1$ and x_2 by 2 adjacent particles is $(1+p)/16$, so that the relative probabilities of other combinations are $(1-p/3)/16$ each. If $p = q$, which is not a priori necessary, a value of $p = 0.29$ makes $D_{yy}/D_{xx} = 1$. This should be contrasted with $q = 0.333$ required to make $D*_{yy}/D*_{xx} = 1$. The reasoning leading to Eq. 12 is of course very qualitative and heuristic, although illuminating. It should be possible to carry out a Monte Carlo calculation to obtain values of D_{yy}/D_{xx} for the interaction parameters appropriate to the O/W(110) system and also for the double site occupancy hypothesis for H/W(110). Such a calculation would also show if $D_{yy}/D_{xx} = 1$ can result from sufficiently different next-nearest and next-next nearest neighbor interactions for H, without invoking double site occupancy.

Anisotropy was also observed in the mean square fluctuations, both for oxygen and hydrogen. For oxygen $f(0)$ with the narrow dimension of the slit along the y axis in Fig. 1 was approximately twice that observed with the narrow dimension along the x axis. For hydrogen and deuterium the trend was qualitatively similiar but the ratios observed were larger, 2 -3, and showed some temperature dependence, although this varied unsystematically with coverage. The details of this variation are probably beyond present understanding, but it is possible to give a general explanation why an orientation dependence can occur (5). At first blush no such dependence is expected of course because Eq. 4 indicates that $f_n(0)$ should be proportional to K, the 2-dimensional compressibility, a thermodynamic quantity, independent of slit dimensions or orientation. However, when $f_n(0)$ is expressed in terms of the correlation lengths of fluctuations in the x and y directions, an effect can be seen to occur if at least one of the correlation lengths is comparable to the small slit dimensions. The equal time density-density correlation function can be written

$$\langle \delta c(\vec{r}) \, c(\vec{r}') \rangle = C \, e^{-|x-x'|/\xi_x} \cdot e^{-|y-y'|/\xi_y} \qquad (14)$$

where $c(r)$ is the density at point $\vec{r}$, C a constant and ξ_x and ξ_y correlation lengths along the x and y directions. Integration of Eq. 14 over the rectantular probed region of dimensions $2a$ x $2b$ then yields

$$f_n(0) = C \, 2 \, \xi_x^2 (2a/\xi_x - 1 + e^{-2a/\xi_x}) \cdot 2 \, \xi_y^2 (2b/\xi_y - 1 + e^{-2b/\xi_y}) \qquad (15)$$

If a correlation length is comparable to a $f_n(0)$ is not invariant to interchanging a and b, i.e. to a rotation by $90°$. In the limit $\xi_x/a \to 0$, $\xi_y/b \to 0$ i.e., for sufficiently large slit dimensions

$$f_n(0) \rightarrow C \ a \ \zeta_x \zeta_y. \tag{16}$$

and there is no orientation dependence, as required. Unfortunately C cannot readily be determined (5) so that absolute values of ζ_x and ζ_y cannot be found, but the fact that an anisotropy is seen experimentally is interesting and points to correlation lengths of the order of $\geqslant 40$ Å, the magnitude of dimension $2\underline{a}$.

3. TRITIUM DIFFUSION ON W(110)

We turn next to the diffusion of hydrogen isotopes on the W(110) plane. Fig. 2 shows values of log D vs. 1/T for ^{1}H and ^{2}H, obtained by

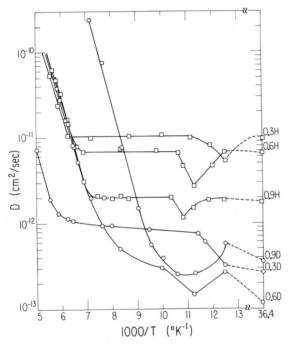

Fig. 2   Log D vs. 1000/T for the diffusion of ^{1}H and ^{2}H on W(110). The numbers and letters, (e.g. 0.9 D) refer to relative coverage and isotope, respectively: D stands for ^{2}H, H for ^{1}H.

DiFoggio and the author (9). Two features stand out. First there is temperature independent, or nearly independent diffusion at T $\leqslant$ 160-140 K, depending on coverage and isotope. This strongly suggests tunneling. From the time these results were obtained in 1982 to almost the present the small difference between the ^{1}H and ^{2}H tunneling diffusion coefficients presented a major puzzle, in that attempts to explain this in terms of a simple tunneling model ran into the following difficulty. If the barrier was sufficiently transparent to account for the small

mass effect, the observed diffusion coefficients were orders of magnitude smaller than any reasonable calculation would indicate. Alterna-
tely, if the experimental diffusion coefficient for, say, ^{1}H was used to
estimate barrier dimensions the value then calculated for the diffusion
coefficient of ^{2}H was many orders of magnitude smaller than observed.
The resolution of the difficulty has recently been obtained by Muttalib
and Sethna (10) who argue that adsorbate-substrate interactions lead to
a lattice distortion which is carried along in tunneling. This has two
effects. First there is a strong mass renormalization so that the effective masses of ^{1}H and ^{2}H are much closer than 1:2. Second, the perturbation of the substrate when an ad-atom moves from one adsorption
site to another by tunneling leads to the emission of phonons; this in
turn means that the transition occurs from a discrete initial state to
a continuum of final states and thus can be treated as tunneling in the
conventional sense, i.e. by the WKB approximation or equivalently by
Fermi's golden rule. Thus the tunneling probability will be proportional to the square of the appropriate hopping matrix element. If there
were no emission of phonons, tunneling would correspond to resonance of
amplitudes between wells; in the case of an extended periodic structure
a band picture would thus apply, and the diffusion coefficient would be
found in terms of a group velocity and a mean free path (9). The group
velocity, however, is proportional to the first power of the hopping
matrix element, not its square. By using the path integral method to
calculate tunneling probabilities (essentially extending the WKB method
to include phonon coordinates) Muttalib and Sethna (10) were able to
come up with values of the actual diffusion coefficients close to the
experimental ones.

Fig. 3 shows results for ^{3}H, obtained by S.-C. Wang and the author
(11). Again there is an activated regime followed by a tunneling one.
The values of D at low temperature are roughly 1-2 orders of magnitude
smaller than those for ^{1}H, and this again is predicted by the work of
Muttalib and Sethna (10). It is quite clear that tunneling of bare ^{3}H
would be many orders of magnitude less probable than tunneling of bare
^{1}H, so that the observed ratio $\sim 10^2$ is very strong support for the dres-
sed particle model.

The second striking feature of the ^{1}H and ^{2}H results is the following. The slopes of the curves in Fig. 2 are nearly identical, meaning
that the activation energies are nearly independent of isotope or coverage, but the position of the curves varies considerably with coverage
and is different for the two isotopes. In other words, if D is expressed as

$$D = D_o \, e^{-E/kT} \tag{17}$$

D_o varies widely with coverage and isotope. Values of E and D_o plotted
against coverage θ are shown in Figs. 4 and 5. It was speculated (9)
that the increase observed for ^{2}H might be the result of diffusion
chains. To first approximation

$$D_o \propto \langle \lambda^2 \rangle \tag{18}$$

where $\langle \lambda^2 \rangle$ is the mean effective jump length squared. If a ^{2}H atom

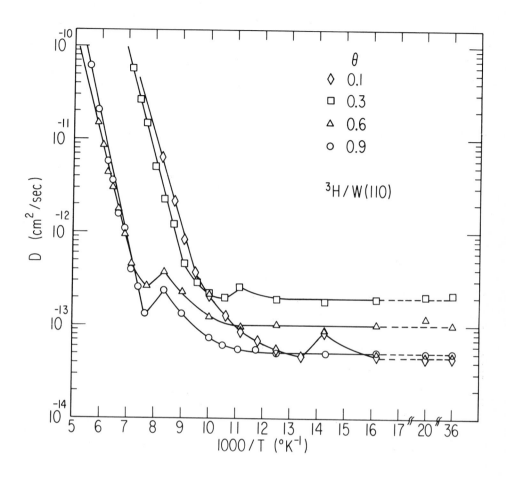

Fig. 3   Log D vs. 1000/T for ^{3}H.   Relative coverages are indicated on the figure.

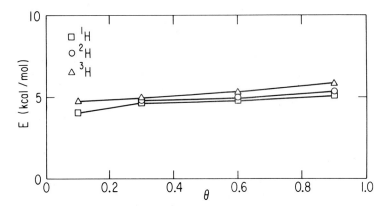

Fig. 4   Activation energy for diffusion of the 3 hydrogen
isotopes on W(110) vs. relative coverage θ.

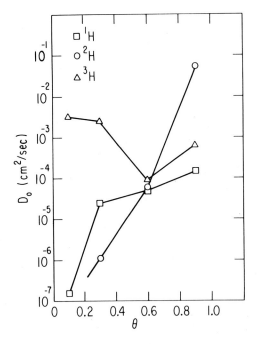

Fig. 5   Prefactors D_o vs. relative coverage for the diffusion
of the 3 hydrogen isotopes on W(110) vs. relative coverage θ.

entering an occupied well could be forward scattered or eject the occu-
pant and the scattered atom in turn entered an occupied well, continuing
the process, chains would result. It is not hard to show (9) that

$$\langle \lambda^2 \rangle = a_o^2 / (1-\theta)^2 \tag{19}$$

where a_0 is a single jump distance. However, if successful forward scattering were to occur only in a fraction of cases, no appreciable chain length and hence no appreciable increase in D_0 with coverage would occur. It was now postulated that ^1H behaved quantum-mechanically so that statistics were important and prevented overlap of the space parts of the nuclear wave function in 3 out of 4 encounters, while ^2H was heavy enough to behave classically, with no restrictions on collisions. On this basis one would predict that ^3H, which has the same statistics as ^1H but is even heavier than ^2H and hence more "classical" should behave like ^2H. The results shown in Fig. 3 for ^3H differ from both ^1H and ^2H. The values of D_0 for tritium are higher than for ^1H but the coverage dependence is unlike that seen with ^2H. At the moment no good explanation can be offered. It is possible that diffusion chains occur with ^2H but are quenched by excessive friction for ^3H, resulting from less mass mismatch with the tungsten substrate. This might also account for the high values of D_0 seen even at low θ. However, the anisotropy results discussed earlier speak against diffusion chains. If such chains occurred for ^2H but not ^1H, the anisotropy would be expected to be different for the two isotopes.

At this point it is not possible to offer definitive explanations for the behavior of the 3 hydrogen isotopes in the activated diffusion regime. It is clear only that this system is much more complicated than might have been expected, and that it offers considerable challenges to theorists.

ACKNOWLEDGMENT

This work was supported in part by NSF Grant CHE83-16647 and its predecessor grants. We have also benefitted from the Materials Research Laboratory of the National Science Foundation at the University of Chicago.

REFERENCES

1) R. Gomer, Surf. Sci. 38, 373 (1973).
2) G. Mazenko, J. R. Banavar, and R. Gomer, Surf. Sci. 107, 459 (1981).
3) B. Bell, R. Gomer, and H. Reiss, Surf. Sci. 55, 494 (1976).
4) M. Tringides and R. Gomer, to be published.
5) D. R. Bowman, R. Gomer, K. Muttalib, and M. Tringides, Surf. Sci. in press.
6) M. Tringides and R. Gomer, Surf. Sci. in press.
7) J.-R. Chen and R. Gomer, Surf. Sci. 79, 413 (1979).
8) V. C. Gonchar, Y. M. Kagan, O. V. Kanash, A. G. Naumovets, and A. G. Fedorus, Zh. Eksp. Teor. Fiz. 84, 249 (1983) (Sov. JETPS 57, 142 (1983)).
9) R. DiFoggio and R. Gomer, Phys. Rev. B25, 3490 (1982).
10) K. Muttalib and J. Sethna, to be published.
11) S.-C. Wang and R. Gomer, to be published.

ON ENERGY PATHWAYS IN SURFACE REACTIONS

Uzi Landman and Robert H. Rast
School of Physics
Georgia Institute of Technology
Atlanta, Georgia   30332

Molecular dynamics simulations allow detailed investigations of energy pathways and the dynamical evolution of interaction processes at surfaces. Studies of surface diffusion of single particles and clusters reveal the microscopic mechanisms of energy transfer, excitation, and mode coupling in activated rate processes. The phenomena of dynamical rate enhancement is defined and stochastic evolution is discussed in terms of nonlinear dynamical processes.

The evolution of chemical reactions is often[1,2] described by a set of kinetic equations which express the rate of change with time of the concentrations of reactants, products and possibly reaction intermediates. The kinetic approach represents a phenomological, most contracted description and in general there is no simple correspondence between the order of the reaction (the exponents of the concentrations as they appear in the kinetic equations) and the microscopic reaction mechanism. In addition the rate coefficients are treated as phenomenological, empirical constants.

 Basic understanding of surface reactions requires studies of the states (electronic, vibrational and rotational) of atoms and molecules in the vicinity of the solid and of the interaction dynamics. The dynamical evolution is governed by the potential surface of the interacting system and by dynamic couplings and response characteristics. The interphase-interface nature of gas-surface systems introduces unique spatial characteristics reflected in the topology of the potential surface and/or boundary conditions. In addition, solid substrates possess unique spectra of elementary excitations (electronic and vibrational) of an extended (collective) nature. The degree to which an extended picture of the substrate rather than a local (cluster or embedded cluster) description should be maintained in a proper formulation of dynamical processes at surfaces depends on the class of phenomena under study. While in claculations of chemical bond formation a representation of the substrate by a finite (often small) cluster may be adequate[3], energy dissipation and redistribution mechanisms, particularly for low-energy (thermal) processes, require an extended picture.

B. Pullman et al. (eds.), Dynamics on Surfaces, 215–230.
© *1984 by D. Reidel Publishing Company.*

In the latter case the validity of the Born-Oppenheimer approximation which is central to calculations of adiabatic potential energy surfaces is seriously questioned due to the availability of easily excited conduction electrons in metal substrates. Moreover, since the substrate is characterized by a continuous spectrum of electron-hole pair excitations, a description of the gas-surface interaction potential for metals in terms of a single adiabatic potential is incorrect and should be replaced by a continuous band of close-lying hypersurfaces. The strength of nonadiabatic coupling which causes transitions between these levels govern the degree of departure from the adiabatic description. The above discussion indicates the difficulties in constructing a proper potential hypersurface let alone a formulation of the evolution of the system.[4]

Energy dissipation and redistribution via coupling to substrate phonons is likely a dominant process.[5,6] The substrate phonons provide a "momentum" and energy reservoir (although the energy transfer if only single-phonon processes are considered is limited to $\lesssim 10^{-2}$ eV). The phonons in turn may dissipate energy via electron-phonon coupling, or scattering processes.

Of paramount importance for the understanding of reaction dynamics on a microscopic level are the mechanisms of coupling and energy exchange and redistribution between the reactants.[7] In the case of gas phase reactions certain general conclusions with regard to the relative importance of translational, vibrational and rotational energies in surmounting the reaction barrier have been drawn. For example, when the barrier is encountered early in the approach of the reactants translational energy is the key factor. While for reactions in which the potential energy barrier for reaction occurs when the reactants have come to close proximity, the vibrational and rotational energy content is dominant in dictating the probability of reaction. As indicated above the formulation and theoretical analysis of surface reactions are much more complex due to the heterogeneous nature of these systems. Consequently, at this stage of development, progress may be made via model theories (employing, critically, simplifying assumptions) of coupling and excitation mechanisms of molecular species near surfaces and of the dynamical evolution leading to bond formation (sticking), bond rupture (desorption, dissociation) rearrangement and migration. The purpose of such theoretical models is to identify and elucidate the dominant mechanisms and relevant characteristic parameters which govern the dynamical physical processes. It should be noted that even though in certain cases theories can be stated in most general terms, their implementation often requires input information which for the systems under study is not yet available (such as a complete surface phonon spectrum, accurate interaction potential hypersurfaces, localized state wave functions etc., although progress in these directions is realized).

The scope of this paper precludes a comprehensive review of the issues raised in the above introductory remarks (for a recent review see Reference 5). Instead, we will confine ourselves to the presentation of some new results pertaining particularly to the subjects of the dynamics and energy pathways in certain surface processes. Many

studies[8] of activated rate processes have been limited to descriptions
of the system evolution along the reaction coordinate ignoring non-
reactive modes. More recently[9], the importance of the multidimensional
nature of interactive systems and effects due to coupling between the
"reaction coordinate" and manifolds of non-reactive modes have been
noted. Such couplings may modify the potential energy surface and
introduce entropic effects which can be included in a transition state
theory description. Our focus here is to demonstrate and emphasize
the role of nonreactive modes and their couplings in the excitation
dynamics. These modes, which provide coupling channels between the
molecular system and the solid substrate, acting as "doorway states" in
the reaction dynamics[5,6] are essential for the proper dynamical description of certain surface processes. The case study which we describe in
some detail is the diffusion of an adparticle cluster on a solid surface. We chose this case since the conclusions which we reach impact
our understanding of the dynamics of other activated processes (such
as desorption, sticking and certain scattering phenomena) and because
it allows us to demonstrate the potential of molecular dynamics simulations in investigations of the microscopic dynamics of reaction processes.

On The Dynamics of Surface Diffusion

Diffusion processes on or in the vicinity of surfaces are of importance
in many surface controlled, or driven, physical and chemical phenomena.
Such phenomena include crystal growth, surface phase transformations,
annealing and recovery of damage, faceting, surface and interfacial
(grain-boundary) segregation and chemical processes, heterogeneously
catalyzed by surfaces.

   Several experimental techniques provide valuable information
about surface diffusion (such as Field Emission Spectroscopy (flicker
noise), photoemission monitoring as the light beam traverses across the
surface, changes in contact potential and scanning Auger methods). The
most direct observations are provided via Field Ion Microscopy (FIM)
techniques (for a review see Reference 10). FIM provides data on the
atomic scale concerning the local structure of the system and configurations of the diffusing species which when analyzed properly provide
information about the kinetics and energetics of surface diffusion
processes (for recent reviews see References 5 and 10).

   The nature of current experimental probes limits the temporal
resolution with which diffusion processes can be studied. Valuable
information about diffusion systems can be obtained via computer dynamical simulations[11] which enable investigations on arbitrarily short time
scales and refined spatial resolution. In the following we describe
our recent molecular dynamics studies of surface diffusion which
demonstrate the potential of such theoretical experiments in revealing
the microscopic dynamical mechanisms of certain activated processes.

   FIM studies of the diffusion of adatom clusters on surfaces have
shown that the migration of the clusters proceeds via distinct alternat-
ing cluster configurations (most easily observed for diffusion on
"channeled" substrates such as the (211) surfaces of bcc solids).

Motivated by these observations we have developed[5,12] a general
stochastic theory of diffusion on ideal and defective lattices in
which the diffusing system performs transitions over a manifold of
internal states, which may correspond to energy levels and/or distinct
spatial configurations. Proper analysis of experimental data in terms
of such multi-state migration mechanisms allows determination of rate
coefficients (frequency factors and activation energies) for transitions among the states participating in the migration mechanism thus
extending the information content which may be extracted from the data
beyond the mere determination of global diffusion coefficients.

A particularly curious FIM observation concerns the diffusion of
Re dimers on the (211) surface of tungsten, where the dimer rate of
diffusion was found to be over a factor of five larger than that of
single Re adatoms. The enhanced dimer diffusion was attributed[10] to an
incompatibility between the equilibrium distance between the adsorbed
dimer atoms and the distance between the cross-channel minima in the
adsorption potential surface dictated by the geometrical structure of
the substrate. As a consequence the adsorbed dimer is never found in
its natural equilibrium state and this accounts for a decrease in the
energy needed to bring the system to a saddle region of the potential
surface which may result in a migration event.

In our study we wish to focus on dynamical aspects of dimer
diffusion and therefore have chosen the parameters of our system such
as to eliminate the geometrical incompatibility. Employing generic
6-12 Lennard-Jones potentials whose parameters (σ and ε) were fitted[13]
to materials properties we found that for a system of a lead dimer
adsorbed in a cross-channeled configuration on the (110) face of
copper, no such geometrical mismatch exists, (see potential energy
curves in Figure 1). In order to achieve a faithful simulation of the
substrate our simulations were performed on a thick slab consisting of
14 layers (with 70 atoms in each layer) exposing the (110) surface of
the fcc copper substrate. The numerical integration of the equations
of motion was performed using Gears' predictor-corrector algorithm
with an integration time step $\Delta t = 0.0075\ t_{Cu}$, where $t_{Cu} =$
$(M_{Cu}\sigma_{Cu}^2/\varepsilon_{Cu}) = 2.9668 \times 10^{-13}$ sec (in the following reduced units
expressed in terms of the Cu-Cu parameters[14] are used, i.e., distance
and energy in terms of σ_{Cu} and ε_{Cu}, respectively).

Following equilibration we have performed extensive studies of
dimer and single particle diffusion at various temperatures in which
we found enhanced dimer diffusion rates (factor of 2-3) compared to the
single particle diffusion rates. Since "geometrical mismatch" is
absent in our system (by construction) we have set to investigate the
origin of the observed enhancement (which as we will demonstrate in the
following is dynamical in nature).

In Figure 2(a-c) sample trajectories of the system evolution over
a time span of $5 \times 10^4\ \Delta t$, for a cross-channel adsorbed dimer, calculated at reduced system temperature $T^* = 0.2$ (corresponding to 2/7 of
the substrate melting temperature), are shown. In these figures the
crosses correspond to time averaged positions of the top layer substrate
atoms and the dots are the actual instantaneous locations of the atoms
comprising the adsorbed dimer as a function of time. A plot of the

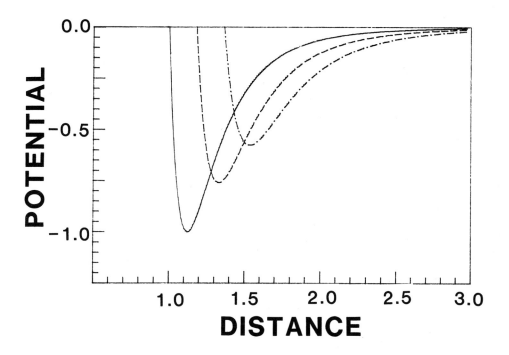

Figure 1.   Interatomic interaction pair-potentials in units of ε_{Cu},
            distance in units of σ_{Cu}. Solid line Cu-Cu, dashed line
            Cu-Pb and dash-dotted line corresponding to Pb-Pb potential.
            The minima in the Pb-Pb potential is at a distance equal to
            the cross-channel distance in the (110) face of Cu.

Pb-Pb bond length as a function of time shown in Figure (2d) allows a
clear identification of the jump events. We observe that at t = 0
the dimer started at a vertical configuration (bond length $\sim$ 1.5 σ_{Cu}).
At t $\simeq$ 4 x 10^3 Δt the bottom particle migrated to a neighboring site in
the channel, returning to the original site at t $\simeq$ 8 x 10^3 Δt, and so
on. Similar information is contained in Figure (2f) were the time
evolution of the angle ϕ (in radians) the inplane angle between the
dimer axis and the positive x direction, along the channel is shown,
along with the corresponding dimer configurations. A plot of the
angle θ, between the dimer axis and the normal to the surface, shows
that the orientation of the dimer axis remained almost parallel to the
surface plane throughout the course of the sample simulation (see
Figure (2e)).
    To investigate details of the migration mechanism we focus on the
time interval indicated by arrows on the time axis of Fig. (2f), which
spans the time between two jump events culminating in the migration of
the bottom adparticle (see Fig. (2c)). A close-up view of the variation in the inplane angle ϕ with time is shown in Fig. (3a). We

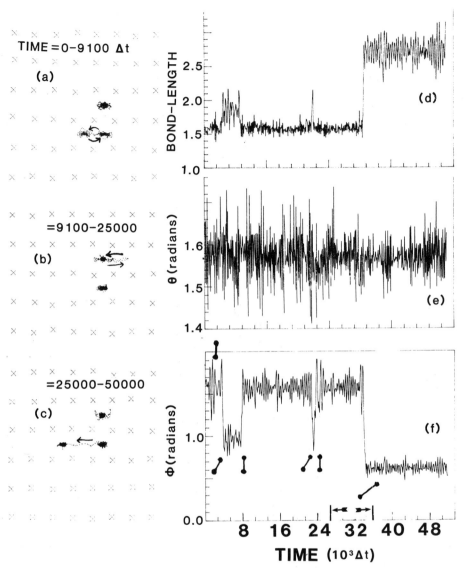

Figure 2. (a–c): Sample trajectories of the system evolution over a time
span of 5×10^4 Δt, where $\Delta t = 0.0075$ $t_{Cu} = 2.25 \times 10^{-15}$ sec.
Crosses indicate averaged positions of the first layer sub-
strate atoms and dots are the instantaneous locations of the
dimer atoms. The system was equilibrated with the dimer at
a cross channel configuration along the Y direction. (d–f):
Time evolution of the dimer bond length, out-of-plane angle
(θ) between the dimer axis and the normal to the surface,(z)
and in-plane angle between the dimer axis and the positive X
direction, respectively. Bond length in units of σ_{Cu}.
Corresponding dimer configurations are shown in (f).

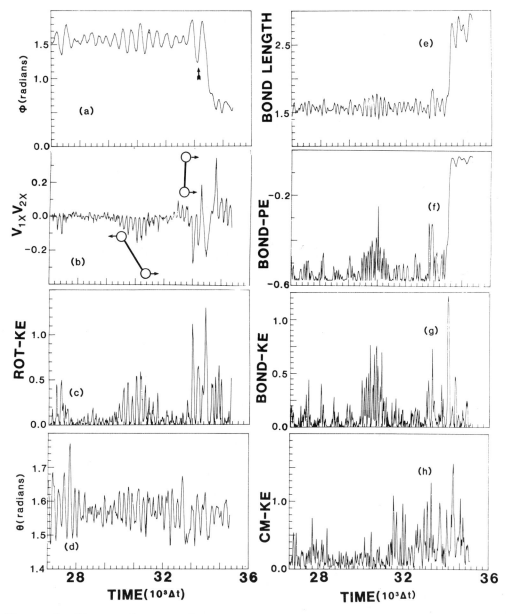

Figure 3. Time evolution of the angles ϕ and θ (a,d), X component of
the equal time velocity correlation function $V_{1x}(t)V_{2x}(t)$ in
(b), dimer rotational energy (c), dimer bond length (e), bond
potential and kinetic energies (f and g) and dimer center of
mass kinetic energy. Energy in units of ε_{Cu}, and length in
units of σ_{Cu}.

observe a regular oscillatory variation about the value $\pi/2$ (with frequency $\sim 1 \times 10^{12}$ sec^{-1}) in time. The amplitude of the dimer libration is modulated, achieving a maximum and decreasing prior to the jump event (marked by an arrow, compare Figs. (2d) and (2f)). In order to interrogate further the type of motion which the dimer is engaged in we show in Fig. (3b) the equal-time correlation of the x-components of the velocities of the dimer atoms, $V_{1x}(t) V_{2x}(t)$. Librational motion in which the two particles move in opposite directions corresponds to a negative value of this correlation function, while a rocking mode in which both particles oscillate in phase yields a positive value. We note that during the increase in the ϕ-oscillation amplitude $V_{1x}V_{2x}$ (as well as $V_{1y}V_{2y}$, see Fig. (4c)) takes the negative values, while during the decrease in ϕ prior to the jump event a reversal in sign occurs, indicating a change in the nature of the dimer mode of motion. This conclusion is corroborated by the rotational kinetic energy shown in Fig. (3c) which exhibits a similar sequence. We note also that the same pattern occurs in the time evolution of the adparticle-adparticle bond length (Fig. 3e)), bond potential energy (Fig. (3f)) and bond kinetic energy (Fig. (3g)). These observations indicate clearly that the librational, out-of-phase, motion of the dimer, upon decrease prior to the jump event, exchanged energy with a mode characterized by an in-phase motion of the dimer atoms which did not involve significant vibrations or stretch of the dimer bond. An additional clue to the nature of the pre-jump mode is provided by a plot of the kinetic energy of the dimer center-of-mass motion as a function of time, given in Fig. (3h), which exhibits low values while the dimer is librating, and increases in magnitude in coincidence with the noted decrease in the librational motion. Further information concerning the dimer motion is given in Fig. (3d) and Figs. (4c) and (4d) where the time development of the out-of-plane angle, θ, and the correlation functions $V_{1y}(t)V_{2y}(t)$ and $V_{1z}(t)V_{2z}(t)$ are shown, respectively. The oscillations in θ are of a much higher frequency than the in-plane ϕ librational mode and the out-of-plane velocity correlation (Fig. (4d)) does not show a clear pattern. In fact the out-of-plane motion appears rather irregular, while coupled to the in-plane degrees of freedom. Inspection of the time sequences for the dimer particles kinetic energy (Fig. (4a)) and dimer total energy (kinetic plus inter-dimer and dimer-substrate potential energies, Fig. (4b)) reveals that the energy content of the dimer increased pursuant to the excitation of the librational mode and remained so until the eventual jump event.

From the above observations we conclude that the evolution of the system leading to the reactive (jump) event is characterized by distinct, though coupled, modes which exchange energy with the substrate and among themselves. The in-phase, rocking mode, which is characterized by a high-degree of center of mass motion is coupled directly to the reaction-coordinate (jump event). This mode is excited by the librational mode which preceeds it, whose excitation is caused by coupling to the substrate vibrations and perhaps partially via high-frequency out-of-plane modes (and possibly dimer vibrations), which are orthogonal to the reaction coordinate. The librational mode is therefore identified as the "door-way state" for the reaction.

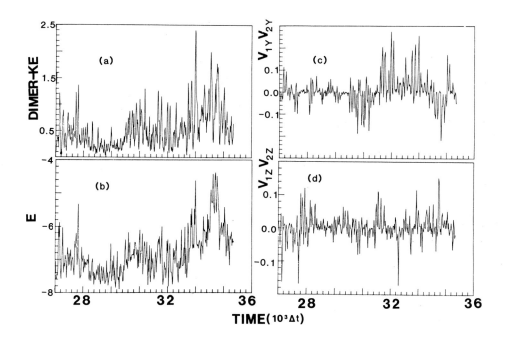

Figure 4. Time evolutions of dimer kinetic energy (a), total energy
of the dimer (kinetic energy plus adparticle-adparticle and
dimer-substrate potential energies) in (b), and equal-time
correlation of functions of the velocity components,
$V_{1y}(t)V_{2y}(t)$ in (c) and $V_{1z}(t)V_{2z}(t)$ in (d). Energies in
units of ε_{Cu}, and $\Delta t = 0.0075\ t_{Cu}$.

It is intuitively obvious and rigorously known in the theory of
nonlinear oscillators and parametric modulations that the excitation
dynamics in such systems depends critically on the frequencies and
coupling parameters of the system. To further investigate the activation dynamics and energy pathways in our system we calculated the
density of vibrational states (Eq. 1a) and velocity autocorrelation
functions (Eq. 1b), for our system, defined by

$$\Phi_{A\alpha}(\omega) = \int \Phi_{A\alpha}(t)\ \cos(\omega t)dt, \quad \alpha = X,\ Y,\ Z \qquad (1a)$$

$$\Phi_{A\alpha}(t) = <A_{\alpha}(0)A_{\alpha}(t)>/<A_{\alpha}(0)A_{\alpha}(0)> \qquad (1b)$$

where A is a dynamical variable, and the angular brackets denote
averaging over time origins and particles (when applicable). The
density of vibrational states for the X, Y and Z components of the

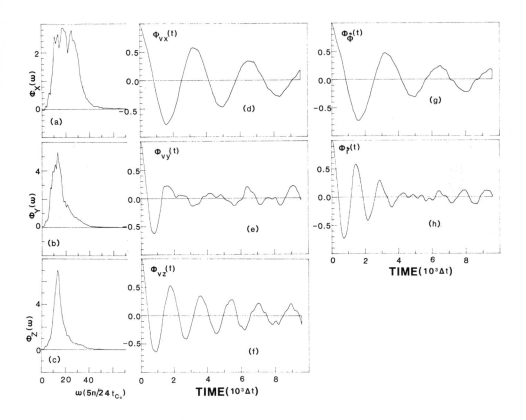

Figure 5.   (a-c): Cosine transforms ($\Phi_\alpha(\omega)$, α = X,Y,Z) of the velocity
            autocorrelation function for first-layer substrate particles,
            as a function of ω expressed in units of $(5\pi/24)$ $t_{Cu}^{-1}{\approx}2.2\times10^{12}$
            sec. Note anisotropy in the vibrational spectra which
            disappears as one moves into the substrate slab. (d-f):
            Velocity autocorrelation functions ($\Phi_{V\alpha}(t) \equiv \langle V_\alpha(0)V_\alpha(t)\rangle/$
            $\langle V_\alpha(0)V_\alpha(0)\rangle$, α = X,Y,Z) of dimer particles. Data for these
            correlation functions and those shown in (g) and (h) was
            accumulated in intervals between jump events. Note the
            difference in frequency between the X and Z modes and the
            fast damping of the Y motion. (g) Autocorrelation function
            of the angular rate of change, $\Phi_{\dot\phi}(t) \equiv \langle\dot\phi(0)\dot\phi(t)\rangle/\langle\dot\phi(0)\dot\phi(0)\rangle$.
            (h) Autocorrelation function of the dimer bond, r, rate of
            change, $\Phi_{\dot r}(t) \equiv \langle\dot r(0)\dot r(t)\rangle/\langle\dot r(0)\dot r(0)\rangle$. Note the higher
            frequency compared to the one shown in (g).

vibrations of the top-layer atoms of the substrate slab are shown in
Figs. (5a-5c), respectively. We observe a marked anisotropy in the
vibrational spectrum, which disappears when calculated for deep layers
in the slab. This emphasizes the importance of employing sufficiently
extended system in simulations in order to faithfully represent the

system. Components of the velocity autocorrelation functions for
motions of the adsorbed dimer atoms, where data has been gathered
between jump events, are shown in Figs. (5d-5f). We observe that
motion along the surface channel, X (Fig. (5d)), is of lower frequency
than that in the normal direction, Z (Fig. (5f)). In addition the
velocity autocorrelation in the Y direction (Fig. (5e)) is damped
rapidly indicating that motion in this direction is coupled strongly to
the substrate, colliding inelastically with the substrate potential
(which rises steeply). Finally the velocity autocorrelations for the
rate of change of the inplane angle, ϕ, and dimer bond length, r, are
shown in Figs. (5g) and (5h), respectively. The frequency of the bond
vibrations is about double that of the angle variations. Analysis of
the dimer modes of motion and their frequencies suggests that the librational mode is excited via coupling to the substrate phonons, and to
the dimer rocking and bond-stretch (see Figs. (5f) and (5h)), subsequently channeling energy to the in-phase mode which leads to the jump
event. This sequence of activation and energy flow is governed by
frequency resonances of the dimer modes with the substrate phonons and
by anharmonic couplings. The observed enhancement of the dimer diffusion rate compared to that of single particles is dynamical in nature,
originating from the dimer additional degrees of freedom (bond stretch,
librations and rocking) which provide (resonance) channels to energy
flow from the substrate.

     While our focus in this paper has been surface diffusion, it is
obvious that the concepts which we introduced can be applied in a
straight-forward manner to other activated rate processes such as desorption and dissociation and perhaps even processes involving energy
dissipation such as sticking. Indeed, our present investigations
supplement our earlier[6] door-way state model for thermal desorption, in
which excitations of the vibrational ladders leading to bond-rupture
occurs via coupling of low-frequency non-stretch modes of the adsorbate
to substrate phonons. In that case the introduction of the low-frequency door-way modes was necessary due to the frequency mismatch between
the frequencies of transitions between vibrational levels of the bond-
rupture stretch mode and the maximum substrate phonon frequency[6] (see
Fig. 6).

Nonlinear Dynamics of Activated Rate Processes

In this section we treat activated rate processes, diffusion in particular, as a problem in nonlinear dynamics. The problem of motions
generated by the Hamiltonian

$$H(\underline{I},\underline{\theta},t) = H_o(\underline{I}) + \mu H_1(\underline{I},\underline{\theta},t) \; ; \; \mu \ll 1 \tag{2}$$

where H_o is an integrable Hamiltonian depending on the actions $\underline{I} =$
$(I_1,\ldots,I_N)$ and μH_1 is a perturbation periodic in the angle variables
$\underline{\theta} = (\theta_1,\ldots,\theta_N)$ and in time t, was considered by Poincare in 1892 as
"the fundamental problem of dynamics".[15] Studies of this problem have
since led to significant developments in applied mathematics and in
many areas of physics ranging from celestial mechanics and analysis of

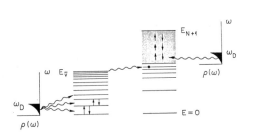

Figure 6. Schematic picture of the door-way state model for thermal surface desorption or dissociation reaction mechanism. A characteristic Debye phonon density of states, $\rho(\omega)$, is shown on the left and right. Excitation of a low-frequency, door-way, mode of vibration (typically a nonstretching mode) occurs via an incoherent multiphonon mechanism. Upon achieving the level $E_{\overline{v}}$ the excitation is transferred to the high-lying levels of a stretching mode (or combination of such modes) via anharmonic coupling. Further excitation in the dense vibrational manifold corresponding to the bond-rupture coordinate can occur via direct incoherent multiphonon excitations induced by thermal coupling to the substrate. The predissociation level is denoted by E_{N+1}. Having achieved this level the reaction proceeds through coupling to possible final state channels, such as dissociation, desorption or migration.

colliding beam devices to Josephson junctions and chemical dynamics[16]. It is well established, that under certain conditions, the behavior of classical systems may exhibit statistical behavior up to a limiting true randomness[16,17], where a statistical description applies (for both closed, conservative, systems and those exhibiting unbounded diffusion in phase space under the influence of an external regular perturbation). Consider a system described by the Hamiltonian

$$H = \frac{p^2}{2M} + f(q + \lambda \sin\Omega t),\qquad(3)$$

where M is the mass, Ω is a modulation frequency, and f is an arbitrary function. Applying to Eq. (3) two canonical transformations, in sequence, with generating functions $F^{(1)}$ and $F^{(2)}$ given by $F^{(1)} = (I - M\lambda\Omega\cos\Omega t)(q + \lambda\sin\Omega t)$, $(z = \partial F/\partial I$ and $P = \partial F/\partial q)$, and $F^{(2)} = Pz + M\lambda^2\Omega\sin\Omega t/2$, $(I = \partial F/\partial z = P$ and $q = \partial F/\partial P = z)$, $(F^{(1)}$ effects a transformation to an oscillatory coordinate frame), we obtain

$$H = \frac{p^2}{2M} + f(q) + qM\lambda\Omega^2\sin\Omega t.\qquad(4)$$

We model now the one-dimensional motion of a particle on a substrate lattice whose potential is represented by a periodic in space potential, by specifying f in Eq. (3) as

$$H = \frac{p^2}{2M} - \frac{V_B}{2}\cos(\frac{2\pi}{a}(x - x_o(t))\qquad(5)$$

where a is the substrate unit cell dimension, V_B is the potential barrier for migration on the lattice and $x_o(t)$ is given by $\sum_i \lambda_i \sin\Omega_i t$, where the λ_i and Ω_i are chosen to model the spectrum of vibrations of the substrate. The above Hamiltonian describes the motion of a particle dynamically modulated by the substrate vibrations. Transforming the Hamiltonian as in Eq. (4) yields a most appealing picture of the dynamics of the system. The time modulation (last term in Eq. (4)) which is linear in the coordinate tilts the spatially periodic cosine potential up and down. Consequently, to each set of modulation amplitudes and frequencies there corresponds a critical energy of the particle (determined by the initial conditions in Eq. (5), with $x_0(t) = 0$), for which upon modulation we may reach an unstable equilibrium (separatrix motion for a pendulum, see refs. 17(a,c)), causing the appearance of a stochastic layer (breaking-up of the KAM tori, in the jargon of nonlinear dynamics) which leads to migration (diffusion) of the particle. We illustrate the above, by choosing the parameters in Eq. (5) to correspond to the channeled (1D) motion of a Pb particle on a Cu(110) surface (expressing energy, mass, distance and time in units of ε_{Cu}, M_{Cu}, σ_{Cu} ($a/\sigma_{Cu} \sim 1.1$) and t_{Cu}, respectively). The barrier for diffusion in the x direction was determined via a relaxation calculation on our molecular dynamics sample to be 1 ε_{Cu}. We show results for individual modulation frequencies selected from the spectrum of substrate vibrations (see Figs. 5(a-c)) and the amplitude of modulation $(2\pi/a)\lambda$ was taken as 0.5 to correspond to the observed (in the simulation) typical rms vibrational amplitude of the substrate atoms. In Figs. (7a-7d) we show surfaces of sections (the intersections of the particle trajectory obtained via integration of Hamilton's equations of motion with the p-x plane at times $n(2\pi/\Omega)$, for different modulation frequencies. For $\Omega = 4\ t_{Cu}^{-1}$ and initial conditions which correspond to an unperturbed energy of -0.4206 (near the bottom of the well which lies at -0.5) we observe a regular motion with the particle oscillating in the well, see Fig. (7a) and the corresponding time record of the particle position shown in Fig. (7e), (the maxima of the potential at $\pm a/2\sigma_{Cu} = \pm 0.55$). For an initial energy of -0.0427875 (only slightly above the critical energy) we observe a dense chaotic region (Fig. (7b)) and a migration of the particle (Fig. (7f)). The dependence of the critical initial energy on the system parameters is demonstrated in Figs. (7c) and (7d) for $\Omega = 5\ t_{Cu}^{-1}$ and an initial energy of -0.20119179 and $\Omega = 10\ t_{Cu}^{-1}$ and the same initial energy as in Fig. (7b). It is seen that stochastic behavior is obtained for $\Omega = 5$ for a lower starting critical energy than for $\Omega = 4$, and that the high-frequency modulation is ineffective in producing stochastic behavior, all of which are characteristics of a resonance phenomenon. The data obtained via the integration can be used to calculate diffusion coefficients employing well-known techniques[17a,c 18]. It is of importance to note that the Hamiltonians given in Eqs. (3-5) possess an infinite sequence (multiplet) of evenly spaced primary resonances, with the amplitude of nth resonance given by $J_n(\lambda)$, where J_n is a Bessel function of the first kind. Global stochasticity, and the associated diffusion rates, depend upon resonance overlap criteria which provide analytical estimates of the dependence of the degree of stochasticity on system parameters. It can be shown that low-frequency modulations are more effective in bringing about resonance overlap[17a,c,16].

We may now consider a more complex situation in which we add to the Hamiltonian (Eq. 5) other degrees of freedom such as motions of the

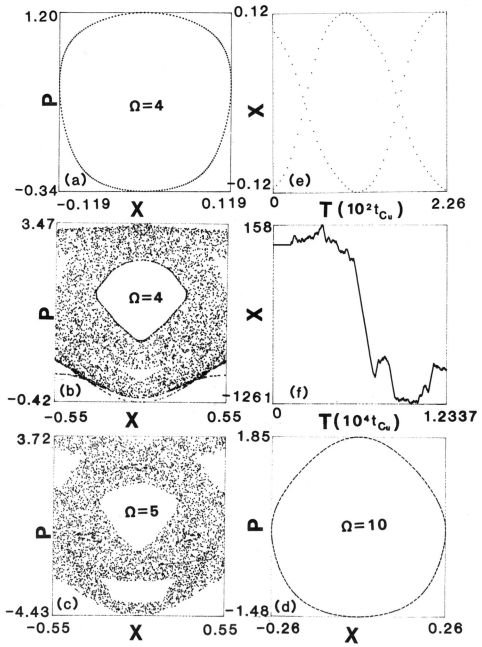

Figure 7. (a–d): Surfaces of section corresponding to the Hamiltonian
in Eq. (5) all quantities are in Cu units, and for modulation frequencies, Ω, as indicated; (e,f) Time records of
the particle position vs. time corresponding to the trajectories in (a) and (d), respectively.

particle in other coordinate directions subject to the appropriate
potentials or another particle (making a dimer), and coupling terms
between the various degrees of freedom, all of which can be modulated.
Thus we have studied: (i) a single particle which in addition to the
x motion given in Eq. (5) is allowed to move with a time modulation in
the y direction subject to a potential which presents a higher barrier
to migration in that direction, and the motions in the two directions
being coupled, and (ii) two particles whose individual Hamiltonians
are as given in Eq. (5), coupled via a Lennard-Jones potential, where
the coupling parameter may also be time modulated. Analysis of these
complex systems[18] is beyond the scope of this article and thus we give
only a schematic description of the mechanism. The presence of the
additional degrees of freedom and their coupling to the "reaction
coordinate degree of freedom" (called x-motion in our discussion) introduces "driving resonances" which "pump" stochastically the reaction
coordinate. In order to achieve effective transfer of stochasticity
into the reaction coordinate motion it is obviously necessary that the
pumping modes themselves experience stochastic motion, i.e., obey the
resonance overlap condition, which as mentioned above, favors low modulation frequencies. It is thus possible to start the system at a low-
energy content (below the critical initial energy) and let the reaction
coordinate motion diffuse in frequency space towards the critical frequency, at which time the reaction coordinate itself experiences stochastic, diffusive, motion leading to reaction. The rate is often limited
by the "stochastic pumping". Numerical experiments verify this description and the dynamical enhancement of dimer diffusion, thus corroborat-
ing the conclusions reached from our simulation studies.

We gratefully acknowledge the help of F. Vivaldi and E. Kuster.
Work supported by the U. S. DOE Contract No. EG-S-05-5489.

REFERENCES

1. K. J. Laidler, Reaction Kinetics, Vols. 1,2 (Pergamon, N.Y., 1963).
2. D. Menzel in Topics in Applied Physics, V.4, (Springer, Berlin, 1975).
3. C. F. Melius et al., J. Vac. Sci. Technol. 20, 559 (1982).
4. J. C. Tully, Ann. Rev. Phys. Chem. 31, 319 (1980).
5. U. Landman, Israel J. Chem. 22, 339 (1982).
6. G. S. De, U. Landman, M. Rasolt, Phys. Rev. B 21, 3256 (1980).
7. J. C. Polanyi, Acc. Chem. Research 5, 161 (1972).
8. H. A. Kramers, Physica 7, 284 (1940).
9. B. Carmeli and A. Nitzan, Chem. Phys. Lett. (1984) and refs. therein.
10. G. Ehrlich and K. Stolt, Ann. Rev. Phys. Chem. 31, 603 (1980).
11. G. De Lorenzi, J. Jacucci and V. Pontikis, Surf. Sci. 116, 391 (1982).
12. M. F. Shlesinger and U. Landman, Applied Stochastic Processes, G.
    Adomian, ed. (Academic, New York, 1980).
13. T. Halicioglu and G. M. Pound, Phys. Stat. Sol. (a) 30, 619 (1975).
14. σ_{Cu}=2.338Å, σ_{Pb}=3.197Å, ε_{Cu}=4750K, ε_{Pb}=2743K; σ_{Cu-Pb}=(ε_{Cu}+ε_{Pb})/2,
    ε_{Cu-Pb} = ($\varepsilon_{Cu}\varepsilon_{Pb}$)$^{1/2}$; after reference 13.
15. H. Poincare, "Les Methodes Nouvelles de la Mecanique Celeste", (1892)

16. A. J. Lichtenberg and M. A. Lieberman, "Regular and Stochastic
    Motion", Springer, N.Y. (1983).
17. (a) B. V. Chirikov, Physics Reports 52, 263 (1979); (b) J. Ford, AIP
    Conf. Proc. No. 46, (1978); (c) F. Vivaldi, Rev. Mod. Phys. (1984).
18. U. Landman and R. H. Rast (Phys. Rev., to be published).

COLLISION INDUCED DISSOCIATION OF MOLECULAR IODINE ON SINGLE
CRYSTAL SURFACES

E. Kolodney and A. Amirav
Department of Chemistry
Tel-Aviv University
69978 Tel-Aviv, Israel

ABSTRACT

Iodine molecules have been aerodynamically accelerated in the energy
range of 1-10 eV and collided with single crystals sapphire (0001) and
MgO (100) surfaces. The dissociation probabilities have been measured
and compared with R. Elber's and R.B. Gerber's theoretical prediction
emerging from classical trajectory calculation. The surface temperature
effect as well as the energy transfer mechanism were studied. An unexpectedly large iodine-MgO energy transfer was observed. The comparison
with theory resulted in a detailed dynamical insight. The dissociation
and the large molecular energy loss are described as a single collision
event where the molecule is highly rotationally excited and then either
rotationally dissociates or loses energy to the MgO surface by creation of a travelling compression wave in the MgO.

INTRODUCTION

Molecular dissociation induced by a high kinetic energy collision with
a solid surface is a process of fundamental importance and interest
[1-4]. We have addressed ourselves to studying the problem of dissociation dynamics of a diatomic molecule colliding with a solid surface in
order to obtain a better understanding and insight into the various
aspects of high kinetic energy reactive gas surface collisional encounters. In order to get a coherent picture, the following questions are
being asked:

a) What is the dissociation probability in a single high kinetic energy
   collision (1-10 eV) and what is its energy dependence.
b) What is the importance of the surface identity (or orientation).
c) How does the surface temperature affect the dissociation process.
d) What are the angular distributions of both the scattered molecules
   and atoms
e) What are the energy transfer processes involved in this reactive
   gas-surface collision.

B. Pullman et al. (eds.), Dynamics on Surfaces, 231–242.
© *1984 by D. Reidel Publishing Company.*

These questions are addressed in this paper. We have chosen iodine
since its large molecular weight and low dissociation energy made it a
good candidate both experimentally [4,5] and theoretically [3,5,6].

EXPERIMENTAL

The iodine molecules were seeded in a hydrogen or helium supersonic
beam and accelerated to a high kinetic energy in the range of 1-10 eV
[7]. The beam was then skimmed and collimated through two differential
pumping chambers into an Ultra High Vacuum chamber (UHV) (base pressure
5×10^{-10} torr). The accelerated beam was either square-wave modulated
for phase sensitive detection or mechanically chopped for kinetic
energy measurements using Time of Flight (TOF) technique. In the UHV
chamber the beam collided with a single crystal MgO (100) slab either
in its <100> or <110> azimuth, or with a single crystal sapphire (0001).
The MgO and sapphire surfaces were commercially available (Adolf-Meller
and Union Carbide) and X-ray analyzed. They were mounted on an UHV
manipulator and annealed. The MgO was annealed at about 900°C to give
highly reproducible sharp specular and first order diffraction peaks.
A detailed study of the energy dependence of helium and hydrogen diffraction from MgO will be published elsewhere [8] . The sapphire surface
annealing required a higher temperature of ∿1100°C and resulted in a
doubly modulated diffraction pattern of helium or hydrogen. During the
I_2 scattering the MgO and sapphire were held at 275°C or 650°C respectively, ensuring a clean and unadsorbed surface as determined from He or
H_2 diffraction. Two quadrupole Mass Spectrometer (QMS) heads were used
for detection. One was in line with the direct molecular beam and
served for kinetic energy measurements [7]. The second was at 45° to
the beam direction and mounted either 2.5 cm from the surface for high
solid angle integration or remotely from the surface (37 cm) for higher
angular resolution and scattered beam TOF experiments.

EXPERIMENTAL RESULTS

In Fig. 1 we show some experimental results demonstrating iodine dissociation following a single high kinetic energy collision with a
sapphire surface [4]. Fig. 1a shows the QMS signal vs. the beam-surface
angle, at a fixed beam detector angle of 45°. The two I and I_2 curves
are identical in shape except for a constant electron energy dependent
peak height ratio of two, indicating no collision induced dissociation.
In Fig. 1(b) however, the kinetic energy is increased to 9.8 eV and
iodine dissociation is evident. The ratio of atomic to molecular iodine
is increased by more than a factor of 2 as compared to Fig. 1(a). In
spite of the low resolution of ±8°, the atomic iodine angular distribution is broader than the molecular one. The two curves are centered
near the specular reflection angle indicating a single collision condition without adsorption to the surface.

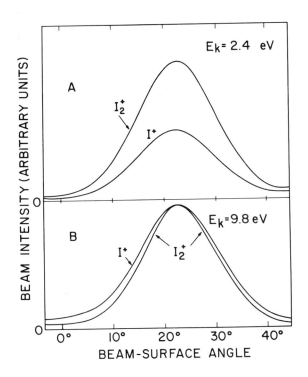

Figure 1: I_2-sapphire. The molecular and atomic iodine signal at the
QMS detector vs the surface-beam axis angle. The QMS detector is fixed at 45° to the beam axis while the sapphire
surface was rotating. The sapphire was held at 650°C. The
30μ nozzle temperature was 75°C corresponding to 12 Torr
iodine partial pressure. The hydrogen nozzle backing
pressure was in Fig. 1(a) 700 Torr resulting in 2.4 eV
molecular iodine kinetic energy, while in Fig. 1(b) it was
13,500 Torr resulting in 9.8 eV molecular iodine kinetic
energy. The QMS ionizing electron energy was 50 eV.

    The major problem in the quantification of results such as in Fig.1
both for I_2 on sapphire and on MgO was the decoupling of the surface
induced dissociation from the electron induced dissociation in the ionizer of the QMS. If one assumes that the electron induced dissociation
does not (or only slightly) depends on the iodine internal energy, then
the calculations are straightforward. We have carefully checked this
issue both for I_2 on sapphire and on MgO and did not find any internal
energy effect on the electron induced dissociation. The following
theoretical and experimental aspects are pertinent to this problem.

a) The dissociation energy of I_2^+ ion is 60.1 kcal [9], much larger than that of the neutral I_2 molecule which is 36.1 kcal.

b) We believe rotational excitation to be dominant in the dissociation mechanism [6]. Rotational excitation by electron impact usually obeys $\Delta J = \pm 1$ selection rules [10]. (Small deviation at 100–25 eV).

c) The dissociation probabilities calculated (assuming decoupling) using electron energies of 25, 50 and 70 eV were the same within our uncertainty range. Most of the data was collected using a reduced electron energy of 25 eV.

d) The angular distributions of the scattered molecular I_2 were exactly the same at 25 and 70 eV electron energy. Since we know from TOF measurements that the molecular energy loss is larger outside the specular reflection, this strongly suggests an internal energy independent electron induced dissociation.

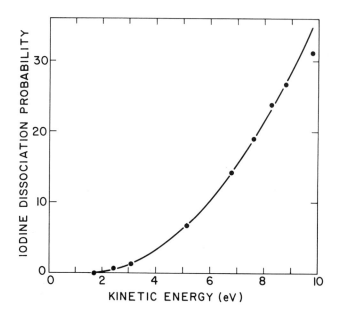

Figure 2. Molecular iodine dissociation probability after a single collision with a sapphire single crystal vs. its kinetic energy. Iodine is expanded through a 30μ nozzle at 75°C seeded with hydrogen. The full points are experimental results and the solid line is given by $P = 0.514 \ (E_k - 1.54)^2$. P is the dissociation probability and E_k is the molecular iodine kinetic energy. The uncertainty in the absolute values is ±35% of the probabilities. The uncertainty in the relative probabilities is about ±10% of their value.

The extraction of the dissociation probability is therefore calculated assuming energy independent electron induced dissociation, from
the angle integrated I^+ and I_2^+ signals. This was measured when the QMS
head was close (2.5cm) to the surface, for ±8 out of plane integration.
The absolute dissociation probabilities were measured for I_2 on MgO
using the surface temperature effect to calibrate the QMS relative
sensitivity to I^+ and I_2^+ ions. Raising the surface temperature resulted in reduced I_2^+ signal and correspondingly increased I^+ signal. In
the case of I_2 dissociation on sapphire, no surface temperature effect
was detected, therefore calibration was performed by a comparison
between the molecular iodine signal vs. kinetic energy probed in the
direct beam (without collision with the surface), and that of the
scattered beam, integrated over the surface angle and normalized to low
kinetic energy (where no dissociation occurrs). The details of the
calculation will be published elsewhere.

Figure 2 summarizes the iodine-sapphire dissociation probability
dependence on the molecular kinetic energy. Figure 2 shows the dissociation threshold to be below 2.5 eV and the probabilities rise up to
31% at 9.8 eV. The threshold behavior can be fitted to a simple formula
$P=K(E_k-E_D)^2$. E_D is the iodine bond energy (1.54 eV), E_k is the kinetic
energy, and the free parameter K was fitted to K=0.514.

In Figure 3, we show some experimental results demonstrating iodine
dissociation and angular scattering following a single high kinetic
energy (indicated in Fig. 3) collision with an MgO surface. The results
exhibit the following features: a) the angular scattering of I_2 molecules is centered near the specular reflection angle. b) The width of
the scattering angular distribution increases with increased kinetic
energy. This aspect is qualitatively different in comparison with
iodine scattering and dissociation on sapphire, where the angular distribution was narrower and reduced itself with increased kinetic energy.
c) Molecular dissociation is evident from the substantially increased
ratio of I^+ signal to I_2^+ signal at the higher kinetic energy and also
from the slight asymmetry in the I^+ curve as compared with the I_2^+
curve in Fig. 1b.

In Fig. 4, both the theoretical and experimental dissociation probabilities are plotted against the incident kinetic energy. The main
features are a fast probability rise with a maximum slope at 4 eV and
approaching saturation at higher energies. The match between the trajectory calculation and experimental results is appealing. Two aspects
should be noted: a) The relative experimental probabilities are accurate within 15%, but the uncertainty in the absolute value is ∿ 40%
for P/(1-P) calibrated at 4.3 eV. b) The threshold for dissociation is
at 2.9 eV for the calculation without invoking velocity distribution,
and as shown in Fig. 2 with 15% width for Gaussian velocity distribution. The experimental values do not drop to zero because of surface
temperature effect which is not treated by the theoretical calculation.

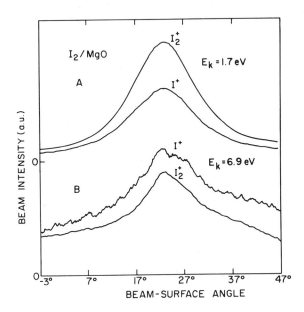

Figure 3. I_2–MgO. The molecular and atomic iodine signal at the QMS
detectors vs. the surface-beam axis angle. The QMS detector
is fixed at 45° to the beam axis while the MgO surface was
rotating. The MgO was held at 275°C. The 30μ nozzle temperature was 75°C corresponding to 12 Torr iodine partial
pressure. The hydrogen nozzle backing pressure was in Fig.
1(a) 500 Torr resulting in 1.7 eV molecular iodine kinetic
energy   while in Fig. 1(b) it was 3200 Torr resulting in
6.9 eV molecular iodine kinetic energy. The QMS ionizing
electron energy was 25 eV and it was 37 cm from the surface.

    A quite interesting feature is the qualitative   difference between
the dissociation probabilities of I_2 on sapphire and MgO. In the former
case it could be fitted to $P=K(E_k-E_D)^2$, while on MgO the dissociation
probabilities are higher, with different curvature and in good agreement with the theory. This difference probably represents some change
in the dissociation mechanism stressing the role of surface identity
even in the case of high kinetic energy single collision dissociative
encounter. Additionally, we have found no surface orientation dependence for <100> and <110> azimuthes in MgO, and the dissociation probabilities were essentially azimuth independent.

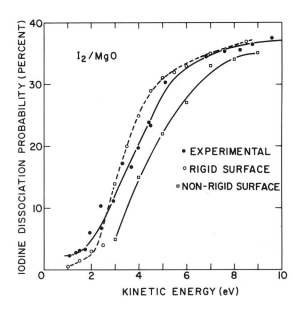

Figure 4. Molecular iodine dissociation probability after a single
collision with MgO surface vs. kinetic energy. MgO surface
temperature is 275°C, nozzle diameter is 30μ and nozzle temperature is 75°C. Both hydrogen and helium (full circles ●●●)
are used as carrier gases. The upper dashed curve (- - -)
is the classical trajectory calculation results for rigid
surface with only little gas-surface energy transfer. The
lower curve is the trajectory calculation for a more realistic MgO surface which also exhibits a substantial gas surface
energy transfer.

Next we have considered the effect of surface temperature on dissociation probabilities. In the case of an I_2 dissociative collision on
sapphire the surface temperature had little or no effect. Using 9.8 eV
I_2 kinetic energy, we have found only up to 1.7% increase in the dissociation probability of I_2 on sapphire, in the temperature range 100-
1100°C. The large MgO surface temperature effect represents another
major difference in comparison with sapphire.

In Figure 4, where the surface temperature is 275°C, one can see
that even at the lowest kinetic energy used below the dissociation
threshold there is about 3-4% dissociation. The role of surface temperature is shown in Fig. 5, where the dissociation probabilities are
plotted vs. the surface temperature at three different kinetic energies
(indicated in Fig. 5). The picture emerging from Fig. 5 is that of a
cooperative effect where the dissociation probabilities largely depend
on the surface temperature coupled to the kinetic energy. This means

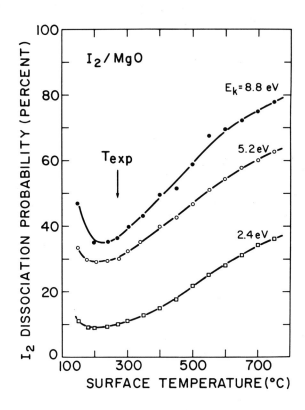

Figure 5. MgO surface temperature effect. The dissociation probabilities vs. the surface temperature in the range 150°C-800°C.
The real temperature is probably slightly higher than indicated. Probabilities are given for three different kinetic
energies indicated.

that at increased surface temperatures the kinetic energy is much more
efficient in promoting dissociation. In addition we have found that the
angular distribution of undissociated I_2 molecules is rather independent on the surface temperature while that of the atomic iodine is
somewhat broader. The low temperature side of Fig. 5 is due to I_2
adsorbed on the surface as evident from He diffraction and should be
disregarded. Probably the major reason to the difference between MgO
and sapphire resides on their different surface Debye temperature
which is significantly higher in sapphire.

At this point we note that the surface temperature of 275°C was
chosen in Fig. 4 as a compromise between a minimum surface temperature
effect and a clean and unabsorbed surface as determined by He and H_2
diffraction. The dissociation probabilities vs. kinetic energy can be
plotted from Figs. 4 and 5 for any surface temperature in this range

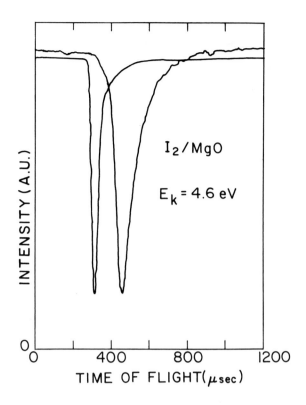

Figure 6: Time of Flight spectra of iodine with and without scatterring
from the MgO surface. The time resolution is 20 μsec. The
path length is 50 cm for the direct beam, 25 cm to the surface and 37 cm from the surface to the detector for the
scattered beam. Beam-surface angle is 22.5° which is the
specular angle to the QMS at 45° to the beam direction.
Surface temperature is 275°C. I_2 is seeding in hydrogen
using 30μ nozzle at 75°C. Hydrogen backing pressure is 1600
Torr and the measured kinetic energies are 4.6 eV incident
kinetic energy and 2.4 eV scattered kinetic energy.

of 275-750°C. The temperatures are probably slightly higher than indicated in Fig. 5 due to experimental difficulties in the measurements
of the exact surface temperature.

The most important question concerning the dissociation probabilities is the energy transfer mechanism involved. We have performed an
extensive study of the Time of Flight of scattered I_2 molecules from
MgO, and   surprisingly found that a very large amount of kinetic
energy is transferred from the molecule to the surface. A typical TOF
spectrum is shown in Fig. 6. The narrow trace is the direct unscattered

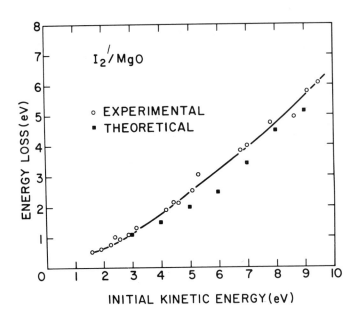

Figure 7: Energy loss of I_2 molecules after a high kinetic energy
          collision with MgO surface. Experimental (open circles ooo)
          and classical trajectory calculations (full circles ●●●).
          Each experimental point was taken from TOF spectra as in
          Fig. 6 at various helium or hydrogen nozzle backing pressure.
          The solid curve arbitrarily connects the experimental points.

beam and the wider and longer trace is that of the scattered beam at
the specular angle. The direct beam travelled 50 cm and the scattered
beam travelled 25 cm to the surface and then 37 cm from it to the
detector. The calculated peak kinetic energies (corrected for TOF as
ions in the QMS [7]) are 4.6 eV and 2.4 eV. It is important to note
that in spite of the large energy loss of 48%, the reflected beam is
definitely supersonic with a narrow velocity distribution. This is the
main proof for the single collision condition prevailing in our experiments (No adsorption). In addition we have also observed the following
features:

     a) The surface temperature had only a small effect on the amount
of energy transferred to the molecules that did not dissociate, although it significantly increased the dissociation probabilities.

     b) The degree of energy loss was somewhat larger at both sides of
the specular reflection.

     c) The reflection was supersonic over all the energy range with a
small energy dependent TOF width.

     Our study of the I_2 energetics is summarized in Fig. 7. The mole-
cular energy losses are plotted against the incident kinetic energy,

and compared with the classical trajectory calculation. It is seen that
above ∿5 eV incident energy, the molecules lose more than 50% of its
kinetic energy in a single collision! At 9.5 eV the molecule loses more
than 6 eV. The energy loss increases supralinearly with the initial
energy and correspondingly the final energy increases in a sublinear
way. This large energy transfer cannot be interpreted in terms of internal vibrational-rotational excitation since the molecule cannot hold
more than 1.6 eV rotational energy without dissociating. Electronic
excitation is ruled out since we have carefully checked for light
emission and found no collisional induced fluorescence with an estimated sensitivity of better than 10^{-6} of the colliding molecules. Thus,
we end up with an extremely large single collision energy transfer to
the MgO surface.

Discussion

From the experimental results described, and from the classical trajectory calculations and modelling of R. Elber and R.B. Gerber, a better
insight into the dissociation dynamics of I_2 on MgO surface is obtained. The iodine molecules approach the surface and collide with it. On
the first moment of collision with the surface, the molecule experien-
ced a tremendous torque acquiring a high degree of angular momentum
implying a large rotational excitation. The degree of vibrational
excitation is low [6], and rotational excitation above the dissociation
energy, leads to rotational induced dissociation. The molecule after
being rotationally excited is smashed into the surface and causes an
irreversible travelling compression wave, therefore losing a large
amount of kinetic energy. It is important to note that there are two
separable time scales. On the first step of the collision dynamics the
surface imparts a large torque on the molecule but very little energy
is transferred to the surface. Only then the molecule will either
rotationally dissociates or loses translational energy to the surface.
Rotationally excited molecules start dissociating before the second
smashing collision (which stops the I_2 center of mass movement) with
the surface. Therefore, as shown in Fig. 4, this secondary collision
effects only to a small degree the dissociation probabilities. The
theoretical prediction shown in Fig. 4 for the "Non rigid surface" are
a lower limit since it does not take into account the dissociation
initiated between the first impact with the surface and the secondary
"smashing" into it. Invoking this time separation and rotational induced
dissociation principles we feel that we can also rationalize the MgO
surface temperature effect. Going back to Fig. 4 one of the main features of both the experimental and theoretical results is the saturation of the dissociation probabilities at high kinetic energy. This is
another manifestation of the rotational involvement in the dissociation
mechanism. The impulsive torque exerted on the molecule by the surface
at the collision instant is strongly angle dependent. The available
angle space which leads to effective rotational excitation resulting in
dissociation is limited and easily saturated. If the surface is heated
this restriction is partially removed, large temporal corrugation is

created and atoms on the surface are moving fast on the initial collision time scale so that the molecule can be rotationally excited in any
angle. It is important to note that this rationalization can also
explain the fact that in spite of the large change in the dissociation
probabilities, the energy transferred to the surface is only slightly
effected since the movement of an atom on the surface is unlikely to
effect the creation of an irreversible travelling compression wave in
the MgO by the iodine collision.

The case of I_2 dissociation on sapphire is probably different in
several important aspects and we intend to continue studying it.

Acknowledgements

This work was done with full theoretical collaboration with R. Elber
and R.B. Gerber whom we thank for useful discussions throughout this
work. We would like to thank C. Horwitz for skillful technical assistance. This work was supported by the Petroleum Research Fund adminis-
tered by the American Chemical Society and the United States-Israel
Binational Science Foundation Grant No. 3209.

References

1. M.S. Connolly, E.F. Greene, C. Gupta, P. Marzuk, T.W. Morton,
   C. Parks and G. Staker, J. Phys. Chem. 85, 235 (1981).
2. G. Prada-Silva, D. Loppler, B.L. Halpern, G.L. Haller and J.B. Fenn,
   Surf. Sci. 83, 453 (1979).
3. R. Elber and R.B. Gerber, Chem. Phys. Lett. 4, 97 (1983).
4. E. Kolodney and A. Amirav, J. Chem. Phys. 79, 4648 (1983).
5. E. Kolodney, A. Amirav, R. Elber and R.B. Gerber, submitted to
   Chem. Phys. Lett.
6. R.B. Gerber and R. Elber, Chem. Phys. Lett. 102, 466 (1983).
7. E. Kolodney and A. Amirav, Chem. Phys. 82, 269 (1983).
8. E. Kolodney and A. Amirav, submitted to Surf. Science.
9. H.M. Rosenstock, K. Craxt, B.W. Steiner and J.T. Herron, J. Phys.
   and Chem. Reference Data 5, supplement 1 (1977).
10. B.M. DeKoven, D.H. Levy, H.H. Harris, B.R. Zogarski and T.A. Miller,
    J. Chem. Phys. 74, 5659 (1981).

ROTATIONALLY INELASTIC SCATTERING OF NITROGEN FROM Fe(111)

J. Levkoff, A. Robertson, Jr. and S.L. Bernasek
Department of Chemistry
Princeton University
Princeton, NJ 08544

Abstract

The electron beam-induced fluorescence technique has been used to measure the angular distribution and internal energy of rotationally cold (~ 19 K) nitrogen scattered from a clean Fe(111) surface. The scattering is found to be direct inelastic. Rotational state distributions are found to be non-Boltzmann with excess energy in higher rotational states. As the surface temperature is increased from 400 to 700K, the measured rotational temperature of the specular beam increases from 148 ± 30 to 172 ± 30 K. Rotational temperatures measured at several exit angles and incident beam energies show increasing temperatures as the exit angle moves toward the surface normal. These results are interpreted as evidence for a surface corrugation in the nitrogen molecule/iron interaction potential.

I. Introduction

The interaction of nitrogen on iron surfaces has been the subject of numerous investigations.[1-4] These have been prompted by a strong interest in understanding the catalytic synthesis of ammonia on iron catalysts. These studies have concluded that the rate limiting step in the ammonia synthesis is the dissociative adsorption of nitrogen, even though the apparent activation energy for the dissociation is very low or even slightly negative. The slow rate of dissociative adsorption results from a very low sticking coefficient into the molecular precursor state ($\sim 10^{-2}$) and a competition on the surface between desorption and dissociation ($k_{des}/k_{dis} \approx 10^4$) of the molecularly adsorbed nitrogen.

We have been prompted by this earlier work to attempt to obtain a better understanding of the dynamics of the iron-nitrogen interaction by studying the scattering of nitrogen molecules from the iron surface. The goal of these experiments is to derive information about the interaction potential which could provide insight into the dynamics of adsorption at the surface. Ultimately, we desire an understanding of the complex mechanism by which molecules become trapped on a surface.

B. Pullman et al. (eds.), Dynamics on Surfaces, 243–256.

Measurements of velocity and angular distributions of atoms scattered from surfaces have provided enough data to generate numerous theoretical treatments of the atom-surface interaction.[5] With the advancement of techniques to measure internal state distributions of molecular species, experiments have now been extended to the scattering of diatomics from surfaces. Measurements of the internal state distributions of NO scattered from Ag(111)[6,7] and Pt(111)[8,9] have shown that transfer of translational energy into molecular rotation may be an important channel for the loss of molecular translational energy leading to trapping at the surface. The effect of well depth on the measured rotational distribution has been discussed by Zare[7] for the case of NO on Ag(111). Direct inelastic scattering results in incomplete rotational accommodation to the surface, but under conditions where trapping becomes more important the rotational temperature of the NO molecules approaches that of the surface. Results are also available for HF[10] and CO[11] scattering from LiF. Once again rotational accommodation is found to be incomplete.

We report here measurements of the rotational temperature and angular distribution of a supersonic, rotationally cold (~ 19 K) beam of nitrogen molecules scattered from a clean Fe(111) surface. The electron beam induced fluorescence technique was used to measure the rotational temperature and number density of scattered nitrogen molecules. The effect of varying incident beam energy and surface temperature was investigated. Auger Electron Spectroscopy (AES) and Low Energy Electron Diffraction (LEED) were used to characterize the iron surface.

II. Experimental

The apparatus used to measure the rotational temperature of scattered nitrogen is depicted in Figure 1. It consisted of a high pressure beam source chamber pumped by a Varian VHS-6 diffusion pump, a differential pumping chamber pumped by a LN_2 trapped Varian M4 diffusion pump and a scattering chamber pumped by a LN_2 trapped Varian VHS-6 diffusion pump.

The beam was generated from a solenoid valve with an effective nozzle diameter of 0.1 mm. at a backing pressure of 60 atmospheres. Molecular beam pulses of approximately 800 μs FWHM were generated by a high voltage spike to the solenoid valve. The expansion was skimmed 3.5 cm from the source by an electroformed skimmer with a 0.5 mm diameter orifice. A 1.0 mm diameter opening located 12 cm from the skimmer collimated the beam before it entered the scattering chamber. The distance from skimmer to crystal was 24 cm and the resulting beam had a diameter of 1.0 mm at the crystal. The base pressure of the scattering chamber was $2x10^{-10}$ torr and rose to $5x10^{-9}$ torr with the nozzle valve operating at 20 Hz.

The translational energy of the incident beam was varied by seeding the nitrogen with helium. A value of 7/2 kT was used for the translational energy of the pure N_2 beam. Energies of the mixtures were calculated based on the mass average composition of the N_2/He mixtures, and an energy of 5/2 kT for the helium expansion.

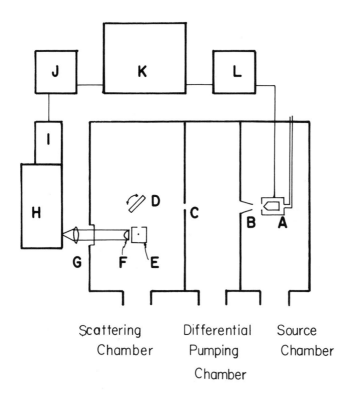

Figure 1. Schematic diagram of experimental apparatus:
A) pulsed nozzle, B) skimmer, C) collimating orifice,
D) sample, E) electron gun, F) light collection lens,
G) focusing lens, H) monochromator, I) PMT
J) amplifier/discriminator, K) signal averager
L) valve trigger

The measured rotational temperature of the incident beam varied from 14 to 19 K at the three beam energies. An Auger electron spectrometer allowed determination of surface impurity levels during a scattering experiment, and an argon ion gun mounted above the sample at a 45° angle to the crystal face facilitated cleaning of the surface.

An electron gun capable of producing 5 mA of current into a 2 mm diameter beam at 1000 eV was mounted perpendicular to the scattering

plane. The gun could be rotated to vary the angle of detection at the
crystal surface. A 2.4 mm collimating orifice located 2.3 cm from the
center of the crystal defined the acceptance angle of the electron
gun, giving an angular resolution of 6°. A lens mounted on the
electron gun collected the nitrogen fluorescence. The collimated
light passed through a 1" sapphire window and was focused by an f/5
biconvex lens into the entrance slit of an f/4.9, 0.6 meter Jobin-Yvon
monochromator equipped with a 2400 line/mm grating operated in second
order. Entrance and exit slit widths of 300 μ were chosen to give a
reasonable signal to noise ratio while providing moderate resolution
(2.0 Å FWHM) of the rotational lines in the R branch of the 0,0 vibrational band of the first negative system. An f/2 biconvex lens
focused the light from the exit slit onto the cathode of a Hamamatsu
R585 photomultiplier tube. The photon counting system consisted of a
Pacific Precision A/D6 amplifier/discriminator and a Hewlett-Packard
5308A counting system.

The spectrometer slit function was measured using the helium
$4\,^1P_1 \to 2\,^1S_0$ transition at 3964.7Å which could be excited by
the electron beam. Wavelength calibration was completed by adjusting
the maximum in the K'=7 → K"=6 line of the N_2^+ emission to 3906.04Å [12]

The signal from 20,000 valve pulses was added at each wavelength
to provide usable signal to noise ratio. Rotational temperatures were
determined by fitting calculated line intensities, convoluted with the
monochromator slit function, to the experimental data. The calculation of line intensity in the theoretical spectrum is taken from the
analysis of Muntz.[13] Rotational temperatures computed by this
method are based upon a Boltzmann rotational state distribution in the
ground state molecule.

The iron single crystal discs were cut from a 1/4" diameter
single crystal boule of iron of 99.9% purity (Atomergic Chemetals).
The boule was mounted on a goniometer and oriented to within 1/2° of
the desired plane using the Laue back reflection method. The
goniometer was transferred to a low speed, wafer-thin, diamond saw and
several discs were cut. Each disc was mounted in plastic and polished
using a succession of 180, 240, 320, 400, and 600 grit emery papers
followed by 15μ, 6μ, and 1μ diamond polishes. A final .05μ Alumina
Corunda polish was used to give a mirror finish.

The crystal was mounted on the front of two 1/16" diameter
stainless steel rods which could be heated resistively with a low
voltage, high current power supply. The crystal temperature was
measured by a W/5% - W/26%Re thermocouple pair spotwelded to the top
edge of the crystal.

After placing in vacuum, an AES spectrum of the crystal showed
carbon to be the major surface impurity. The crystal was heated to
900 K for 10 minutes which removed most of the carbon. A few percent
of a monolayer of sulfur remained. The crystal was then argon ion
bombarded for 15 minutes at 1000 eV with a current of 1.5μA at the
sample and finally annealed for 15 minutes at 700 K. This technique
produced samples with sharp, well defined LEED patterns and impurity
levels of less than 1% sulfur and carbon.

Measurement of rotational temperature of scattered nitrogen was performed at several exit angles, surface temperatures, and beam energies. The sample was cleaned and annealed between each measurement.

The apparatus was modified in order to measure the angular distribution of scattered molecules. For angular distribution measurements a fiberoptic cable was used to allow rotation of the electron gun in the scattering plane. Two identical f/1.6 aspherical lenses mounted on the electron gun holder collected the fluorescence and focused it onto the polished end of the fiberoptic cable. The other end of the fiberoptic cable was held fixed at the center of the 1" sapphire window. A small f/1.6 aspherical lens collimated the output beam and a f/5 lens focused the light onto the monochromator slit. The signal was measured at the maximum in the P branch, $\lambda = 3914.50\text{Å}$ with the monochromator slits fully open, FWHM = 10.2Å. The resulting signal is proportional to the number density of nitrogen in the scattered beam. Angular distributions were measured at several incident angles with beam energies of 90, 200 meV and 340 meV at a surface temperature of 400 K.

III. Results

The rotational temperature of nitrogen molecules scattered from the Fe(111) surface was measured at several surface temperatures and exit angles. In all cases the rotational state distributions had excess intensity in the high K region of the spectrum, i.e., the high K region of the spectrum could always be fit to a higher temperature than the low K region. Even though the distributions exhibit non-Boltzmann behavior it is convenient to characterize the rotational state distribution with a rotational temperature. Measured rotational temperatures of molecules scattered from the surface at $\theta_i = \theta_f = 45°$ at surface temperatures of 400 and 700 K are shown in Table I.

TABLE I

$E \diagdown T_s$	400 K	700 K
.09 eV	148 ± 30 K	172 ± 30 K
.20	190 ± 25	192 ± 25
.34	263 ± 40	269 ± 40

Table I. Measured rotational temperatures of molecules scattered into $\theta_f = 45°$ at several beam energies and surface temperatures ($\theta_i = 45°$)

The rotational temperature of scattered molecules shows a dependence on both beam energy and surface temperature over the range of
measurements made.   The rotational temperature is found to vary
approximately linearly with beam energy.   The variation of the rotational temperature with surface temperature is rather weak.
Angular distributions at T_s = 400 K were measured at several
angles of incidence and at all three beam energies.   The angular
distribution of the 90 meV nitrogen scattered from the surface at
θ_i = 25° and θ_i = 45° is shown in Figures 2 a) and b).

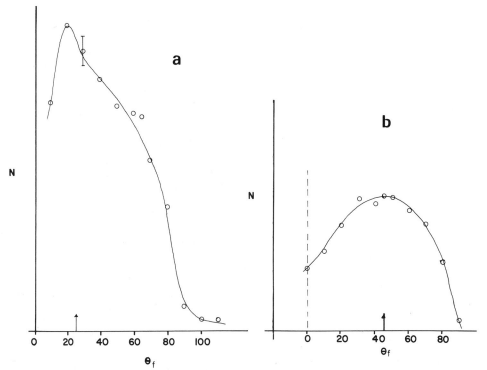

Figure 2.   Angular distribution of N_2 scattered from Fe(111);
a) θ_i = 25° b) θ_i = 45° (E = 90 meV, T_s = 400 K).

Note that the scattering occurs in broad peaks centered near the
specular angle.   The FWHM of the distribution at θ_i = 45° is 70°.
The angular distribution for the 200 meV beam at θ_i = 45° is
shown in Figure 3.   The 340 meV beam gave a qualitatively similar
result. Note the drastic shift in the maximum in the angular distribution toward the surface tangent as compared to the 90 meV beam.   The

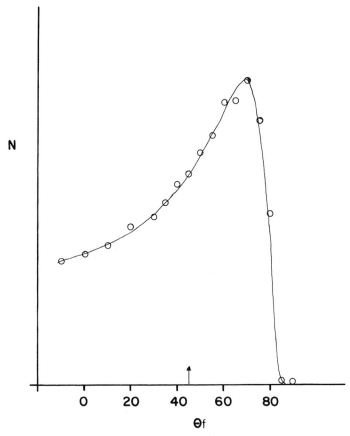

Figure 3.   Angular distribution of N_2 scattered from Fe(111),
            $\theta_i = 45°$, E = 200 meV, T_s = 400 K.

scattering maximum occurs at $\theta_f = 70°$ and the FWHM of the distribution
has decreased to 30°. An angle of incidence of $\theta_i = 60°$ was also
selected to provide a comparison between our results and those of
Boheim, et. al.[14] on Fe(110). The angular distribution is shown in
Figure 4. While the distributions are qualitatively similar, there
are several differences. For an incident beam energy of 176 meV at a
surface temperature of 550 K̇, Boheim's angular distribution peaked at
the specular angle $\theta_m = 60°$ and had a FWHM = 26°. Under our experimental conditions the measured maximum falls closer to the surface
tangent, $\theta_m = 70°$ and is slightly broader, FWHM = 32°. We also
observed a slightly greater intensity at the surface normal relative
to the peak intensity in the distribution. Whether the disagreement
is due to differences in experimental conditions or surface structure
is not clear.

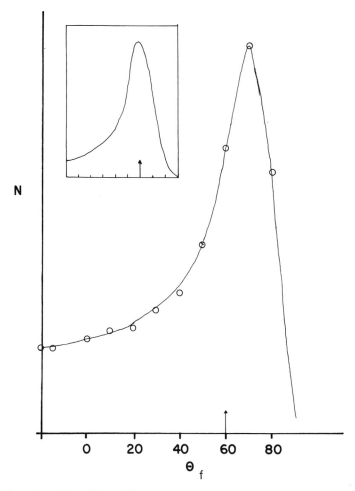

N

Figure 4.   Angular distribution of N_2 scattered from Fe(111),
            $\theta_i = 60°$, E = 200 meV, $T_s = 400$ K.

        The rotational temperature of scattered nitrogen was measured at
exit angles of $\theta_f = 20°$ and $\theta_f = 70°$ at $T_s = 400$ K and 700 K at all
three incident energies. Rotational temperature calculations are
based on the best fits to lines K' = 3, 5, 7, 9, 11, and 13 in the
rotational spectrum. Measured rotational temperatures are shown in
Table II.
        These distributions were also found to be non-Boltzmann with
excess energy in the higher K states. Note that the rotational temperature increases as the exit angle moves toward the surface normal. Again there is a strong dependence on the incident beam energy
and a weaker dependence on surface temperature for both exit angles.

Table II

	E \Ts	400 K	700 K
$\theta_f = 20°$	.09 eV	224 ± 30 K	234 ± 30 K
	.20	214 ± 35	240 ± 35
	.34	280 ± 50	402 ± 50
$\theta_f = 70°$	0.09 eV	91 ± 30 K	88 ± 30 K
	.20	91 ± 25	113 ± 25
	.34	144 ± 40	162 ± 40

Table II. Measured rotational temperature of molecules scattered into $\theta_f = 20°$ and $\theta_f = 70°$ at several beam energies and surface temperatures, ($\theta_i = 45°$).

IV. Discussion

The lobular nature of the scattering distributions, combined with the dependence of the measured rotational temperature of scattered molecules on the incident beam energy is evidence that molecules have undergone direct inelastic scattering. Ertl, et. al[15] have measured a sticking coefficient of ~ 10^{-2} for molecular adsoprtion of nitrogen on Fe(111). His measurement of a low sticking coefficient is in agreement with our observation of direct inelastic scattering.

To estimate the sticking coefficient for the 200 meV N_2 at $\theta_i = 60°$ incident on a 400 K Fe(111) surface, the angular distribution in Figure 3 was divided into a trapping-desorption fraction and a direct-inelastic scatttering fraction. Integrating the angular distribution with the assumption that the intensity at the normal is due to molecules which trap and desorb with a $\cos\theta$ distribution while direct-inelastic scattering is cylindrically symmetric about the peak in the lobe yields a sticking coefficient of 0.8. Accounting for the fact that our angular distributions are a measure of number density rather than flux with the assumption that the velocity of trapped-desorbed molecules is equilibrated with the surface while the directly scattered molecules retain their initial velocity only reduces the estimated sticking coefficient by a factor of 3. The discrepancy between this calculated value of the sticking coefficient and that of Ertl's experimentally measured value suggests that a large fraction of the intensity at the surface normal is due to molecules that are directly scattered.

Previous scattering experiments from well characterized single crystal surfaces have been restricted to the (111) planes of silver and platinum[16] which can apparently be described as "flat". Asada found the angular distributions of N_2 scattered from epitaxially grown

Ag(111) films to be qualitatively similar to Ar scattering.

Janda, et. al.[17] measured angular and velocity distributions of Ar and N_2 scattered from Pt(111). For an incident N_2 beam energy of 1050 K (90 meV) the peak in the scattering distribution was shifted only a few degrees toward the surface tangent relative to the peak in the argon distribution which peaked within a few degrees of specular. The distribution had a FWHM of 30°, only slightly broader than that observed for a similar Ar scattering experiment. The intensity of scattering at the surface normal was < 10% of the peak intensity. The velocity distribution of molecules scattered at the normal was in equilibrium with the surface, indicating that these molecules trapped and desorbed. Based on the angular distribution, he estimated the sticking coefficient of molecular nitrogen on Pt(111) to be between 0.4 and 0.6. Calculated differential translational energy accommodation coefficients based on measurement of velocity distributions of molecular nitrogen and argon scattered at the specular angle were 0.79 and 0.87, respectively. The fact that these coefficients differ by less than 10%, and the similarity of the N_2 and Ar scattering distributions suggest that transfer of translational energy into rotations has little effect on the scattering of nitrogen from the Pt(111) surface.

Our results for scattering of nitrogen from Fe(111) indicate that there are large differences between the iron surface and either silver or platinum surfaces. At a similar beam energy (90 meV) the angular distribution for nitrogen scattered from iron peaks near the specular angle but is very broad, FWHM = 60°. At higher energy (200 meV) the maximum in the distribution shifts drastically to $\theta_f = 70°$ and narrows to a FWHM = 30°.

We have attempted to apply the simple hard cubes theory[18] to explain these angular distributions. The basic assumptions of the theory are

  1) surface atoms are modelled by cubes having a Maxwellian distribution of velocities normal to the surface,
  2) collisions with the cubes are impulsive,
  3) the surface is flat, i.e., parallel momentum is unchanged by the collision,
  4) only single collision events are considered.

This model successfully predicted the qualitative behavior of the scattering maxima and width of the angular distributions of rare gas atoms scattered from single crystal metal surfaces.[19] The hard cubes theory of atom scattering predicts only slight shifts (~ 10°) toward the surface tangent as the energy of the incident beam is increased. The observed changes in the maximum in our scattering distributions are much larger than those predicted by this theory.

The hard cubes theory also predicts that a fixed fraction of the incident normal translational energy will be lost upon collision. A comparison of Figures 3 and 4 will show that this is not the case for scattering of nitrogen from Fe(111). A calculation of the fraction of normal energy lost by molecules incident at $\theta_f = 45°$ when applied to the $\theta_i = 60°$ distribution predicts that the distribution be peaked much closer to the surface tangent than was observed. These results

suggest that the assumptions of the hard cube theory are not applicable to the scattering of nitrogen from Fe(111).

Nichols and Weare[20] have modified the hard cube theory to include
homonuclear diatomic scattering from flat surfaces. The molecule is
modelled as an ellipse with the ratio of major to minor axis, the
ellipticity, being an adjustable parameter.

A comparison of rotational excitation of scattered molecules may
provide insight into the rotational anisotropy of the scattering
interaction. Zare[7] and Auerbach[6] have measured rotational state
distributions of NO scattered from Ag(111). Their results indicate
that for similar incident beam energies, NO molecules scattered at the
specular angle had significantly higher rotational temperatures than
our measured rotational temperature of N_2 molecules scattered from
Fe(111). Considering the relatively large well depth of N_2 - Fe,
7.5 kcal compared to ~ 2 kcal for NO - Ag and its effect on increasing
the translational energy of the incoming molecule prior to collision,
this is a rather low measured rotational temperature for the nitrogen.
It is likely that the smaller rotational anisotropy of N_2 relative to
NO manifests itself in a lower rotational temperature of scattered
molecules.

Some interesting observations about the angular dependence of the
rotational temperature of scattered molecules can be made based on
this model. The model predicts that those molecules which lost the
largest component of their normal momentum to rotations would exit at
angles nearer the surface tangent. Likewise, those exiting nearer to
the surface normal should have smaller rotational energies.

Auerbach[6] has measured state selective angular distributions of
NO molecules scattered from Ag(111). He found that the angular
distribution of higher rotational states was peaked farther toward the
surface tangent than for lower rotational states. This observation is
in agreement with the predictions of the hard cube theory for diatomic
scattering. For nitrogen scattering from Fe(111), the measured rotational temperature of scattered molecules monotonically increases as
the detector is rotated toward the surface tangent.

Apparently, differences in the N_2/Fe(111) and NO/Ag(111) interaction
potential result in opposite trends in the angular dependence of the
rotational temperature, and the hard cube theory of diatomic scattering does not adquately describe the N_2/Fe(111) interaction. The
angular dependence on the measured rotational temperature can be
interpreted in terms of a model with a surface corrugation effect.

Consider a one-dimensional periodic potential in the plane of
incidence and the three trajectories indicated in Figure 5.

We can apply the assumptions of Nichols and Weare's model to a
collision using the micronormal to the surface rather than the
normal to the macrosurface. Those molecules with trajectories indicated by A have a larger component of normal momentum and are consequently scattered with larger rotational energies than those of
trajectory B or C. These molecules are also scattered in a direction
located farther toward the surface normal from that predicted by a
flat surface. Molecules of trajectory C will be scattered toward the

surface tangent with lower rotational energies. This model successfully predicts the qualitative trends in the angular dependence of the
measured rotational temperature of N_2 scattered from Fe(111).

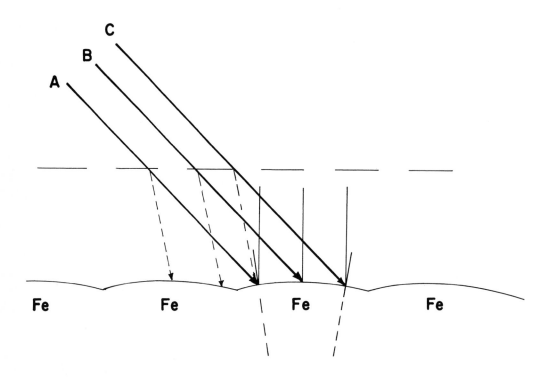

Figure 5. Corrugation model for N_2/Fe interaction.

    We have attempted to use the corrugated surface model to predict
the observed trends in the measured angular distribuions. We note
that the measured signal in the angular distribution is proportional
to number density not flux. Without a knowledge of the velocity
distribution of scattered molecules, we were unable to present the
data as a flux distribution.

    Referring to Figure 5, trajectory A results in scattering toward
the surface normal relative to Trajectory B, which is unchanged from
that predicted by the flat surface model. The large probability of
striking the front of the corrugated potential produces a shift in the
scattering maximum toward the surface normal when compared to the flat
surface model. Figure 4 indicates that the shift toward the surface
normal was not observed in our experiments.

The presence of a large potential well at the surface reduces the "shadowing" effect by altering the trajectories of the incoming molecules (dotted lines in Figure 5). This effect becomes less pronounced as the incident energy increases relative to the well depth, predicting a shift in the distribution toward the surface normal with increasing beam energy.

Though the corrugation effect accounts for the angular dependence on the measured rotational temperatures, the model inadequately describes the observed angular distributions. The analysis of angular distributions and rotational temperature of scattered nitrogen would benefit from detailed trajectory calculations such as those applied to $N_2/Ag(001)$ scattering.[21] In this way a more realistic potential, including well depth, rotational anisotropy and surface corrugation can be applied to the $N_2/Fe(111)$ scattering interaction.

If in fact this corrugation does exist for molecular adsorption, it suggests that there are preferred sites of adsorption on the Fe(111) surface. This has been postulated by Kishi and Roberts[2] based upon the existence of two peaks in the XPS spectrum of molecularly adsorbed nitrogen on iron films.

They proposed the presence of bridged and on-top molecularly adsorbed nitrogen. Whether or not these oriented molecules exist on the surface could be determined by a detailed analsysis of the rotational structure in the EELS spectrum of molecularly adsorbed nitrogen on iron. If these species do exist, then the preferred site or orientation effects could be responsible for the unusually low sticking coefficient of nitrogen on iron.

V. Conclusions

Measurements of angular distributions and rotational temperature of nitrogen molecules scattered from Fe(111) have shown that the scattering results in direct inelastic events. Rotational distributions of scattered molecules measured at various scattering angles indicate that the molecules are leaving the surface in non-Boltzmann population distributions, the high K portions of the distribution being characterized by higher temperatures than the low K region.

Conversion of translational energy into rotations of the molecule is significant but the rotational temperature used to characterize the distribution lags behind the surface temperature, especially at higher temperatures. A comparison of the measured rotational temperature of molecules scattered from the surface in the NO/Ag(111) and $N_2/Fe(111)$ systems concluded that the rotational anisotropy of the $N_2/Fe(111)$ interaction potential is smaller than for the NO/Ag(111) system. The rotational temperature of scattered molecules increases monotonically as the scattering angle moves from the surface tangent toward the surface normal. A model which assumes a surface corrugation accounts for the measured rotational temperatures of scattered nitrogen as a function of exit angle. Angular distributions are not accurately described by this model. Experiments are currently underway in our laboratory to measure the velocity distribution of scattered nitrogen molecules. This data would be useful in developing a full detailed balance of the energy exchange occuring at the surface.

We plan to apply these techniques to the Pt(111) surface, where velocity distribution measurements of in-plane scattered N_2 are already available.[17] We are currently proceeding with measurements of N_2 scattering from Ag(111). It is hoped that these measurements will elucidate the differences in the interaction of N_2 with the Ag(111) and Fe(111) surfaces.

Acknowledgement

We gratefully acknowledge the support provided for this research by the National Science Foundation and for fellowship support provided by SOHIO.

References

1. F. Bozso, G. Ertl, M. Grunze, M. Weiss, J. Catalysis 49, 18 (1977).
2. K. Kishi, M. W. Roberts, Surf. Sci. 62 252 (1977).
3. K. Yoshida, G. A. Somorjai, Surf. Sci. 75, 46 (1978).
4. See G. Ertl, Catalysis Rev. 21, 201 (1980) and referenences therein.
5. See J. C. Tully, Ann. Rev. Phys. Chem. 31, 319 (1980).
    W. H. Weinberg, Adv. Colloid and Interface Sci. 4, 301 (1975).
6. A. W. Kleyn, A. C. Luntz, D. J. Auerbach, Surf. Sci. 117, 33 (1982).
7. G. D. Kubiak, J. E. Hurst, Jr., H. G. Rennagel, G. M. McCleland and and R. N. Zare, J. Chem. Phys. 79, 5163 (1983).
8. J. Segner, H. Robota, W. Veilhaber, G. Ertl, F. Frenkel, J. Häger, W. Walther, Surf. Sci. 131, 273 (1983).
9. Q. L. Guthrie, T. H. Lin,. S. T. Ceyer, and G. A. Somorjai, J. Chem. Phys. 76, 6398 (1982).
10. D. Ettinger, K. Honma, M. Keil and J. C. Polanyi, Chem. Phys. Lett. 87, 413 (1982).
11. J. W. Hepburn, F. J. Northrup, G. L. Ogram, J. C. Polanyi and J. W. Williamson, Chem. Phys. Lett. 85, 127 (1982).
12. G. Herzberg, Spectra of Diatomic Molecules, Van Nostrand, New York (1950).
13. E. P. Muntz, NATO AGARDorgraph 132 (1968).
14. J. Böheim, W. Brenig, T. Engel, U. Leuthäusser, Surf. Sci. 131, 258 (1983).
15. G. Ertl, S. B. Lee, M. Weiss, Surf. Sci. 144, 515 (1982).
16. H. Asada, Jap. J. Appl. Phys. 20, 527 (1981).
17. K. C. Janda, J. E. Hurst, J. Cowin, L. Wharton, D. J. Auerback, Surf. Sci. 130, 395 (1983).
18. R. M. Logan and R. E. Stickney. J. Chem. Phys. 44, 195 (1966).
19. F. O. Goodman, H. Y. Wachman, Dynamics of Gas-Surface Scattering, Academic Press, New York (1976).
20. W. L. Nichols and J. H. Weare, J. Chem. Phys. 62, 3754 (1975).
21. C. W. Muhlhausen, J. A. Serri, J. C. Tully, G. E. Becker, and M. J. Cardillo, Isreal J. Chem. 22, 315 (1982).

DYNAMICAL PROCESSES AT SURFACES: EXCITATION OF ELECTRON-HOLE PAIRS AND
PHONONS

B.N.J. Persson
Institut für Festkörperforschung
Kernforschungsanlage Jülich GmbH
Postfach 1913
5170 Jülich
Federal Republic of Germany

ABSTRACT

I discuss the response of a metal surface to an external electric field
which varies slowly in space and time. Both electron-hole pair and
phonon production is considered and the theoretical results are compared
with inelastic electron scattering measurements from Cu(100). Excellent
agreement between theory and experiment is obtained.

1.   INTRODUCTION

The response of a metal surface to an external electromagnetic wave enters many important problems in surface science.[1] Despite this, our
present understanding of this problem is still rather limited.[2] Here, I
would like to report on a very simple theory for the response of a metal
surface to an external electric field which varies slowly in space and
time.[3] The theory agrees remarkably well with electron energy loss measurements[4] from Cu(100), and has also several important applications.
    The basic quantity for the following discussion is the surface re-
sponse function $g(q_\parallel, \omega)$ defined as follows: Consider a semi-infinite
metal occupying the half space $z > 0$ and let

$$\phi_{ext} = e^{-q_\parallel z} \, e^{i(\vec{q}_\parallel \cdot \vec{x} - \omega t)} \tag{1}$$

be an external potential which polarizes the metal. The polarization
charges will give rise to an induced potential which, outside the metal
(where $\nabla^2 \phi_{ind} = 0$), can be written as

$$\phi_{ind} = -g(q_\parallel, \omega) \, e^{q_\parallel z} \, e^{i(\vec{q}_\parallel \cdot \vec{x} - \omega t)} \quad . \tag{2}$$

It has been implicitly assumed that the external potential is so weak
that the metal responds linearly to it and also that the metal can be
treated as translational invariant parallel to the surface. A very useful equation[5] is the relation between Img and the power absorption in
the metal, $\hbar\omega w$, caused by an external potential of the form (1):

257

B. Pullman et al. (eds.), Dynamics on Surfaces, 257–269.
© *1984 by D. Reidel Publishing Company.*

$$\text{Img} = \pi\hbar w \; / \; Aq_{\parallel} \quad , \tag{3}$$

where A is the surface area. The energy absorption in the metal is due to excitation of electron-hole pairs and phonons so that in general

$$\text{Img} = (\text{Img})_{\text{e-h pair}} + (\text{Img})_{\text{phonon}} \quad . \tag{4}$$

In this paper, we will not consider $\omega > \omega_p$, where also plasmons can be excited. Since g is a causal response function, Reg can be obtained from Img via a Kramers-Kronig equation.

In the standard textbook treatment of the dielectric response of a metal surface, one assumes that the solid is characterized by a local dielectric function which jumps discontinuously at the surface, i.e. $\in = \in(\omega)$ for $z > 0$ and $\in = 1$ for $z < 0$. From the continuity of ϕ and $\in \partial\phi/\partial z$ at the solid-vacuum interface, one obtains the standard textbook result

$$g(q_{\parallel},\omega) = \frac{\in(\omega) - 1}{\in(\omega) + 1} \quad . \tag{5}$$

A real metal does not have a step like surface profile but a profile which varies smoothly on a microscopic scale. In addition, the bulk dielectric function depends on the wave vector $q_{\parallel}$. Thus (5) is in general not correct although one can prove that it is exact for $q_{\parallel} = 0$.

Within the jellium model, $g(q_{\parallel},\omega)$ depends on $q_{\parallel}$ via the dimensionless parameters[6] $q_{\parallel}/k_F$ and $(q_{\parallel}/k_F)(2\omega_F/\omega) \equiv 1/\eta$. Thus, if $q_{\parallel}/k_F \ll 1$ and $1/\eta \ll 1$, it is possible to expand $g(q_{\parallel},\omega)$ in a power serie in $q_{\parallel}$ to obtain $g(q_{\parallel},\omega) = (\in(\omega)-1)/(\in(\omega)+1) + q_{\parallel}A(\omega) + O(q_{\parallel}^2)$. An expansion of this type has been given by Feibelman[7]. However, in the present work, we will be interested in both small $q_{\parallel}$ and ω so that although $q_{\parallel}/k_F \ll 1$, the parameter $\eta^{-1} = (q_{\parallel}/k_F)(2\omega_F/\omega) > 1$ for small enough ω. There are many interesting applications where the last inequality is valid. Consider, for example, a low energy electron incident upon a metal surface. The electric field of the incident electron penetrates into the metal where it can excite e.g. electron-hole pairs. In the so-called dipole scattering regime, the momentum transfer to the excitation in the metal is small, $q_{\parallel} \sim k(\hbar\omega/2E)$, [$E = \hbar^2k^2/2m$ is the kinetic energy of the incident electron], so that in this case $(q_{\parallel}/k_F) \times (2\omega_F/\omega) \sim k_F/k > 1$ if $E < E_F$. We will show in the next section that with a few plausible assumptions, it is possible to determine the dielectric response even when $\eta^{-1} > 1$.

2.   EXCITATION OF ELECTRON-HOLE PAIRS

We treat the metal within the jellium model, i.e., the metal conduction electrons are assumed to move in a semi-infinite positive background obtained by smearing out the positive metal-ion cores. The electron wavefunctions are assumed to satisfy a one-particle Schrödinger equation where the effective one-particle potential $V_{eff}(z)$ is obtained, e.g., from the functional density scheme within the local density approximation.

The contribution to Img from electron-hole pair excitation can be calculated from (3) as follows:[3] The rate w at which the external potential (1) excites electron-hole pairs in the metal is obtained from the Golden rule formula

$$w = \frac{2\pi}{h} \int d^3k \; d^3k' \; f_k(1-f_{k'}) \; |<k'|e\phi|k>|^2 \; \delta(\epsilon_{k'} - \epsilon_k - \hbar\omega) \tag{6}$$

where, at zero temperature, $f_k = 1$ if $\epsilon < \epsilon_F$ and $f_k = 0$ if $\epsilon > \epsilon_F$. $\phi = \phi_{ext} + \phi_{ind}$ is the total potential in the metal induced by the external potential. As discussed in Ref. 3, it is convenient to decompose $\phi = \phi_{surf} + \phi_{bulk}$ into a component ϕ_{surf} which is non-vanishing only in the surface region of the metal plus another component $\phi_{bulk} = \phi - \phi_{surf}$ which vanishes in the surface region of the metal (see Figure 1). We then have

$$<k'|\phi|k> = <k'|\phi_{surf}|k> + <k'|\phi_{bulk}|k> \quad , \tag{7}$$

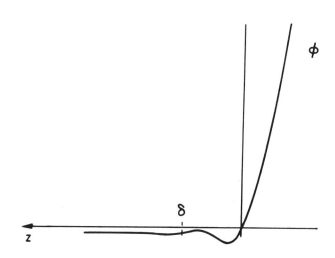

Fig. 1.   The screened (i.e. total) potential as a function of the distance z into the jellium ($z=0$ is the jellium edge). We define $\phi_{bulk} = \phi$ for $z > \delta$ and $\phi_{surf} = \phi$ for $z < \delta$.

and the rate w will have three contributions $w = w_{surf} + w_{bulk} + w_{int}$, where the last term is the interference term between bulk and surface excitation of electron-hole pairs. In the $q_\parallel \ll k_F$ and $\omega \ll \omega_F$ limits the matrix elements in (7) have been calculated under the assumptions (a) that the bulk potential ϕ_{bulk} is well approximated by its classical form

$$\phi_{bulk} = \frac{2}{\mathcal{E}(\omega)+1} \, e^{-q_{\parallel} z} \, e^{i(\vec{q}_{\parallel} \cdot \vec{x} \, -\omega t)} \quad , \tag{8}$$

and (b) that the surface potential ϕ_{surf} is well approximated by the potential obtained from a static $(\omega=0)$ calculation of the type presented by Lang and Kohn.[11] Under these assumptions, one obtains

$$<k'|\phi_{surf}|k> = - \frac{q_{\parallel}^2}{\pi^2} \, \delta(\vec{k}_{\parallel} +\vec{q}_{\parallel} -\vec{k}_{\parallel}') \, <\Psi_{k_z'}|A(z)|\Psi_{k_z}> \quad , \tag{9}$$

$$<k'|\phi_{bulk}|k> = - \frac{2}{\pi} \, \frac{1}{1+\mathcal{E}} \, \delta(\vec{k}_{\parallel} +\vec{q}_{\parallel} -\vec{k}_{\parallel}') \, \frac{q_{\parallel}}{q_{\parallel}^2+(k_z-k_z')^2} \quad . \tag{10}$$

In (9) $A = A(z)$ is the induced potential in the metal for $\omega = 0$, due to an uniform charged sheet located outside the metal surface and carrying one unit of charge per unit area.

From (3), (6), (7), (9), and (10), one can now calculate the contribution to Img from electron-hole pair production:[3]

$$Img = (Img)_{surf} + (Img)_{bulk} + (Img)_{int} \quad , \tag{11}$$

where

$$(Img)_{surf} = 2\xi \frac{q_{\parallel}}{k_F} \frac{\omega}{\omega_p}$$

$$(Img)_{bulk} = \frac{1}{2} \left(\frac{m^*}{m}\right)^2 \left(\frac{\omega}{\omega_p}\right)^2 \eta^3 \, G(\eta) \tag{12}$$

$$(Img)_{int} = - \frac{8}{\pi^2} \frac{m^*}{m} \frac{1}{k_F a_o} \left(\frac{\omega}{\omega_p}\right)^2 \eta \, H(\eta)$$

where $\xi = \xi(r_s)$ is a function of the electron gas density parameter r_s tabulated in Ref. 8 and where

$$G(\eta) = \begin{cases} 8 & \text{if} \quad \eta < 1 \\ 8(1-[1+\frac{1}{2} \eta^2][1-\eta^{-2}]^{1/2}) & \text{if} \quad \eta > 1 \end{cases}$$

and

$$H(\eta) \approx h(r_s) \, / \, (1+\eta^2)$$

where $h(r_s)$ is tabulated in Ref. 8. To the expression (11) should be added the classical Drude contribution

$$(Img)_{Drude} = Im \frac{\mathcal{E}(\omega)-1}{\mathcal{E}(\omega)+1} = \left(4 \frac{\omega_F}{\omega_p}\right) \frac{m^*}{m} \frac{1}{k_F \ell} \frac{\omega}{\omega_p}$$

where ℓ is the mean-free path due to scattering aginst phonons. Note that $\ell \to \infty$ as $T \to 0$ so that by performing experiments at low temperatures one can eliminate the Drude contribution to Img. We will discuss this in more detail in Sec. 4.

3.   EXCITATION OF PHONONS

In this section, we will discuss the contribution to Img from excitation of bulk phonons.[9] Again we focus on the case $q_{||} \ll k_F$. One can show that the contribution to Img from excitation of phonons is derived almost entirely from ϕ_{surf} , the contribution from ϕ_{bulk} being about a factor 10^{-6} times smaller.

The discussion below is based on the following result, first noticed by Budd and Vannimenus:[10] The total force which acts on the metal ion-cores from an external charge is zero, within the assumption of linear response. This implies that we have a force-sum-rule $\Sigma \, \bar{F}_i = 0$ where $\bar{F}_i$ are the forces which act on the individual ion cores. From the force sum rule it follows that the screened potential induced by an external potential of the form (1) must exhibit oscillations. Figure 2 shows the z-variation of the screened electric field E as calculated by Lang and

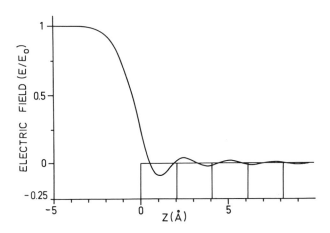

Figure 2.

Kohn[11] for $\omega = 0$ and $q_{||} = 0$ and $r_s = 3$. The monovalent ion core layers are represented by slabs of uniform positive background along the z-axis as depicted in the figure. The Friedel oscillations decay rather rapidly with increasing z ($E_z \sim z^{-2} \cos(2k_F z - \gamma)$ as $z \to \infty$) and in the simplest possible model one therefore assumes that the screened electric field is non-zero only at the two-topmost layers of ions. To satisfy the force-sum-rule, opposite forces of identical magnitudes, F and -F, must

act on the two layers of ions. Within linear response theory, these
forces are linearly related to the electric field E_o just outside the
metal, i.e. $F = \lambda |e| E_o \equiv e^* E_o$ where $|\lambda| < 1$ is a screening parameter
and $|e|$ the charge of the ion. We have defined an effective charge
$e^* = \lambda |e|$.

For $q_\parallel \ll 1/a$, where a is the lattice constant, only bulk
phonons which propagate normal or almost normal to the metal surface can
be excited. These phonons correspond to rigid displacement of the lattice planes against each other. Thus we have a simple one-dimensional
problem for the phonon dynamics:

$$M\ddot{u}_o + M\Omega^2(u_o - u_1) = e^* E_o$$

$$M\ddot{u}_1 + M\Omega^2(2u_1 - u_o - u_2) = - e^* E_o$$

$$M\ddot{u}_i + M\Omega^2(2u_i - u_{i-1} - u_{i+1}) = 0 \quad , \quad i = 2, 3, \ldots$$

Here u_i is the displacement from the equilibrium position of an ion
(mass M) in the i-th lattice plane and Ω^2 is determined by the force
constant between the lattice planes. The system of equations above is
easily solved for the dipole moment $p = e^*(u_o - u_1)$ and hence polarizability $\alpha \equiv p/E_o$

$$\alpha(\omega) = \frac{4e^{*2}}{M\omega_o^2} f(\omega)$$

where

$$f(\omega) = 1 + 4\bar{\omega}^{-2} - 8\bar{\omega}^{-4} + i8\bar{\omega}^{-3}(1-\bar{\omega}^{-2})^{1/2} \quad .$$

We have introduced $\bar{\omega} = \omega/\omega_o$ where $\omega_o = 2\Omega$ is the maximum longitudinal phonon frequency in the [100] direction. Since there are $2A/a^2$ ions
(a = lattice constant = 3.61 Å for Cu) on the surface area A, the total
time averaged (denoted by a bar), power absorption is $\hbar\omega w = (2A/a^2) \times$
$\times \overline{p \cdot E_o} = 16(A/a^2) \omega q^2 |\phi_o|^2 \,\text{Im}\alpha(\omega)$. Substituting this into (3) gives

$$(\text{Im}g)_{phonon} = 4q_\parallel a \, (e^*/e)^2 \, (\omega_{ion}/\omega_o)^2 \, \text{Im}f(\omega) \tag{13}$$

where $\omega_{ion} = (4\pi e^2/(Ma^3/4))^{1/2}$ is the unscreened ion-plasma frequency.
For copper we have $\hbar\omega_o = 30$ meV and $\hbar\omega_{ion} = 32$ meV.
We will discuss the value of the screening parameter $\lambda = e^*/e$ in
the next section where we compare theory with experiment.

4.   INELASTIC SCATTERING OF ELECTRONS FROM Cu(100)

It has recently been shown that $\text{Im}g(q_\parallel, \omega)$ can be measured directly
using inelastic electron scattering from clean metal surfaces.[12,4] Such
measurements have been performed by Andersson on Cu(100) and we will here
compare his experimental data with the theory for $\text{Im}g$ presented in the
last two sections.

The experimental results discussed below show that the long-range dipole interaction dominates the inelastic electron scattering process. Hence the probability $P(\vec{k},\vec{k}')d\Omega_k, d\hbar\omega$ that an incident electron of wave vector $\vec{k}$ is scattered inelastically into the solid angle $d\Omega_k$, around the direction of $\vec{k}'$ (the wave vector of the scattered electron) losing energy in the range $\hbar\omega$ and $\hbar(\omega+d\omega)$ is given by[13]

$$P = \frac{2}{(ea_0\pi)^2} \frac{1}{\cos\alpha} \frac{k'}{k} \frac{q_{\parallel}}{(q_{\parallel}^2+q_{\perp}^2)^2} (n_\omega+1) \ \mathrm{Img}(q_{\parallel},\omega) \qquad (14)$$

valid for small momentum transfer, $q \ll k$. Here α is the angle of incidence, and $\hbar\vec{q} = \hbar(\vec{k}-\vec{k}')$ and $\hbar q_\perp = \hbar(k_\perp-k_\perp')$ are the changes in the parallell and normal components of momentum, respectively, $a_0 \simeq 0.53$ Å the Bohr radius and $n_\omega = (\exp(\hbar\omega/k_BT)-1)^{-1}$ is the Bose-Einstein factor. Eq. (14) gives the scattering probability on the loss side; on the gain side $(n_\omega+1)$ must be replaced by $n_{|\omega|}$. In EELS one does not measure $P(\vec{k},\vec{k}')$ directly, but rather P integrated over the angle of detection $\Delta\Omega$

$$\Delta P = \int_{\Delta\Omega} P \ d\Omega \qquad .$$

The experimental data discussed below were obtained with the analyzer in the specular direction (angle of incidence, $\alpha = 54^0$) and with an acceptance angle (full width at half maximum) of 0.9^0.

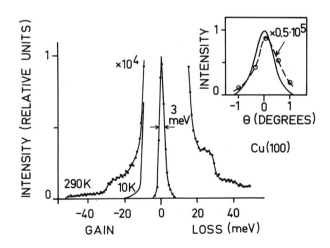

Figure 3.

Figure 3 shows electron energy loss data from Cu(100) at room temperature (T = 290K) and at T = 10 K. The intensity on the loss side is proportional to $n_{|\omega|}+1$ while the intensity on the gain side is proportional to $n_{|\omega|}$ and thus smaller. For $|\omega| > \omega_o$, where ω_o is the highest phonon frequency in the metal, it is only possible to excite electron hole pairs while for $|\omega| < \omega_o$ phonons also can be excited. This explains the rather sharp drop in the loss and gain intensities when $|\omega|$ increases beyond ω_o = 30 meV, which is the highest longitudinal phonon frequency in the [100] direction of Cu(100). We will discuss Figure 3 in more detail below.

The loss data in Figures 3-5 involve excitation of very low energetic electron-hole pairs, $\hbar\omega \sim 0.1$ eV. The momentum transfer, according to standard dipole scattering theory, is therefore also very small, $q_{||} \sim (\hbar\omega/2E_o)k \sim 10^{-2}$ Å^{-1}, and we can use the expression for Img derived in Sec. 2 which we now write

$$\text{Img} = \left(\frac{a}{k_F \ell} + b \frac{q_{||}}{k_F} + c \, \eta^3 \, G(\eta) \, \frac{\omega}{\omega_p} + \frac{d\eta}{1+\eta^2} \, \frac{\omega}{\omega_p} \right) \frac{\omega}{\omega_p} \quad , \tag{15}$$

where (we treat Cu as a r_s=2.67 jellium with m^* = 1.5 m): a = 3.89, b = 0.98, c = 1.125, and d = 3.10.

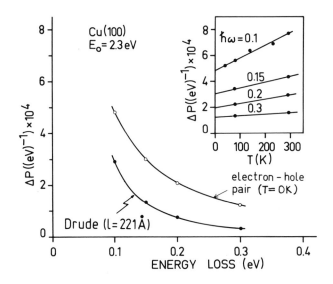

Fig. 4.   The theoretical predicted electron-hole pair contribution (T = 0 K) is scaled with a factor 0.88 .

Figure 4 shows how the inelastic scattering probability, ΔP, for Cu(100), depends on the loss energy ħω. The inset shows the measured

data for ΔP at $\hbar\omega$ = 0.1, 0.15, 0.2, and 0.3 eV and for several temperatures. We note that ΔP varies linearly with temperature which is
also expected from optical data for copper. This is also the prediction
from standard theory of phonon-resistivity for $T > 0.2\, T_D$ (where T_D
is the Debye temperature). The open circles in Figure 4 correspond to
the ΔP values obtained by extrapolating the data in the inset to $T=0$.
The filled circles show the Drude contribution to ΔP at room temperature as given by $\Delta P(T=293\ K) - \Delta P(T=0\ K)$. The solid curves are theoretically predicted results for ΔP using $Img = (Img)_{surf} + (Img)_{bulk} +$
$+ (Img)_{int}$ for the upper curve and $Img = (Img)_{Drude}$ with
$\ell=221\ \text{Å}$ for the lower curve. We note that there is an almost perfect
agreement between theory and experiment with respect to the dependence
of ΔP on the loss energy $\hbar\omega$.

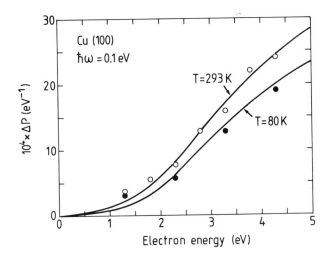

Figure 5.

The absolute value of ΔP in Figure 4 deviates only by 10% from
the theoretical result and on the average, over a range of impact energies, with less than this as can be seen in Figure 5. The open and
filled circles in this figure are experimental data for $\hbar\omega$ = 0.1 eV obtained at 293 K and 80 K substrate temperature, respectively. The
solid curves correspond to the calculated contribution using the expression for Img given by (15) for $T = 293\ K$ ($\ell = 221\ \text{Å}$) and $T = 80\ K$
($\ell = 809\ \text{Å}$). The agreement between theory and experiment is very good,
what concerns the dependence of ΔP on the incident electron energy,

the absolute value of ΔP , and the relative magnitude of the Drude contribution.

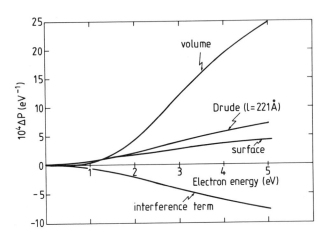

Figure 6.

Figure 6 shows separately the various contributions to ΔP for the case shown in Figure 5.

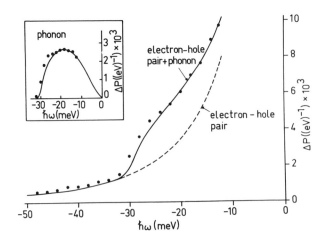

Fig. 7.   The theoretical predicted electron–hole pair contribution is scaled with a factor 1.2 .

Figure 7 shows the function $\Delta P(\omega)$ on the gain side of the electron energy loss spectrum. The black dots are the experimental data points (the same as in Figure 3), and the dashed line is the calculated contribution to ΔP using (14) but with $n_\omega + 1$ replaced by $n_{|\omega|}$. The full line is the result obtained by adding a contribution caused by excitation of longitudinal bulk phonons (see below). With the same parameters as above ($m^* = 1.5\ m$ and $\ell = 221\ \text{Å}$), the contribution from excitation of electron-hole pairs deviates only by about 20% from the measured data points.

In Figure 7, calculating the phonon contribution to ΔP, we have used $e^* = 0.0155\ e$ which is the only adjustable parameter in the theory. The agreement between theory and experiment is excellent. The solid line in the inset shows the phonon contribution separately which also agrees very well with the experimental data (black dots).

The screening parameter $\lambda = e^*/e \simeq 0.016$ shows that the force on the lattice ions in the first layer is reduced to 1.5% of its unscreened value. This result agrees well with the prediction of the jellium calculation (Figure 2) from which an estimate of λ can be obtained by averaging $E(z)/E_0$ over a layer of ion cores, taken as a slab of positive background, centered at an ion nucleus and of thickness given by the inter-planar separation in the [100] direction of a fcc crystal. For $r_3 = 3$, this gives $\lambda_1 = -0.011$, $\lambda_2 = 0.008$, $\lambda_3 = 0.005$, $\lambda_4 = -0.0003$, $\lambda_5 = -0.002$, $\lambda_6 = -0.0002$, ..., which should be compared with our result $\lambda_1 = -0.016$, $\lambda_2 = 0.016$, $\lambda_3 = \lambda_4 \ldots = 0$. The agreement between the results of the jellium calculation and the experiment is even more striking if one accounts for the extended nature of the Friedel oscillations by using a more detailed model for the phonon dynamics. We then find that the calculated electron scattering cross section for Cu (interpolation to $r_s = 2.7$) is only 10% larger than the experimental value.

5.   SUMMARY AND OUTLOOK

In this work, I have discussed the response of a metal surface to an external electric potential which varies slowly in space and time. We focussed the discussion on the surface response function $g(q_\parallel, \omega)$ which determines the induced electric potential outside the metal. It was shown that $g(q_\parallel, \omega)$ can be measured directly using inelastic electron scattering and that such measurements performed on Cu(100) agree remarkably well with the theoretical results. It is therefore meaningful to apply the theory to other dynamical surface processes involving small $q_\parallel$ and ω. Here, I note that there are several such interesting applications, including
(a) the lifetime of excited states at surfaces,[1,14]
(b) the Van der Waals interaction between an atom and a surface,[1,15]
(c) the surface power absorption,[1]
(d) the friction force on a charged particle moving above a metal
     surface.[1,16]
We refer the readers to the quoted references for detail concerning these applications.

I close this work by breafly mentioning how the formalism presented above can be extended to larger ω and $q_\parallel$. Assume first that $q_\parallel$ is small, $q_\| \ll k_F$, but ω so large that $(q_\|/k_F)(2\omega_F/\omega) = \eta^{-1} \ll 1$. Then it is possible to expand $g(q_\|,\omega)$ in a power serie in $q_\|$ and Feibel-man has shown that[1]

$$g(q_\|,\omega) = \frac{\in(\omega)-1}{\in(\omega)+1} + 2q_\| \, d_\perp(\omega) \, \frac{\in(\omega)(\in(\omega)-1)}{(\in(\omega)+1)^2} + 0(q_\|^2)$$

where $d_\perp(\omega)$ is the centroid of the induced charge density. Feibelman calculated $d_\perp(\omega)$ for $\omega \gtrsim \omega_p/2$ and from the formalism presented in Sec. 2, one can obtain $\mathrm{Im}\, d_\perp(\omega)$ for $\omega \ll \omega_p$. In addition, it has been shown that $\mathrm{Im}\, d_\perp(\omega)$ and $\mathrm{Re}\, d_\perp(\omega)$ must satisfy the Kramers-Kronig equations and a set of sum rules[17] which gives useful con-straints on $d_\perp(\omega)$. Thus the $q_\| \ll k_F$ limit of $g(q_\|,\omega)$ is now quite well understood for all frequencies ω .

Some progress has also been achieved in calculating $g(q_\|,\omega)$ for arbitrary $q_\|$ and ω. Eguiluz[18] has calculated the response of a thin jellium slab to an external time varying electric field, in the non-retarded limit. His calculation is in principle valid for arbitrary $q_\|$ and ω but if the thin slab is to represent a semi-infinite jellium then the film thickness L must satisfy $L \gg 1/q_\|$. Thus from a prac-tical point of view, it is hard to study the small $q_\|$ limit using this method - this, however, might be an unimportant restriction since the small $q_\|$ limit is now quite well understood from the works discussed above.

We note finally that by studying inelastic scattered electrons off-specular in EELS, it is possible to measure $\mathrm{Img}(q_\|,\omega)$ for arbitrary $q_\|$. This is probably the most accurate and straightforward way to obtain information about this important surface response function.

ACKNOWLEDGMENTS

The author is grateful to his collaborators S. Andersson, N.D. Lang, M. Persson, and E. Zaremba who have contributed to various parts of the work presented here.

REFERENCES

1. See, e.g., P.J. Feibelman, *Prog. Surf. Sci.* 12,   (1982).
2. K.L. Kliewer, *Surf. Sci.* 101, 57 (1980); G. Mukhopadhyay and S. Lundqvist, *Phys. Scr.* 17, 69 (1978); F. Forstmann and R.R. Gerhards, *Festkörperprobleme* 22, 291 (1982); P.J. Feibelman, *Phys. Rev. B* 22, 3654 (1980); 12, 1319 (1975); 14, 762 (1976); P. Apell, *Phys. Scr.* 24, 795 (1981).
3. B.N.J. Persson and N.D. Lang, *Phys. Rev. B* 26, 5409 (1982); B.N.J. Persson and S. Andersson, *Phys. Rev. B* (April 1984); B.N.J. Persson and E. Zaremba, to be published.
4. S. Andersson and B.N.J. Persson, *Phys. Rev. Lett.* 50, 2028 (1983); see, also, the last two works in Ref. 3.
5. See the last two works in Ref. 3.

6. This follows, e.g., from the fact that the RPA bulk dielectric function $\in(q,\omega)$ contains the factor $(q/q_F)(\omega_F/\omega)$.

7. P.J. Feibelman, *Phys. Rev. B* $\underline{22}$, 3654 (1980).

8. See the last work in Ref. 3.

9. S. Andersson, B.N.J. Persson, M. Persson and N.D. Lang, submitted to *Phys. Rev. Lett.*

10. H.F. Budd and J. Vannimenus, *Phys. Rev. B* $\underline{12}$, 509 (1975).

11. N.D. Lang and W. Kohn, *Phys. Rev. B* $\underline{7}$, 3541 (1973).

12. B.N.J. Persson, *Phys. Rev. Lett.* $\underline{50}$, 1089 (1983).

13. See, e.g., H. Ibach and D.L. Mills, *Electron-Energy-Loss-Spectroscopy and Surface Vibrations*, Academic Press, New York (1982); W.L. Schaich, *Phys. Rev. B* $\underline{24}$, 686 (1981); R.F. Willis, A.A. Lucas and G.D. Mahan, in *The Chemical Physics of Solid Surfaces and Heterogeneous Catalysis*, Vol. $\underline{2}$, eds. D.A. King and D.P. Woodruff, Elsevier (1983).

14. H. Morawitz, *Phys. Rev.* $\underline{187}$, 1792 (1969); R.R. Chance, A. Prock and R. Silbey, *Adv. Chem. Phys.* $\underline{37}$, 1 (1978); B.N.J. Persson and M. Persson, *Surf. Sci.* $\underline{3}$, 609 (1980).

15. E. Zaremba and W. Kohn, *Phys. Rev. B* $\underline{13}$, 2270 (1976); B.N.J. Persson and P. Apell, *Phys. Rev. B* $\underline{27}$, 6058 (1983); E. Zaremba and B.N.J. Persson, to be published.

16. W.L. Schaich, *Phys. Rev. B* $\underline{14}$, 1488 (1976).

17. See the last two works in Ref. 15.

18. A.G. Eguiluz, *Phys. Rev. Lett.* $\underline{51}$, 1907 (1983).

PROBE-HOLE FEM AND TDS STUDIES OF NO ADSORBED ON INDIVIDUAL TUNGSTEN
PLANES.

R.E. Viturro and M. Folman
Department of Chemistry, Technion I.I.T.
Haifa 32000, Israel.

ABSTRACT

The adsorption, migration and desorption of NO on single tungsten
planes was studied by means of probe-hole field emission and thermal
desorption techniques over a wide range of coverages and temperatures.
A molecular beam system that enabled deposition of strictly controlled
doses of gas was employed. The system was connected to a probe-hole
field emission tube in which the electron emission from individual
planes was measured by means of a photoelectric method. Work function
changes were measured for the (211), (100) and (111) planes as a
function of coverage and temperature. Thermal desorption spectra were
obtained for polycrystalline ribbon and single W (100) plane. It has
been concluded from the results obtained that at 77K NO is adsorbed as
molecular species, being a precursor to dissociative adsorption at
higher temperatures. The adsorbed nitrogen resulting from dissociation
exists mainly in the β_1 state on the (100) plane. The corresponding
desorption peak is very narrow and well defined as compared with the
much weaker β_2 peak. The desorption spectrum of the β_1 state was
satisfactorily compared with the computed one from the Wigner-Polanyi
first order desorption kinetics and a lateral interaction dependent
activation energy.

1. INTRODUCTION

The adsorption of NO on transition metal surfaces has received
considerable attention in recent years. Few studies were reported on
the NO/W system and the interpretation and conclusions reached are
controversial, mainly for the low temperature results (1-8). In order
to elucidate the nature of this adsorption system we have carried out
the study of adsorption and desorption of nitric oxide on single
crystal planes of tungsten, at different temperatures, by means of the
probe-hole field emission technique (FEM). The study on macroscopic
crystal planes was accomplished by means of flash desorption and
molecular beam techniques. Desorption of nitrogen from adsorbed layers

B. Pullman et al. (eds.), Dynamics on Surfaces, 271–285.
© 1984 by D. Reidel Publishing Company.

of nitric oxide on polycrystalline tungsten and W(100) single crystal
plane were measured over a large range of coverages.

2. EXPERIMENTAL

The method employed consisted of depositing NO by means of a
molecular beam onto a tungsten field emitter, held at liquid nitrogen
temperature. Emission currents from individual planes were measured as
a function of applied voltage for different NO coverages and
temperatures using the probe-hole method. The apparatus consisted of a
U.H.V. stainless steel chamber (300 cm^3 in volume). It was used in
both FEM and FD measurements. The chamber had six exits connected to
different systems, each of them with specific functions: two were
connected to the pumping and molecular beam systems, one to a Bayard-
Alpert ionization gauge, another to a manipulator and the last two to
a window and to a probe-hole FEM tube or to a mass spectrometer. The
whole system, after baking, could be pumped down to a pressure lower
than $2x10^{-10}$ torr.

The probe-hole FEM tube, a modification of the one described by
Gomer et al (9), is shown in Figure 1.
The field emitter was spot welded to a tungsten loop 0.1 mm in
diameter and was mounted on a manipulator. Electrons emitted from the
tip struck a nickel anode coated with a fluorescent material
(willemite). A fraction of the emitted electrons, from a particular
single plane, pass through a small hole (1 mm in diameter) in the
anode. This emission current was measured by means of a photomultiplier detector. The emitted electrons, after passing through the
probe hole are collimated and pass through a transparent tungsten mesh
grounded electrode. The electrons are accelerated up to 10 KeV and
struck a glass anode coating with a fast responding phosphor,
producing light that is photomultiplied. The second anode is coated
with a thin aluminum film to avoid direct light from reaching the
photomultiplier (EMI 9558). The performance of the FEM tube was tested
by plotting the emission current from individual tungsten faces as a
function of the applied voltage to the second screen for constant
total emission current. A gain factor of about 10^5 as compared to the
Faraday cage collector method was obtained.

Thermal desorption measurements were done on a polycrystalline
tungsten ribbon and on a (100) oriented tungsten crystal. The W sample
supplied by Material Research Corporation had impurities of a few ppm.
After polishing the crystal, its orientation was determined to within
0.5^o of the (100) plane using X-ray back scattering. The crystal was
heated to any desired temperature by electron bombardment from a W
filament positioned behind the crystal. The temperature was monitored
with a W/W-Re 26% thermocouple. For TDS measurements the crystal
temperature was raised in a linear form by controlling the electron
emission and the potential difference between the filament and the
crystal. Typical heating rate β was 25 K s^{-1}. The characteristic
pumping time of the system was about 0.03 sec. After dosing the
adsorbent by means of the molecular beam the desorption mass spectra

were recorded using a UTI 100 C Quadrupole Mass Analyser mounted in line of sight to the crystal. Calibration of the mass analyser was done against a Bayard-Alpert ionization gauge.

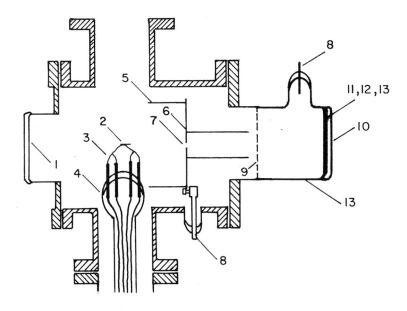

Fig. 1: The probe-hole field emission tube.

1. Glass window. 2. Emitter. 3. Heating loop. 4. Support loop.
5. Anode. 6. Screen. 7. Probe hole. 8. High voltage. 9. Grounded electrode. 10. Window. 11. SnO_2 coating. 13. Fluorescent screen. 14. Al coating.

    Research grade O_2, N_2, and NO were further purified by repetitive distillation. They were stored at pressures of about 500 torr in an external manifold and admitted to the Knudsen cell through a metal leak valve.
    Cleaning of the sample was done by exposure of the crystal, held at about 1500K, to oxygen flux and flashing in UHV to 2500K. The sample was considered clean when no CO signal was detected in the mass analyser.

3. RESULTS

3.1. Field emission results.

    As mentioned above, the NO layer was obtained by depositing gas onto the emitter by means of a molecular beam. Similarly to the results reported by Madey and Yates (1) the spreading of the adsorbate over the emitter was met with serious difficulties. Only at temperatures above 600K migration of the chemisorbed species occurred.

To avoid this difficulty the emitter was positioned with the 110 axis parallel to the impinging beam. This permitted measurements of work function dependence on coverage at low temperatures. The dose magnitude was maintained constant by fixing the pressure in the effusion cell and the exposure time. Correction was introduced for each single face by multiplying the dose size by cos γ_i, where γ_i is the angle between the normal to the face and the beam axis. The work function values were determined from the current-voltage dependence

$$\ln \frac{I}{V^2} = \ln A - \frac{\alpha \phi^{3/2}}{V}$$

relying on the Fowler-Nordheim equation (10) where I is the emission current, V the applied voltage and ϕ the work function. The change in work function of single crystal faces caused by adsorption was found by taking the ratio of the slopes of the Fowler-Nordheim plots before and after adsorption, S^o and S^i respectively. Then

$$\phi^i = \phi^o \left(\frac{S^i}{S^o} \right)^{2/3} \left(\frac{\alpha^o}{\alpha^i} \right)^{-2/3}$$

α^o and α^i are functions of the field constants for the crystal face before and after adsorption, respectively. Their ratio is usually taken as 1.

The emission patterns at low temperatures for increasing NO doses show small changes in the relative emission of different portions of the tip. A decrease in contrast between the bright and shadow regions was noticed, followed by a decrease in the apparent surface area of the shadow zones (110), (211) and (100). This being similar to the behaviour of CO on W (10). Figure 2 shows the change in $\Delta\phi$ as a function of a number of doses at 77K.

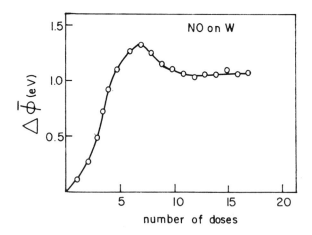

Fig. 2: Variation of $\Delta\phi$ with number of doses. 2×10^{14} molecules. cm^{-2} of NO in each dose.

Figure 3 shows the dependence $\Delta\phi$ of three different faces (211), (100) and (111) on the number of normalized doses. It may be seen that for the initial stage of adsorption, up to n=5-6, a linear dependence of $\Delta\phi$ on n is obtained. For higher doses a change in the slope is seen and for n=15-18, the curves show a maximum. The $\Delta\phi$ values in increasing order are $\Delta\phi(211) < \Delta\phi(100) < \Delta\phi(111)$. Further dosing up to n=23 causes a decrease in $\Delta\phi$, this being again larger for the (111) face. For n=25 the curves level off. Finally to ensure that full coverage was achieved a dose corresponding to about 50L was applied. No additional change in $\Delta\phi$ was observed.

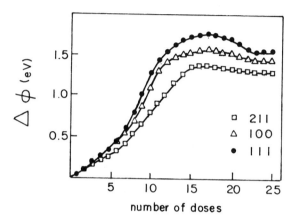

Fig. 3: Work function changes of single planes versus number of NO doses deposited. Each dose 1.2×10^{14} molecules.cm^{-2}.

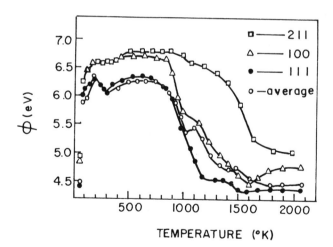

Fig. 4: Work function changes of single planes with temperature.

Figure 4 shows the change in work function when the emitter,

kept at 77K, was saturated with NO and subsequently heated to various
temperatures for 40 sec intervals.
At first an increase in $\bar{\phi}$ is seen by about 0.46 eV at 170K. On
subsequent heating the work function decreases and reaches a minimum
at 300K. In parallel the field emission pattern shows a number of
bright spots localized between the (110) and (121) faces. Most
probably this being the reason for the drop in the $\bar{\phi}$ value. On further
heating up to 550K the bright spots disappear and a change in the
pattern occurs. The pattern resembles the original clean one, but
without sharp contrast between bright and shadow regions, and
enlargement of the shadow areas is observed. The work function
practically remains constant up to 900K. From that temperature onwards
the $\bar{\phi}$ drops sharply on further heating and reaches the value of a
clean emitter at about 1800K. The work function change for the (111)
plane resembles quite closely the behaviour of the $\bar{\phi}$ up to 1000K.
Above that temperature the decreases in ϕ (111) is sharper than that
of $\bar{\phi}$. The ϕ versus T curves for the (211) and (100) planes show a
steady increase up to 700K with no local minimum at 300K. Above 300K
the work function of the (100) plane drops sharply and reaches a
minimum of 4.45 eV at 1600K. Above that temperature the ϕ (100) raises
to the value of the clean plane. The high temperature behaviour of the
NO covered (211) plane shows a slow decrease of the work function from
900K to 1800K, where the $\Delta\phi$ equals zero.

3.2. Thermal desorption results.

    Figure 5 shows thermal desorption spectra of nitrogen for differ-
ent coverages of nitric oxide adsorbed on polycrystalline tungsten.

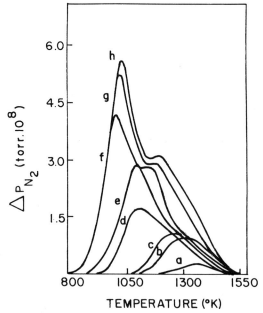

a. $\Theta = 0.05$
b. $\Theta = 0.15$
c. $\Theta = 0.20$
d. $\Theta = 0.35$
e. $\Theta = 0.50$
f. $\Theta = 0.83$
g. $\Theta = 0.92$
h. $\Theta = 1.00$

Fig. 5: Thermal desorption of N_2 from NO/W (poly) $\beta = 40$ K s^{-1}.

The TDS spectra in the temperature range 800K-1600K resemble quite
closely those reported by Yates and Madey (1). The only difference
being the better resolution attained in our experiments. The main
reason for that, most probably, was the high pumping speed of our
system and the relatively low heating rate used. The full coverage
spectrum shows two overlapping bands with maxima at 1040K and 1250K.
The two bands were further resolved by computer deconvolution. The
fractional surface coverages were calculated by measuring the areas
under the thermal desorption curves and dividing them by the area
under the saturation curve. At saturation the relation between the
areas of the two high temperature modes was 1:1. We did not find the
broad band reported by Yates and Madey (1) for mass 30, with a maximum
at ~ 450K. However, in the low temperature region, 100K-300K, we found
a peak for mass 30 with a maximum at 140K, Figure 6.

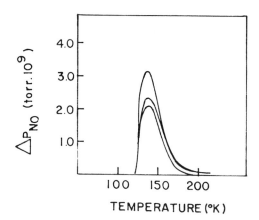

Fig. 6: Low temperature desorption of N_2 from NO/W (poly).

Figures 7 and 8 show thermal desorption spectra of nitrogen from
nitric oxide adsorbed on W(100).

The development of two peaks even at the low coverages attainable
is observed. The high temperature peak, assigned to the β_2 state of
nitrogen grows slowly and widens with increasing dose. The low
temperature peak assigned to the β_1 nitrogen state increases rapidly
and shifts to higher temperatures with coverage. Another feature of
the TDS spectra is the diminution of the full width at half maximum
(FWHM) of the β_1 peak with increasing coverage. At saturation the
ratio between the β_1 and β_2 bands was about 5:1. The experimental
value of the saturation coverage was 10×10^{14} NO molecules.cm^{-2}.

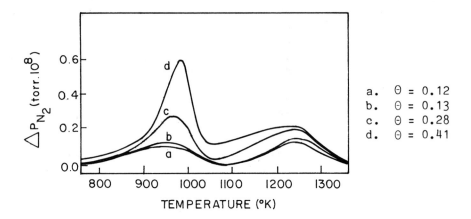

a. $\Theta = 0.12$
b. $\Theta = 0.13$
c. $\Theta = 0.28$
d. $\Theta = 0.41$

<u>Fig. 7</u>: Thermal desorption of N_2 from NO/W(100).
Low coverage range. $\beta=27$ K s^{-1}.

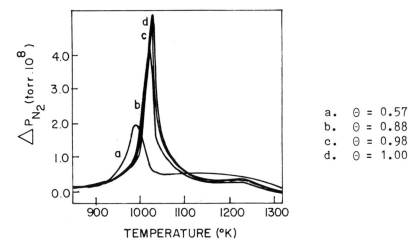

a. $\Theta = 0.57$
b. $\Theta = 0.88$
c. $\Theta = 0.98$
d. $\Theta = 1.00$

<u>Fig. 8</u>: Thermal desorption of N_2 from NO/W(100).
High coverage range. $\beta=27$ K s^{-1}.

4. DISCUSSION

4.1. Changes of work function with coverage.

To present the results on a quantitative basis the work function
changes were depicted as a function of coverage. To do this, one has
to know the dependence on coverage, Θ, of the sticking probability, S,
for nitric oxide on single crystal planes of tungsten. Up to now very
few measurements of S for NO/W were done (3,11). However, these do not
seem to be very accurate, because of the method used. We have measured

S for NO/W (polycrystalline) and NO/W(100) using the molecular beam
and desorption techniques. The results indicate that, within the
resolution of this method, the dependence of S on Θ is very similar
for the different crystal planes. Thus an average value for S(Θ) was
adopted. This dependence is shown in Figure 9. The Δφ values replotted
as a function of Θ are shown in Figure 10.

    In the curves three different regions can be distinguished.
First, for all the planes studied there is a linear dependence between
Δφ and Θ up to Θ ≈ 0.4. Also the three slopes are similar. From that
coverage onwards a second linear portion is seen with a larger slope
as compared to the first one. The three curves reach a maximum at
about Θ =0.9. For values of Θ between 0.9 and 1 there is also a
decrease in Δφ. Similar behaviour of Δφ vs. exposure was found by
Klein and Yates (3) in NO/W(110) at 120K using an electron gun and the
retarding potential method.

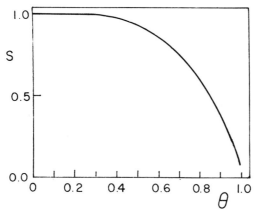

Fig. 9: Sticking probability as a function of surface
        coverage for NO/W(100).

    In order to explain these changes of work function on adsorption
a number of possibilities should be taken in account. The first one is
the existence of adsorbed molecular species. In such a case the binding and orientation of the adsorbed molecules on the three planes have
to be quite similar, and the change produced in Δφ vs. Θ curves could
be explained by the different bonding states of the nitric oxide molecule at each stage of the adsorption process. The bonding states of
the two first regions must comply with the fact that there is an
increase of the work function, which means a formation of a surface
dipole moment with negative end pointing outwards, while in the third
stage the surface dipole moment will have an opposite direction.

    It is assumed that molecular adsorption results in an NO molecule
bound to the surface via its N end. There is some experimental evidence for that through XAES results which show a stronger perturbation
of the N KLL than the O KLL spectrum on adsorption (12). Also that is
the only observed bonding geometry in nitrosyls (13). Three different
geometries can be assigned to bonded NO:linear, bent and bridged. The
last two are usually assumed to be negatively charged (increase in Δφ)

while linear species are supposed to be positively charged, NO⁺,
through charge transfer from the antibonding π^* orbital to the metal
(decrease in $\Delta\phi$). To discuss the bonding of the two electronegative
states we must compare the relative strength of the two surface dipole
moments formed in bridged and bent configurations.

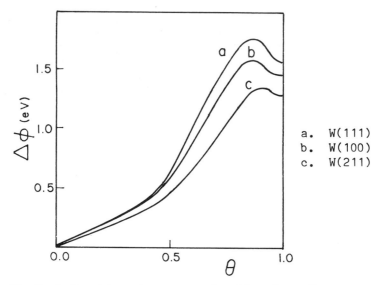

a. W(111)
b. W(100)
c. W(211)

Fig. 10: Work function change vs. fractional surface coverage.

A qualitative explanation may be given based on the XPS spectra of
these configurations. It was found that the binding energy difference
ΔE of O1s and N1s peaks is in the range 132 eV for linear nitrosyls
and 128 eV for bent nitrosyls (14). There is no data of ΔE for bridged
nitrosyls, but it is believed that it lies between those of linear and
bent structures; i.e. $\Delta E=130$ eV; an assumption which is supported by
some experimental evidence (12). The lower value of ΔE of the bent
structure may be explained by applying the equivalent cores
approximation (15), which shows a diminution in the O1s binding energy
of the bent configuration caused by an electron flow from the metal to
the molecule. The opposite occurs in the linear structure. Therefore
we assume that the bent structure is more electronegative than the
bridge configuration. According to our results we propose that
molecular NO is adsorbed in a bridge configuration at lower coverages
followed by adsorption in a bent configuration for coverages larger
than $\Theta=0.4$. The existence of a transient molecular NO state on the
W(110) plane at 100K was assumed by other workers relying on results
from XPS and UPS measurements (7). It is believed that in our case the
lowering of the temperature down to 77K stabilizes these molecular
structures.

The decrease in $\Delta\phi$ found at high coverages, $\Theta > 0.9$, may be
caused by adsorption of an electropositive species, such as linearly
bound NO. There also exists the possibility that mutual depolarization
of surface dipoles will diminish the dipole moment of the adsorbed

molecules. The strength of this depolarization will depend on the
distance of the neighbouring dipoles. There exists still the
possibility of formation of N_2O on the surface as a result of the following reaction: $2NO(ads) \longrightarrow N_2O(ads) + O(ads)$. In many cases such a
reaction has been found even at low temperatures and low surface
coverages. In case of tungsten adsorption of NO at 100K and at very
high coverages, this led to the formation of N_2O as deduced from UPS
spectra (7). It is not easy to predict the influence of possible N_2O
formation on the work function.

The assignment of the changes in ϕ to complete dissociation of
the molecule and presence of O and N adsorbed on the surface could
cause some difficulty in interpreting of the results. It is known that
both O and N raise the work function of the (111) and (211) planes
whereas on the (100) plane nitrogen lowers by 0.8–0.9 eV (16) and oxygen raises the work-function by 1.6 eV at saturation (17). Therefore
the $\Delta\phi$ vs. Θ dependence should be much different for the (100) plane
as compared to the two remaining planes, which is not the case.

Another possibility is partial dissociation of the NO molecule.
In such a case three species should be present on the surface, and in
order to explain the linear dependence of $\Delta\phi$ on Θ the concentration
ratio of the different species has to stay constant, which is not very
likely to occur.

The last possibility is the existence of physisorbed molecules at
77K at low coverage as single species and at higher coverages as
dimers, although well known to occur on ionic crystals and semiconductors (18), cannot be accepted here since the adsorbed layer
shows a complete absence of mobility at this and higher temperatures,
as shown in our measurements and other workers (19,1).

Therefore we conclude that NO is adsorbed as molecular species at
77K being a precursor to dissociative adsorption.

4.2. Changes of work function with temperature.

The initial increase in the work function on heating from 77K up
to about 170K on all the planes studied, may be explained as due to
partial desorption of the NO molecules, mainly those adsorbed in the
electropositive state (linear configuration), and the dissociation of
the remaining NO into atomic N and O. In addition, a decrease in the
preexponential factor, not shown, is also observed on all the planes,
which can be interpreted as a change in the adsorbed state and a
decrease in the amount of adsorbed species. The XPS and UPS results
obtained on W (polycrystalline) (2) and on W(110) (7) at about 100K,
as well as results obtained in these systems and in our XPS work on
W(100) also indicate complete dissociation at room temperature.

The results for the (100) plane above 200K show that the work
function increases slightly by about 0.1 eV up to 300K. It is accepted
that adsorbed nitrogen may exist in two surface states, β_1 and β_2. The
β_1 state raises the work function by about 0.15 eV and desorbs at
temperatures above 800K, whereas β_2 lowers the work function of this
plane by −0.9 eV (16). Since our results give a total $\Delta\phi$ of about
1.8 eV in the region considered, it is assumed that on that plane

nitrogen in the β_1 state coexists with atomic oxygen. For temperatures above 900K the work function drops sharply and reaches a minimum at 1600K. This decrease is caused by desorption of the nitrogen and by rearrangement of the remaining oxygen. Such a decrease in ϕ when only oxygen is adsorbed has been noticed by Bell et al. (17). The fact that mainly one nitrogen desorption peak was obtained in our TDS spectrum at 1030K points out that mainly the β_1 nitrogen state is present on the (100) plane as result of dissociation of NO.

After the initial increase in $\Delta\phi$ already explained and present in all the surface planes studied, the $\Delta\phi$ of the (211) plane reaches the value of ~ 1.7 eV and remains constant up to 550K. This value is made up from contributions from adsorbed N and O. In this temperature range nitrogen raises the work function by 0.4 eV, and oxygen by 1.4 eV at 200K increasing to 1.6 eV at 550K. Above this temperature nitrogen leaves this plane by surface migration onto other regions (preferentially the (111) plane) (16). On the contrary, oxygen most probably diffuses into this plane. The work function remains constant, and begins to decline slowly above 900K. The clean plane value is reached at high temperature, above 1900K, which may be due to the higher binding energy of oxygen in the (211) plane (17).

The interpretation of the results obtained on the (111) plane presents some difficulty. The initial portion of the $\Delta\phi$ curve shows not much difference from the results on the other planes. Also in this case it is assumed that dissociation of NO begins above 80K and the observed $\Delta\phi$ values represent contributions from atomic N and O. Oxygen on this plane raises the work function by 1.6 eV and this value remains constant up to 1200K. Nitrogen only slightly affects the work function, causing an increase of ~ 0.1 eV. Therefore, mainly the adsorbed oxygen contributes to the measured ϕ value. The occurrence of a local minimum at 300K is difficult to explain. It coincides with the appearance of the bright spots mentioned in the "results" section. The decrease in $\Delta\phi$ begins at a temperature 200 degrees lower than that observed in adsorption of a complete oxygen monolayer. However, results obtained with fractional coverages of oxygen on tungsten show changes in $\Delta\phi$ already at 800K (17). The descending portion of the curve is caused by desorption of nitrogen with simultaneous migration of oxygen out of this plane. These two effects lead to a value close to the characteristic one for clean (111) plane already at 1150K.

4.3. Thermal desorption spectra.

One of the most striking results of the TDS measurements on W(100) is the relatively high intensity of the low temperature peak, assigned to the β_1 state of the adsorbed nitrogen. This is in agreement with our field-emission measurements, where as it was stressed above, the work function changes with temperature on the (100) plane are interpreted as due to the presence of N(β_1) state as a result of NO dissociation. The TDS spectra of adsorbed nitrogen on W(100) show clearly that the N(β_2) state is the dominating one and the N(β_1) state is obtained only for surface coverages approaching saturation (20). These findings may give some support to the belief that the adsorption sites for nitrogen and oxygen on the W(100) plane are not the same. It is believed that nitrogen in the β_1 state is adsorbed on top of the outer tungsten atom layer while in the β_2 state the nitrogen is bound in a C_5 site - in a well between four tungsten atoms. Most probably in our case this site is occupied preferentially by oxygen atoms, while the sites for nitrogen are those in the top position.

Another finding which supports this, is connected with the saturation concentration of nitrogen on the (100) plane. From numerous measurements it is known that in adsorption of nitrogen the ratio of surface tungsten atoms to adsorbed nitrogen atoms is 2:1, and the LEED pattern shows a C 2x2 structure at saturation. In case of NO adsorption the saturation ratio becomes 1:1, as deduced from the present work and from XPS studies of this system (21). Thus the presence of O in the C_5 position causes an electron charge redistribution in the adjacent tungsten atoms that facilitates the formation of N-W bonds on each atom.

The broad peak with a maximum at ~ 1230K at saturation is obtained most probably from a fraction of nitrogen atoms adsorbed in the β_2 state. As stressed this fraction is only ~ 15% of the total desorbed amount. Despite the fact that the C_5 sites are preferentially occupied by oxygen atoms, due to equilibrium requirements some of these sites should be still available for nitrogen adsorption. A similar result is obtained in decomposition of NH_3 on W(100) as studied by the probe-hole field emission method (22). Another possible contribution, may be surface heterogeneity, and simultaneous surface reconstruction due to the remaining oxygen, which may change the surface symmetry.

To explain the dependence of both the peak temperature T_m and of FWHM on Θ, we assume that the activation energy for desorption depends on the surface coverage. This dependence may be due to the interaction between the ad-species. On assuming that $E_d = E_o + c\Theta$, where E_d is the activation energy for desorption, and c is a constant, and supposing first order desorption kinetics, the computed dependence of T_m and FWHM on Θ shows poor agreement with the experimental results. On the other hand, when attractive lateral interaction between adsorbate atoms was taken into account the agreement was quite satisfactory. A computer fitting, using the Runge-Kutta-Verner method, was carried out relying on the Wigner-Polanyi equation for first order desorption kinetics and a lateral interaction dependent activation energy

$$- \frac{d\Theta}{dt} = \nu\Theta \exp \left[- \frac{E(\Theta,w)}{kT} \right]$$

where ν is the preexponential factor and w is the pairwise attraction energy (23)

$$E(w,\Theta) = E_o + \frac{zw}{2} \left\{ 1 - \frac{1-2\Theta}{[1-4\Theta(1-\Theta)(1-\exp\frac{w}{RT})]^{1/2}} \right\}$$

z=4 for a square lattice.

The computed curves are shown in figure 11, and the numerical values of the fitting parameters were:

$$E_0=51.6 \text{ kcal/mol, } w=2.5 \text{ kcal/mol, } \nu = 1.45\times10^{12} \text{ s}^{-1}$$

The E_0 value compares well with the experimental one for desorption of the β_1 state of $N_2/W(100)$ (20), where a value of ~ 50 kcal/mol was obtained assuming $\nu = 10^{13} \text{ s}^{-1}$.

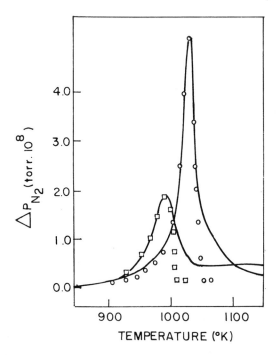

Fig. 11: Theoretical fit to thermal desorption trace for the β_1 state. Points indicate the computed spectra.

For intermediate and high coverages the high temperature tail of the β_1 experimental curve does not reach the base-line and the agreement between the two curves is not too good. This may be explained as due to variable background caused by desorption of nitrogen from the walls as result of filament radiation. The striking feature of the $N(\beta_1)$ peak is its low FWHM value. This is adequately explained by the theory presented above and by the value of the w parameter. Comparison of the TDS results from polycrystalline ribbon and the (100) single crystal indicates that the resolution of the two peaks ascribed to the $N(\beta_1)$ and $N(\beta_2)$ surface states is much better for the latter system. This is not surprising since the adsorption energy and the concentration ratio of the two states differs for the different planes; therefore the desorption spectrum from the ribbon represents a superposition of a number of spectra characteristic for the different planes. The overall spectrum shown in Fig. 5 is similar to the one obtained with the W(110) single plane (3) where the surface packing is the closest one and differs from the W(100).

ACKNOWLEDGEMENTS

The authors wish to thank Mr. A. Levy and I. Lior for constructing the experimental apparatus.

REFERENCES

1. J.T. Yates, Jr., and T.E. Madey, J. of Chem. Phys., 45, (1967) 1623.
2. J.T. Yates, Jr., T.E. Madey and N.E. Erickson; Surf. Sci., 43, (1974) 526.
3. R. Klein and J.T. Yates, Jr., Japan. J. Appl. Phys. Suppl. 2, pt.2 (1974) 461.
4. S. Asami and T. Nakagima, Japan. J. Appl. Phys. Suppl. 2, pt.2 (1974) 237.
5. J.C. Fuggle and D. Menzel, Surf. Sci., 79 (1979) 1.
6. A.K. Bhattacharya, J.Q. Broughton and D.L. Perry, Surf. Sci., 78 (1978) L689.
7. R.I. Masel, E. Umbach, J.C. Fuggle and D. Menzel, Surf. Sci., 79 (1979) 26.
8. K.J. Rawlings, S.D. Foulias and B.J. Hopkins, Surf. Sci., 108 (1981) 49.
9. R. Gomer and R. Diffoggio. The 25th International Field Emission Symposium (1978).
10. R. Gomer. Field Emission and Field Ionization (Harvard University Press - 1961).
11. T. Tamura and T. Hamamura, Bull. Chem. Soc. Japan, 49 (1976) 1780.
12. E. Umbach, S. Kulkarni, P. Feulner and D. Menzel, Surf. Sci., 88 (1979) 65.
13. L.Y. Chan and F.W.B. Einstein, Acta Cryst., B 26 (1970) 1899.
14. C.C. Su and J.W. Faller, J. Organomet. Chem., 84 (1975) 53.
15. W.L. Jolly, Electron Spectroscopy: Theory, Techniques and Applications, Volume 1, Ed. by Brundle and Baker (Academic Press, 1977), page 119.
16. M. Wilf and M. Folman, Surf. Sci., 52 (1975) 10.
17. A.E. Bell, L.W. Swanson and L.C. Crouser, Surf. Sci., 10 (1968) 254.
18. A. Lubezky and M. Folman, Trans. Faraday Soc., 67 (1971) 58.
19. R.E. Viturro, Ph.D. Thesis, Israel Institute of Technology, 1983.
20. L.R. Clavenna and L.D. Schmidt, Surf. Sci., 22 (1970) 365.
21. E. Pelach, R.E. Viturro and M. Folman, to be published.
22. M. Wilf and M. Folman, J. Chem. Soc., Faraday Trans. I, 72 (1976) 1165.
23. C.G. Goymour and D.A. King, J. Chem. Soc., Faraday Trans. I, 69 (1973) 736.

SURFACE AMPLIFICATION OF PHOTOIONIZATION SIGNAL – A PROBE FOR ADSORBED MOLECULES

A. Petrank, D. Lubman[*]   and R. Naaman

Department of Isotope Research
Weizmann Institute of Science
Rehovot, Israel

Abstract

The surface enhanced ionization (SEI) effect, in which adsorbed molecules are ionized by near UV light, is discussed. Following previous studies on some aromatic molecules adsorbed on metal, we present in this work recent results obtained with small molecules, and with semiconductor surfaces.

All the data obtained so far can be explained by a two stage model. In the first step adsorbed molecules are ionized thus forming an ionic layer which reduced the surface work function. In the second stage, photoelectrons are emitted very efficiently due to the lowered work function. Fast recombination (neutralization) is responsible for the structural spectra obtained.

----

[*]Permanent address: Department of Chemistry, The University of Michigan, Ann Arbor, MI 48109, U.S.A.

B. Pullman et al. (eds.), Dynamics on Surfaces, 287–296.
© 1984 by D. Reidel Publishing Company.

Introduction

The effect of surfaces on the interaction of light with molecules
is a very complicated, yet important subject. Recently, several related phenomena are under intensive study, among others, the Surface
Enhanced Raman Scattering (SERS) [1], and second harmonic generation
on metal surfaces [2]. One would expect that other multiphoton processes can also be enhanced by surfaces in a similar way. In previous
papers [3] we introduced the new technique called Surface Enhanced
Ionization (SEI). The near UV spectra of adsorbed molecules was obtained, and it was found that with only 10^{12} molecules sampled, a current of 10^{-5} A of electrons could be observed. This mean amplification by several orders of magnitude of the number of electrons
produced in the initial multiphoton ionization process. The results
could not be explained by simple enhancement of the electromagnetic
field by itself, and a more elaborate model had to be used.

The surface effect on generation of photoelectrons is an established phenomenon. Ultraviolet photoelectron spectroscope is a useful
tool for probing surfaces and adsorbance, and its theory is well understood. Adsorbed molecules modify the work function (WF) of metals
and semiconductors, therefore, decreasing or increasing the photoelectron yield. Using this idea Langmuir explained the observed alkali
induced WF [4]. In his model the adatoms are ionized by spontaneous
charge transfer to the surface, and the ionic layer reduces the WF at
the surface. Our results could not be explained by any one of the
known theories by itself, and only a combination of processes could
rationalize all the observations. The SEI effect was discovered first
using metal surfaces and molecules with low ionization potentials, so
that resonance two photon ionization (R2PI) could occur. In this work
the studies were expanded to some small molecules and semiconductor
surfaces. This was done to gain better understanding of the mechanism
and the processes involved. In particular, effects of ion recombination rate, and surface states were studied.

Experimental

The experimental set-up consisted of a Pyrex cell pumped by a 2"
diffusion pump with liquid N_2 trap. In one configuration - i.e. external reflection, the cell contained two parallel plates 1 cm apart
(Fig. 1a).
One electrode served as the surface under study, and the second plate
was the collector. Several different metallic surfaces were used including aluminum, silver and platinum. The thin films were produced
using vapor decomposition under vacuum. The thickness of the metal
films was varied from 50-1000Å.

Semiconductor surfaces such as In doped SnO_2, CdS and SnO_2 were
also investigated, but we concentrated in studying n and p type sili-

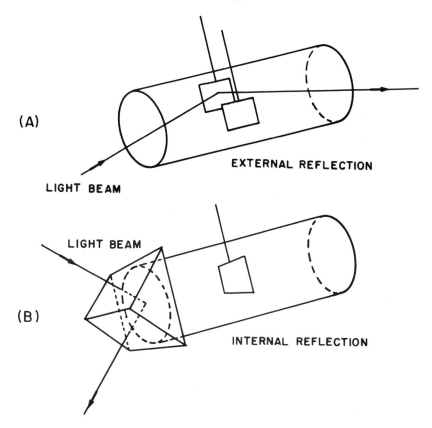

Figure 1: Experimental set up. (a) External reflection configuration. (b) Internal reflection configuration.

con. The silicon surfaces were polished on microcloth with polishing
alumina with particle sizes of 5μ, 1μ and 0.3μconsecutively until a
shining surface was obtained. Etching was done by putting the surfaces for one minute in a solution with volumetric composition of 8 parts
70% nitric acid, 3 parts 100% acetic acid, and 1 part 50% hydrochloric
acid. The laser was introduced at an angle almost tangent to the
electrode.

In a second configuration, a metal or semiconductor coated quartz
prism served as the window and as the excitation surface (Fig. 1b).
The laser light was introduced into the cell by total internal reflection (IRF) through the back of the metal surface. In this configuration the transmission of the evanescent wave is such that we obtain
excitation of molecules within about 500Å of the coated prism surface.
In this case, the light interacts mainly with adsorbed molecules near
the surface. Thus, it is possible to eliminate gas phase processes
which may be induced in the external reflection technique.

In the IRF configuration, the same metals were investigated as in the external reflection technique. The metal film thickness was varied from 50 to 500Å and a collector faced the coated surface at a distance of one cm. The pressure in the cell could be varied from the vapor pressure of the molecules at room temperature down to 10^{-6} torr using the diffusion pump and a cold finger for determining the partial pressure of the probed molecules.

The coated surfaces were biased between -400 and -600V typically and the current collected at the second electrode. The light source was a Chromatix CMX-4 flashlamps pumped dye laser, which is frequency doubled into the UV. This laser has a long pulse length (1μs) and thus provides low peak power of up to one KW and with a beam diameter of 0.3 cm. The signal was processed using a PAR model 162 boxcar integrator with model 164 plug-in unit. The signal detected at the collector was electron current measured through a MΩ resistance. The current could be monitored as a function of wavelength. The measurements were performed on the molecules Maphthalene, aniline, NO_2, NO and SO_2 on metal films, and naphthalene and aniline on the semiconductors.

Results

SEI was applied to examine several small molecules, such as NO, NO_2, and SO_2 on metal surfaces. In these cases the ionization potential is greater than twice the photon energy, so that R2PI does not occur. Using an unfocused laser beam in the external reflection configuration no signal could be observed on an aluminum surface for any of these molecules. Using a focused beam we obtained a spectrum only for NO_2 as shown in Fig. 2.

The IRF configuration was used to further investigate the process. Initially no signal could be observed using this technique. However, if the pressure of the gas was greater than 10 m torr and the bias voltage was raised above -600V than a weak violet breakdown would occur, that was independent of the laser light.

As the bias voltage was lowered back to -400V, no discharge was present and the laser induced signal was on the order of 2×10^{11} electrons/sec. No structural spectrum could be observed as a function of wavelength, and no signal was observed when the collector was positively biased.

The signal obtained was found to decay slowly on the aluminium surface as a function of time. Signal was still present even ten minutes after the disappearance of any discharge. If the laser was blocked or the bias voltage turned off for several minutes, a signal was observed again when they were turned back on, even without the creation of a new discharge. In the case of Al the charge could be leaked off the

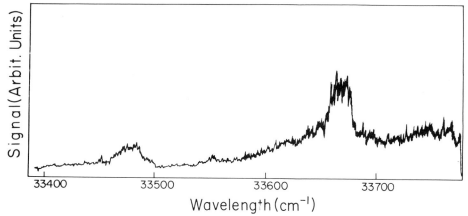

Figure 2:   SEI spectrum of NO_2 on Al surface.

surface using a piece of paper to accept the static charge and the signal would disappear completely until new discharge was formed.

This is evidence of a long lifetime for ions remaining on the aluminium surface after the initial discharge. This is not surprising because Al is readily oxidized to form Al_2O_3 and the recombination rate is very slow on Al_2O_3, so that the ions can live for minutes [5]. The same experiment was performed in the IRF configuration on a Pt surface and the recombination rate was essentially instantaneous on the time scale of our experiment.

The necessary conditions for observing the SEI effect are that the surface can be negatively biased and that positive ions are produced which reduce the work function of the surface. In principle, semiconductor surfaces should produce a SEI signal. Therefore, we have studied the SEI effect for the case of naphthalene on several semiconductor surfaces, i.e. In doped SnO_2, CdS (n-type semiconductors) n and p-silicon. We concentrated on the naphthalene/silicon system because we have characterized naphthalene behavior on metal surfaces, and silicon is one of the most studied semiconductors - its photoelectric parameters are well known, enabling comparison of n and p doped semiconductors. Before describing the experimental results one has to emphasize the fact that the experiments were performed at relatively low vacuum. For this reason and due to the high reactivity of silicon no claims on its purity or the coverage of the surface can be made. However, our surface preparation procedure ensured us reproducible results. As the observations are quite different from those of metals some important conclusions can be reached. For all semiconductor surfaces used, a sharp and intense naphthalene spectrum was obtained as demonstrated in Fig. 3.
In general polished surfaces were found to have a smaller enhancement effect than etched ones. This fact was established both by comparing

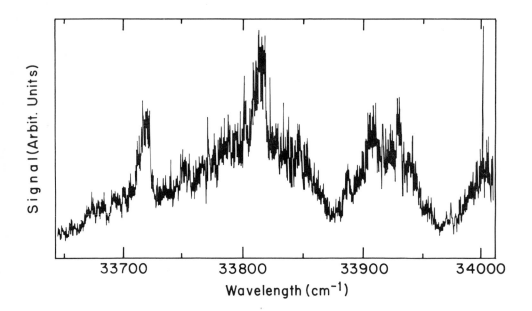

Figure 3:   Napthalene spectrum taken on CdS surface.

the gas signal with the surface enhanced signal, and by observing the
ratio between negative and positive particles arriving at the collector. As was established in the case of metal surfaces, here also only
the electron current arriving at the collector is enhanced while no
such effect could be detected for the positive ion current. For
etched p and n Si the ratio between the negative and positive charges
was 17:1 and 10:1 respectively, which means that a larger enhancement
factor was found for p silicon. Using the near UV light at about 300
nm the signal dependence on laser power was found to be quadratic. As
visible light (600nm) was combined with the near UV beam, a further
increase in the signal was observed and the signal dependence on the
total light intensity was found to be cubic.

Discussion

All our observations fit the model given before, in which two stages are assumed. In the first one ions are formed on the surface and they reduce the effective WF. In the second stage photoelectrons are emitted due to the near UV laser light. When the laser is tuned to a molecular resonance an ion is formed, either by two photon ionization or by absorption of one photon followed by spontaneous charge transfer to the surface. Due to the fast recombination process the lifetime of the ion sheet formed is short, and all the charge is neutralized in the interval between two laser pulses. Therefore, only as the laser is tuned on resonance the WF is reduced and efficient photoelectron current can be obtained.

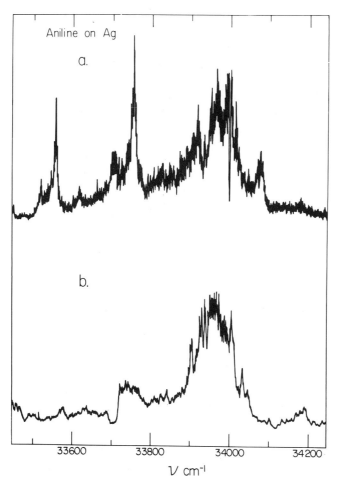

Figure 4: Aniline spectrum taken on silver surface in the (a) external reflection and (b) internal reflection configuration.

When the laser is off resonance, no ionic layer is formed, therefore, no reduction in the WF occurs, which results in almost no emission of photoelectrons. The process can therefore be viewed as a type
of "resonance photomultiplier" which amplifies the photoelectrons current only at molecular resonances.

Spectral broadening depends on the recombination rate. For surfaces on which this rate is high, sharp spectra are expected, as in the
case of silver (Fig. 4). For low neutralization rates, broad and diffuse spectra are seen as was the case for the Al surface (Fig. 5).

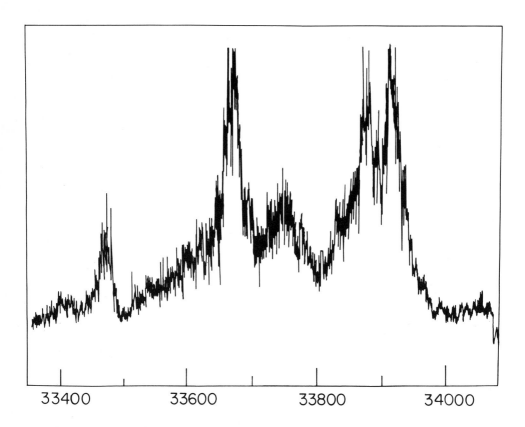

33400   33600   33800   34000

Figure 5: Naphthalene spectrum taken on Al surface in the
external reflection set-up.

The bigger enhancement observed on p-Si versus n type samples are
consistent with former theoretical [6,7] and experimental results
[9,10], in which it was established that cesium coated p-silicon exhibit a higher photoelectric quantum efficiency than n type samples.
This is due to the band bending with a net positive charge in the sur-

face states [11]. The contribution of the 600 nm light indicates that the WF of the surface covered with the ionic layer is reduced to about 2eV instead of being about 4.7 eV in the uncovered surface. Using the Helmholtz equation [12]

$$\phi' = \phi - 4\pi\sigma X_\circ$$

where $X_\circ = 3\times10^{-8}\,cm$ is the distance between the ions and the surface, σ is the charge density and ϕ and ϕ' are the initial and modified work functions respectively, we can obtain a coverage of about 4×10^{13} molecules/cm^2, which corresponds to almost a monolayer of ions.

The technique described here can be used both for detection of adsorbed molecules, and for probing different surface properties following the absorption process. Its simplicity and the ability to use it both at UHV and at high pressure conditions make it a very promising method.

Acknowledgements

The authors are grateful to V. Yakhot for many helpful suggestions and discussions regarding the model for the SEI process. The work is supported by the Fund for Basic Research administrated by the Israel Academy of Sciences and Humanities.

References

1. For recent reviews:
    a. R.P. Van Duyne, in Chemical and Biological Applications of Lasers 4, C.B. Moore, Ed., Academic Press, New York (1979).
    b. T.E. Furtak and J. Reyes, Surface Sci., 93 351 (1980).

2. a. C.H. Lee, R.K. Chang and N. Bloembergen, Phys. Rev. Lett., 18 167 (1967).
    b. C.K. Chen, A.R.N. de Castro and Y.R. Shen, Phys. Rev. Lett., 46 145 (1981).
    c. T.F. Heinz, C.K. Chen, D. Richard and Y.R. Shen, Phys. Rev. Lett., 48 478 (1982).

3. a. D.M. Lubman and R. Naaman, Chem. Phys. Letts., 95 325 (1983).
    b. R. Naaman, A. Petrank and D.M. Lubman, J. Chem. Phys., 79 4608 (1983).

4. I. Langmuir, J. Am. Chem. Soc. 54, 2798 (1932).

5. M. Even-Or, Tel-Aviv University, Department of Chemistry - private communication, Aug. 1983.

6. W.E. Spicer, J. Appl. Phys. 31, 2077 (1960).

7. L. Apker, E. Taft, and J. Dickey, Phys. Rev. 74 1462 (1948).

8. D. Redfield, Phys. Rev. 124, 1809 (1961).

9. J.J. Scheer, Philips Research Repts. 15, 584 (1960).

10. G.W. Gobell and F.G. Allen, Phys. Rev. 127, 141 (1962).

11. S.R. Morrison in "Treatise on Solid State Chemistry", ed. N.B. Hannay, Plenum Press, New York 1976.

12. a. C. Herring and M.H. Nichols, Rev. Mod. Phys. 21, 185 (1949).
    b. R.V. Culver and F.C. Tanphins, Adv. Catal. 11, 68 (1959).

VALENCE AND CORE-EXCITATIONS OF ADSORBATES: SPECTROSCOPY AND RELAXATION DYNAMICS

Phaedon Avouris
IBM Thomas J. Watson Research Center
P. O. Box 218
Yorktown Heights, New York 10598
USA

Abstract: We present a brief discussion of the spectroscopy and relaxation dynamics of adsorbates on metal surfaces. First we consider the excitations of physisorbed species. We discuss non-radiative decay mechanisms involving energy or electron transfer to the substrate and compare the predictions of the simple models with experiment. We then discuss the valence and core-excitations of chemisorbed species using CO as a prototype. We propose a new assignment for its excitations and discuss the bonding of the excited molecule to the surface. We close by briefly considering photochemistry at surfaces.

Introduction

A large number of surface dynamical processes involve or produce electronically excited states. These processes include photon- and electron-stimulated desorption[1], photo-chemistry of adsorbates[2], sputtering[3], surface Penning ionization[4] and chemiluminescence at surfaces.[5] The quantitative understanding of surface optical spectroscopies such as surface Raman scattering[6], second harmonic generation[7], resonance photoemission[8] and photoluminescence[9] also requires knowledge of the location and nature of the electronic resonances of the adsorbates. A large fraction of the chemical reactions at surfaces involves charge transfer to empty adsorbate levels. Study of the adsorbate electronic excitations provides valuable information regarding these levels. Finally, the electronic excitations can be used to test proposed ground state bonding schemes since in many cases a unique set of filled bonding and empty antibonding levels can be associated with a particular scheme.

Despite their importance in surface science, however, much less is known about excitations at surfaces than about the excitations of gases or bulk solids. This is partly due to the fact that the appropriate techniques such as synchrotron or EELS spectroscopy under UHV conditions have only recently become more generally available[10], and partly because the mixing of the discrete adsorbate and the continuous substrate (metal) states results in broad featureless bands which are difficult to assign.

Here we first discuss the new decay paths opened up to the excited adsorbate by the presence of the metal substrate. We then compare experimental results on the broadening of the excitations of weakly bound adsorbates with the predictions of the simple models of non-radiative decay. Special emphasis is then placed on the valence and core excitations of chemisorbed CO, a prototype chemisorption system. It is shown that the excitation spectra and recent inverse photoemission results agree that, unlike the generally accepted picture, the virtual $2\pi_a^*$ level of chemisorbed CO is at lower energies than the $2\pi^*$ level of free CO. The bonding and relaxation of ground, valence and core-excited chemisorbed CO is discussed on

297

B. Pullman et al. (eds.), Dynamics on Surfaces, 297–311.
© *1984 by D. Reidel Publishing Company.*

the basis of these results and ab-initio cluster calculations.   Finally, we discuss briefly the effects of the metal surface on the photochemistry of adsorbates.

I. Excited States of Physisorbed Species

Physisorbed adsorbates are weakly bound to the substrate primarily by dispersion forces.   In the absence of any specific chemical interaction with the substrate, the excited adsorbate-substrate interaction is treated as the dipole-dipole interaction between a dynamic dipole on the adsorbate with the image dipole it induces in the substrate (see next section). To a first approximation (ignoring Franck-Condon factor effects) changes in the excitation energies between gas and adsorbed phases reflect differences in the interaction between the ground and excited state adsorbate with the substrate.   Unlike ionization based spectroscopies such as photoemission (PES) and inverse photoemission (IPES), excitation spectroscopy is largely free from the complications of many body effects.[11] This is due to the fact that for a purely intraadsorbate excitation the charge state of the adsorbate does not change, and therefore there is no monopole adsorbate-substrate coupling to trigger a many body screening response. Here we discuss electronic excitations from the point of view of orbital levels as is customary in surface science.   To establish the connection between excitation spectroscopy and the observables in PES and IPES consider a simple two level adsorbate.   Level A has two electrons while level B is initially empty.   The free adsorbate has an excitation energy $E_{EXC}=IP-EA-U$ where IP and EA are respectively the ionization potential and electron affinity and U is an electron-hole attraction term.   When this system is placed on a metal surface (no chemical interaction), the ionization level (A) is raised due to the attractive interaction, v, of the hole in A with the free electrons of the metal and, similarly, the affinity level B is lowered by v.   Thus PES measures IP-v while IPES measures EA+v with v in the classical limit been given by $e^2/4d$.   However, despite the fact that the one electron levels have moved closer together, the excitation energy is not changed because the electron-hole interaction U has also been reduced (screened) by 2v due to the repulsive hole-image hole and electron-image electron interactions.   In principle, the effect of electron spin can be included by the use of the analogous concept of "image spin".

II. Non-Radiative Decay of Excitations at Surfaces

The adsorbate-substrate interaction besides shifting the excitation frequency, also quenches the excitation.   This enhanced non-radiative decay of electronic (and also vibrational) excitations at surfaces has very profound influence on many surface dynamical processes and we will consider it here in more detail.   In principle we can define two distinct types of quenching mechanisms: (A) damping by electromagnetic field coupling to substrate excitations or (B) damping by charge transfer[12] as shown in Fig. 1.

A. Electromagnetic Field Coupling

Consider an electronically excited molecule at a distance d from a metal surface.   The excited molecule can decay radiatively by the emission of a photon or non-radiatively by either internal conversion[13] or energy transfer to the excitations of the substrate.[12] The excitations of the solid which can accept energy from the excited molecule are electron-hole pairs, plasmons and phonons.   Since the Debye frequencies of metals are on the order of 10^{-2}eV, energy transfer to the phonons will be a high order multiphonon process, very much slower than transfer to the electronic degrees of freedom.   The relative importance of electron-hole pair vs. plasmon excitations depends on the excitation frequency Ω of the molecule. Plasmon excitation becomes important when $\Omega \sim \omega_p$ (bulk plasmon) or ω_{sp} (surface plasmon) frequencies.

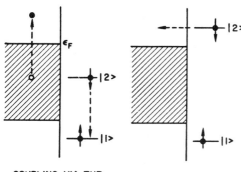

COUPLING VIA THE
ELECTRIC FIELD

RESONANT ELECTRON
TRANSFER

Figure 1   Quenching mechanisms for excitations at surfaces.

In the classical description[14] of the energy transfer process, the excited molecule is treated as a vibrating dipole characterized by a dynamic dipole moment $\mu = <A|\hat{\mu}|B>$, where A and B are the ground and excited state, respectively. Thus the energy transfer problem translates into the problem of the damping of an oscillating (with a frequency Ω) dipole coupled to the metallic electronic excitations. The energy transfer process is subject to conservation of momentum as well as energy. In the bulk of the metal the momentum required can be supplied by electron-phonon or electron-impurity scattering (in intraband transitions) or by the crystal potential (interband transitions). By virtue of the termination of the transverse periodicity at the surface, the surface potential can act as a "source" of momentum. (The spatial variation of the near field of the excited molecule can also contribute momentum components up to $\hbar/d$).[12]

In general, the non-radiative lifetime τ_{NR} can be expressed as[14]:

$$\frac{1}{\tau_{NR}} = \frac{\mu_\parallel^2/2 + \mu_\perp^2}{4d^3\hbar} F(\omega,d) \tag{1}$$

where the damping function F is given by

$$F = \int_0^\infty dq_\parallel q_\parallel^2 \, \text{Img}(q_\parallel,\omega)e^{-2q_\parallel d} \tag{2}$$

In Eq. (2) $g(q_\parallel,\omega)$ is the linear response function of the substrate (assumed translationally invariant parallel to the surface) to the external potential due to the excited adsorbate.

Causality requires that damping of the oscillator be associated with a shift (renormalization) of its frequency. The physical origin of this shift is the induced polarization (determined by μ and $g(q_\parallel,\omega)$) acting back on the oscillator. The renormalized frequency Ω' is complex; i.e. $\Omega' = \Omega + \Delta\Omega \pm i\hbar\tau_{NR}^{-1}$, with $\Delta\Omega$ and τ_{NR} being determined by the real and imaginary parts of $g(q_\parallel,\omega)$, respectively.

Classical electrodynamics gives a proper account of the processes in the bulk, $g(q_\parallel,\omega) = [\varepsilon(\omega) - 1]/[\varepsilon(\omega) + 1]$ and thus[14]

$$F_{bulk} = 2Im\frac{\varepsilon(\Omega) - 1}{\varepsilon(\Omega) + 1} \tag{3}$$

where $\varepsilon(\omega)$ is the bulk dielectric function of the substrate. To account for surface processes one must use a dielectric function which varies continuously in the surface region of the metal. This has been done by using a jellium model for the metal, and thus[15]

$$F_{surface} = 6\xi(r_s)\frac{1}{k_F d}\frac{\Omega}{\omega_p} \tag{4}$$

where ω_p is the bulk plasmon frequency, k_F is the Fermi wavevector and ξ is a function of the electron gas density. By comparing Eqs. (3) and (4) we see that bulk and surface processes result in a different functional dependence of τ_{NR} on distance from the surface - $\tau_{NR} \sim d^3$ and $\tau_{NR} \sim d^4$ respectively. Thus it is clear that surface processes will be important only close ($<10\text{Å}$) to the surface. Additional insight into the question of the relative contribution of bulk and surface e-h pair processes to the non-radiative decay comes by using a Drude form for $\varepsilon(\omega)$ in Eq. (3) which can then be recast in the form:

$$F_{bulk} \approx 8\frac{\omega_F}{\omega_p}\frac{1}{k_F \ell}\frac{\Omega}{\omega_p} \tag{5}$$

Here ℓ is the mean free path of the free electrons in the metal. Bulk processes will be important provided that ℓ is small. This condition $(k_F \ell \sim 1)$ is usually true for transition metals and $\hbar\Omega \sim$ a few eV. On the other hand, $k_F \ell$ is large (~ 100) for free electron metals and for noble metals away from interband (e.g. d$\rightarrow$sp) transitions and therefore surface e-h pair processes may dominate at small distances.

B. Resonance Ionization

We consider now an excited state quenching mechanism involving charge transfer, rather than energy transfer, to the substrate. This mechanism is illustrated in Fig. 1. The adsorbate is again pictured as a two level system. When the energy of the excited electron is higher than the Fermi energy (ε_F) then the metal-adsorbate coupling allows that electron to tunnel to the empty band states $|k\rangle$ of the substrate. In effect then the excited level becomes autoionizing with respect to the continuum of the conduction band states. If we denote by ε the one electron energies and by U the electron-electron Coulomb repulsion energies in the two level system, and if v is the image interaction, then the energetic requirement for the resonance ionization process can be written as: $\varepsilon_2 - \varepsilon_1 - U_1 + U_{12} > \varepsilon_F - \varepsilon_1 - U_1 - v$. For weak adsorbate-substrate coupling the resonance ionization rate can be expressed by the Golden Rule formula:

$$\begin{aligned}\tau^{-1} &\cong \frac{2\pi}{\hbar}\sum_k |V_k|^2 \delta(\varepsilon_2 + U_{12} + v - \varepsilon_k)\\ &= \frac{2\pi}{\hbar}<|V_k|^2>\rho(\varepsilon_2 + U_{12} + v)\end{aligned} \tag{6}$$

where $\rho(\varepsilon)$ is the density of unoccupied levels of the metal and $<|V_k|^2>$ the average (in k − space) of $|V_k|^2$. Since V_k depends on the overlap of the substrate and adsorbate wavefunctions and since the metal wavefunctions decay roughly exponentially away from the metal surface we expect $<|V_k|^2>\sim e^{-\beta d}$, where d is the separation between the metal surface and the orbital $|2>$, and $\beta\sim2(2m|\varepsilon_F + \Phi-\varepsilon_k|/\hbar)^{1/2}$, where ϕ is the metal work function.

While the above simple description of the process allows some qualitative predictions to be made, quantification of the problem is very difficult. A two level adsorbate-metal system has been treated with an Anderson-Newns Hamiltonian and expressions for the spectral function (renormalized frequency and self-energy) correct to order V^2 have been given.[16] However, the problem of determining V_k still remains. Density functional studies of the chemisorption of atoms on jellium[17] give widths Δ_0 of the adsorbate-induced density of states at equilibrium chemisorption distance of the order of $1eV$ ($\tau\sim1\times10^{-15}s$). The distance dependence of the width is given approximately by $\Delta(d) = \Delta_0 e^{-\gamma d}$ with $\gamma\sim1.0-1.5\text{Å}^{-1}$.[17] Note that the width Δ is found not to be completely a monotonic function of the distance of the atom from the surface. The reason is that as the atom approaches the surface, the width of the resonance increases but at the same time the energy of the resonance changes so that not only V_k but also $\rho(E)$ in Eq. 6 changes.

A different approach to the evaluation of the resonance ionization rate is provided by the analysis of experiments measuring the survival probability of high energy ions scattered by surfaces[19] or the survival probability of excited species produced in sputtering.[3] In both cases, the functional form for the rate R of resonance electron transfer is taken to be: $A\exp(-ad)$. There is agreement that $a\simeq2\text{Å}^{-1}$, but values given for A vary in the range $10^{14}-10^{17}s^{-1}$.

C. Relaxation Dynamics from Spectral Widths

From the discussion on the non-radiative decay mechanisms of excitations at surfaces we expect extremely short lifetimes and small fluorescence yields. Elegant experiments measuring the luminescence lifetimes of excited molecules placed (by the use of noble gas spacer layers) at different distances d from the surface have confirmed the validity of the classical electromagnetic field model of energy transfer for $d\gtrsim10\text{Å}$.[20] The most interesting case of course is that of an excited species in direct contact with the surface. In this case however the fluorescence yields become too small to measure. One would therefore like to obtain dynamical information from the broadened linewidths of the electronic excitations which can be observed easily in optical or EELS experiments. For this purpose it is important to consider closely the different broadening mechanisms contributing to the observed linewidth.

The spectral lineshape is the Fourier transform of the transition dipole correlation function[21]

$$I(\omega)\propto \int_{-\infty}^{+\infty}dt e^{i\omega t}<\mu(t)\mu(0)> \tag{7}$$

Therefore, the time behavior of the correlation function determines the observed lineshape. So, for example, if $<\mu(t)\mu(0)>\sim\exp(-i\Omega t-|t|/\tau)$ then the lineshape is Lorentzian with a linewidth (FWHM)$=2/\tau$.

There are two fundamentally different mechanisms by which $<\mu(t)\mu(0)>$ can decay in time. (A) Electronic relaxation (radiative and non-radiative) which as discussed earlier leads to the decay of the amplitude $<\mu^2(t)>$ as $t\to\infty$. (B) Elastic collisions with, for example, phonons or electron-hole pairs which can lead to dephasing, i.e. distraction of the definite phase relationship between $\mu(t)$ and $\mu(0)$ and decay of $<\mu(t)\mu(0)>$ while $<\mu^2(t)>$ remains constant.

One instructive way of looking at the dephasing is to consider the excitation as an oscillator whose fundamental frequency is modulated by the fluctuations of the surrounding heat bath.[21] To quantify the expression for the lineshape however one must know both the coupling strength and time structure of these fluctuations. Dephasing is expected to be more important for delocalized excitations (2D-adsorbate exciton systems) since the elastic scattering events (e.g. phonon scattering) modulate not only the excitation frequency but also the through the field and through the substrate inter-adsorbate coupling. In addition, impurity or defect scattering may become important in the dephasing of these delocalized excitations.

From an experimental point of view, the most important property of the dephasing process is its temperature dependence. Dephasing shows a strong T dependence and (except for defect scattering) vanishes as $T \rightarrow 0$.[21] Therefore, if one is interested in obtaining the decay rate of the surface electronic excitation, then the Lorentzian component of the lineshape at low temperatures should provide a good measure of the lifetime broadening Γ.

Finally, we note that besides the two <u>homogeneous</u> broadening mechanisms discussed above, <u>inhomogeneous</u> broadening due to disorder should be considered. Such broadening will be most important in incomplete layers, when multiple binding sites are occupied and near domain walls. Random inhomogeneities lead to Gaussian lineshapes.

D. Experimental Studies of Electronic Relaxation

As discussed in Section IIA, classical electrodynamics appears to describe very accurately the decay of electronic excitations as close as ~ 10Å away from metal surfaces. Here we would like to investigate to what extent the same models can be applied in the case of molecules in direct contact with the metal surface (Fig. 1a). To test this we have studied two systems: N_2/Al(111) at 15K[22] and pyrazine/Ag(111) at 140K.[23]

Nitrogen on Al(111) at 15K is physisorbed with minimal perturbation of its ground state electronic structure. This is evidenced by the rigid shift of its valence photoemission spectrum on going from the gaseous to the adsorbed phase.[22] The excited state of interest is the $C^3\Pi_u$ state, which is coupled by a dipole transition to the lower $B^3\Pi_g$ state. The excited $1\pi_g^*$ orbital of N_2 is located below the Fermi level of Al, and therefore quenching via resonant electron transfer is not present. The homogeneous linewidth of the $C^3\Pi_u \leftarrow X^1\Sigma_g^+$ transition of a monolayer of N_2 on Al(111) is found to be $\Gamma \sim 140$meV ($\tau \sim 5 \times 10^{-15}$s).[22] Classical electrodynamic theory predicts a much smaller width $\Gamma \sim 15$meV. However, when the surface contributions to the relaxation via e-h pair generation are included according to the model of Persson and Lang[15], a total width of ~ 155meV is obtained.

In the case of pyrazine on Ag(111), on the other hand, bulk quenching alone overestimates (~ 3 times) the width of the $^1B_{2u} \leftarrow {}^1A_1$ transition (~ 120meV) while surface contributions are ~ 5 times smaller.

The differences between the two systems can be understood along the lines of discussion in Section IIA. The pyrazine transition at 4.8eV is in the midst of strong d$\rightarrow$sp interband transitions, so that ℓ is small due to scattering by the crystal potential, and bulk quenching dominates. On the other hand, Al at 3.7eV ($C^3\Pi_u \rightarrow B^3\Pi_g$) behaves like a free electron metal and the requirement of momentum conservation is satisfied by the termination of the transverse periodicity at the surface.

The above two examples suggest that the electrodynamic theory, with the effects of the substrate surface included, can predict to better than an order of magnitude the lifetime of these weakly adsorbed species. To the extent that this conclusion is a general one the non-radiative lifetimes of dipole allowed excitations on metal surfaces are expected to be $\sim 10^5$-10^6 times shorter than the free molecule radiative lifetimes. The conclusion that dipole coupling with substrate e-h pair excitations is important even for adsorbed molecules is supported by our observations that transitions to excited states not dipole coupled to lower states are not appreciably broadened. This is the case for O_2/Ag at 20K, where no significant homogeneous broadening of the $^1\Delta_g \leftarrow {}^3\Sigma_g^-$ and $^1\Sigma_g^+ \leftarrow {}^3\Sigma_g^-$ transitions was observed by resonant electron scattering.[24]

Noble gas atoms adsorbed on metal surfaces provide a prototype system for studying the resonant electron transfer mechanism of non-radiative decay (Fig. 1b). In their ground state the noble gases have a full np^6 outer shell and interact with the metal surface primarily via dispersion forces. Their electronically excited states are of a Rydberg nature, the lowest energy one having the alkali-like configuration $np^5(n+1)s$; so for example, an excited Xe atom looks, in terms of electronic structure, like a ground state Cs atom. The alkalis are known to adsorb ionically on most metals, and excited noble gases may behave analogously so that the excited states of the adsorbed noble gases could interact strongly with the surface and may be ionic. In fact, using optical reflectance techniques Flynn and coworkers[25] concluded that when the excited level is above E_F, the excitonic transitions of mono- and sub-monolayer noble gases on metals are unobservable. To avoid local field screening effects which can complicate optical studies we have used low energy EELS to study the spectroscopy of adsorbed noble gases.[26,27] In Fig. 2 we show the electronic spectra of one and two layers of Xe on Cu^{27} at ~15K, a case where $E(6s)>E_F$.[27] Also shown diagrammatically are the free Xe atomic spectra and the lineshape of the lowest energy $5p^6 \rightarrow 5p^5_{3/2}6s$ excitation after background subtraction. From these spectra we see that the excitonic absorption of the Xe monolayer is clearly visible (detailed EELS studies indicate that these absorptions are seen for as low as 0.08 layers and that the integrated intensity per atom is approximately constant).[26]

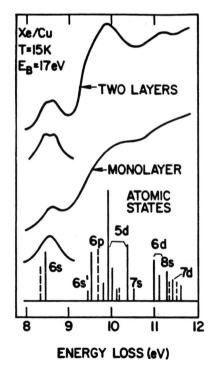

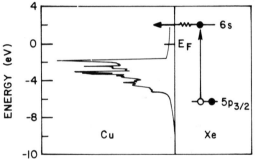

Figure 2 Left: Electron energy loss (EEL) spectra of Xe on Cu at 15K. Also shown are the lineshape of the $5p^6 \rightarrow 5p^5_{3/2}6s$ transition ans the free Xe excitations.
Above: Energy level diagram of the Cu DOS and the Xe atomic levels on Cu. From Ref. 27.

There is also an obvious correlation of the bands in the adsorbed phase with groups of atomic excitations. We also observe that the lineshape of the lowest $5p^6 \rightarrow 5p^56s$ transition of monolayer is symmetric with a linewidth (FWHM) ~0.5eV. The intensity maximum is shifted by ~0.2eV to a higher energy than the gas phase value. By deconvolution of the spectral profile the Lorentzian (homogeneous) width was found to be ~0.2eV, which translates to an electron transfer rate of ~$3 \times 10^{14} s^{-1}$. We find similar results for Xe or Ar on Al, Ag and Au.[26] To compare with theory, density functional calculations predict four to five times higher relaxation rates[28], while classical electrodynamics predict about ten times slower relaxation. Our results agree with the studies of quenching of excited states produced in sputtering with A~$10^{14} - 10^{15} s^{-1}$ (see Section IIB). We expect that, whenever resonant electron transfer is energetically possible, it will dominate over the decay path involving field coupling of the adsorbate and substrate excitations. Resonant electron transfer can be an important decay path for both physisorbed and chemisorbed adsorbates.

III. Valence Excitations of Chemisorbed Molecules

Chemisorption involves significant modification of the electronic structure of the adsorbate. The valence molecular orbitals of the adsorbate and the metallic bands are mixed to produce new bonding and antibonding orbitals and the electronic excitations involve transitions between these hybrid orbitals.

Carbon monoxide has long been a prototype molecule in studies of chemisorption. Ultraviolet photoemission spectroscopic (UPS) studies of its valence orbitals have shown that, while in free CO the 5σ, 1π and 4σ orbitals are spaced ~3eV apart, upon chemisorption on transition metals or Cu the 5σ and 1π levels become essentially degenerate.[29] This observation has been interpreted as evidence for the involvement of the 5σ orbital in the bonding with the metal; CO donates electrons to the metal via its 5σ orbital, whose binding energy thus is increased. The frequency of the C-O stretch mode is also decreased upon chemisorption; this has been interpreted to be the result of charge backdonation from the metal to the $2\pi^*$ orbital of CO. In Fig. 3, we show an orbital correlation diagram for free and chemisorbed CO taken from Ref. 30 which shows the formation of bonding and antibonding $\widetilde{5\sigma}$ and $\widetilde{2\pi}^*$ orbitals (the ~ denotes chemisorption). From this diagram the energy difference between the $\widetilde{5\sigma}_b$ and $\widetilde{2\pi}_a$ orbitals (only chemical shifts are shown) is larger than the corresponding difference of the 5σ and $2\pi^*$ orbitals of free CO. Thus, while the singlet-coupled $5\sigma \rightarrow 2\pi^*$ excitation of free CO is at ~8.5eV, the corresponding $\widetilde{5\sigma}_b \rightarrow \widetilde{2\pi}_a$ excitation has been placed generally at much higher energies (~14eV).[31]

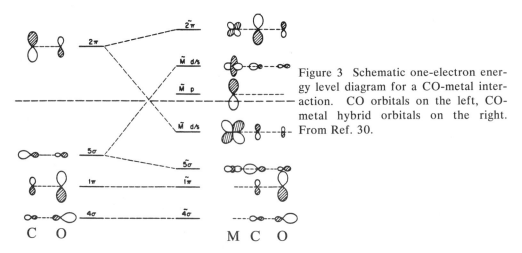

Figure 3 Schematic one-electron energy level diagram for a CO-metal interaction. CO orbitals on the left, CO-metal hybrid orbitals on the right. From Ref. 30.

    In Figs. 4 and 5 we show the low energy EELS electronic spectra of CO on Cu.[11] The clean Cu surface spectrum shows interband transitions at ~2.2eV ($L_3 \rightarrow FS$) and at ~4.2eV ($L_{2'} \rightarrow L_1$ and $X_5 \rightarrow X_{4'}$) which are further enhanced upon CO adsorption.[11] The transitions at ~6 and ~8.5eV are chemisorbed CO excitations as can be verified by comparing them with the spectra of other adsorbates on Cu and with the Cu bulk ($-\mathrm{Im}[\varepsilon(\omega)]^{-1}$) and surface ($-\mathrm{Im}[\varepsilon(\omega)+1]^{-1}$) loss functions.[11] At higher than monolayer CO coverages the two excitations at ~6 and ~8.5eV develop vibronic structure and can be identified as $^3(5\sigma \rightarrow 2\pi^*)$ and $^1(5\sigma \rightarrow 2\pi^*)$ excitations of solid CO.[11] In the case of Ni, the clean surface spectrum shows broad interband ($W_3 \rightarrow L_1$, $X_5 \rightarrow W'_1$) and plasmon excitations between 8-12eV, and a surface excitation of ~0.7eV, all of which are quenched upon CO adsorption. The spectrum at a CO exposure of 2L shows again two broad bands at ~6 and ~9eV which are assigned as excitations of chemisorbed CO.[11] The coincidence of the excitation energies of chemisorbed CO

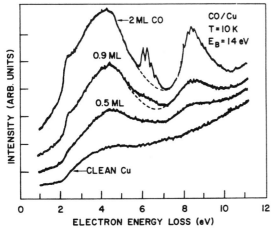

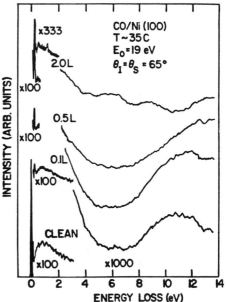

Figure 4   EEL spectra of CO on Cu at 10K. Coverage in monolayers calibrated by UPS. From Ref. 11.

Figure 5   EEL spectra of CO on Ni(100) at 35C. Exposure in Langmuirs. From Ref. 11.

with the $^{1,3}(5\sigma \rightarrow 2\pi^*)$ excitation energies of free CO raises the possibility of a common origin. UPS shows the binding energies (with respect to E_F) of the $5\tilde{\sigma}$ levels of CO/Cu and CO/Ni to be about the same.[29] How about the $2\tilde{\pi}_a^*$ energies? Recently, the technique of inverse photoemission (IPES) has been applied to chemisorbed CO. In Fig. 6 we show the spectra of the $2\tilde{\pi}_a^*$ levels of CO/Ni(111)[32], CO/Cu(110)[33] and the $2\pi^*$ shape resonance of free CO. We see that the IPES resonances of both CO/Cu and CO/Ni are at lower energies than the free CO resonance. As we discussed in Section I the experimentally observed shifts of $2\tilde{\pi}^*$ by IPES are the sum of chemical and relaxation shifts. By ab-initio cluster calculations, however, we confirm that the resonances of chemisorbed CO are of $2\tilde{\pi}_a^*$ nature and that the <u>chemical</u> shift of $2\tilde{\pi}_a^*$ is to <u>lower</u> energies.[34] Therefore, both the filled $5\tilde{\sigma}$ valence level and the $2\tilde{\pi}_a$ virtual level are pulled down in energy by chemisorption, not pulled apart as Fig. 3 suggests. On the basis of the IPES results, the 5σ binding energies on the two metals[30] and the work functions of the CO covered surfaces we find that for the $5\tilde{\sigma} \rightarrow 2\tilde{\pi}_a^*$ excitation IP-EA$\simeq$11eV

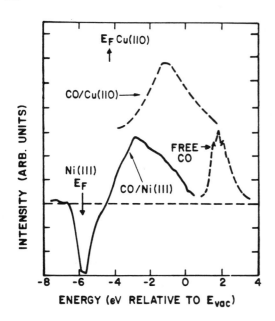

Figure 6   Inverse photoemission spectra of CO on Ni(111) (Ref. 32) and on Cu(110) (Ref. 33).

for both metals thus explaining why this excitation has the same energy on both Cu and Ni. The traditional picture of the CO-metal interaction expressed in Fig. 3 fails to predict correctly the location of the $2\tilde{\pi}_a^*$ level and the excited states because it overemphasizes the interaction of the CO with the d-band of the metal which indeed will push $2\tilde{\pi}_a^*$ to a higher energy. There are, however, strong interactions between virtual $2\tilde{\pi}_a^*$ and virtual p (and s) metal bands which push the $2\tilde{\pi}_a^*$ level to lower energies. The maximum of the p,s density of states is at higher energies than the $2\pi^*$ level of CO. Finally, note that the 14eV electron loss peak, traditionally associated with the $^1(5\tilde{\sigma} \rightarrow 2\tilde{\pi}_a^*)$ transition of chemisorbed CO[31], correlates well with the 5σ ionization energy and also the energy threshold for electron stimulated desorption (ESD) of CO^+.[35] The 8-9eV loss, on the other hand, correlates with a resonance for ESD of neutral CO.[36] On the basis of the chemisorption bond model of CO it is easy to see why the $5\tilde{\sigma} \rightarrow 2\pi^*$ and $5\tilde{\sigma}^{-1}$ states could lead to unbound CO and CO^+, respectively.

In the discussion of the non-radiative decay rate of excited noble gases we have used the spectral width of the $np^6 \rightarrow np^5(n+1)s$ excitation to infer the broadening of the (n+1)s resonance. This was made possible by the fact that the np level does not mix with the metallic band and thus remains narrow. The same analysis cannot be used in the case of the valence excitations of chemisorbed species (e.g. $5\tilde{\sigma} \rightarrow 2\tilde{\pi}^*$ of CO) when both orbital levels mix with band states.

IV.   Core Excitations of Chemisorbed Molecules

Core $\rightarrow$ valence excitations of chemisorbed species in principle are easier to interpret than valence $\rightarrow$ valence excitations since the core levels do not participate in the chemisorption bond, and thus remain relatively sharp. The recent availability of tunable radiation in the soft x-ray region by synchrotrons has stimulated new interest in the spectroscopy and photochemistry of core excitations.[1] In Fig. 7 we show high resolution $C1s \rightarrow 2\tilde{\pi}_a^*$ and $O1s \rightarrow 2\tilde{\pi}_a^*$ synchrotron spectra of solid CO and CO on Ni(111) obtained by the technique of photoelectric yield spectroscopy.[37] Three important observations can be made: (A) both core excitations of chemisorbed CO appear at lower energies than the corresponding excitations of solid (free)

CO, (B) both excitations of chemisorbed CO are strongly broadened and (C) the $O1s \rightarrow 2\tilde{\pi}_a^*$ transition of CO/Ni(111) is more red-shifted and broadened than the $C1s \rightarrow 2\tilde{\pi}_a^*$ transition.

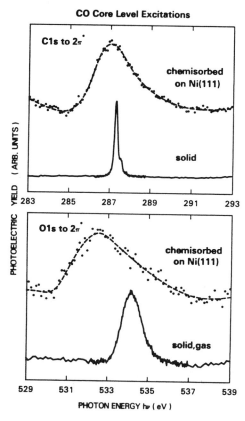

CO Core Level Excitations

Figure 7 $C1s \rightarrow 2\tilde{\pi}_a^*$ and $O1s \rightarrow 2\tilde{\pi}_a^*$ synchrotron spectra of solid (gas) CO and CO on Ni(111) obtained by photoelectric yield spectroscopy. From Ref. 37.

The red shift (A) can be explained on the basis of the discussion in the previous section: the $2\tilde{\pi}_a$ level energy is lowered as a result of the mixing with virtual s and p metallic levels. The broadening (B) reflects the mixing of the $2\tilde{\pi}_a$ level with band states and implies lifetimes of the order of $\sim 10^{-16}$s. Observation (C) is particularly interesting and suggests a stronger bonding interaction of the $O1s \rightarrow 2\tilde{\pi}_a^*$ state with Ni(111) than either the CO ground or $C1s \rightarrow 2\tilde{\pi}_a$ excited state. Thus it appears that, although the localized core levels do not directly participate in bonding, core level occupancy does affect bonding.

The core level occupancy dependence of the binding energy can be explained in terms of the different charge distribution in the $2\pi^*$ orbital and the extent of metal $\rightarrow 2\pi^*$ backbonding in the two core-excited states. The change in the charge distribution can be easily predicted on the basis of "equivalent core" considerations. In the equivalent core picture[38], an atom with a core hole is considered equivalent to an atom with a unit higher atomic number; so then $C^+O(C1s^{-1})$ and $CO^+(O1s^{-1})$ are equivalent to NO and CF, respectively. Thus, while in the NO $2\pi^*$ orbital the charge is almost evenly distributed between the N and O atoms, in the CF case most of the $2\pi^*$ amplitude is on the C atom. Since CO bonds with the C end to the metal we expect the backbonding to be optimized when the overlap of the $2\pi^*$ orbital and the metal states is the highest, which, from the preceding discussion, will occur for the $O1s \rightarrow 2\pi^*$ state, thus giving the highest excitation energy red-shift and spectral width, in agreement with experiment (Fig. 7). If indeed the extent of backbonding determines the energy of core excitations of chemisorbed CO we may expect a binding site dependence of this energy. From vibrational spectroscopic studies[31] it is known that metal→$2\pi^*$ backdonation is stronger at higher ligancy sites. Thus, the much smaller red shifts of the core excitations of CO on Ni(100)[39], where CO is primarily in a top site, compared to CO on Ni(111)[37], where the CO is primarily in a three-fold hollow site, can, at least partially, be ascribed to changes in backdonation. These general ideas about the nature and bonding of the core-excited states of chemisorbed CO are supported by ab-initio cluster calculations.[34] For example, the notion that the red-shift of the O1s→$2\pi^*$ CO excitation upon chemisorption is due to stronger backbonding in the excited than in the ground state is clearly seen in the orbital plots of Fig. 8.[34] Here we show the backbonding contributions of Cu sp-electrons (left) and Cu d-electrons (right) in the ground state (top) and O1s→$2\tilde{\pi}^*$ excited state (bottom). Also shown is the atomic arrangement of the Cu$_5$CO cluster. It is clear that both sp- and d→$2\pi^*$ backbonding is enhanced in the excited state. The calculations[34] also find that upon core ionization a full metal electron is transferred in the $2\tilde{\pi}^*_a$ orbital to screen the core-hole. The resulting "shake-down" states have a very similar charge distribution to that of the neutral core excitations. Indeed, experiment shows that the screened ionic states have binding energies with respect to E_F which are about equal to the corresponding excitation energies.[34]

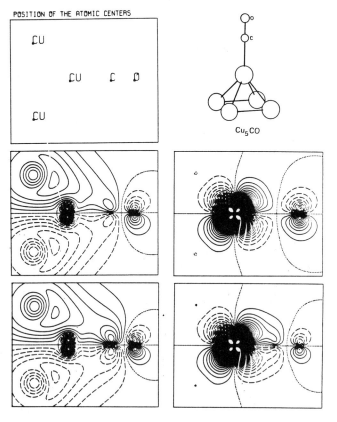

Figure 8   Orbital plots of the sp→$2\pi^*$ and d→$2\pi^*$ backbonding interaction in a Cu$_5$CO cluster (see text). From Ref. 34.

V. Photochemistry at Metal Surfaces

Even in the absence of specific adsorbate-substrate interactions, the enhanced non-radiative decay rate of the excited molecule on the surface will greatly reduce photochemical yields. However, under conditions such as rough surfaces or gratings, where the incident light couples to the plasmon resonances of the substrate, the optical fields are amplified, which could enhance photochemistry.[2] Since the field enhancement is of longer range than the damping interaction, the photochemical yield should peak for a molecular position at some distance from the metal surface.[2] This apparently has been observed recently by the use of inert spacer layers.[40] For chemisorbed molecules more drastic and varied modifications of their photochemical behavior are expected. Charge-transfer and rehybridization modifies the strength of intramolecular bonds so that transitions which were not dissociative in the free molecule may become so in the adsorbed phase. For example, strong metal$\rightarrow 2\tilde{\pi}^*_a$ backdonation in chemisorbed CO combined with optical excitation to the $2\tilde{\pi}^*_a$ orbital could lead to the cleavage of the C-O bond. Excitation between orbitals directly involved in the adsorbate-substrate bonding such as $5\tilde{\sigma}\rightarrow 2\tilde{\pi}^*_a$ in chemisorbed CO could lead to desorption.

Adsorbate-substrate charge-transfer (CT) excitations[41,42] can be particularly photochemically active. As a simple example, consider the adsorption of halogens on alkali metal surfaces; for example, Cl on Na. The chlorine adsorbs as Cl^-, and upon Franck-Condon excitation in the $Cl^-/Na\rightarrow Cl^\circ/Na^-$ CT-band the Cl° finds itself in the repulsive part of the Cl°-Na potential curve. Unless it recaptures an electron fast enough, it is ejected from the surface.[43] CT excitation can lead to desorption even in the absence of a chemisorption bond. This is illustrated by the case of noble gas atoms on metal surfaces. For example, Kr on $W(110)$[44] shows a desorption threshold at the onset of the $Kr(4p_{3/2})\rightarrow W(E_F)$ CT-transition. We can envision the desorption mechanism as involving the acceleration of the positive ion towards its image, neutralization, and, finally, desorption if the kinetic energy gained is higher than the neutral binding energy at the neutralization position.

Core excitation of adsorbates can lead to interesting photochemistry. This photochemistry can be a direct effect of the excitation to an unbound excited state or the result of the non-radiative decay of the excited state by an Auger process. Let us take CO again as an example. The C1s and O1s$\rightarrow 2\pi^*$ excited states are bound, but fast ($\sim 10^{14}s^{-1}$) CVV Auger decay can put two holes in a valence orbital(s) (e.g. $3\sigma^{-2}$) resulting in the "Coulomb explosion"[45] of the molecule to give C^+ and O^+ ions. In the Auger decay of the free molecule the electron in the $2\pi^*$ orbital could be a spectator so that the final state is $V^{-1}V'^{-1}2\pi^{*1}$, or a participant, i.e. $V^{-1}2\pi^{*-1}$. The core-excited states of chemisorbed CO according to the cluster results[34] are bound. Unlike free CO, however, the fastest non-radiative decay path is now resonance ionization (Fig. 1B), so that the $2\tilde{\pi}^*_a$ electron leaves the adsorbate in $\sim 10^{-15}-10^{-16}$s. The subsequent Auger decay again can be of the CVV type, but the two hole state need not Coulomb explode in this case. In fact, no O^+ is observed upon C1s or O1s$\rightarrow 2\tilde{\pi}^*_a$ excitation of CO/Ni(100).[46] The difference between free and chemisorbed CO is twofold. First, the hole-hole repulsion which provides the driving force for Coulomb explosion is largely screened by the substrate. Second, the holes can be reneutralized by electron hopping from the substrate. Two hole reneutralization occurs at $\sim 10^{-14}$s while one hole reneutralization is faster $\sim 10^{-15}$s. Finally, note that by using tunable synchrotron radiation the potential exists for site specific core photochemistry. The dynamics of core-excitation decay are illustrated in Fig. 9.

PHOTOCHEMISTRY OF CORE EXCITATIONS

FREE MOLECULE

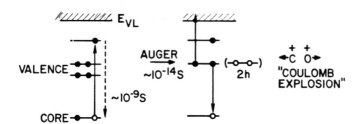

ADSORBED MOLECULE

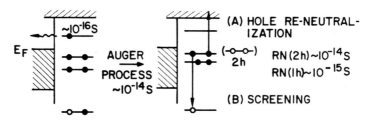

Figure 9 The effects of the metallic substrate on the decay of core excitations (see text).

Acknowledgements

It is a pleasure to acknowledge my coworkers P. S. Bagus, J. E. Demuth, N. J. DiNardo, F. J. Himpsel, B. N. J. Persson and A. R. Rossi who made valuable contributions to the work described here.

References

1. Desorption Induced by Electronic Excitations, Springer Series in Chemical Physics, edited by N. H. Tolk, M. M. Traum, J. C. Tully, and T. E. Madey (Springer, New York, 1983).
2. A. Nitzan and L. E. Brus, J. Chem. Phys. 75, 2205 (1981); G. M. Goncher and C. B. Harris, J. Chem. Phys. 77, 3767 (1982); C. J. Chen and R. M. Osgood, Appl. Phys. A31, 171 (1983).
3. Inelastic Ion-Surface Collisions, N. H. Tolk, J. C. Tully, W. Heiland and C. W. White, editors (Academic, New York, 1977).
4. H. Conrad, G. Ertl, J. Küppers, W. Sesselmann and H. Haberland, Surf. Sci. 121, 161 (1982); F. Bozso, J. Arias, G. Hanrahan, R. M. Martin, J. T. Yates, Jr. and H. Metiu, Surf. Sci. 136, 257 (1984) and references cited therein.
5. B. Kasemo and L. Wallden, Surf. Sci. 53, 393 (1975).
6. A. Otto, in Light Scattering in Solids, edited by M. Cardona and G. Guntherodt (Springer, New York, 1983).
7. T. F. Heinz, H. W. K. Tom and Y. R. Shen, Laser Focus 19, 101 (1983).

8.   G. Loubriel, T. Gustafsson, L. I. Johansson and S. J. Oh, Phys. Rev. Lett. 49, 571 (1982).

9.   G. Ritchie and E. Burstein, Phys. Rev. B24, 4843 (1981).

10. J. Demuth and Ph. Avouris, Physics Today 36, 62 (1983).

11. Ph. Avouris, N. J. DiNardo and J. E. Demuth, J. Chem. Phys. 80, 491 (1984).

12. For a review see: Ph. Avouris and B. N. J. Persson, J. Phys. Chem. 88, 837 (1984).

13. Ph. Avouris, W. M. Gelbart and M. A. El-Sayed, Chem. Rev. 77, 793 (1977).

14. For a review see: R. R. Chance, A. Prock and R. Silbey, Adv. Chem. Phys. 37, 1 (1978).

15. B. N. J. Persson and N. D. Lang, Phys. Rev. B26, 5409 (1983).

16. B. N. J. Persson and Ph. Avouris, J. Chem. Phys. 79, 5156 (1983).

17. N. D. Lang and A. R. Williams, Phys. Rev. B18, 616 (1978).

18. N. D. Lang, Phys. Rev. B27, 2019 (1983).

19. H. D. Hagstrum, in Electron and Ion Spectroscopy of Solids, L. Fiermans, J. Vennik and W. Dekeyser, editors (Plenum, New York, 1978).

20. P. M. Whitmore, H. J. Robota and C. B. Harris, J. Chem. Phys. 77, 1560 (1982) and references cited therein.

21. R. Kubo, Adv. Chem. Phys. 15, 101 (1969).

22. Ph. Avouris, D. Schmeisser and J. E. Demuth, J. Chem. Phys. 79, 488 (1983).

23. Ph. Avouris and J. E. Demuth, J. Chem. Phys. 75, 4783 (1981).

24. D. Schmeisser, J. E. Demuth and Ph. Avouris, Phys. Rev. B26, 4857 (1982).

25. J. E. Cunningham, D. Greenlaw and C. P. Flynn, Phys. Rev. B22, 717 (1980).

26. J. E. Demuth, Ph. Avouris and D. Schmeisser, Phys. Rev. Lett. 50, 600 (1983).

27. Ph. Avouris, J. E. Demuth and N. J. DiNardo, J. Phys. (Paris) C10 44, 451 (1983).

28. N. D. Lang, A. R. Williams, F. J. Himpsel, B. Reihl and D. Eastman, Phys. Rev. B26, 1728 (1982).

29. E. W. Plummer and W. Eberhardt, Adv. Chem. Phys. 49 (1982).

30. M. J. Freund and E. W. Plummer, Phys. Rev. B23, 4859 (1981).

31. For a review see: Ph. Avouris and J. E. Demuth, Ann. Rev. Phys. Chem. 35, 49 (1984).

32. Th. Fauster and F. J. Himpsel, Phys. Rev. B27, 1390 (1983).

33. J. Rogozik, H. Scheidt, V. Dose, K. Prince and A. Bradshaw, Surf. Sci. 1984, to be published.

34. A. R. Rossi, Ph. Avouris and P. S. Bagus, to be published.

35. D. Menzel, Ber. Bunsenges. Phys. Chem. 72, 591 (1968).

36. J. Rubio, J. M. López-Sancho, P. M. López-Sancho, J. Vac. Sci. Technol. 20, 217 (1982).

37. Y. Jugnet, F. J. Himpsel, Ph. Avouris and E. E. Koch, to be published.

38. W. L. Jolly, in Electron Spectroscopy, edited by D. A. Shirley (North-Holland, New York, 1972).

39. J. Stöhr and R. Jaeger, Phys. Rev. B26, 4111 (1982).

40. G. M. Goncher and C. B. Harris, J. Phys. Chem. to be published.

41. Ph. Avouris and J. E. Demuth, in Surface Studies with Lasers, Springer Series in Chemical Physics, edited by F. Aussenegg, A. Leitner and M. E. Lippitch (Springer, New York, 1983).

42. E. Burstein, A. Brotman and P. Apell, J. Phys. (Paris) 44, C10-429 (1983).

43. N. D. Lang, J. K. Norskov and B. I. Lundquist, to be published.

44. Q.-J. Zhang, R. Gomer and D. R. Bowman, Surf. Sci. 129, 535 (1983).

45. T. A. Carlson and M. O. Krause, J. Chem. Phys. 56, 3206 (1972).

46. R. Jaeger, R. Treichler and J. Stöhr, Surf. Sci. 117, 533 (1982).

47. D. E. Ramaker in Ref. 1.

MOLECULE–SURFACE INTERACTIONS STIMULATED BY LASER RADIATION

T. J. Chuang and Ingo Hussla
IBM Research Laboratory
5600 Cottle Road
San Jose, California 95193 USA

ABSTRACT: Laser photons can interact with a gas-solid system to promote heterogeneous reactions and to stimulate desorption. Two classes of photo-enhanced surface processes have been extensively investigated in our studies.   The first one involves laser-induced chemical etching of solids examplified by the silicon-fluorine reaction affected by visible photons.   Recent time-resolved mass spectrometric measurements have revealed interesting surface interaction characteristics quite different from steady-state reactions.   The second one involves infrared laser photodesorption initiated by excitation of internal molecular vibrations. New results on ammonia photodesorbed from Cu(100) and NaCl surfaces are presented to elucidate desorption mechanisms.   Both single-photon and multi-photon excitation processes are examined and the velocity distributions of desorbed particles are analyzed. Various energy transfer and relaxation processes related to desorption are evaluated and a model for infrared photodesorption with particular emphasis on assessing the quantum and thermally-assisted effects is also discussed.

I. INTRODUCTION

It is by now quite well established that photon beams, in particular lasers, can be used to induce or enhance chemical interactions between a gas and a solid surface. In a recent review given by Chuang [1], it is shown that heterogeneous interactions can be affected by the laser radiation due to the excitation of the gaseous species, the adsorbates and the solid substrates involving electronic, vibrational and lattice phonon activations.   The enhanced surface processes include photon-induced adsorption, adsorbate-adsorbate and adsorbate-absorbent reactions, and desorption.   Whereas there exist numerous examples demonstrating the photon radiation effects, detailed studies on a given phenomenon or on a set of chemical systems under well-controlled conditions in order to elucidate the basic interaction steps and mechanisms remain relatively few.   Even less is known about the dynamics of molecule-surface interactions stimulated by laser beams.

In the continuing effort to gain better understanding of the reaction pathways and surface dynamics, we have performed further studies on two classes of photochemical processes, namely, direct gas-solid reactions resulting in chemical etching of solids and vibrationally activated photodesorption. For the first class of reactions, we investigate the radiation effects on a semiconductor surface exposed to an active halogen-containing gas.   The band gap excitation by a photon beam to

313

B. Pullman et al. (eds.), Dynamics on Surfaces, 313–327.
© *1984 by D. Reidel Publishing Company.*

create electron-hole pairs can result in the enhancement of surface sticking probability, the product formation and the vaporization of the surface species. Specifically, we have chosen to study Si-XeF$_2$ reactions influenced by visible light pulses. This gas-solid system has been rather extensively investigated as a model for understanding basic mechanisms involved in halogen-semiconductor interfacial reactions [2]. Although the steady-state reaction of the system is relatively well understood, important information about the time-dependent behavior, in particular, study on the effects due to pulsed laser irradiation has been lacking. The pulsed laser time-resolved study can also provide important insight into the etching mechanisms of other gas-solid systems, since many surface photochemical reactions including metals [3,4] are initiated by surface halogenation and the photon interaction with the halogenated surface layers. For the second class of photon-surface interactions, we have investigated ammonia desorption from clean metal and NaCl surfaces excited by a tunable infrared laser [5,6]. Although molecular desorption from ionic crystals [7-9] and metal surfaces [1,10-12] induced by resonant vibrational excitation of adsorbed species has been reported, prior experimental studies have been carried out exclusively with rather high power CO$_2$ lasers involving multi-photon absorptions. Multi-photon excitation complicates the rate analyses and render the desorption kinetics and mechanisms difficult to deduce. Furthermore, questions such as surface accommodation in translational and internal modes of motions associated with the photodesorption process need to be answered. In order to elucidate the surface molecular dynamics, we have applied time-resolved mass spectrometry to investigate these gas-solid systems. These studies have revealed new and interesting surface interaction characteristics which allow us to assess the relative importance of the various competitive processes involved in the observed desorption phenomenon. A model of infrared photodesorption is then discussed in light of the new experimental results and the existing theoretical considerations [13-17].

II. EXPERIMENTAL

A typical experimental setup for present studies is shown in Fig. 1. The apparatus consists of a UHV chamber with an ESCA/Auger spectrometer, an ion gun, a quadrupole mass spectrometer and an rf induction heater as described previously [10]. A Q-switched Nd:YAG laser (Quanta-Ray) is frequency doubled to 532 nm and used to pump a dye laser containing Exciton LDS 765 and 867 dyes. The tunable IR pulses are generated from the frequency difference between the 1.064 μm radiation and the dye laser in a LiNbO$_3$ crystal. The polarized IR beam at 10 Hz (6 nsec pulse width) is focused and incident at 75° from the surface normal covering a surface area of about 5 mm^2 (17 mm^2 for 532 nm light). The laser beam enters and leaves the UHV chamber through sapphire windows. The visible light pulses at 532 nm are used for Si-XeF$_2$ reactions and to perform laser-induced thermal desorption studies. The mass spectrometer is placed in a line-of-sight arrangement along the surface normal and the ionizer is located about 45 or 80 mm from the sample. The TOF mass signal amplified with a fast preamplifier (3.3 μsec rise time) [18] is recorded with a Tracor signal averager with 2 μsec time resolution which is

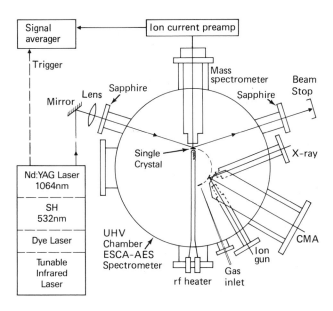

Fig. 1. Experimental setup for time-resolved mass spectrometric study on laser-stimulated gas-surface interactions. Light pulses at 532 nm are used for enhancing Si(111)-XeF$_2$ reaction and for laser-induced thermal desorption studies. The tunable IR laser in 2-4 μm is used for studying vibrationally excited desorption.

triggered by the laser synchronization pulse. The Si(111) and Cu(100) crystals are held separately with a manipulator and can be cleaned by Ar$^+$-bombardment, annealed by rf heating and cooled to about 90K with liquid N$_2$. The gas is dosed through a small copper tubing directly facing the crystal. The amount of surface coverage is determined from ESCA intensity analyses and from the conventional thermal desorption spectra. For ammonia adsorption and photodesorption on a NaCl film deposited in situ on sapphire at 10-90K, the experiments are carried out in a separate UHV chamber described previously [12]. Except for the differences in substrate material and temperature, the basic optical excitation and mass spectrometer detection arrangements are similar to those described above.

III. LASER-ENHANCED SILICON-FLUORINE REACTIONS

When silicon is exposed to gaseous XeF$_2$ at 25°C, surface fluorination to form a layer containing SiF$_x$ (x≤3) species readily occurs [19]. Further exposure of the fluorinated Si surface to the gas can result in the spontaneous formation of volatile products and chemical etching of the solid [20]. The major volatile product in the dark reaction has been identified to be SiF$_4$ with less F-coordinated SiF$_x$ species being the minor products [21,22]. When the gas-solid system is subject to laser irradiation, enhanced reaction is observed and the photo-enhancement can be nonthermal in nature [23,24]. In the reaction enhanced by a CW Ar^+-ion laser at 515 nm due to direct band-gap excitation and the formation of electron-hole pairs, the major product of the Si etching reaction has been found to be SiF_4 also [24]. In contrast, we find that the dominant photodesorbed species is not SiF_4 when the fluorinated Si surface is irradiated by 532 nm light pulses. At the laser intensity about 10 MW/cm^2, the ion yields for SiF_2^+ and SiF^+ observed with a mass spectrometer are about the same. They are, however, much higher than the SiF_3^+ yield which is the dominant component when SiF_4 is cracked by the electron impact ionization of the mass spectrometer. It is clear that less F-coordinated species, *i.e.*, SiF_x ($x \leq 3$), are more readily volatilized than SiF_4 production when the fluorine-exposed surface is irradiated by relatively high power laser pulses in the visible region. It is also interesting to note that at this laser intensity, the Si^+ mass signal can be substantially higher than the combined signals of SiF^+, SiF_2^+ and SiF_3^+, suggesting that some Si atoms are directly desorbed by the pulsed radiation. The probability of Si atom emission apparently increases with the laser intensity.

Figure 2 shows the typical time-of-flight (TOF) signals for SiF_3^+, SiF_2^+, SiF^+ and Si^+ species detected by the mass spectrometer as the fluorinated surface is irradiated by 532 nm light pulses at 9 MW/cm^2. From the maximum of TOF signal (t_m), the translational temperature of the desorbing species (T_d) can be calculated. A detailed description of the theoretical treatment assuming that the desorbed species leave the surface with a Maxwell-Boltzmann distribution has been given by Wedler and Ruhmann [25] among others. Following this approach, T_d can be calculated according to:

$$T_d = \frac{m}{4k} \frac{\ell^2}{t_m^2} \tag{1}$$

where m, k and ℓ are the mass, the Boltzmann constant and the distance between the sample and the middle of the ionization chamber, respectively. The average temperature rise (ΔT_s) in the laser-heated surface layer can also be estimated according to the equation given by Bloembergen [26],

$$\Delta T_s = \frac{(1-R)It_p}{C_v \rho (2Dt_p)^{1/2}} \tag{2}$$

where R is optical reflection coefficient, I is laser intensity, t_p is pulse duration, and C_v, ρ and D are specific heat, density and thermal diffusivity of the material, respectively. At I=9 MW/cm^2, ΔT_s is estimated to be about 400K and the peak surface temperature due to the 6 nsec laser pulse could reach above 700K. The translational temperatures calculated from the observed t_m for SiF_x species are in the 400-560K range. These temperatures are apparently substantially lower than the estimated surface temperature during the laser excitation pulse. A lower T_d than T_s should not be particularly surprising in view of the fact that the surface species can desorb during the rise of surface temperature before reaching the maximum temperature. Also perhaps, a desorbing species has to utilize part of its energy for breaking the surface bond in the desorption process which can result in a lower translational temperature than the substrate. Similar phenomenon of a significantly low T_d relative to T_s has also been observed in pulsed laser-induced thermal desorption of CO from Cu and Fe surfaces [18,25]. It is further observed that irradiation of the fluorinated Si surface at a high laser fluence can induce the formation of plasma resulting in the ejection of both Si atoms and Si^+ ions (the latter are detected without using the ionizer of the mass spectrometer) in addition to SiF_x species. The velocities of these ejected particles are very high as shown in TOF signals in Fig. 3. The threshold of laser fluence for plasma generation is much lower for the fluorinated Si than the clean surface. For instance, at 15 MW/cm^2, plasma formation giving rise to the characteristic optical emission from the surface is readily observed when Si is exposed to XeF_2. Yet, on a clean, well polished and annealed Si surface, plasma generation is not detectable even at the intensity of 20 MW/cm^2. Once the Si crystal is damaged due to either chemical etching or laser physical ablation, the laser threshold for particle emission is also substantially reduced. In short, our pulsed-laser time-resolved mass spectrometric study of the Si-XeF_2 has revealed interesting surface interaction characteristics quite different from the

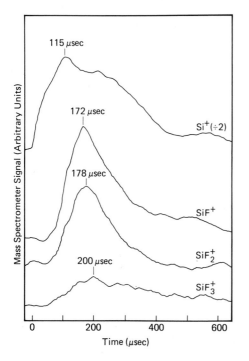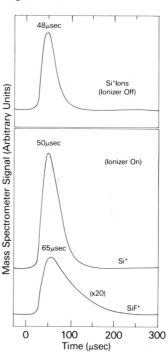

Fig. 2 (left). Time-of-flight (TOF) signals of SiF_3^+, SiF_2^+, SiF^+ and Si^+ detected by the mass spectrometer from XeF_2-exposed Si surface at 300K, irradiated by 532 nm laser pulses at I=9 MW/cm^2 on the surface; actual Si^+ signal a factor of 2 higher than other signals.

Fig. 3 (right). TOF signals of SiF^+ and Si^+ detected by the mass spectrometer from XeF_2 exposed Si surface irradiated by 532 nm light pulses at I=18 MW/cm^2; the top signal for Si^+ ions is detected by the mass spectrometer without using the ionizer.

steady-state reactions. Namely, less F-coordinated SiF_x species are more readily produced and desorbed than SiF_4 due to pulsed laser excitation and the translational temperatures of the desorbed particles can be substantially lower than the Si substrate temperature.

IV. IR LASER INDUCED PHOTODESORPTION

4.1 Ammonia on Cu(100)

The adsorption behavior of NH_3 on Cu(100) at 90K has been studied in detail by thermal desorption and X-ray photoemission spectroscopy. Figure 4 shows the $Cu(2p_{3/2}, 2p_{1/2})$ and N(1s) XPS spectra at monolayer and multilayer surface coverages. Apparently, there are no major chemical shifts in the core level peaks due to NH_3 adsorption. The surface coverages (θ) are estimated according to the formula discussed previously [19], assuming the electron mean-free path length to be about 30Å. They are also determined from thermal desorption spectra. These two types of measurements agree reasonably well. The thermal desorption spectra as shown in Fig. 5 indicate the presence of three different adsorbed states with desorption temperatures of about 143K, 194K and 264K, respectively. The two major chemisorbed states at submonolayer coverages have the estimated heats of adsorption of about 64 kJ/mole and 46 kJ/mole [27]. The corresponding value for the physisorbed species present at a coverage greater than a monolayer is about 34 kJ/mole. At very high NH_3 exposures, solid NH_3 can be condensed on the metal showing a desorption peak around 109K. The vibrational spectrum of ammonia adsorbed on Cu(110) has been investigated by Lackey et al. [28], using electron energy loss spectroscopy (EELS). It was shown that at low NH_3 coverages, the ν_s (N-H) stretching mode had the vibrational frequencies at 3360 and 3430 cm^{-1}, which merged into a broad band centered around 3400 cm^{-1} at more than 1 monolayer coverages. The corresponding symmetric and antisymmetric N-H stretching modes for adsorbed ND_3 appeared at 2420 and 2520 cm^{-1}. No evidence of dissociative adsorption on the Cu surface was observed. It was also suggested that the molecular adsorption involved the nitrogen lone pair electrons interacting with the metal surface. Similar vibrational frequencies and configuration of interactions were determined for NH_3 and ND_3 adsorbed on Ag surfaces [29].

$\underline{NH_3 \text{ on } Cu(100)}$. The laser experiments involve the vibrational excitation of ν_s stretching mode of adsorbed NH_3 molecules. As a first step, the IR laser is tuned to $\nu = 3340$ cm^{-1}. When the p-polarized light with the intensity I=10 mJ/cm^2 is incident at 75° onto the Cu(100) surface with about 1 monolayer of NH_3, molecular desorption is detected by the mass spectrometer. The p-polarized radiation is chosen because it is much more effective in promoting surface vibrational excitation and desorption than s-polarized light [11]. The observed desorption yield (Y) per pulse strongly depends on the laser frequency (ν). As shown in Fig. 6a, major desorption occurs in the laser frequency range between 3320 and 3370 cm^{-1}. The full width at half maximum (FWHM) of the laser absorption-desorption spectrum is about

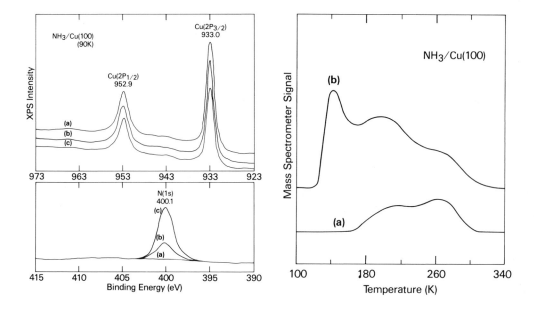

Fig. 4 (left). XPS spectra of NH_3 adsorbed on Cu(100) at 90K: (a) clean surface; (b) surface coverage about 1 monolayer ($\theta=1$), (c) about 4 monolayer coverage ($\theta=4$).

Fig. 5 (right). Thermal desorption spectra of NH_3 adsorbed on Cu(100) at 90K: (a) θ slightly below a monolayer; (b) $\theta=2$. The heating rate is about 8K/sec.

33 cm^{-1}, substantially narrower than the vibrational spectrum of NH_3 adsorbed on Cu(110) obtained by EELS [28] which does not have the comparable high spectral resolution. When the laser is tuned off resonance from NH_3 vibrational bands, no desorption is detected for $I \leq 15$ mJ/cm^2. For the monolayer of adsorbate excited at 3340 cm^{-1}, the laser threshold for desorption is about 5 mJ/cm^2, *i.e.*, about 0.85 MW/cm^2 or 7.5×10^{16} photons/cm^2. At this monolayer coverage ($\theta=1$), the desorption yield increases nonlinearly with the laser intensity (see Fig. 7a). A plot of $\ell n Y$ *versus* $\ell n I$ yields a slope close to 2, indicating that two-photon absorption may be involved in inducing the desorption of chemisorbed species. In great contrast to the chemisorbed states, resonantly excited desorption is readily detected at a rather low laser intensity for the physisorbed species at $\theta>1$. Figure 6b shows the photodesorption spectrum of a multilayer NH_3 ($\theta=3.4$) adsorbed on Cu(100) excited at $I=4$ mJ/cm^2. The peak of the spectrum clearly shifts to a higher laser frequency ($\nu=3370$ cm^{-1}) and the band is substantially broadened (FWHM=47 cm^{-1}). There is a significantly lower laser threshold for desorbing the multilayer adsorbate. Figure 7b shows the desorption yields as a function of laser intensity at $\nu=3370$ cm^{-1}. The essentially linear dependence suggests that single-photon excitation is involved in promoting the desorption. It is further observed that at a given laser fluence, the photodesorption yield increases with the thickness of the NH_3

overlayer. When the laser intensity is raised to $I>20$ mJ/cm^2, laser-induced thermal desorption due to direct substrate heating can overtake the resonant photodesorption effect for the multilayer adsorbate. Apparently at the laser fluence higher than 20 mJ/cm^2, the increase in surface temperature can be greater than 35K above T_s at 90K to thermally desorb physisorbed molecules due to substrate absorption of laser photons.

The absolute photodesorption yields are difficult to determine. We have irradiated the adsorbate at both the monolayer (with ν at 3340 cm^{-1}) and the multilayer (with ν at 3370 cm^{-1}) coverages with 550 laser pulses at $I=10$ mJ/cm^2 covering a sample area of 5 mm $\times$ 5 mm. This is done by translating the sample during the laser irradiation period and examining the sample with XPS which probes about the same surface area. No measurable changes in XPS intensities due to laser photodesorption at these laser frequencies are observed although typically, no desorption signals by mass spectrometer are detected after 50 pulses. Clearly, the

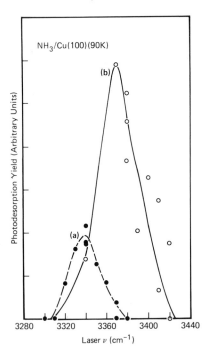
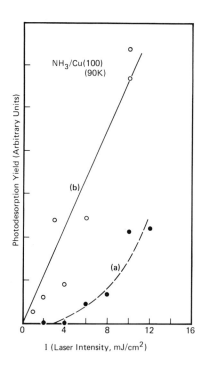

Fig. 6 (left). IR photodesorption yields of NH$_3$ adsorbed on Cu(100) at 90K as a function of laser frequency: (a) surface coverage about a monolayer, $\theta=1$ and laser intensity, $I=10$ mJ/cm^2; (b) multilayer coverage, $\theta=3.4$ and $I=4$ mJ/cm^2. The laser is p-polarized and at 75° angle of incidence. Each data point is an average of mass spectrometer signals due to 20 laser pulses.

Fig. 7 (right). Photodesorption yields of NH$_3$/Cu(100) at 90K as a function of laser intensity: (a) $\theta=1$, $\nu=3340$ cm^{-1} and (b) $\theta=3.4$, $\nu=3370$ cm^{-1}, sampling average of 20 laser pulses.

amount of molecules desorbed form the surface is very small, possibly less than 10^{11} molecules/cm^2 per pulse, i.e., $<6 \times 10^{-7}$ molecules per incident IR photon. The desorption quantum yields per absorbed IR photon is estimated to be about 5×10^{-4} or less. In great contrast, decrease of NH_3 coverage on the Cu(100) surface by a single laser pulse at 532 nm and I=20 mJ/cm^2 which is quite strongly absorbed by the Cu substrate to cause laser-induced thermal desorption is readily detected by XPS.

Figure 8a shows the NH_3 photodesorption signal as a function of time following a laser pulse at the surface coverage of about a monolayer. The laser excites the adsorbate at $\nu=3340$ cm^{-1} and I=10 mJ/cm^2. From the time-resolved mass spectrometer signals, the translational temperatures of desorbing molecules can be calculated according to Eq. (1). For the observed t_m at 128 μsec corrected with the drift time in the mass spectrometer, T_d for the photodesorbed NH_3 is estimated to be about 80K±25K. The large temperature spread is due to the finite length of the ionizer in comparison with the relatively short TOF distance. The maximum rise in the surface temperature due to laser heating of the substrate which is highly reflective in the IR is estimated to be about $\Delta T_s \lesssim 15K$ at the laser intensity of 1.7 MW/cm^2. The observed translational temperature of the photodesorbed molecules apparently is about the same as or slightly below the substrate temperature (T_s). Namely, the desorbing translational temperature seems to be quite accommodated to the temperature of the underlying substrate. Figure 8b shows the TOF signal for the multilayer adsorbate excited by laser pulses at $\nu=3370$ cm^{-1} and I=10 mJ/cm^2. The desorption temperature calculated from $t_m=120$ μsec is about 90K±25K, also very close to T_s. From the line shapes of the TOF spectra as shown in Figure 8, it appears

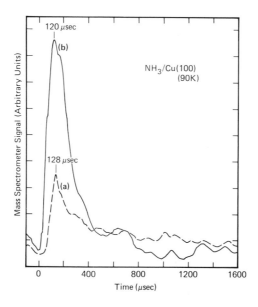

Fig. 8. Time-of-flight mass spectrometer signals of NH_3 photodesorbed from Cu(100) at 90K: (a) $\theta=1$, $\nu=3340$ cm^{-1} and I=10 mJ/cm^2; (b) $\theta=3.4$, $\nu=3370$ cm^{-1} and I=10 mJ/cm^2, sampling average of 20 laser pulses.

that the velocity distributions of the desorbed particles do not deviate far from the Maxwell-Boltzmann distribution. It should be pointed out, however, that since the observed T_d is often substantially lower than T_s as discussed in previous section, the actual peak surface temperature during the photodesorption process can be substantially higher than the original substrate temperature at 90K.

$\underline{ND_3}$ on $\underline{Cu(100)}$. The N-D stretching vibrational modes for adsorbed ND_3 on $\overline{Cu(100)}$ are expected to be in the 2400-2550 cm^{-1} region. When the adsorbate at a monolayer or a multilayer coverage is excited with laser pulses in this spectral region at $I=4$ mJ/cm^2, no significant photodesorption is detected by the mass spectrometer. We cannot produce more intense laser pulses because the laser approaches its tuning range and the IR generating efficiency is quite low at or near 4 μm. The result suggests that IR photons at 4 mJ/cm^2 near 2500 cm^{-1} can excite adsorbed ND_3 molecules but cannot induce desorption by single-photon absorption. In contrast, physisorbed NH_3 molecules can be readily desorbed by IR photons at 3370 cm^{-1} with the same laser fluence. This should not be surprising because an IR photon at 2500 cm^{-1} contains 30 kJ/mole of energy which is smaller than the adsorption energy of 34 kJ/mole for physisorbed ammonia. Whereas the IR photon at 3370 cm^{-1} possesses 40 kJ/mole of energy, enough to break the physisorption bond by single-photon absorption. The result further suggests that at 4 mJ/cm^2, the laser fluence is apparently inadequate in promoting multi-photon excitation to enhance desorption. This is consistent with the observation that at this fluence, neither chemisorbed NH_3 nor ND_3 can be photodesorbed at $\nu=3340$ cm^{-1} or near 2500 cm^{-1}. As discussed previously, chemisorbed NH_3 can be desorbed via a two-photon excitation at a higher laser fluence as shown in Fig. 7a.

$\underline{NH_3}$-$\underline{ND_3}$ coadsorbed on $\underline{Cu(100)}$. A fresh NH_3 and ND_3 gas mixture is prepared and dosed onto the Cu crystal at 90K. The composition of the gases coadsorbed on the Cu surface is directly determined by laser induced thermal desorption with 532 nm light pulses. ND_3 is detected after being ionized at 20 amu and NH_3 at 17 amu by the mass spectrometer. Figure 9 shows the TOF signals for a NH_3/ND_3 (1:1) mixture thermally desorbed by a 532 nm light pulse. From the observed t_m, T_d is estimated to be about 300K. When about a monolayer of this gas mixture is adsorbed on Cu at 90K, the photodesorption signals due to resonant excitation in 3320-3350 cm^{-1} range are very weak. Therefore, the surface coverage is increased to about 2~3 monolayers. As shown in Fig. 10, both NH_3 and ND_3 desorptions are detected at $\nu=3370$ cm^{-1} and $I=6$ mJ/cm^2. The relative desorption yields are about the same. Repetitive measurements show that within the relatively large experimental uncertainties, i.e., about $\pm20\%$, there is no apparent isotope selectivity in the photodesorption of the isotopic coadsorbates. Clearly, although only NH_3 molecules are initially excited by IR photons, either by direct vibrational energy transfer or by thermally-assisted processes (further discussed below), the coadsorbed ND_3 can be desorbed with almost equal probability as NH_3. It should be noted that our earlier study on C_5H_5N and C_5D_5N at about 2 monolayer coverages coadsorbed on a KCl surface excited by CO_2 laser pulses also showed the lack of molecular selectivity in the multi-photon excited desorption [1,10]. The new results demonstrate that even a

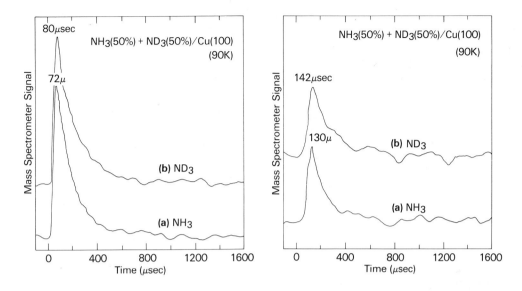

Fig. 9 (left).   TOF signals due to laser-induced thermal desorption of a NH_3-ND_3 (1:1) mixture adsorbed on Cu(100) at 90K excited by 532 nm light pulses at 20 MW/cm^2, θ=4.

Fig. 10 (right).   TOF signals of NH_3 and ND_3 photodesorbed from Cu(100) at 90K for a (1:1) mixture in the adsorbed phase: θ=2.5, ν=3370 cm^{-1} and I=6 mJ/cm^2, sampling average of 20 laser pulses.

single-photon absorption cannot promote efficient separation of different isotope species by IR photodesorption.   The translational temperatures of the photodesorbed species as estimated from TOF signals shown in Fig. 10 are about 75K$\pm$25K, again reasonably close to the substrate temperature.

4.2 Ammonia on NaCl

The experimental results for this system can be briefly summarized below.   For the NH_3 surface coverage of a monolayer or a few monolayers adsorbed on NaCl at 90K, no significant photodesorption is observed for ν in the 3330-3450 cm^{-1} range with I as high as 14 mJ/cm^2.   Comparison of this result with the NH_3/Cu(100) system at 90K leads us to conclude that direct excitation of the N-H stretching mode does not induce desorption even for physisorbed molecules with a single-photon excitation.   The apparently higher photodesorption efficiency for Cu substrate is thus most likely due to the contribution from the direct laser substrate heating effect. Although the polished Cu crystal is highly reflective to IR photons, there is still significant IR absorption to heat up the surface of the substrate and thereby enhance the desorption probability when the adsorbed molecules are also resonantly excited. The Cu substrate heating effect is clearly evident at a higher laser fluence.   As mentioned earlier, for I just above 20 mJ/cm^2, laser-induced thermal desorption which is insensitive to ν is readily observable for a NH_3 multilayer on a Cu substrate. In contrast, the NaCl film on sapphire is sufficiently transparent so that it does not cause severe substrate heating effect in the laser power range of interest.

Resonantly excited desorption of NH_3 molecules begin to be observable as the thickness of the adsorbate is increased to about 15 monolayers with the laser tuned to $\nu=3370$ cm^{-1} and I=6 ~10 mJ/cm^2. No desorption is detected at $\nu=3500$ cm^{-1} under the same exposure conditions. As the surface coverage is increased, the observed desorption signal increases and the required laser fluence for desorption decreases. At a very high surface coverage, photodesorption is detectable even at 35K NaCl substrate temperature. It is interesting to note that under this condition, the observed translational temperature (T_d~70K$\pm$25K) of the photodesorbed NH_3 appears to be substantially higher than the original T_s at 35K. The desorption yield dependence on the overlayer thickness clearly suggests that "indirect heating" [1,10] or "resonant heating" [15,16] of the molecular layers and the substrate due to the transfer of the absorbed photon energy to the surrounding phonons plays a crucial role in the photodesorption phenomenon. Similar photodesorption yield dependence on the adsorbate layer thickness has also been observed for C_5H_5N on KCl and Ag surfaces [1,10].

4.3 Model of IR Photodesorption

The possible desorption processes activated by photoexcitation of internal molecular vibrational modes have been outlined, theoretically evaluated and quite extensively discussed by Gortel, Kreuzer, Lin, George and their coworkers [13-16]. Basically, the processes involve the excitation of a molecular vibrational mode by a single-photon or multi-photon absorption. The photoactive mode usually has a much higher vibrational frequency than the modes directly associated with the molecule-surface bonds and it is generally considered to be decoupled from its interaction with the surface. Once excited, the vibrational energy can be transferred to energy levels in the molecule-surface potential via the bound state-bound state transitions with the transition probability $P_{i'i}^{v'v}$, following the notations used by Gortel, Kreuzer $et\ al.$ [15,16], where v and v' are the vibrational levels in the internal molecular mode and i and i' are the bound states in the surface potential. The process is followed by rapid thermalization of the levels in the surface potential via $P_{i'i}^{vv}$, and the eventual decay of the surface energy into lattice phonons resulting in the heating of the solid. We had called this process "indirect substrate heating" [1,10], while Gortel, Kreuzer $et\ al.$ used the term "resonant heating" [15,16]. If the rate of "indirect" or "resonant" heating is sufficient rapid, it is then possible that the levels of excitation within the surface potential can be so high as to reach the desorption continuum and the molecule can depart from the surface with the probability $P_{ci}^{vv}>0$, where c represents the level of desorption continuum. In the picture just described, direct energy transfer from the photoactive mode to the surface and lattice phonons is considered highly unlikely because of the large difference in the frequencies of these two types of vibrations. In the initial photoexcitation step, if the molecule can be pumped into such a high vibrational level that its total energy is degenerate with the continuum states of momentum, then tunneling from the vibrational level into these continuum states may occur via both elastic $Q_{ci}^{v'v}$ and inelastic $P_{ci}^{v'v}$ processes causing the molecular to desorb. The desorption resulting from these elastic [30] and inelastic [14] tunneling processes, the latter process being the bound state to continuum state transitions mediated by phonons, has been referred as genuine "photodesorption" due to resonant laser-molecule vibrational coupling [15]. In previous experimental [7-12] and theoretical [13-16] studies, it was generally considered that photodesorption via direct vibrational coupling and tunneling could be very important in the observed IR-laser-excited desorption phenomenon.

In addition to the basic photoexcitation, energy transfer and tunneling processes described above, Lin and George [13] also considered the effects of intra-molecular vibrational energy transfer and relaxation on the desorption efficiency. For many molecule-solid systems, depending on the mode of excitation, these factors definitely have to be taken into account. In fact, under most practical conditions, even the inter-molecular energy transfer and relaxation processes, such as intermolecular vibrational-vibrational coupling, also have to be considered and the overall picture can become very complex indeed. However, regardless of the exact details of the energy flow through the internal molecular vibrational-rotational states and even coupling into other molecules, the essential question as far as desorption is concerned is still related to the relative importance among the various coupling processes between the excited internal modes and the bound states as well as the continuum levels in the surface potential. Namely, what are the relative contributions of such processes as elastic and inelastic tunneling and the "indirect" or "resonant" heating to the IR excited desorption? Or, more generally, is the observed photodesorption a genuine quantum effect or a basically thermally-assisted effect? Also, could there be molecular selectivity in IR photodesorption?

In our earlier study on multi-photon excited C_5H_5N and C_5D_5N desorption from KCl and Ag surfaces [1,9-12], it was observed that the photodesorption yields per laser pulse were generally very low and they were strongly dependent not only on the thickness of the molecular overlayer but also on the optical and thermal properties of the substrate. Furthermore, there was no significant molecular selectivity in the photodesorption of C_5H_5N and C_5D_5N coadsorbates when either type of isotropic molecules was initially excited. These observations had led us to conclude that thermally-assisted processes including the effects of direct laser substrate heating and indirect (or "resonant") localized heating of the molecular layer and the substrate played a major role in photo-activated desorption of those molecule-solid systems [1,10]. Our new experimental results on NH_3 and ND_3 adsorbed on both Cu(100) and NaCl as presented in this paper suggest that pure quantum processes such as the elastic and inelastic tunneling leading directly to molecular dissociation from the surface cannot be by themselves very important. This is apparently true even for single-photon excitation in which the energy of the initially excited vibrational level is degenerate with levels in the desorption continuum. If the tunneling processes are very important, i.e., $P_{ci}^{v'v}$ and $Q_{ci}^{v'v} >> P_{i'i}^{v'v}$, then from the experimental results, we have to suggest that such tunneling do not necessarily lead to direct or instantaneous breaking of the surface bonds. In this context, it is possible that once the molecule-surface complex is suddenly in the continuum states of the surface potential, the energy can relax so rapidly by phonon-assisted cascade down to the lower levels of the surface potential and consequently very few molecules can escape from the solid surface. This process is essentially the reversal of the P_{ci}^{vv} process followed by $P_{i'i}^{vv}$ which thermalizes the bound states in the surface potential. In any event, we believe that bound state to bound state transitions, i.e., both $P_{i'i}^{v'v}$ and $P_{i'i}^{vv}$ processes, play essential roles in channeling the adsorbed photon energy into the levels in the surface potential via phonon-mediated steps resulting in the thermal excitation of the surface potential. In favorable cases, the levels of excitation can approach the desorption continuum and desorption can take place. In other cases, this thermal excitation in the molecule-surface potential can enhance the desorption probability when coupled with the elastic and inelastic tunneling processes. Without the thermal activation of the surface potential, tunneling processes alone cannot be effective in inducing molecular desorption. It is interesting to note that the theoretical calculations performed by Gortel et al. [15] also illustrate the importance of the processes involving bound state to bound state transitions. In these calculations, it is shown that the photodesorption rate is drastically reduced if $P_{i'i}^{v'v}$ is set to 0.

This model of IR photodesorption which emphasizes the importance of the thermally-assisted processes appears to be consistent with the important experimental observations including the low photodesorption efficiency and the strong desorption yield dependences on the adsorbate overlayer thickness and the substrate optical and thermal properties for the pyridine and ammonia-surface systems that we have investigated. For CH_3F-NaCl system studied by Heidberg et al. [7,8], a rather high desorption cross section (2×10^{-19} cm^2) approaching the IR absorption cross section was suggested. The estimation, however, was not obtained from direct measurements of the change in surface coverages due to the CO_2 laser irradiation. Thus, it is not clear whether a very high desorption efficiency due to purely molecular vibrational excitation exists for this particular system. If the photodesorption model just described can be generalized, it suggests that molecularly selective desorption would be very difficult to achieve because of the thermal effects. For mixed molecular isotope species coadsorbed on surfaces, ultrafast intermolecular energy transfer via, e.g., dipole-dipole coupling, can further delocalize the absorbed photon energy and render the hope to separate isotope species effectively by means of photodesorption difficult to realize. Nevertheless, it should be recognized that IR photodesorption is an interesting phenomenon and photodesorption studies, as we have demonstrated, can open up a new area for exploring the dynamics of molecular-surface interactions.

ACKNOWLEDGMENTS

We would like to thank J. Goitia for his technical assistance with the experiments and H. Seki and A. C. Luntz for many helpful discussions, and also to acknowledge that this work is supported in part by San Francisco Laser Center, a National Science Foundation Regional Instrumentation Facility, NSF Grant No. CHE 79-16250, awarded to University of California at Berkeley in collaboration with Stanford University.

REFERENCES

1. T. J. Chuang, *Surf. Sci. Reports* **3**, 1 (1983).
2. H. F. Winters, J. W. Coburn and T. J. Chuang, *J. Vac. Sci. Technol.* **B1**, 469 (1983).
3. T. J. Chuang, *J. Vac. Sci. Technol.* **21**, 798 (1982).
4. T. J. Chuang, *Mat. Res. Soc. Symp. Proc.* **17**, 45 (1983).
5. T. J. Chuang and I. Hussla, to be published.
6. I. Hussla, H. Seki and T. J. Chuang, to be published.
7. J. Heidberg, H. Stein, E. Riehl and A. Nestmann, *Z. Phys. Chem. (N.F.)* **121**, 145 (1980).
8. J. Heidberg, H. Stein and E. Riehl, *Phys. Rev. Lett.* **49**, 666 (1982); also, *Surf. Sci.* **126**, 183 (1983).
9. T. J. Chuang, *J. Chem. Phys.* **76**, 3828 (1982).
10. T. J. Chuang, *J. Electr. Spectr. Relat. Phenom.* **29**, 125 (1983).
11. T. J. Chuang and H. Seki, *Phys. Rev. Lett.* **49**, 382 (1982).
12. H. Seki and T. J. Chuang, *Solid State Commun.* **44**, 473 (1982).
13. J. Lin and T. F. George, *Surf. Sci.* **100**, 381 (1980); also, *J. Phys. Chem.* **84**, 2957 (1980); J. Lin, Ph.D. Thesis, University of Rochester (1980).
14. H. J. Kreuzer and D. N. Lowy, *Chem. Phys. Lett.* **78**, 50 (1981).
15. Z. W. Gortel, H. J. Kreuzer, P. Piercy and R. Teshima, *Phys. Rev.* **B27**, 5066 (1983).
16. Z. W. Gortel, H. J. Kreuzer, P. Piercy and R. Teshima, *Phys. Rev.* **B28**, 2199 (1983).
17. C. Jedrzejek, K. K. Freed, S. Efrima and H. Metiu, *Surf. Sci.* **109**, 191 (1981).
18. D. R. Burgess, Jr., R. Viswanathan, I. Hussla, P. C. Stair and E. Weitz, *J. Chem. Phys.* **79**, 5200 (1983).
19. T. J. Chuang, *J. Appl. Phys.* **51**, 2614 (1980).
20. H. F. Winters and J. W. Coburn, *Appl. Phys. Lett.* **34**, 70 (1979).
21. Y. Y. Tu, T. J. Chuang and H. F. Winters, *Phys. Rev.* **B23**, 823 (1981).
22. H. F. Winters and F. A. Houle, *J. Appl. Phys.* **54**, 1218 (1983).
23. T. J. Chuang, *J. Chem. Phys.* **74**, 1461 (1981).
24. F. A. Houle, *Chem. Phys. Lett.* **95**, 5 (1983); also, *J. Chem. Phys.* **79**, 4237 (1983).
25. G. Wedler and H. Ruhmann, *Surf. Sci.* **121**, 464 (1982).
26. N. Bloembergen, in MRS Symposium on Laser-Solid Interactions and Laser Processing – 1978, AIP Conf. Proc., No. 50, 1979, pp. 1-9.
27. I. Hussla and T. J. Chuang, *Mat. Res. Soc. Symp. Proc.* (1984, in press).
28. D. Lackey, M. Surman and D. A. King, *Vacuum* **33**, 867 (1983).
29. J. L. Gland, B. A. Sexton and G. E. Mitchell, *Surf. Sci.* **115**, 623 (1982).
30. D. Lucas and G. E. Ewing, *Chem. Phys.* **58**, 385 (1981).

DESORPTION BY RESONANT LASER-ADSORBATE VIBRATIONAL COUPLING

J. Heidberg, H. Stein, Z. Szilágyi, D. Hoge and H. Weiß

Institut für Physikalische Chemie und Elektrochemie der Universität Hannover, D-3000 Hannover 1, F. R. Germany

Dedicated to Professor Hermann Hartmann

Features of the desorption induced by resonant laser-adsorbate vibrational interaction on ionic crystal, semiconductor and metal surfaces are reviewed. The description of the multiphoton excitation process is founded on quantum level schemes and stepwise incoherent transitions. Rate equations express the basic mechanism and relate the desorption rate and yield to the intensity, the duration of the laser-surface interaction, and to spectral and thermal properties of the adsorption system. The potential of adsorbate photochemistry induced by monochromatic infrared pulses of tunable lasers in combination with high resolution linear infrared spectroscopy is indicated for a one-component adsorbate with several phases within a single monolayer.

INTRODUCTION

Desorption induced by resonant laser-adsorbate vibrational coupling has been observed recently. This process has been investigated in the systems CH_3F on NaCl, SF_6 on NaCl (1), pyridine on KCl and silver, NH_3 on Cu (2) and other systems (3). A quantum statistical theory has been advanced and in particular applied to the system CH_3F on NaCl (4). The primary process has been shown to be the excitation of an internal normal vibration of the adsorbate.

In the following, some remarks on the basic mechanism of laser-adsorbate interaction, direct photodissociation and predissociation are intended to render the subsumption of the new phenomenon into the general frame of spectroscopy and kinetics. The importance of the evaluation of the statics of the adsorbate, such as two-dimensional phases and their transitions, isotherms, metastables, linear spectra and the particular role of the infrared spectroscopy of adsorbates thereby are indicated. As example, the system CH_3F-NaCl has been chosen. The excitation of adsorbates with coherent monochromatic laser radiation is treated and the rate equations relating the desorption rate with laser and adsorbate properties are reviewed.

329

B. Pullman et al. (eds.), Dynamics on Surfaces, 329–344.
© *1984 by D. Reidel Publishing Company.*

LASER-ADSORBATE INTERACTION

A laser pulse striking a surface covered with an adsorbate can deposit
energy into
        1) the adsorbate
        2) the adsorbate-adsorbent bond
        3) the adsorbent,
for a given adsorption system depending upon wavelength, intensity and
pulse length. Laser polarization and angle of incidence can be important
also for the strength of interaction. Processes 1) and 2) are qualitatively different from process 3) due to the characteristic structures
of adsorbates with a finite number of strongly bound atoms as distinct
from the solid adsorbents. The structure of the adsorbate, when isolated
on the surface, can be specified in terms of approximate point group
symmetry, the fundamental vibrational excitations are discrete, being
classified according to normal and local modes. The electronic excitations are also discrete below the first ionization potential. These
discrete features of low-lying excitations in adsorbates, which generate
sharp resonances in light-adsorbate coupling and yield substance specificity and localization in the initial energy deposition, differ markedly from the continuous level structure in solids.

    In the ordered lattices with the approximate symmetry of a periodic
space group the density of states of elementary excitations, e.g. phonons, plasmons, excitons, is continuously extending over a finite energy range. However, the characteristic length of elementary excitations,
e.g. the free paths of phonons and electrons, and the coherence length,
is in general appreciably smaller than the solid dimensions, rendering
possible certain localized excitations in the adsorbent also. In two-
dimensional, ordered adsorbate layers nonlocalized elementary excitations, spanning a continuous energy range, may be generated. However,
the width e.g. of the infrared absorption by the collective internal
vibrations of ordered CO monolayers on metals (Pt, Pd, Cu) are only
$\lesssim 20$ cm^{-1}, being mainly determined by homogeneous and inhomogeneous
line broadening. (5)

PHOTODISSOCIATION

Dissociation of a chemical bond upon photoexcitation may occur in
either of two ways (6):
        a) Direct dissociation,
by single photon excitation from the ground state 1) to a state above
the dissociation limit of an excited state 2), yielding a continuous
absorption,
by multiphoton excitation, giving also a continuous absorption spectrum.
        b) Predissociation
by single or multiphoton excitation to a bound excited state 2) followed
by transition to a dissociative state 3) above the dissociation limit of
the ground state 1) or to a repulsive state 3'), yielding a discrete
absorption spectrum but broadened due to the lifetime shortening of the
state 2). Such processes were first discovered in the x-ray region by

Auger.

Whereas direct dissociation has been considered only for electronic (vibronic) stimulation, the excited state being a higher electronic state, three cases of predissociation have been distinguished: electronic, vibrational and rotational. Vibrational predissociation cannot arise for diatomic molecules in the gas but is of considerable importance for polyatomic molecules, radicals and ions. Most unimolecular decompositions belong to this case, in which no change of the electronic state occurs.

Direct Photodesorption

Direct photodissociation of the adsorbate-adsorbent bond has not yet been shown experimentally.

Though this bond can be excited vibrationally – for example the hydrogen-tungsten bond in H-W(100) by CO_2-laser radiation (6)

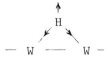

ν_s (fundamental) 1046 cm^{-1}
half width   14 cm^{-1}
adsorption energy $\gtrsim$80 kJmol^{-1}

similarly the C-Cu bond vibration by electron impact (EELS) in (7,8)

Cu(100)-CO ν_{Cu-C} (fundamental)   339 cm^{-1}
                           half width   30 cm^{-1}
                  adsorption energy   70 kJmol^{-1}

– its direct photodissociation is unlikely. To understand this we bear in mind that the adsorbate-adsorbent bond will be very anharmonic, particularly at high excitation, so that desorption induced in a stepwise multiphoton process by monochromatic radiation with a frequency close to that of the fundamental will be rather improbable, except at enormous laser intensities, calculated to be in the order of $\gtrsim 10^{12}$ Wcm^{-2}. In addition there must be considerable broadening of the energy levels. Then the process is expected to be of little resonant character, yielding a broad continuous absorption spectrum.

Direct photodesorption by excitation to a momentum state above the desorption limit of an excited internal vibrational or electronic state might occur under certain conditions e.g. if the equilibrium adsorbate-adsorbent distance changes strongly in going from the lower to the upper state, since then the Franck-Condon maximum could be in the continuum.

Predesorption

Electronic predesorption. In the desorption of molecules on metal and semiconductor surfaces (ZnO, TiO_2, V_2O_5), stimulated by ultra-violet and visible photons (h$\nu \lesssim 6$ eV), the effective light is usually absorbed by the electrons in the solid adsorbent, up to a depth of about 10^2 Å. In metals fast relaxation of the electronic excitation into the lattice phonons within 10^{-13} s takes place, causing phonon-mediated or thermal desorption with a quantum yield $\lesssim 10^{-7}$. For laser pulse lengths of $\gtrsim 10^{-8}$s ordinary bulk thermal conduction laws are applicable (9). Photodesorption from oxide and sulfide semiconductor surfaces involving band gap electronic excitation and subsequent migration of the excitation to the desorbing molecule (CO_2, H_2O) can have quantum yields of 10^{-3}. Apparently studies on electronic predesorption induced by adsorbate photo-excitation have not yet been reported, except for very high photon energies $\gtrsim 10$ eV (10).

In photodesorption it could be favorable that for electronic excitation the absorption cross section is in general higher, the energy absorbed in a single photon process larger, and the degradation of energy proceeds more slowly than for vibrational excitation. Moreover, focusing of light and local macroscopic control of reactions is the sharper the shorter the wavelength. In electronic excitation, however, substance specificity is less pronounced than in vibrational excitation.

Vibrational predesorption. In vibrational predesorption by resonant laser-adsorbate interaction, an infrared laser pulse striking an adsorbate-covered surface of a solid is coupled into some internal vibrational normal mode of the adsorbate. A marked dependence of the desorption yield upon laser wavelength is observed. On the surfaces of either highly transparent insulators and semiconductors or strongly reflecting metals, the primary energy deposition can be localized in the adsorbate. Even vibrational predesorption of molecules from supported thin metal films, which themselves absorb infrared radiation continuously over the whole spectral range, exhibits distinct resonance features. Quantum efficiencies as high as 10^{-2} have been measured just above the threshold intensity.

Vibrational adsorbate excitation shows a rich structure rendering high specificity in infrared-adsorbate coupling, as is well established in surface analysis. Linear infrared absorption spectra discriminate by the frequency shifts between different isotopes and different bonding coordination of a single adsorbed species, such as on top of an adsorbent atom and bridged over two adsorbent atoms. Phase equilibria between a two- dimensional adsorbate "gas" and a two-dimensional condensed island phase have been determined, the frequency shifts between the signals of the various phases being $10 - 20$ cm^{-1}. Also kinetic effects such as the formation of metastable adsorbates upon low temperature adsorption due to hindrance to surface diffusion are detectable from the infrared frequencies of adsorbates. Though monochromatic infrared laser pulses can be coupled selectively into a single one of the adsorption phases, the high selectivity of excitation can in general not be fully preserved in the subsequent reaction steps due to fast energy relaxation.

It is the intricate competition between the excitation, the energy transfer into continuum states above the desorption limit and relaxation which determines the yield and selectivity of the photon induced process. The excitation rate is determined by the laser intensity. Certain threshold intensities are required in order to promote the process at observable rates and yields. For desorption of neutrals with CO_2 laser pulses

threshold intensities of about 10^5 Wcm^{-2} were measured. At high intensities the desorption rate becomes independent of intensity, i.e. "saturation" takes place. In general most efficient photochemistry can be
performed in the intermediate intensity range between the threshold and
saturation level. No matter how judiciously one has selected the laser
wavelength and intensity, specificity may suffer, if the time of interaction between the radiation and the sample is too long. Energy relaxation, whose rate is temperature dependent, will heat the whole system
relatively uniformly, causing less specific processes, even though the
energy absorption occurs initially at certain localized sites.

Simple rate equations relate the photoreaction yield and rate to
the intensity, the duration of the laser-substrate interaction, spectral properties of the adsorbate and adsorbent, thermal properties of
the adsorbent and to the temperature. For optimizing laser wavelength,
intensity and pulse duration, as well as sample preparation, the use of
these simple rate equations could be expedient.

INSTRUMENTATION

Only the main components of the apparatus are described here.

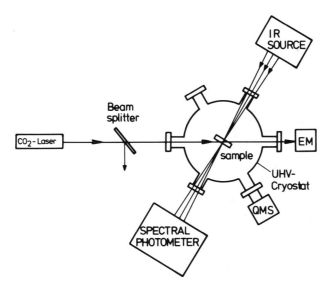

Fig. 1. Block diagram of apparatus for desorption induced by resonant
laser-adsorbate coupling.

An ultrahigh vacuum (uhv) cryostat with flanged analyzer cell suitable
for linear infrared absorption spectroscopy as well as laser stimulated
processes is equipped with a mass spectrometer. The entire uhv system
is pumped to a base pressure in the 10^{-10}mbar range by an ion getter
pump and a titanium sublimation pump after bakeout at 420 K.

Considering an experimental arrangement for the study of desorption

by resonant laser-adsorbate interaction, we have to deal with three
general problems:

1) Small linear infrared absorption bands. For a monolayer of CO on
single crystal surfaces of transition metals, the linear absorption is
typically in the order of 1% to a few percent of the incident intensity.
The absorption is determined by the number of adsorbed molecules, their
surface concentration, the adsorbate-adsorbent distance, the vibrational frequency, the electronic and vibrational polarizability, α_e and α_v,
respectively. The values for α_e and α_v were estimated to be appreciably
larger for CO-metal systems than the gas phase values. The corresponding values for α_e and α_v for CO-ionic crystal systems are not known,
but definitely smaller than for molecule-metal systems. No linear infrared absorption signals from molecules at monolayer coverage on single crystal cleavage planes of alkali halides have been detected as
yet. Therefore NaCl films are evaporated in situ onto optically polished or cleaved NaCl(100) single crystals at ca. 100 K and then annealed
at room temperature. The air-cleaved NaCl (100) surface was cleaned by
heating up to 420 K under uhv conditions. In order to prevent gas, e.g.
water, adsorption we filled up a refrigerant vessel inserted in the uhv
cryostat with liquid nitrogen before cooling down the sample to 100 K.
Sharp infrared absorption signals with half widths of $\lesssim 20$ cm^{-1} at submonolayer coverage on (100) planes of the crystallites of 1μm size are
obtained.

2) The second general problem refers to the time- and mass
resolved detection of the desorbed particles at high sensitivity of the
quadrupole mass analyzer operated in time-of-flight mode. Time resolution of a few microseconds, mass resolution of at least 1:100 for desorption of much less than a monolayer from an area of about 1 mm^2 into
a small solid angle are desirable. Momentum and time-of-flight spectra,
respectively, of molecules desorbed from single crystal cleavage planes
of NaCl (100) can be measured. Due to the high sensitivity of the
secondary electron multiplier in the mass spectrometer, infrared vibrational spectroscopy of adsorbates at monolayer coverage on single crystal planes by desorption due to resonant laser-adsorbate interaction
can thus be carried out.

3) At this stage a third general problem arises, the crystalo-
graphic characterization of ordered ionic surfaces, clean and covered
with adsorbates, e.g. CO, CH_3F, by atom beam diffraction. Not much work
has been performed in this field.

A carefully purified CH_3F (or other gas) stream is directed towards
the sample surface and controlled by an uhv leak valve. The beam of a
transversely excited atmosphere CO_2 laser is focused onto the sample.
The laser is furnished with an intracavity aperture to generate infrared pulses of ca. 200 ns (FWHM) duration and approximate TEM$_{oo}$ spacial
distribution. Oscillation occurs simultaneously on several longitudinal
modes at a single rotational vibrational line. The wavelength is determined with a grating spectrum analyzer, the pulse energy measured with
a pyroelectric energy meter. The time-profile of the laser pulse is
measured with a fast photon drag detector. Application of the lens formula for TEM$_{oo}$ radiation gives the area of the focus required for the
estimation of the fluence and intensity, in approximate accordance with

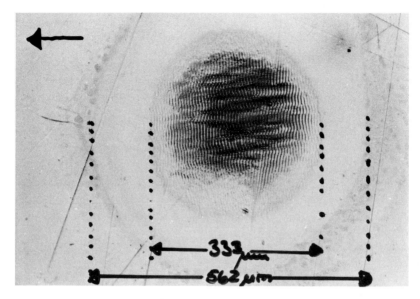

100 μm ├─────┤

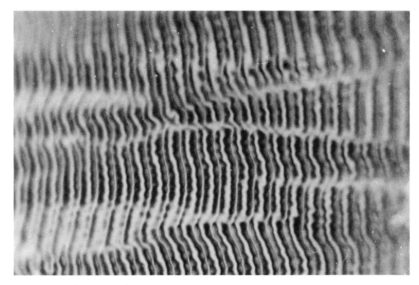

20 μm ├─────┤

Fig. 2. Focal spot on NaCl (100) cleavage plane generated by a CO_2 laser pulse. Surface microstructure, wavelength 6.7 μm, wavefront mainly normal to laser polarization, incidence of radiation normal to NaCl surface.

the measured area: Fig. 2. The focal spot was obtained on the back side cleavage plane of a NaCl disc, 4 mm thick, as used as substrates in the desorption experiments at much lower intensities, with a CO_2 laser pulse

of 100 mJ energy, 250 ns (FWHM) duration, wavelength 9.47 μm, employing a lens with 15 cm focal length. The circular structural area has a
diameter of ∿330 μm, the diameter of the total spot is∿560 μm. (11)

THE STATIC ADSORBATE

Infrared photodesorption is best attacked by splitting the problem into
two parts to be solved separately:
            1) The determination of the static properties of the adsorbate:
existence conditions of adsorption phases, isotherms, heats of adsorption, stationary molecular (spectroscopic) energy states of the particular phases, structure.
            2) Coupling of the energy states by the laser field.
      Obviously, the static problem is a vast field of research by itself.
We confine our attention here only to a few characteristic points relevant to the infrared photodesorption in the adsorption system CH_3F-NaCl.

Two-Dimensional "Gas" and Island Phases in the One-Component Adsorbate
CH_3F-NaCl

Linear infrared absorption spectroscopy shows the existence of several
phases in the one-component adsorbate CH_3F-NaCl at submonolayer coverage. At low coverage ($Θ ≲ 0.2$) and sufficiently high temperature ($≳ 190K$)
only phase α is present. Its $ν_3$ band frequency (CF-stretch) is observed
at 946 cm^{-1} and shifts only slightly with increasing coverage up to
saturation (+2 cm^{-1} not shown in fig. 3 , to be published), being consistent with the assumption that lateral interactions in phase α are

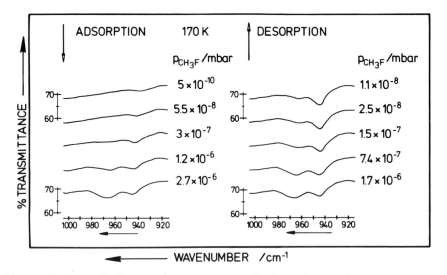

Fig. 3. Linear infrared absorption spectra of the adsorption system
CH_3F-NaCl (film) on NaCl (100) with phase α (947 cm^{-1}, two-dimensional
"gas") and phase ß (967 cm^{-1}, island).

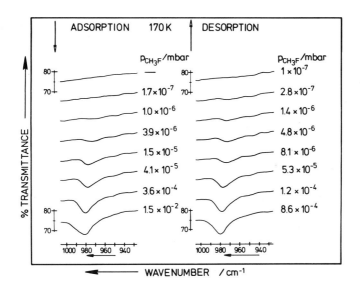

Fig. 4. Linear infrared absorption spectra of the adsorption system CH$_3$F–NaCl (film) on NaCl (100) with phase α (947 cm^{-1}, two-dimensional "gas") and phases ß and γ (967 – 980 cm^{-1}, island). Thickness of NaCl film in fig. 3 larger than in fig. 4. NaCl crystal size ca. 1 μm.

practically absent. The frequency of the α-phase remains constant even as a second phase ß is formed and the coverage is further increased. Therefore we ascribe approximately constant intermolecular distances to the α-phase after the appearance of the ß-phase as the coverage increases. These distances are much larger in the α- than in the ß-phase, the saturation coverage of α being appreciably smaller than those of ß. The ß-phase is described as an island phase. The islands grow continuously larger to phase γ (980 cm^{-1}) with increasing coverage up to saturation. Fig. 4. At further increasing the coverage (at sufficiently low temperature $\lesssim$ 100 K) multilayer adsorbate formation takes place (996 cm^{-1}). Application of the isotopic mixing method should further illucidate the interesting phase boundaries in CH$_3$F–NaCl.

Apparently the α, ß transition is not a first order two-dimensional phase transition because in that case the isotherms should show a vertical step where large amounts of gas are adsorbed at constant and definite pressure, which is not observed in these experiments, though kinetic effects could interfere.

On admission of CH$_3$F to the NaCl surface at 77 K a multilayer phase is formed even at submonolayer coverage probably because of hindrance to surface migration at that temperature. Raising the temperature to $\gtrsim$ 100 K leads to the formation of the phases α, ß, γ. If α- or coexistant α- and ß-phase are prepared at temperatures above 150 K and cooled down to 77 K without further gas admission no changes in intensity and frequency of the α- and ß-IR bands are observed. γ-phase shows a similar behaviour.

Desorption by resonant laser adsorbate vibrational coupling has been observed in all adsorbates, including α.

In the photodesorption only one internal vibration of the adsorbate is considered to play an active part, namely that vibration into which the laser is coupled resonantly, in case of CH_3F-NaCl the CF-stretching vibration ν_3, which has a frequency lower than those of all internal adsorbate vibrations, but much larger than the fundamental frequency of the adsorbate-adsorbent bond.

EXCITATION OF ADSORBATES WITH COHERENT, MONOCHROMATIC LIGHT

Excitation of a two-state system with coherent intense, monochromatic radiation induces the well-known Rabi oscillations of the state occupation functions. For systems with many states an oscillatory (complicated) time dependence of the state occupations can still be obtained. To describe the evolution in time of such systems, the von Neumann-Liouville equation for the density matrix $\rho_{\nu\nu'}$ is used and rate equations may not be applicable.

The systems considered are characterized by a separable IR-active

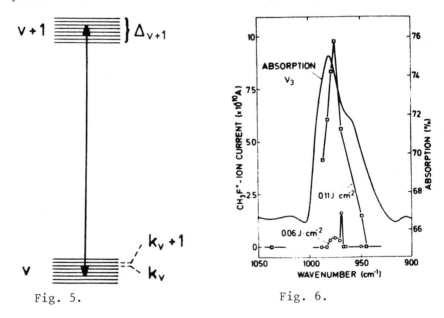

Fig. 5.   Fig. 6.

Fig. 5. Closely-spaced states within the widths Δ_ν of the energy levels v and v+1 of the infrared active vibration coupled to the laser field (reproduced with permission of Heidberg 1g).

Fig. 6. Measured linear infrared absorption band ν_3, width 45 cm^{-1} of α-CH_3F-NaCl (film) on NaCl (100) and yield of CH_3F desorption (CH_3F^+ ion current) at constant coverage and fluence vs laser excitation wavenumber. Sample temperature 77 K, pressure after gas admission 1.6×10^{-9} mbar, (reproduced with permission of Heidberg, Stein, Riehl 1b).

(possibly harmonic) vibration with energy levels E^V having many closely-spaced states (sublevels) k_v and/or true continua within their widths Δ_v. It is this vibration, which is resonantly pumped by the laser in the desorption process. We should bear in mind, however, that the absorption bands in the high resolution linear infrared spectrum strictly correspond to molecular eigenstates or spectroscopic states, and not to separable modes. Even for the fundamental transition the coupling energies can be of the order of cm^{-1}. Nevertheless, the concept of one IR active vibration strongly coupled to the laser field appears to be useful and justified for the desorption systems considered. The occupation functions n^V of the levels v develop in time according to a simple rate equation

$$\dot{n}^V (t) = \sum_{v'} (L^{vv'} n^{v'} - L^{v'v} n^v) \tag{1}$$

with time-independent rate coefficients $L^{v'v}$, if the widths Δ_v of the levels in frequency units are large compared with the laser-pumping band width $L^{v'v}$:

$$\Delta_v, \Delta_{v'} \gg L^{v'v} > g_v, g_{v'} \tag{2}$$

n^v is the sum of the occupations of all states k_v within the level v. The use of rate equations involves the assumption that coherence in the excitation decays very fast compared with energy dissipation, which in general can be expected for systems with many states within the level width Δ_v. Then the pumping rate coefficients are given in accordance with Fermi's golden rule

$$L^{v'v} = 2\tilde{\pi} < |H_{v'v}|^2 > g_{v'} , \tag{3}$$

where $<|H_{v'v}|^2>$ stands for the average square coupling matrix element between the two levels v and v', and g_v denotes the density of states in level v. As a starting point g_v can be estimated by a combinatorial algorithm (12). the important parameter being the level width Δ_v. Because $|H_{v'v}|^2 = |H_{vv'}|^2$, the long-time relaxed distribution is given by

$$\left(\frac{n^v}{n^{v'}}\right)_{relaxed} = \frac{g_v}{g_{v'}} = \frac{L^{vv'}}{L^{v'v}} . \tag{4}$$

For a system with broad energy levels, (2, the rate coefficients $L^{v'v}$ are proportional to the laser intensity I with a constant absorption cross section σ as one factor of proportionality. σ can be measured at low intensity. For z-polarized monochromatic light with intensity I one has

$$<|H_{v'v}|^2> = <|M^z_{v+1,v}|^2> \frac{2 I}{4\tilde{\hbar}^2 c \varepsilon_0 \eta} c(\vartheta) , \tag{5}$$

$$<|M^z_{v+1,v}|> \cong <\varphi_{v+1} |\vec{\mu} \cdot \vec{e}_z| \varphi_v> . \tag{6}$$

$M_{v+1,v}^z$ is the z component of the electric dipole transition moment, $\vec{e}_z$ the unit vector in the z-direction, being perpendicular to the adsorbent surface. $\mathcal{E}_0$ is the vacuum dielectric constant, η the refractive index of the medium, c the velocity of light, h = $2\pi\hbar$ Planck's constant. $C(\vartheta)$ is a parameter, which is dependent upon the surface conditions, fig. 2, and upon the angle of incidence ϑ of the radiation, assumed to be polarized parallel to the plane of incidence.

As an approximation to begin with, the level width is taken to be of the order of the band width 10^1 cm^{-1} of the observed fundamental vibrational spectrum. For adsorption systems the number of states in the wavenumber interval 1 cm^{-1} is expected to be high, e.g. due to coupling with low frequency hindered rotations and translations of the adsorbed molecule.

No spectral structures caused by coupling with the high frequency (translational) vibration of the adsorption bond whose fundamental frequency was calculated to be $\gtrsim 150$ cm^{-1} have been observed. Individual translational-vibrational and rotational-vibrational states will not be resolved, but rather the overall envelope is recorded. Kreuzer et al. calculated the broadening of the internal vibrational adsorbate states to be of the order 1 cm^{-1}. The broadening is mediated by phonon and internal-external vibrational interaction. Time-resolved hole burning experiments could give experimental evidence for the homogeneous vibrational band width. Even at zero coverage, broadening due to inhomogeneous molecule- adsorbent interactions should appear also.

The level width plays an additional role, since small anharmonicity of the internal adsorbate vibration by the laser field is compensated in the multiquantum excitation with monochromatic laser radiation. For the system CH$_3$F in argon at a temperature of 9 K the first anharmonicity constant was reported to be 8 cm^{-1} (13). The ν_3 band width in CH$_3$F-NaCl is 20 to 50 cm^{-1} depending on coverage.

RATE EQUATIONS FOR DESORPTION BY RESONANT LASER-ADSORBATE VIBRATIONAL COUPLING

The interaction between a molecule and the solid can be represented by an adsorption potential V(x) which is attractive a few Angströms away from the surface and becomes strongly repulsive close to contact. We assume that the motions of the molecule parallel to the surface are weakly hindered. The corresponding frequencies should be small compared to the frequency of the vibration perpendicular to the surface. Then the adsorption potential is a function of the distance x from the surface only. We bear in mind, that these features of the static molecule-solid interaction are consistent with the level structure being excited by coherent light. In general V(x) will have several bound states with vibrational energies $E_0,\ldots,E_i$, etc. and a continuum above. Molecules occupying the bound states form the adsorbate, molecules in the continuum states the gas.

Desorption of a molecule takes place, when the instantaneous vibrational energy in the adsorption potential exceeds the binding energy of adsorption. In thermal desorption phonon energy is transferred from the adsorbent to the vibration of the adsorptive bond. In desorption by resonant laser-adsorbate vibrational coupling, the energy is transferred from the pumped internal vibration of the adsorbed molecule into the adsorptive vibration. In a first approximation the energy of an adsorbed molecule can be expressed as

$$E_i^v = E_i + (v+1/2)\omega\hbar , \tag{7}$$

where i labels the states in the adsorption potential and v counts the number of quanta in the internal vibration of the adsorbate. The occupation functions of the levels n_i^v change in time at random. In photodesorption there are two forces driving the walk of the molecule up and down the energy levels. One is due to the already mentioned thermal vibration of the lattice atoms, which the adsorbed molecule perceives as a time-dependent force. If the molecule is in a level (i,v), this force induces transitions to other levels (i', v') with a transition probability $P_{i'i}^{v'v}$ per unit time. For $(E_i^v - E_{i'}^v) < 0$ the transition occurs via absorption of a phonon, otherwise via emission. If the spacing of the energy levels $(E_i^v - E_0^v)$ is larger than the Debye energy and if relaxation into the ground level is to be included, multiphonon processes must be considered. The other driving force is the field of the laser which induces transition $v \to v\pm 1$ with a transition probability $L_{i'i}^{v'v}$ per unit time, where

$$L_{i'i}^{v'v} = L^{v'v} \delta_{i'i} . \tag{8}$$

As stated above the $L^{v'v}$ are linear functions of the intensity. The level populations n_i^v evolve in time according to the rate equation

$$\frac{d}{dt} n_i^v(t) = \sum_{\substack{i'=0 \\ v'=0}}^{\substack{i_{max} \\ v_{max}}} R_{ii'}^{vv'} n_{i'}^{v'}(t) - R_{i'i}^{v'v} n_i^v(t) - \sum_{v'=0} (P_{ci}^{v'v}+Q_{ci}^{v'v}) n_i^v(t) \tag{9}$$

where it is assumed that

$$R_{i'i}^{v'v} = L_{i'i}^{v'v} + P_{i'i}^{v'v} . \tag{10}$$

The $P_{ci}^{v'v}$ and $Q_{ci}^{v'v}$ are the rate coefficients for transitions from a bound level (i,v) into all continuum states (c, v'), the first being mediated by phonons, the second caused by coupling between the pumped internal and the external vibration of the whole molecule normal to the surface. These two terms represent desorption. The rate equation (9 is similar to Kreuzers. If readsorption can be neglected, as is appropriate for the experiments considered, the adsorbate concentration $\Theta(t)$ on the surface changes in time t according to

$$\frac{\Theta(t)}{\Theta(0)} = \frac{\sum_{v=0}^{v_{max}} \sum_{i=0}^{i_{max}} n_i^v(t)}{\sum_{i'} \sum_{v'} n_{i'}^{v'}(0)} = \sum_k A_k e^{\lambda_k t} , \tag{11}$$

starting with a thermal occupation

$$n_{i'}^{v'}(0) = B \exp\left[-E_{i'}^{v'} / kT\right].$$ (12

The $\lambda_k < 0$ are the eigenvalues of the matrix $(L + P + Q)$. In general the largest eigenvalue λ_1 is the only one close to zero. Then after a sufficiently long time the contributions from all other terms $\exp[\lambda_k t]$ may be neglected, which defines a "steady state", where we have

$$\frac{-d\ln[\Theta(t) / \Theta(0)]}{dt} = -\lambda_1 = k_u.$$ (13

The calculated and measured k_u values are of the same order of magnitude: $k_u \approx 10^6$ s^{-1} at a laser intensity 0.5 MW cm^{-2}, 77 K. The experimental k_u values are obtained from measurements of the desorption yield dependence on the laser fluence. Fig. 7. Further information should be gained from the time-of-flight spectra. Fig. 8.

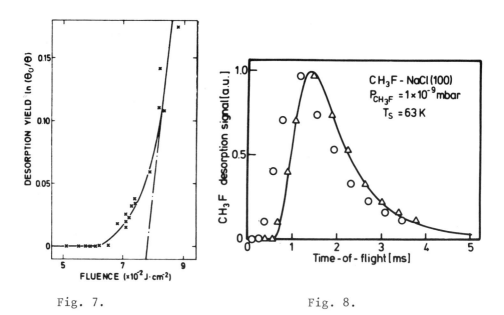

Fig. 7.   Fig. 8.

Fig. 7. Yield of CH_3F desorption from γ-CH_3F-NaCl (film) on NaCl (100) vs. fluence ϕ at constant laser excitation, wavenumber $\tilde{\nu}$ =975.9 cm^{-1}. Sample temperature 77 K, pressure 1.6 x 10^{-9}mbar. Reproduced with permission of Heidberg, Stein, Riehl 1b.
Fig. 8. Measured time-of-flight spectrum in desorption induced by resonant laser-adsorbate vibrational coupling. CH_3F on NaCl (100) cleavage plane. Reproduced with permission of Heidberg, Stein, Riehl, Hussla 1f.

We conclude that the simple rate equation approach, as proposed by H.J. Kreuzer (4), and others (1d), is useful for obtaining a general view of the basic mechanism of desorption by resonant laser-adsorbate vibration-

al coupling. It has practical importance in the optimization of
laser intensity and pulse length
for photodesorption in a given
substrate. We shall attempt to
measure the infrared absorption
of the adsorption bond vibration
in order to determine the range
γ^{-1} of the adsorption potential
for the two-dimensional "gas"
phase α. (14)

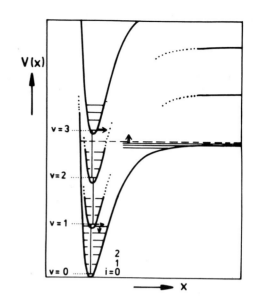

Fig. 9. Adsorption potentials V(x)
of four levels of the internal CF
vibration in the adsorbate CH$_3$F-
NaCl. Multiquantum internal vibrational predesorption in CH$_3$F-NaCl.
Reproduced with permission of J.
Heidberg 1g.

ACKNOWLEDGEMENTS

"Gefördert mit Hilfe von Forschungsmitteln des Landes Niedersachsen",
der Deutschen Forschungsgemeinschaft und des Fonds der chemischen In-
dustrie. We should like to thank E. Riehl for his scrutiny in the experimental work, Prof. H. J. Kreuzer, Prof. M. Quack and Dr. Z. W.
Gortel for stimulating discussions and sending preprints.

REFERENCES

1.a. J. Heidberg, H. Stein, A. Nestmann, E. Hoefs, I. Hussla, Sympos-
      ium 'Laser-Solid Interactions and Laser Processing', 1978, Mat-
      erials Research Society, Boston. - Laser-Solid Interactions and
      Laser Processing, S.D. Ferris, H.J. Leamy and J.M.Poate, eds.,
      American Institute of Physics, New York 1979, pp.49-54.
   b. J. Heidberg, H. Stein, E. Riehl, Phys. Rev. Letters 49, 666
      (1982).
   c. J. Heidberg, H. Stein, E. Riehl and A. Nestmann, Z. Physikali-
      sche Chem. N. F. 121,145 (1980).
   d. J. Heidberg, H. Stein, E. Riehl in 'Vibrations at Surfaces', R.
      Caudano, J.-M. Gilles, A.A. Lucas, eds. (Plenum Press, New York
      1982) pp.17-38.
   e. J. Heidberg, H. Stein, E. Riehl, Surface Sci. 126,183 (1983).
   f. J.Heidberg, H. Stein, E. Riehl and I. Hussla in Surface Studies
      with Lasers, F.R. Aussenegg, A. Leitner, M.E. Lippitsch (Sprin-
      ger Verlag, Berlin 1983) pp. 226-229.
   g. J. Heidberg, Materials Research Soc. Annual Meet, Boston (1983)
      in press.

1.h. J.Heidberg, I. Hussla and Z. Szilagyi, J. Electron Spectrosc. 30,
      53 (1983). - Vibrations at Surfaces, Part B, C.R. Brundle and
      H. Morawitz, eds. (Elsevier, Amsterdam 1983) pp. 53-58.
2.a. T.J. Chuang, J. Chem. Phys. 76, 3828 (1982).
  b. T.J. Chuang and H. Seki, Phys. Rev. Lett. 49, 382 (1982).
  c. T.J. Chuang, J. Electron Spectrosc. 29, 125 (1983), Vibrations at
      Surfaces, Part A, C.R. Brundle and H. Morawitz, eds. (Elsevier,
      Amsterdam 1983) pp. 125-138.
  d. T.J. Chuang, Surface Sci. Reports 3, 1-105 (1983).
  e. I. Hussla and T.J. Chuang, Materials Research Soc. Annual Meet.
      Boston 1983, in press.
3.a. J. Heidberg, D. Hoge and H. Stein to be published.
4.a. Z.W. Gortel, H.J. Kreuzer, P. Piercy and R. Teshima, Phys. Rev. B
      27, 5066 (1983). ibid. 28, 2119 (1983).
  b. H.J. Kreuzer and D.N. Lowy, Chem. Phys. Lett. 78, 50 (1981).
  c. Z.W. Gortel and H.J. Kreuzer, Surf. Sci. 131,L359 (1983).
  d. D. Lucas and G.E. Ewing, Chem. Phys. 58, 385 (1981).
  e. C. Jedrzejek, K.F. Freed, S. Efrima and H. Metiu, Surf. Sci. 109,
      191 (1981).
  f. W.C. Murphy and T.F. George, Surf. Sci. 102,L46 (1981).
  g. J.A. Beswick and J. Jortner, J. Chem. Phys. 68, 2277 (1978).
5.a. F.M. Hoffmann, Surf. Sci. Reports 3, 107-192 (1983).
  b. D.A. King, in 'Vibrations at Surfaces', C.R. Brundle and H. Mora-
      witz eds. (Elsevier, Amsterdam 1983) pp. 11-24 (Part A).
6.a. Y.J. Chabal, A.J. Sievers, Phys. Rev. Lett. 44, 944 (1980).
  b. R.F. Willis, in 'Vibrational Spectroscopy of Adsorbates'.
      Springer, Berlin 1980.
7.   B.A. Sexton, Chem. Phys. Lett. 63, 451 (1979).
8.   J.C. Tracy, J. Chem. Phys. 56, 2748 (1972).
9.a. J.H. Bechtel, J. Appl. Phys. 46, 1585 (1975).
  b. G. Wedler and H. Ruhmann, Surf. Sci. 121, 464 (1982).
  c. J.P. Cowin, D.J. Auerbach, C. Becker and L. Wharton, Surf. Sci. 78,
      545 (1978).
10.a. D. Menzel, J. Vac. Sci. Technol. 20, 538 (1982).
  b. D. Lichtman and Y. Shapira, CRC Critical Rev. Solid State
      Materials Sci. 8, 93 (1978).
11.   P.A. Temple and M.J. Soileau, J. Quantum Electronics, IEEE
      Vol. QE-17, NO 10, 2067 (1981).
12.a. M. Quack, Ber. Bunsenges. Phys. Chem. 83, 757 (1979).
  b. M. Quack, Adv. Chem. Phys. 50, 395-473 (1982).
13.   V.A. Apkarian and E. Weitz, J. Chem. Phys. 76, 5796 (1982).
14.   G. Herzberg, The Spectra and Structures of Simple Free Radicals,
      Cornell University Press, Ithaca, New York 1971.

CHARGE TRANSFER EXCITATIONS IN SERS: COMPARATIVE STUDY OF BENZENE,
PYRIDINE AND PYRAZINE

J. Thietke, J. Billmann, A. Otto
Physikalisches Institut III,
Universität Düsseldorf
Universitätsstr. 1
D-4000 Düsseldorf 1,
Fed. Rep. of Germany

ABSTRACT

We discuss a model of resonant Raman scattering by photon driven charge
transfer between an electrode and an adsorbate as well as the influence
of solvation and Coulomb relaxation on the energetic position of the
electronic affinity levels. The potential dependence of the surface
enhanced Raman (SER) intensity of the total symmetric skeleton modes
at different laser photon energies is in qualitative agreement with the
model. The observed energetic shift of the affinity levels is bigger
than the applied shifts of potential, this is explained by local potential effects. The reversible static charge transfer to pyrazine is observed by a reversible change of the Raman spectrum. The experimental
results cannot be understood with the "ground state CT model of SERS".
More advanced models of SERS by CT should comprise both aspects,
"coherent tunneling" CT and "vibrational driven hopping" CT.

1.   INTRODUCTION

The proposed explanations of surface enhanced Raman scattering (SERS)
/1-5/ have been controversial for years. The concpet of "SERS active
sites" (e.g. /6/ and references therein, /7,8/) finds now more acceptance than some years ago – however, there is disagreement on the
enhancement mechanism at active sites. Whereas Albano et al. /9/ and
Seki and Chuang /10/ envision local electromagnetic resonances in the
porosities /9,11/ of coldly deposited silver films – several experimental indications against the applicability of this concept have been
pointed out /12,13/.
    Most researchers in the field of SERS admit in general terms a
"chemical contribution" to SERS.
    More specific models of this "chemical contribution" comprise a
charge transfer (CT) between metal and adsorbate. There are two classes
of CT discussed.

B. Pullman et al. (eds.), Dynamics on Surfaces, 345–364.
© *1984 by D. Reidel Publishing Company.*

a) "Ground state charge transfer" (e.g. /14/, /15/ and references therein, /16/, /17/).

A vibrating adsorbate pushes charge forth and back between metal and partially filled molecular orbitals at an energy near the Fermi energy. In this way, the Raman polarizability is enhanced due to the extension of the oscillating charge from the molecule into the metal. In the words of Lippitsch /15/ ...."the enhancement is caused by ground state coupling to the metal state and does not involve any resonance of the exciting light with the charge transfer transition". We will discuss this model further in section 6.

b) "Photon driven charge transfer".

Among several hypothetical mechanisms Burstein et al. /18/ described the following process: "The CT process involves the excitation of an e-h-pair in the metal followed by hopping (e.g. tunneling) of the excited electron into (and back from) the virtual bound state (of the adsorbed molecule). The process in effect corresponds to the inelastic scattering of the electron via a negative ion resonance (e.g. shape resonance) of the adsorbed molecule". This Raman process is resonant when the energy of the incident or emitted photon matches the CT excitation energy. Calculations along these lines based on the Newns-Anderson-model /19/, but different in details have been presented by Gersten et al. /20/, Ueba (/21/ and references therein), Arya and Zeyher /22/, Persson /23/, and Adrian /24/.

On the experimental side, the first indication for CT excitations were observed by electron energy loss spectroscopy on smooth /25,26/ and "rough" /27/ silver surfaces. In the latter case, the loss intensity was bigger - indicating the importance of atomic scale roughness /4/. The first indications, that CT excitations are indeed involved in SERS were obtained by Billmann and Otto /28,29/ and Furtak and Macomber /30/, see also further work in refs. /7,31,32/. Here we want to present further evidence for the "photon driven" CT-SERS mechanism by a comparative study of SERS of benzene /33/, pyridine /34/, and pyrazine /33/ adsorbed at silver electrodes.

2.   THE RESONANCE RAMAN MODEL OF "PHOTON DRIVEN CT".

Based on many experimental observations discussed in /12/ we have presented a model in ref. /12/ which we repeat here with a minor modification to accomodate the solvation energy of ions in electrolytes.

The resonant Raman scattering process is depicted in fig. 1. In step 1, a metal electron localized in quasi Ag5s states at sites of atomic scale roughness (ρ(E,ASR) is the corresponding density of states) is excited by annihilation of a laser photon $\hbar\omega_L$. In the second step, the excited electron tunnels to the electron-affinity level A(e,M,Q) of the adsorbate. This level as well as the ground state of the adsorbate (assumed neutral throughout this work, in the case of the negatively charged ground state of cyanide adsorbed to silver see /31/) are plotted versus the normal coordinate Q of an adsorbate vibration (frequency ω_Q)- only totally symmetric vibrational modes of the free molecules are considered throughout this paper. (For the excitation of non totally

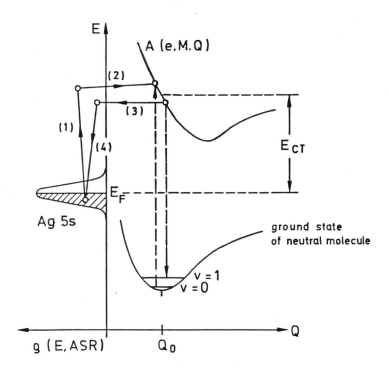

Figure 1. Resonant Raman scattering process, see text.

symmetric modes, see refs. /29/ and /4/.)
        After times of the order of 10^{-14} to 10^{-13} seconds, the electron
tunnels back to the metal (3. step). If the adsorbate was initially in
the vibrational ground state $v = 0$, there is a probability that it is
found in the first vibrationally excited state $v = 1$ after steps 2 and
3. Steps two and three correspond to "shape resonances" /18,4,12/ – for
this case Gadzuk /35/ has described the vibrational excitation in a
very transparent way. Finally, in the fourth step, the metal electronic
excitation is annihilated by emission of a photon $\hbar\omega_S$. According to
perturbation theory, energy is only conserved between the initial and
the final state – which means in our case $\hbar\omega_L = \hbar\omega_S + \hbar\omega_Q$ and hence
Raman scattering. This Raman process is resonant when $\hbar\omega_L$ or $\hbar\omega_S$ is
about equal to the CT excitation energy E_{CT}. E_{CT} is given by the energy
difference between the affinity level $A(e,M,Q \approx Q_0)$ (in the Franck-Condon
region $Q \approx Q_0$) and the Fermi energy E_F.
        A qualitative picture of the energetic position of $A(e,M,Q \approx Q_0)$ is
given in fig. 2. One has to differentiate between adsorbed molecules
and molecules in the bulk electrolyte. The energy of electronic states
is given with respect to the electrostatic potential within the working
electrode (the silver electrode), which is assumed to be zero. (This
corresponds to the actual experimental condition, in which the working
electrode is always connected to ground potential.) The Fermi level E_F
is at $-\phi_0$, where ϕ_0 is the work function of clean silver. The electro-

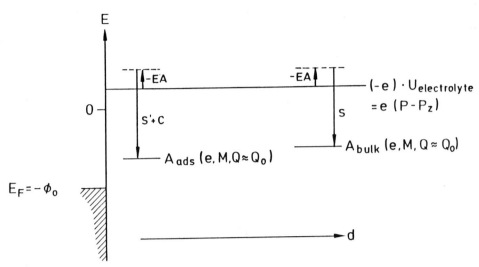

Figure 2. Schematic picture of the energy E of the electron affinity
levels $A_{bulk}(e,M,Q{\approx}Q_0)$ of molecules in the bulk electrolyte and
$A_{ads}(e,M,Q{\approx}Q_0)$ of molecules adsorbed at the working electrode. d is
distance from the electrode surface. E_F: Fermi level, ϕ_0: work function
of clean electrode. EA: electron affinity (assumed negative), S: solvation energy, S'+C: electronic contribution to solvation energy and
metallic screening. U: electrostatic potential, e: positive elementary
charge, P: "potential of the working electrode", P_z: see text.

static potential of the bulk electrolyte $U_{electrolyte}$ is $U_{electrolyte} =$
$-(P-P_z)$. P is the so-called "potential of the working electrode", P_z is
the potential P, where the electrostatic potentials within the working
electrode and the bulk electrolyte are equal. P and P_z are given by
$U_{working\ electrode} - U_{reference\ electrode}$, where the U's are the electrostatic potentials. Note that $-(P-P_z)$ does not depend on the choice
of the reference electrode. The potential quoted in experimental dia-
grams is always $P = U_{working\ electrode} - U_{reference\ electrode}$ (e.g.
potential SCE/V) (SCE: saturated calomel electrode).
   The difference of $A_{bulk}(e,M,Q{\approx}Q_0)$ and $-eU_{electrolyte}$ is given by
the sum of the electron affinity EA of the free molecule and the solvation energy of the negatively charged ion. For benzene, pyridine and
pyrazine the EA's are negative (see below). The solvation energy S, due
to the electronic and orientational polarization of the water surrounding the ion, are of the order of some eV (for instance 4,29 eV for F
and 2,11 eV for J /36/).

$$A_{bulk}(e,M,Q{\approx}Q_0) = e(P-P_z) - EA - S \qquad (1)$$

When P is scanned in negative direction ("cathodic") $A_{bulk}(e,M,Q{\approx}Q_0)$
will eventually be equal to E_F. At values $P<P_b$,

$$P_b = (-\phi_0 + S + EA + eP_z)/e \qquad (2)$$

the free enthalpy of the system is lowered when electrons are transferred from the metal to the affinity level of the molecule and hence
an electrochemical reaction ("reduction" of the molecule and possible
subsequent chemical reactions) sets in, which should be observable by
voltammetry (controlled variation of P).

For an adsorbed molecule, the energetic position of the affinity
level during the excursion of metal electron within $10^{-14} - 10^{-13}$ sec
is given by

$$A_{ads}(e, M, Q \approx Q_o) = e(P - P_z) - EA - S' - C \qquad (3)$$

S' is the polarization energy by electronic polarization of the surrounding water (an orientational polarization is excluded by the short
residing time of the electron in the affinity level). The order of
magnitude of S' may be estimated from the difference between the experimental heats of ion-solvent interaction and the calculated heats of
ion dipole interactions; using tables 2.12 and 2.13 in /36/, one obtains
1,33 eV for F^- and 0,34 eV for J^-. C is the Coulomb interaction between
the negatively charged adsorbate and the positive screening charge at
the metal surface. For pyridine, C is estimated /37/ from measurements
of EA /38/ and charge transfer excitation energies at a silver-vacuum
surface /27/ as $C \approx 2,7$ eV. The charge transfer energy E_{CT} depends on
P and is given by

$$E_{CT}(P) = e(P - P_z) - EA - S' - C + \phi_o \qquad (4)$$

For a given laser frequency ω_L, the maximum Raman intensity according
to the mechanism described above will be obtained at a special $P = P_{max}$
for which $\hbar \omega_L = E_{CT}(P_{max})$. This yields a linear relationship between
$\hbar \omega_L$ and P_{max}:

$$\hbar \omega_L = eP_{max} - eP_z - EA - S' - C + \phi_o \qquad (5)$$

Extrapolation of this relation to $\hbar \omega_L = E_{CT}(P_{max}) = 0$ leads to

$$P_{max} = P_s = (-\phi_o + S' + C + EA + eP_z)/e \qquad (6)$$

P_s is not necessarily equal to P_b. For different molecules i (e.g. benzene and pyrazine), differences in P_{bi} and P_{si} will reflect differences
in EA_i, provided that S_i and $S_i' + C_i$ do not depend on i.

In fig. 3 we have compiled the relevant data of the lowest affinity levels of benzene, pyridine and pyrazine. The negative EA values
have been measured by elastic electron transmission through vapour by
Nenner and Schulz /38/. The positive electron affinity of the $\pi^* - b_{3u}$
level has been obtained indirectly by Nenner and Schulz. They correlated the measured negative EA's with the measured P_b values of various
aromatic molecules dissolved in an electrolyte of 0.1 M Et_4NI (tetra-
ethylammoniumiodide) in DMF (dimethylformamid). A linear relation

$$P_{bi} = \frac{1}{e} \cdot \frac{EA_i}{0.6} + const \qquad (7)$$

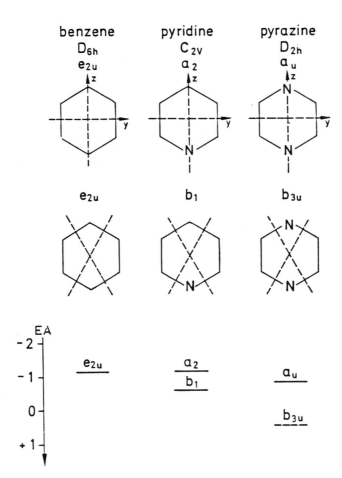

Figure 3. Lowest electron affinity levels EA of benzene,
pyridine and pyrazine, from /38/. Point group symmetry
of the free molecule as well as group theoretical representation (according to the given choice of axis) of one
electron π^* orbitals are indicated. Nodal planes of the
π^* orbitals perpendicular to the molecular plane are
given by dashed lines (from /39/).

was obtained. By extrapolation, the positive value of EA (b_{3u}) for
pyrazine was obtained from the measured $P_{b,pyrazine}$.
      The frequencies ω of the totally symmetric vibrational modes, with
the exclusion of the C-H stretch modes, as well as the vibrational
patterns, are given in fig. 4. According to the selection rules in
shape resonances /41,4/ only totally symmetric modes are excited when
the π^* orbitals corresponding to the affinity levels are nondegenerate.
This is the case for pyridine and pyrazine. As the lowest π^* orbitals
are antibonding with respect to the six-membered ring, but nonbonding
with respect to C-H (see the symmetry of the lowest π^* orbitals in fig.
3) the excitation of C-H stretch modes in low energy shape resonances

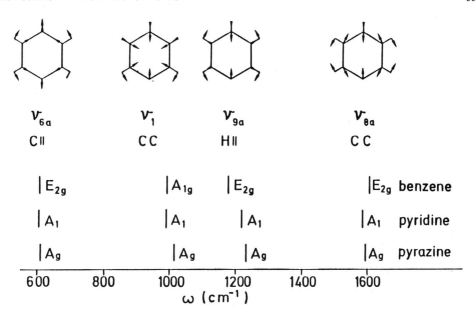

Figure 4. Frequency ω of totally symmetric skeleton modes of pyridine and pyrazine compared to the corresponding A_{1g} and E_{2g} vibrations of benzene. ν_i are Wilson mode numbers. The vibrational pattern are those of the pyridine molecule. CC: ring-stretch, C$\parallel$: in plane ring bend, H$\parallel$: in plane hydrogen bend. Adapted from /39/ and /40/.

is negligible /41,4/. For benzene the lowest π^* orbital of e_{2u} symmetry is degenerate – hence also other than A_{1g} modes are excited in the e_{2u} shape resonance /41,4/. We do not discuss this further, because only the A_{1g} breathing mode of benzene was observed in our experiments (see below).

3. EXPERIMENTAL CONDITIONS AND VOLTAMMETRY.

3.1 Measurements with benzene.

Whereas SER spectra of benzene on coldly deposited silver films are relatively easily obtained /42/, the low solubility of benzene in water prevents strong SERS at silver electrodes in aqueous electrolytes. Nevertheless, Howard and Cooney /43/ found an intense surface benzene line at 995 cm^{-1} for the silver electrode in a 0.1 M KF, 0.01 M C_6H_6 aqueous electrolyte after a single oxidation reduction cycle between −0.2 and +0.4 V_{SCE} at 5 mV sec^{-1} after changing P to −1.4 V_{SCE}.

In our case, the silver electrodes were first etched in an aqueous solution of NH_2 and H_2O_2, then inserted at −1.4 eV into the 0.1 M KCl aqueous electrolyte and cycled for about one hour between P = −1.2 V_{SCE} and −1.4 V_{SCE}. Then benzene was added to the electrolyte and mixed with it by permanent bubbling of N_2 gas through the electrolyte. The silver electrode was activated ("roughened") by a potential jump from P = −1.4 V_{SCE} to +0.3 V_{SCE} for about 5 sec under laser illumination.

3.2 Measurements with pyridine.

The experimental details have been described in /28/.

3.3 Measurements with pyrazine.

Silver electrodes, etched as described above, were inserted in three different electrolytes, namely 0.1 M KCl, 0.1 M KCl with 0.05 M pyrazine, and 0.1 M Na_2SO_4 with 0.05 M pyrazine. Examples of voltammograms are given in fig. 5.

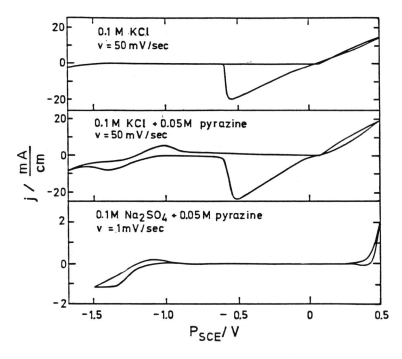

Figure 5. Cyclic voltammograms (current density j versus P) of silver electrodes in three different indicated aqueous electrolytes, triangular sweep rate.

The current above P = −0.6 V_{SCE} in the Cl^- containing electrolytes are due to the formation and reduction of a solid AgCl film, currents in the $SO_4^=$ electrolyte above P = +0.3 V_{SCE} are due to formation an reduction of Ag_2SO_4. Addition of 0.05 M pyrazine to the electrolytes leads to additional charge transfer reactions near P = −1.1 V_{SCE}. In fast cycles (50 mV/sec) one observed a reduction (P ≈ −1.3 V_{SCE}) and oxidation (P ≈ −1.0 V_{SCE}) peak, similar to the ones observed by Wiberg and Lewis /44/ for the reduction ($C_4H_4N_2 \rightarrow C_4H_4N_2^-$) and oxidation ($C_4H_4N_2^- \rightarrow C_4H_4N_2$) of pyrazine dissolved in DMF at a dropping mercury electrode.

In slow cycles (1 mV/sec) a reduction starting at about P = -1.0 V_{SCE}
is observed, whereas the reoxidation is very weak. This reflects the
diffusion of the reduced pyrazine (the negative pyrazine ion) away from
the surface, so that there is little chance for reoxidation. In agreement with ref. /44/, the reduction and reoxidation of pyrazine is a
reversible process, which is also corroborated by our Raman data (see
below).

From our voltammetric measurements follows a value of P_b (the
"halfwave potential") of pyrazine in aqueous Cl⁻ or SO₄⁼ containing
electrolytes of about -1.2 V_{SCE}[*].

At P < -1.5 V_{SCE}, hydrogen gas evolution sets in. For SERS, the
electrode was "activated" ("roughened") by a potential jump from -0.1
V_{SCE} to +0.5 V_{SCE} for 5 sec under laser irradiation. Recording of the
Raman spectra displayed in the next section was started about 30 sec
after activation.

4.   RAMAN SPECTROSCOPIC RESULTS

The spectroscopic equipment including an optical multichannel system
has been described in /31/.

SER spectra of benzene taken with three different laser wavelength
λ_L are displayed in fig. 6. The spectra were recorded during anodic
sweeps with v = 10 mV/sec, the integration time per spectrum was 8 sec.
The ν_1-A_{1g} breathing mode of adsorbed benzene shifts from 987 cm⁻¹ to
991 cm⁻¹ between P = -1.4 V_{SCE}. Similar line shifts have been observed
and discussed for pyridine and cyanide and CO (/12,31,45/ and references therein). In this work, we will not discuss the line shifts and
halfwidths of the SERS lines.

The Raman intensity within the full halfwidth (FWHM) of the ν_1-
line in every spectrum was integrated and the backgrounds on the low
and high energy side integrated over 1/2 FWHM substracted. In this way
one obtains an intensity versus potential profile of the ν_1-line from
which follows the value of $P_{max}(\hbar\omega_L)$. A detailed description of this
procedure will be given elsewhere /46/.

The dependence of P_{max} on $\hbar\omega_L$ is presented in fig. 7, top. The
analogue results for the ν_1, ν_{9a}, and ν_{8a} vibrational lines of pyridine
(see fig. 4) at a silver electrode in 0.1 M KCl from ref. /28/ are also
given in fig. 7, middle. The P_{max} values are different for the different lines.

This is in contrast to the results for pyrazine on silver in 0.1 M
KCl aqueous electrolyte (fig. 7, bottom), where the P_{max} values nearly

[*] For pyrazine in DMF, Wiberg and Lewis /44/ reported P_b = -1.57 V
versus a "mercury pool", which has a potential -0.516 mV versus a SCE
electrode inserted into the DMF electrolyte /44/. This yield P_b =
-2.086 V_{SCE}. The reasons for the discrepancy are not clear. Possible
explanations might be a diffusion potential at the water/DMF junction
between the SCE electrode and the DMF electrolyte or a smaller orientational solvation energy of negative ions in DMF compared to the case
of water.

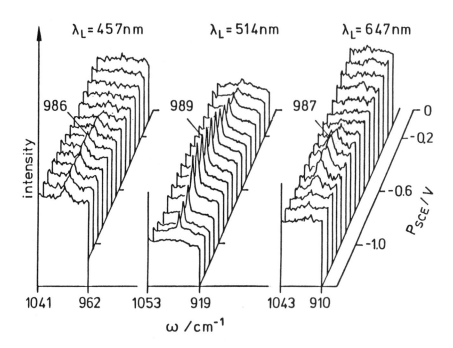

$\lambda_L = 457\,nm$ $\lambda_L = 514\,nm$ $\lambda_L = 647\,nm$

Figure 6. Multichannel Raman spectra of benzene adsorbed
at a silver electrode in 0.1 M KCl electrolyte. Anodic
sweeps with 10 mV/sec for 3 laser wavelengths λ_L.

coincide for the ν_1, ν_{9a}, and ν_{8a} lines.

Fig. 8 gives in its lower part the multichannel spectra of SERS of
pyrazine, showing the resonant appearance of the ν_{6a} (~ 625 cm^{-1}).
ν_1 (~ 1019 cm^{-1}), ν_{9a} (~ 1229 cm^{-1}), and ν_{8a} (~ 1597 cm^{-1}) lines. When
the cathodic scan is extended to -1.45 V_{SCE} (upper half of fig. 8), one
observes a definite change of the spectrum - the ν_{6a}, ν_1, ν_{9a}, and ν_{8a}
lines gradually disappear at the positions of about 618 cm^{-1}, 1015 cm^{-1},
1225 cm^{-1} and 1594 cm^{-1}, respectively, and gradually reappear in the
new positions (at -1.45 V_{SCE}) of about 512 cm^{-1}, 719 cm^{-1}, 939 cm^{-1},
and 1669 cm^{-1}, respectively.

The same transformation is observed for pyrazine in a 0.1 M Na$_2$SO$_4$
aqueous electrolyte (fig. 9). The spectral change is reversible, as
displayed in fig. 9. The spectral change is "centered" around P = -1.1
V_{SCE}, the "transition range" is between -0.9 V_{SCE} and -1.3 V_{SCE}.

5.   DISCUSSION

We find maximum SERS intensity of ν_1 of benzene in the potential range
around -0.7 V_{SCE} (see fig. 7), whereas Howard and Cooney /43/ found for
activated silver in a 0.1 M KCl, 0.01 M benzene aqueous electrolyte
intense surface benzene lines after stepping the potential to -1.2 V_{SCE}.
The reason for this discrepancy is not clear - it might be due to the

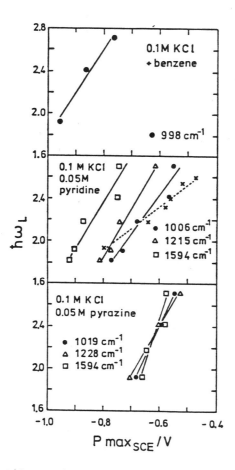

Figure 7. Silver electrode potential of maximum Raman inten-
sity P_{max} (V_{SCE}) versus photon energy $\hbar\omega_L$ for the ν_1 vibrational line of adsorbed benzene (top) and the ν_1, ν_{9a}, and
ν_{8a} vibrational lines of adsorbed pyridine /28/ (middle)
and pyrazine (bottom). Crosses are analogous results for
the ν_1 line of pyridine in the indicated electrolyte by
Furtak and Macomber /30/.

surface carbon layer present on Howard and Cooney's silver electrode
/43/.

Our results for the ν_1 mode of pyridine /28/ (fig. 7) are not
quantitatively in agreement with those of Furtak and Macomber /30/, see
fig. 7. (No data on the modes other than ν_1 were given in /30/.) The
reason for the deviations between the two sets of results are unknown,
but the discrepancy is within the range of the observed local potential
effects (see below).

In table 1 we compare the vibrational line positions of pyrazine
between 600 cm^{-1} and 1800 cm^{-1} obtained in this work with previous SERS
results of pyrazine at silver electrodes by Dornhaus et al. /47/ and
Erdheim et al. /48/ and at a silver vacuum interface by Moskovits and

Sym. a	Wilson mode number a	Bulk (IR+Raman) a	Raman aqu. solut. a	-0.4 V a	-0.4 V b	c	-0.57V this work	-0.78V this work
A_g	6a	596–620	615	635s	636m–s	615m	629w	625s
	1	1015	1017	1018vs	1020vw	1015s	1019vs	1019vs
	9a	1230–1239	1241	1242s	1237m	1233m	1228m	1229s
	8a	1578–1593	1594	1590vs	1597vs	1578s	1607vs	1597s
Raman	2	3054–3074	3060	3050m	3066m	3055w		
A_u	16a	(340–363)	363	362w		352m		
silent	17a	(950–960)		(966)vw		972w		
B_{1g} Raman	10a	754	758	744m	743w	753*w		
B_{3g}	6b	641	(677)	635,662s		*		
	3	1118	1120	1121w	1123*w	*		
	8b	1523–1529	1529	1520w	1519w	1522m	1525vw	1519m
Raman	7b	3040–3060		3031vw				
B_{2g}	4	703	703	700m	698vw	700*w		
Raman	5	920	922	916m–s	897vw	922*w		
B_{1u}	12	1018–1021	–	1038vw		1031w		
	18a	1130–1135	–	(1164?)		–		
	19a	1483–1490	–	1485s	1488m	1484m	1492m	1492w
IR	13	3066						
B_{3u}	16b	416–418	–	436m	440vw	417m		
IR	11	785–804	–	797w	800vw	792m		
B_{2u}	15	1063	–	1069w	1088*w	1088m	1084m	1081w
	14	1346	–	1340vw	1317w	1347*w		1333w
	19b	1411–1418	–	1420vs		1407m	1429vw	1412vw
IR	20b	3066	–	3060		–		

* assignment in ref. a, different from that given in b or c.

Table 1. Comparison and assignment of SERS lines of pyrazine on silver. The first 7 columns have been adapted from ref. /50/. Intensity characterization by vw: very weak, w: weak, m: medium, s: strong, vs: very strong. Ref. a: /47/ and /50/, b: /48/, c: /42/.

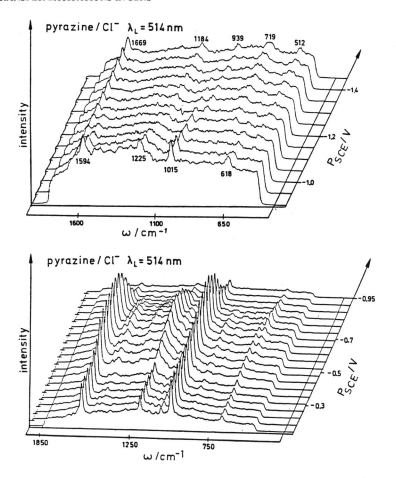

Figure 8. below: Multichannel Raman spectra of pyrazine
adsorbed on a silver electrode in 0.1 M KCl, 0.05 M pyrazine aqueous electrolyte during a cathodic scan from -0.1
V_{SCE} to -1.0 V_{SCE} with 50 mV/sec, 2 seconds integration
time per spectrum. above: Scan extended from -0.9 V_{SCE} to
-1.45 V_{SCE}.

DiLella /42/.
    Discrepancies in the assignment are partly due to different
assignment of some modes for gaseous or liquid pyrazine (see table 2),
however, this does not concern the A_g modes. The main experimental discrepancy is the strong SERS intensity of some IR modes (e.g. 19b) by
Dornhaus et al. /47/ compared to the other results.
    It has recently been reported by Mörschel and Dornhaus /49/ that
the intensity of these modes is enhanced by coadsorbed oxygen. Whereas
nitrogen purging of the electrolyte was not applied by Dornhaus et al.
/47/, has not been mentioned by Erdheim et al. /48/, it was applied
throughout all our experiments.

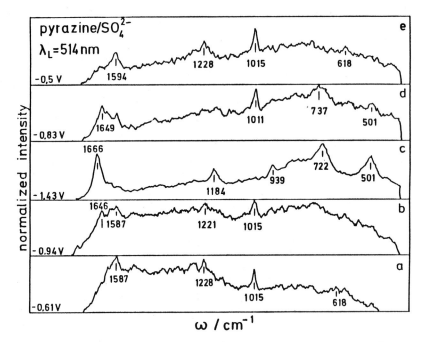

Figure 9. Normalized Raman spectra of the system Ag/0.1 M
Na$_2$SO$_4$, 0.05 M pyrazine aqueous electrolyte. Spectra a and
b were recorded during a cathodic scan (50 mV/sec), spectrum
c during the scan reversal at -1.45 V$_{SCE}$, and spectra d and
e during the anodic return scan.

Wilson mode number	a	b	c	d	e
10a	754	755	927	925	757
6b	641	–	699	641	641?
3	1118	–	785	1118	1118
5	920	922	983	753	918,6
13	3066	–	(3012)	3066	3066
14	1346	–	(1149)	1342	1346
20b	3066	–	3013	3066	3066

Table 2. Controversial assignment of some pyridine
vibrational modes: Ref. a: /50/, b: /48/, d: /40/,
e: /39/.

The dependence of P$_{max}$ on $\hbar\omega_L$ (fig. 7) cannot be explained by
potential dependent coverage or rearrangement of the adsorbates nor on
changes of the electromagnetic surface resonances, as discussed in /29,
4, and 31/. This latter point became very obvious, when a negative

gradient $d\hbar\omega_L/dP_{max}$ was found for SERS from cyanide on silver electrodes /31,12/ (see also section 6). The gradients $d(\hbar\omega_L)/dP_{max}$ of the
linear approximations in fig. 7 are considerably greater than 1 eV/V,
greater than expected from equation (5). These gradients vary with the
chemical nature and concentration of the conducting ions in the electrolyte, as demonstrated for the ν_1-mode of pyridine in fig. 10. They
may also depend on the adsorbate, as gradients for benzene, pyridine
and pyrazine in the same electrolyte are different (see fig. 7).

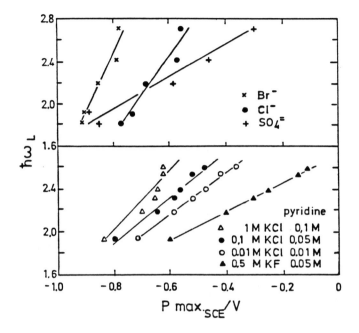

Figure 10. Potential of maximum SERS as function of laser
photon energy for the ν_1 band of pyridine adsorbed on
silver electrodes in various aqueous electrolytes.
top: Br⁻: 0.1 M KBr, 0.05 M pyridine; Cl⁻: 0.1 M KCl,
0.05 M pyridine; $SO_4^=$: 0.1 M Na_2SO_4, 0.05 M pyridine;
from ref. /34/, data below from /30/.

Deviations from the gradient 1 eV/V have also been found for the shift
of delocalized surface states at smooth low index faces of single crystalline silver electrodes /51/, as discussed in /28,29,4/. One should
also note, that the experimentally observed relation (7) does not yield
a gradient dEA/dP_b = 1 eV/V expected from relation (2) but rather a
value of 1.66 eV/V.
     We do not consider the deviations of all these gradients from
1 eV/V as an indication against the concept of a photon driven charge
transfer, but rather as an indication of local potential variations at
the electrode surface due to the necessarily inhomogeneous distribution
of ions at an electrode surface, particularly when it is rough on an
atomic scale /28,29,4/, in agreement with the discussion by Furtak

and Macomber /30/.

It is not clear, why the $\hbar\omega_L$ - P_{max} relations are nearly coincident for the ν_1, ν_{9a}, and ν_{8a} vibrations of pyrazine, but not for pyridine, see fig. 7. A possible explanation might be, that the coupling to these vibrations by CT to the b_{3u} affinity level is much stronger than by CT to the a_u affinity level of pyrazine (see fig. 3), whereas in the case of pyridine coupling is comparable for the b_1 and b_2 affinity levels. Due to the interference between the probability amplitudes (matrix elemements) corresponding to "roundtrips" (see fig. 1) via the a_2 or b_1 affinity levels, excitation profiles for the different A_1 modes could be different. Inspection of the elastic electron transmission spectra of Nenner and Schulz /38/ show indeed, that vibrational substructure in the a_u shape resonance of pyrazine is negligible compared to the accessible vibrational substructure (E > 0) of the b_{3u} positive electron affinity level, whereas this difference is less pronounced for the b_1 and a_2 shape resonances of pyridine.

The reversible change of the pyrazine SER spectrum around -1.1 eV (fig. 8 and 9) is assigned to the reversible reduction and reoxidation of pyrazine in agreement with the electrochemical evidence discussed in section 3. The affinity level being "permanently" filled in this reduction process is the b_{3u} level. The range of transition from SER spectra of neutral pyrazine to those of the negatively charged pyrazine is between -0.9 V_{SCE} and -1.3 V_{SCE}. This reflects inhomogeneous potentials on an atomic scale at the surface.

Extrapolation of the linear approximation to the $\hbar\omega_L$ - P_{max} relation für pyrazine in a 0.1 M KCl aqueous electrolyte (fig. 7) to $\hbar\omega_L$ = 0 yields -0.9 V_{SCE} > P_{max} > -1.2 V_{SCE}. This extrapolation gives the value of P_s in the sense of equations (5) and (6). Surprisingly P_s agrees within its uncertainty limits with $P_b \approx$ -1.2 V_{SCE} (see section 3). Analogous extrapolations of the benzene and pyridine data in fig. 7 to $\hbar\omega_L$ = 0 yields P_s values in the hydrogen evolution range, corresponding P_b values in this range would not be measurable by voltammetry. Analogous extrapolation for SERS of cyanide on silver /31/ gave P_b values in the potential range where the electrochemical reaction of silver dissolution to aqueous $Ag(CN)_2^-$ ions sets in /31/. The approximate agreement between P_b and P_s in the case of pyrazine and CN^- may be just fortuitous. Further experiments, especially of pyrazine in a $SO_4^=$ aqueous electrolyte are needed to clarify the point.

At a given P_{max}, the corresponding $\hbar\omega_L$ for benzene on silver is about 1.2 ± 0.3 eV above the value for pyrazine on silver, as follows from the data in fig. 7. Provided that $S' + C$ has the same value in both cases, this implies that the affinity level of benzene lies 1.2 ± 0.3 eV above that of pyrazine. The difference of the e_{2u} electron affinity of benzene and the b_{3u} electron affinity of pyrazine is 1.54 eV /38/, see fig. 3.

As the different A_1 modes of pyridine do not show the same $\hbar\omega_L$ - P_{max} relation it is not clear how to deduce the position of $A(e,M,Q\approx Q_0)$. However, if one considers only the ν_1 modes, the general trend $\hbar\omega_L$ (ν_1, benzene) > $\hbar\omega_L$ (ν_1, pyridine) > $\hbar\omega_L$ (ν_1, pyrazine) for a given P_{max} is the same as the trend in the position of the affinity levels, see fig. 3.

In summary, our results further support the concept of a SERS process
by "photon driven CT" /12/ - however, we are still far from the quantitative understanding of, for instance, the local potentials and the
different resonances of the different pyridine vibrations.

6.   COMMENTS

Watanabe et al. /8/ find it hard to reconcile the observed gradients
$d\hbar\omega_L/dP_{max} > 1$ with the CT concept alone and propose an additional potential dependent enhancement mechanism. They propose two mechanisms:
1. A surface Ag^+ complex adsorbing at the laser wavelength, "resulting
in a molecular resonance Raman type process, leading to an increased
Raman polarizability". 2. "A charge transfer excitation between a Ag
adatom and an adsorbate".
        Watanabe et al. do not explain why and in which way the first pro-
cess would be potential dependent. Usually, low lying electronic transitions in complexes, giving rise to a resonant Raman effect are CT excitations between the central atom or ion (in this case Ag or Ag^+) and
the ligands. In this sense we do not see a significant difference between our model and the ones by Watanabe et al. /8/, contrary to Watanabe et al. /8/.
        Sandroff et al. /52/ concluded from the frequency of SERS lines of
TTF (tetrathiofulvalene) adsorbed onto silver and gold island films,
that TTF is adsorbed primarily as radical cation TTF^+. The Raman excitation profile of TTF^+ had the shape of the measured island film absorption, in accord with electromagnetic enhancement models. Sandroff et al.
discuss how the static electron transfer from TTF to the metal islands
might give an additional polarizability, leading to an additional
enhancement, which would not have a resonant behaviour in its excitation
dependence. Analogies to the "ground state charge transfer model" of
Aussenegg and Lippitsch /14/ were pointed out /52/. An experimental
separation of the two contributions was not possible /52/.
        One should note, that in many other cases compiled in /4/ no agree-
ment was found between measured absorption and Raman excitation profiles.
It is the advantage of the experiments described here, that the electromagnetic enhancement does change very little, when the potential is
changed (discussed in /29/) whereas the CT Raman mechanism is tuned into
and out of resonance according to the laser photon energy. In this way,
the two contributions are separable.
        Mal'shukov /17/ gives the cross section σ for Raman scattering as

$$\sigma = f(\omega_L,\omega) \, (\Delta n)^2$$

where $f(\omega_L,\omega)$ is determined by the nonlinear metal response and may
include also the local field enhancement factors, and Δn is the charge
pushed back and forth between adsorbate and substrate by the adsorbate
vibrations. Mal'shukov discusses two contributions to $f(\omega_L,\omega)$. One of
these is long range (skin depth) modulation of the metal polarization,
the other is the surface contribution - which varies with applied
potential.

The different presign of the gradients $d\hbar\omega_L/dP_{max}$ for CN^- and pyridine /11/ make Mal'shukov's explanation of the potential dependence very unlikely. As the width of SERS lines should be proportional to $(\Delta n)^2$ /53,17/ and the line shift grows with Δn, Mal'shukov concludes that the width and shift of any line in the SER spectrum increases with its intensity. This is not confirmed by our SERS measurements /54/.

In an advanced theory of CT, both concepts of CT may be reconciled. Perssons and Perssons ansatz /53/ contains terms for "indirect" vibrational excitation of the molecule, namely by charge transfer $(a^+ a_k + a_k^+ a)$ and subsequent electron-vibration coupling within the molecule $((b + b^+) a^+ a)$ and terms of direct vibrational excitation by vibration assisted CT $((b + b^+) (a^+ a_k + a_k^+ a))$.

Here, $a^+ a$, $a_k^+ a_k$, $b^+ b$ are creation and annihilation operators of an electron in $A(e,M,Q \approx Q_o)$, an electron in the metal, and an adsorbate vibrational quantum. The second term of the first two ones may be called "coherent tunneling" the direct term "hopping" in analogy to exciton propagation in molecular crystals.

For reasons of simplicity Persson and Persson /53/ neglected the "hopping term", and consequently it was not further considered in the SERS-CT theory of Persson /23/, and is not included in the model in section 2. The "hopping term" without the "tunneling term" is the fundament of all "ground state CT models" (see section 1), it may become important in some cases. For instance, we conjecture, that the "hopping term" is mainly responsible for SERS from the $\nu_7 B_{1u}$-CH wagging mode of ethylene adsorbed on silver, whereas the "coherent tunneling" is mainly responsible for SERS from the $\nu_3 Ag CH_2$ scissor and ν_2-C-C stretch mode.

REFERENCES

/1/ R.P. VanDuyne, in Chemical and Biological Application of Lasers 4, ed. by C.B. Moore (Academic Press, New York, 1979)

/2/ T.E. Furtak, J. Reyes, Surf. Sci. 93, 351 (1980)

/3/ Surface Enhanced Raman Scattering, ed. by R.K. Chang, T.E. Furtak, (Plenum, New York, 1982)

/4/ A. Otto, in Light Scattering in Solids, Vol. IV, ed. by M. Cardona, G. Güntherodt, (Springer, 1984)

/5/ A.G. Mal'shukov, in Chemical Physics of Solvation, ed. by J. Ulstrup, E. Kálmán, R.R. Dogonadze, A.A. Kornyshev, (Elsevier, to appear)

/6/ Ü. Ertürk, I. Pockrand, A. Otto, Surf. Sci. 131, 367 (1983)

/7/ T.E. Furtak, D. Roy, Phys. Rev. Lett. 50, 1301 (1983)

/8/ T. Watanabe, O. Kawanami, K. Honda, B. Pettinger, Chem. Phys. Lett. 102, 565 (1983)

/9/ E.V. Albano, S. Daiser, G. Ertl, R. Miranda, K. Wandelt, N. Garcia, Phys. Rev. Lett. 51, 2314 (1983)

/10/ H. Seki, T.J. Chuang, Chem. Phys. Lett. 100, 393 (1983)

/11/ J. Eickmans, A. Goldmann, A. Otto, presented at ECOSS VI, to be published

/12/ A. Otto, J. Billmann, J. Eickmans, Ü. Ertürk, C. Pettenkofer, Surf. Sci. 138, 319 (1984)
/13/ C. Pettenkofer, Ü. Ertürk, A. Otto, presented at ECOSS VI, to be published
/14/ F.R. Aussenegg, M.E. Lippitsch, Chem. Phys. Lett. 59, 214 (1978)
/15/ M.E. Lippitsch, Phys. Rev. B 29, 3101 (1984)
/16/ A.G. Mal'shukov, Sol. State Commun. 38, 907 (1981)
/17/ A.G. Mal'shukov, J. de Physique C 10, 315 (1983)
/18/ E. Burstein, Y.J. Chen. S. Lundquist, E. Tossatti, Sol. State Commun. 29, 567 (1979)
/19/ B.I. Lundquist, H. Hjelmberg, O. Gunnarson, in Photoemission and the Electronic Properties of Surfaces, ed. by B. Feuerbacher, B. Fitton, R.F. Willis (Wiley, 1978)
/20/ J.I. Gersten, R.L. Birke, J.R. Lombardi, Phys. Rev. Lett. 43, 147 (1979)
/21/ H. Ueba. Surf. Sci. 131, 347 (1983)
/22/ K. Arya, R. Zeyher, in Light Scattering in Solids, Vol. IV, ed. by M. Cardona, G. Güntherodt, (Springer, 1984)
/23/ B.N.J. Persson, Chem. Phys. Lett. 82, 561 (1981)
/24/ F.J. Adrian, J. Chem. Phys. 77, 5302 (1982)
/25/ J.E. Demuth, P.N. Sanda, Phys. Rev. Lett. 47, 57 (1981)
/26/ J.E. Demuth, Ph. Avouris, Phys. Rev. Lett. 47, 61 (1981)
/27/ D. Schmeisser, J.E. Demuth, Ph. Avouris, Chem. Phys. Lett. 87, 324 (1982)
/28/ J. Billmann, A. Otto, Sol. State Commun. 44, 105 (1982)
/29/ A. Otto, J. Electron Spectr. Rel. Phenomena 29, 329 (1983)
/30/ T.E. Furtak, S.H. Macomber, Chem. Phys. Lett. 95, 328 (1983)
/31/ J. Billmann, A. Otto, Surf. Sci. 138, 1 (1984)
/32/ J.R. Lombardi, R.L. Birke, L.A. Sanchez, I. Bernhard, S.C. Sun, Chem. Phys. Lett. 104, 240 (1984)
/33/ J. Thietke, Diplomarbeit, Universität Düsseldorf, 1984
/34/ J. Billmann, Dissertation, Universität Düsseldorf, 1984
/35/ J.W. Gadzuk, J. Chem. Phys. 79, 3982 (1983)
/36/ J. Bockris, A. Reddy, Modern Electrochemistry, (Plenum Press, New York, 1970)
/37/ A. Otto et al., in preparation
/38/ I. Nenner, G.J. Schulz, J. Chem. Phys. 62, 1747 (1975)
/39/ K.K. Innes, J.P. Pyrne, I.G. Ross, J. Molec. Spectr. 22, 125 (1967)
/40/ F.R. Bollish, W.G. Fateley, F.F. Bentley, Characteristic Raman Frequencies of Organic Compounds, (Wiley, 1974)
/41/ S.F. Wong, G.J. Schulz, Phys. Rev. Lett. 35, 1429 (1975)
/42/ M. Moskovits, D.P. DiLella in /3/
/43/ M.W. Howard, R.P. Cooney, Chem. Phys. Lett. 87, 299 (1982)
/44/ K.B. Wiberg, T.P. Lewis, J. Am. Chem. Soc. 92, 7154 (1970)
/45/ C. Pettenkofer, A. Otto, in preparation
/46/ J. Thietke, A. Otto, in preparation
/47/ R. Dornhaus, M.R. Long, R.E. Benner, R.K. Chang, Surf. Sci. 93, 240 (1980)
/48/ G.R. Erdheim, R.L. Birke, J.R. Lombardi, Chem. Phys. Lett. 69, 495 (1980)

/49/ U. Mörschel, R. Dornhaus, Vhdl. der DPG 1, 153 (HL 122) (1984)

/50/ R. Dornhaus, J. Electron Spectr. Rel. Phenomena 30, 197 (1983)

/51/ W. Boeck, D.M. Kolb, Surf. Sci. 118, 613 (1982)

/52/ C.J. Sandroff, D.A. Weitz, J.C. Chung, D.R. Herschbach, J. Phys. Chem. 87, 2127 (1983)

/53/ B.N.J. Persson, M. Persson, Sol. State Commun. 36, 175 (1980)

/54/ G. Kannen, J. Billmann, A. Otto, in preparation

CATION-INDUCED SUPEREQUIVALENT ADSORPTION OF Cl⁻ IONS:
TEMPORAL AND POTENTIAL EVOLUTION OF SERS

T. T. Chen,[*] J. F. Owen,[+] and R. K. Chang
Yale University
Section of Applied Physics and Center for Laser Diagnostics
P.O. Box 2157, Yale Station
New Haven, Connecticut 06520, USA

ABSTRACT

The cation effect on the SERS intensity of Ag^o-Cl^-, pyridine, and H_2O adsorbates has been investigated by adding high concentration cation-Cl salt to a 0.1 M NaCl + 0.05 M pyridine electrolyte after oxidation and reduction of the Ag electrode. During a potential hold, the slow temporal evolution and the potential dependence of the SERS spectra were detected with an optical multichannel analyzer. Two types of cation effects on the SERS characteristics were observed: one for the lower hydration energy cations (Cs^+, Rb^+, and K^+) and another for the higher hydration energy cations (Na^+ and Li^+). Our data suggest that, even though the Ag electrode is positively charged, the lower hydration energy cations can enter the inner Helmholtz layer and induce changes in the rate of growth and decay of the SERS intensity, which are associated with possible rearrangement of SERS active sites and/or adsorbates on these sites.

1. INTRODUCTION

Since the discovery of surface enhanced Raman scattering (SERS), alkali-chloride salts in aqueous solvent have been most commonly used as the supporting electrolyte for SERS investigations at the electrode-electrolyte interface.[1,2] The role of the Cl⁻ ions as well as of other contact-adsorbed anions in SERS studies is reasonably well understood. During an oxidation-reduction cycle (ORC), the surface coverage of Cl⁻ ions is a single-valued function of the Ag electrode potential and decreases monotonically as this potential is linearly swept to less

---

[*] On leave from the Department of Physics, National Tsing-Hua University, Hsinchu, Taiwan, Republic of China.

[+] Present address: Research Laboratories, Eastman Kodak Company, Rochester, New York 14650.

B. Pullman et al. (eds.), Dynamics on Surfaces, 365–378.
© 1984 by D. Reidel Publishing Company.

positive values.[3] During the oxidation of the Ag electrode, the Ag^+ and Cl^- ions form an insoluble adherent AgCl layer which is subsequently electrochemically reduced and/or photoreduced, producing submicron size surface microstructures, giving rise to electromagnetic enhancement, and Ag^0-adatom related active sites, giving rise to a charge-transfer type enhancement.[4,5] After the reduction of the AgCl layer, the contact-adsorbed Cl^- ions stabilize the Ag^0-adatoms and prevent them from diffusing along the surface and becoming irreversibly lost upon incorporation at surface defects.[6] By sweeping the potential to less positive values, fewer of the Cl^- ions are adsorbed and can therefore stabilize the adatom active sites. Consequently, the SERS intensities are irreversibly decreased even when the potential is swept back to more positive values where the surface coverage of Cl^- ions is again large.

The contact-adsorbed Cl^- ions can coordinate with other coadsorbed molecules such as pyridine and H_2O molecules. Evidence exists that coadsorbed pyridine and Cl^- can form complexes which can stabilize more Ag^0 adatoms and thereby further increase the SERS signals relative to electrolytes containing no pyridine.[7] Furthermore, when Cl^- ions and H_2O molecules are coadsorbed, the Cl^- ions disrupt hydrogen bonds between the H_2O molecules and the rest of the water network and also form weak hydrogen bonds with the H_2O molecules. Such H_2O and Cl^- interactions cause the SERS line shape of interfacial H_2O to be considerably narrower than the normal Raman line shape of bulk water.[8-10] In addition to coadsorption, there can be competition between the Cl^- ions and H_2O molecules for Ag^0-adatom active sites, as evident from the SERS intensity vs potential dependences for the O-H stretching mode of interfacial H_2O and for the Ag^0-Cl stretching mode of the Cl^- adsorbates on the Ag surface.[11]

The role of the cations in SERS studies is much less understood. In electrolytes containing 1 M Cl^- ions and a variety of cations, many researchers have noted that cations affect the SERS characteristics of interfacial H_2O, in particular, the SERS line shape of the O-H stretching mode, and the SERS intensity vs potential dependence.[12-15] New experimental results indicate that the different types of cations can be divided into two groups: one with lower hydration energy (Cs^+, Rb^+, and K^+), and another with higher hydration energy (Na^+, Li^+, Ba^{2+}, Sr^{2+}, Ca^{2+}, and Mg^{2+}).[16]

For cations with lower hydration energy, we have postulated that the interfacial H_2O molecules are preferentially aligned with the oxygen atoms bonded to the positively charged Ag electrode while one or both of the hydrogens are weakly bonded to the coadsorbed Cl^- ions.[15] The narrow symmetric SERS line shape and the rapid decrease of the SERS intensity as the potential is swept toward less positive values are caused by such preferential alignment of H_2O molecules, disruption of hydrogen bonding between H_2O molecules and the rest of the water network, and formation of weak hydrogen bonds between interfacial H_2O and coadsorbed Cl^- ions.[15] For cations with higher hydration energy ($Na^+ \rightarrow Mg^{2+}$), we have postulated that the positively charged Ag surface is not sufficient to preferentially align the interfacial H_2O molecules with their oxygen atoms facing the electrode.

Instead, regardless of the electrode potential, the oxygen atoms are oriented toward the higher hydration energy cations which are located in the outer Helmholtz layer. The broad asymmetric SERS line shape and the increase of the SERS intensity as the potential is swept toward less positive values are associated with such coordination of interfacial H_2O molecules with the cations and the increased hydrogen bonding between these H_2O molecules and the rest of the water network.[15]

In this paper, we present new experimental SERS results in electrolytes containing Cl⁻ which further indicate that different groups of cation effects exist, one with lower hydration energy and another with higher hydration energy. In order to separate the cation effect on the number of active sites produced during oxidation and reduction of a Ag electrode and the cation effect on molecular coordination on SERS active sites, various types of high salt concentration solutions (1 M YCl + 0.05 M pyridine, $Y = Cs^+ \rightarrow Li^+$) were added to the electrolyte after the electrode had been oxidized and reduced in a low salt concentration electrolyte (0.1 M NaCl + 0.05 M pyridine). The SERS spectra of the Ag^o-Cl⁻ stretching mode, the O-H stretching mode, and the 3070 cm⁻¹ pyridine mode were monitored before and after the addition of high concentration YCl salts at a fixed potential, as well as during a voltage sweep toward more negative values. The SERS intensity vs time at a fixed electrode potential and the SERS intensity vs potential were deduced from these SERS spectra which were detected with an optical multichannel analyzer (OMA). Together with the previous results on interfacial H_2O, D_2O, OH⁻, and OD⁻ in 1 M YCl electrolytes,[15,17] these experimental observations provide a clearer picture of the role of cations in SERS studies.

2. EXPERIMENTAL

Our electrochemical cell consisted of a rectangular optical cell containing three electrodes [a Ag working electrode, a Pt counter electrode, and a saturated calomel electrode (SCE)] controlled by a potentiostat and a function generator. All the applied voltages were referenced to the SCE. Figure 1(a) shows the voltage excursion used throughout this work. Specifically, the voltage was swept positively ($-0.2 \rightarrow 0.2$ V_{SCE}, time A→B at 5 mV/s) and swept negatively ($0.2 \rightarrow -0.1$ V_{SCE}, time B→C at 5 mV/s) in a 0.1 M NaCl + 0.05 M pyridine electrolyte, causing the Ag electrode to undergo oxidation and reduction. After the reduction current had reached zero and during the potential hold (-0.1 V_{SCE} for 3 min, time C→D), 1 M YCl + 0.05 M pyridine solution was rapidly added near the Ag electrode, and the electrolyte was thoroughly stirred during time C. The potential was then swept negatively ($-0.1 \rightarrow -0.7$ V_{SCE}, time D→E at 5 mV/s).

The polycrystalline Ag electrode was mechanically polished, ending with 0.05 μm abrasive, and chemically etched in a 1:1 volume mixture of 30% H_2O_2 and 58% NH_4OH. All chemicals were reagent grade and the doubly distilled water was boiled to remove dissolved CO_2. The electrochemical cell was continuously purged with dry N_2 gas in order to prevent O_2 and CO_2 dissolution. The working Ag electrode was placed approximately

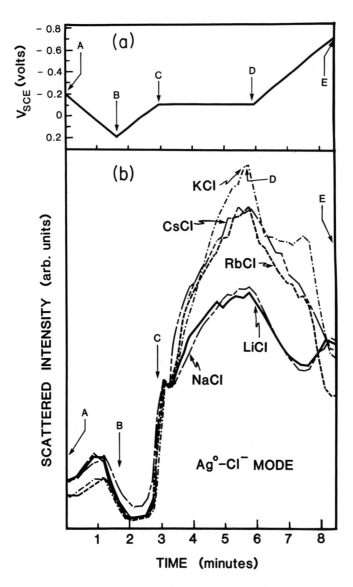

Fig. 1.   (a) The temporal evolution of the applied voltage on the Ag
          electrode, consisting of the oxidation and reduction of the
          Ag electrode (time A→C) at 5 mV/s in a 0.1 M NaCl + 0.05 M
          pyridine electrolyte, a potential hold at -0.1 V_{SCE} (time
          C→D) when the 1 M YCl + 0.05 M pyridine solution was rapidly
          added (time C), and a subsequent voltage sweep from -0.1 to
          -0.7 V_{SCE} (time D→E) at 5 mV/s.   (b) The temporal evolution
          of the peak SERS intensity of the Ag^o–Cl^- stretching mode
          during an ORC shown in (a) above before and after the addi-
          tion of different 1 M YCl + 0.05 M pyridine solutions in
          0.1 M NaCl + 0.05 M pyridine electrolytes.

1 mm from one cell window and the argon-ion laser beam, 50 mW at
514.5 nm, was incident on the electrode surface at ∿45° and focused
to a line within an area ∿0.3 mm^2.

The Raman radiation was collected by a f/1.2 camera lens and
imaged onto the entrance slit of a triple-stage spectrograph (Spex
1877) followed by an intensified photodiode linear array detector
(PAR 1420). Throughout the experiments, the Raman spectra were recorded every 8 s, corresponding to 40 mV intervals when the potential
was swept at 5 mV/s. The integration time of the photodiode array
was 2-5 s. In order to study the diffuse elastic scattering (0 cm^{-1}
shift) and the Ag^o-Cl⁻ mode at 240 cm^{-1}, 1200 g/mm gratings were used
in both the subtractive double-filter stage and the dispersive stage.
The overall spectral coverage of the triple-stage spectrograph was
∿500 cm^{-1} (see Fig. 2). For studying the 0-H stretching mode
(∿3400 cm^{-1}) and the pyridine mode (at 3070 cm^{-1}), 300 g/mm and
600 g/mm gratings were used in the filter and dispersive stages,
respectively, giving an overall spectral coverage of ∿1200 cm^{-1}.
The Ag^o-Cl⁻ mode was measured separately from the 0-H stretching and
pyridine modes, which could be detected simultaneously during an ORC
shown in Fig. 1(a).

3. RESULTS

Ag^o-Cl⁻ mode

Figure 2 shows the evolution of the SERS spectra during an ORC shown
in Fig. 1(a). The Ag electrode was oxidized and reduced (time A→C)
in a 0.1 M NaCl + 0.05 M pyridine electrolyte. The characteristic
valley near the switching voltage (time B) is ascribed to the optical
absorbing AgCl adherent layer, which decreases the reflectivity of
the Ag surface and hence the electromagnetic intensity in front of
the electrode. After the complete reduction of the AgCl layer
(time C), 1 M YCl + 0.05 M pyridine was rapidly added to the electrolyte at a potential hold (time C→D) at -0.1 V_{SCE}. For CsCl and
LiCl salts, both Figs. 2(a) and 2(b) show a gradual increase of the
Ag^o-Cl⁻ 240 cm^{-1} mode during the 3-min potential hold at -0.1 V_{SCE}.
Without the addition of 1 M YCl + 0.05 M pyridine, the SERS intensity
remains unchanged during such a potential hold. After the potential
hold, the voltage is swept toward -0.7 V_{SCE} (time E), causing the
SERS intensity at 240 cm^{-1} to decrease as fewer Cl⁻ ions are adsorbed
on the less positively charged Ag surface. During the subsequent
anodic sweep back toward -0.1 V_{SCE} (time D), some but not all of the
Ag^o-Cl⁻ SERS intensity is restored as more Cl⁻ is adsorbed on the
more positively charged Ag surface.

Figure 1(b) plots the peak intensity of the Ag^o-Cl⁻ mode throughout the ORC for all five 1 M YCl + 0.05 M pyridine additives. The SERS
intensity for various ORCs was set equal just after the reduction
current reached zero in the 0.1 M NaCl + 0.05 M pyridine electrolyte
but before the 1 M YCl + 0.05 M pyridine was added. This normalization
procedure minimizes the slight fluctuation in the SERS intensity

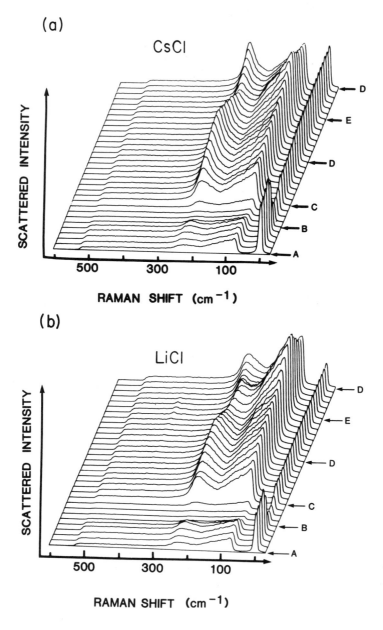

Fig. 2.   The SERS spectra detected with an OMA during an ORC shown
          in Fig. 1(a) with the addition of 1 M CsCl + 0.05 M pyri-
          dine (a) and with the addition of 1 M LiCl + 0.05 M pyri-
          dine (b). The elastic scattering at 0 cm^{-1} and the Ag^{0}-Cl^{-}
          stretching modes at $\sim$240 cm^{-1} are shown. Analogous spectra
          were detected for the 3070 cm^{-1} pyridine mode and the O–H
          stretching mode of interfacial H$_2$O.

introduced by the electrode surface preparation and by the total amount of Ag^0 atoms oxidized and Ag$^+$ ions reduced during the faradaic phase of the ORC. Figure 1(b) shows that both the rate of increase and the final SERS intensities (time C→D) are considerably larger for the lower hydration energy cations than for the higher hydration energy cations. The rate of decrease of the SERS intensity during the cathodic sweep (time D→E) is also shown in Fig. 1(b).

3070 cm^{-1} pyridine mode

Figure 3 shows the SERS intensity of the 3070 cm^{-1} pyridine mode during the ORC shown in Fig. 1(a). These curves are not normalized to each other but are scaled according to the normalization procedure for the Ag^0-Cl$^-$ mode discussed earlier for Fig. 1(b). During the potential hold at -0.1 V$_{SCE}$ (time C→D), the pyridine SERS intensity decreases rapidly for cations with lower hydration energy and the pyridine SERS intensity increases gradually for cations with higher hydration energy. After a 3-min potential hold, the 3070 cm^{-1} SERS intensity is much larger for LiCl and NaCl additives than that for CsCl, RbCl, and KCl additives (see Fig. 3). The cation effect on the pyridine mode is opposite to that for the Ag^0-Cl$^-$ mode.

During the cathodic sweep from -0.1 to -0.7 V$_{SCE}$ (time D→E), the 3070 cm^{-1} pyridine intensity vs potential curves can also be separated into two groups, one common to the lower hydration energy cations and another common to the higher hydration energy cations. Surprisingly, for the CsCl, RbCl, and KCl salts, the 3070 cm^{-1} SERS intensity first decreases and then increases. For the NaCl and LiCl salts, the observed intensity increase is more commensurate with the expected increase of surface coverage for organic molecules as the potential approaches the potential of zero charge (PZC).

O-H stretching mode of H$_2$O

Figure 4 shows the SERS intensity of the O-H stretching mode of interfacial H$_2$O molecules during the ORC shown in Fig. 1(a). Again, these curves are not normalized to each other but are scaled according to the normalization procedure for the Ag^0-Cl$^-$ mode in Fig. 1(b). Two features should be noted in Fig. 4: (1) During the potential hold at -0.1 V$_{SCE}$ (time C→D), the high salt concentration additive caused a gradual increase for NaCl and LiCl salts, while a much steeper increase was noted for CsCl, RbCl, and KCl salts. (2) During the cathodic sweep (time D→E), the O-H SERS intensity for NaCl and LiCl additives increased monotonically even when the voltage reached -0.7 V$_{SCE}$, while that for CsCl, NaCl, and LiCl additives decreased. In the case of CsCl, the intensity rise for voltages negative of -0.5 V$_{SCE}$ (after 7.5 min in Fig. 4) is associated with an overall rise of the continuum throughout the entire inelastic spectrum. No background subtraction was included in our data reduction.

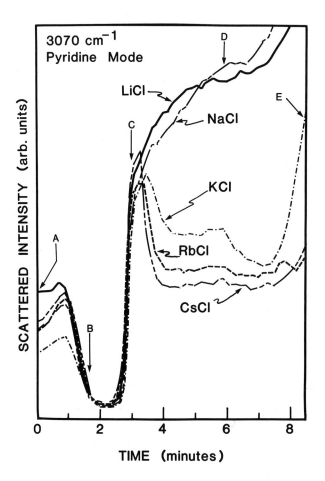

Fig. 3. The temporal evolution of the peak SERS intensity of the 3070 cm^{-1} pyridine mode during an ORC shown in Fig. 1(a) before and after the addition of five 1 M YCl + 0.05 M pyridine solutions in 0.1 M NaCl + 0.05 M pyridine electrolytes.

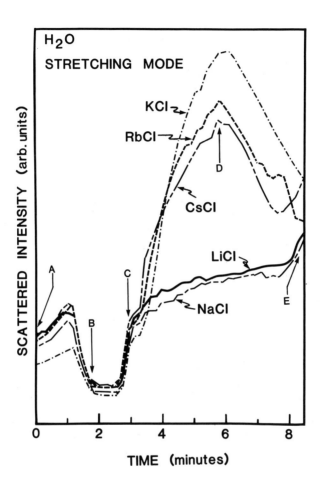

Fig. 4. The temporal evolution of the peak SERS intensity of the O-H stretching mode of interfacial H_2O during an ORC shown in Fig. 1(a) before and after the addition of five 1 M YCl + 0.05 M pyridine solutions in 0.1 M NaCl + 0.05 M pyridine.

4. DISCUSSION

Since, without the addition of a high salt concentration solution, the SERS intensity of the Ag^o-Cl^-, 3070 cm^{-1} pyridine, and O-H stretching modes remained constant during the potential hold, it can be assumed that the surface coverage of the adsorbates and of the SERS active sites on the Ag electrode remained constant. It can therefore be assumed that rearrangement of the surface morphology and adsorbates did not occur on the Ag surface. The small differences in the average surface coverage of SERS active sites, produced by the oxidation and reduction of the electrode, can be minimized by normalizing the Ag^o-Cl^- 240 cm^{-1} mode intensity after the reduction current reaches zero.

Temporal evolution of the SERS intensities occurs after the high salt concentration solution (1 M YCl + 0.05 M pyridine) has been added during the very beginning of the 3-min potential hold portion of the ORC. The rates of growth or decay of the SERS intensity for three adsorbates (Cl^-, pyridine, and H_2O), when different types of cations are used in the high salt concentration additives, can provide information on possible interactions of Y^+ ions with adsorbates located in the inner Helmholtz layer, on possible competition of Y^+ ions with other adsorbates for SERS active sites, and on possible rearrangement of the surface morphology and adsorbates. The extremely slow temporal rates, in the range of minutes, suggest that the growth and decay of the SERS intensities are caused by the possible rearrangement of surface morphology of SERS active sites and/or rearrangement of adsorbates on these sites. Such slow rates are not consistent with the diffusion rates of anions, cations, and molecules into the interfacial double layer. In this paper, we will not discuss the SERS intensity vs voltage dependences which occur after the potential hold and during the negative potential sweep toward the PZC (time D→E). Such voltage dependences have been reported,[11] although not for the ORC used in this paper.

The positively charged Y^+ ions are not expected to contact adsorb on the positively charged Ag electrodes (i.e., for voltages positive of the PZC). However, in the electrochemistry literature, particularly that pertaining to Hg electrodes, evidence exists that superequivalent adsorption can occur where the excess positive charge density on the Ag surface is less than the contact-adsorbed Cl^- density. The net charge density difference is equal to the density of cations in the inner Helmholtz layer.[18,19] We believe that our present SERS data on the Ag^o-Cl^-, pyridine, and H_2O modes support the fact that superequivalent adsorption of Cl^- ions results when the lower hydration energy cations (Cs^+, Rb^+, and K^+) can enter the inner Helmholtz layer. For higher hydration energy cations (Na^+, Li^+, and other Column IIA ions, Ba^{2+} → Mg^{2+}), the hydration sphere cannot be partially shed and thus the cations can remain only in the outer Helmholtz layer. However, these high hydration energy cations can reorient the adsorbed H_2O molecules, causing their oxygen atoms to be coordinated with the cations rather than with the positively charged Ag surface.

In Fig. 1(b), after the addition of 1 M YCl + 0.05 M pyridine, the increases in the Ag^o-Cl^- SERS intensity for all YCl salts are expected

since the Cl$^-$ concentration in the bulk electrolyte has been increased
by 10 times. What is surprising is that the SERS intensity increases
can be separated into two types, one for CsCl, RbCl, and KCl and
another for NaCl and LiCl. After a 3-min potential hold at -0.1 V$_{SCE}$,
the difference between the SERS intensity of the Ag^o-Cl$^-$ mode for the
lower and higher hydration energy cations is significant. Since the
bulk concentrations of Cl$^-$ ions in all these electrolytes are the same,
we believe these SERS results suggest that Cs$^+$, Rb$^+$, and K$^+$ ions can
enter the inner Helmholtz layer and thereby cause the surface coverage
of Cl$^-$ ions to be greater than that of positive charges on the electrode for -0.1 V$_{SCE}$.

After the addition of 1 M YCl + 0.05 M pyridine, the bulk concentration of pyridine remains at 0.05 M. Consequently, in Fig. 3, the
observed decrease and increase of the 3070 cm^{-1} pyridine SERS intensity
is not a result of a pyridine concentration change in the bulk electrolyte but is partially a result of the surface coverage of pyridine.

For NaCl and LiCl salts during time C→D, the rate of intensity
increase of the 3070 cm^{-1} pyridine mode is approximately equal to that
of the 240 cm^{-1} Ag^o-Cl$^-$ mode shown in Fig. 1(b). This near equality
suggests that more pyridine can complex with the more abundant Cl$^-$
adsorbates at the SERS active sites. Ample evidence exists in the
literature that pyridine can complex with halide anions at Ag active
sites.[20]

The rapid increase followed by a rapid decrease of the 3070 cm^{-1}
pyridine intensity after the addition of high concentration CsCl,
RbCl, and KCl solutions (all containing 0.05 M pyridine) shown in
Fig. 3 during time C→D implies that two mechanisms may be operative.
In the beginning, concomitant with the Cl$^-$ coverage increase, the
pyridine and/or pyridine-Cl$^-$ complex coverage also increases, analogous
to the case for the NaCl and LiCl salts. However, as the Cs$^+$, Rb$^+$, and
K$^+$ ions enter the inner Helmholtz layer, the surface coverage of pyridine is decreased either because the Ag-pyridine-Cl$^-$ complexes are dis-
sociated or because the cations directly displace the pyridine molecules from the SERS active sites. We believe that the higher hydration
energy cations displace the pyridine from the SERS sites and attract
more Cl$^-$ ions to cover the SERS sites in order to be consistent with
the larger increase of the Ag^o-Cl$^-$ intensity [shown in Fig. 1(b) for
CsCl, RbCl, and KCl salts].

The temporal increase of the SERS intensity of interfacial H$_2$O
after the addition of 1 M YCl + 0.05 M pyridine solutions near time C
can again be separated into two types, one for NaCl and LiCl salts and
another for CsCl, RbCl, and KCl salts. Even though an increase in the
Cl$^-$ coverage can cause displacement of H$_2$O molecules from SERS active
sites (i.e., Cl$^-$ ions can compete with H$_2$O molecules for SERS active
sites),[10] the increased H$_2$O SERS intensity is still evident for all
the YCl high salt additives. The more rapid rise for the Cs$^+$, Rb$^+$, and
K$^+$ ions can be attributed to a concomitant increase of the Cl$^-$ coverage
because of a ten-fold increase in the bulk Cl$^-$ concentration and
because of superequivalent adsorption with these cations entering the
inner Helmholtz layer. The less rapid rise for Na$^+$ and Li$^+$ ions can be
ascribed only to the increase of Cl$^-$ coverage resulting from a ten-fold

increase in the bulk electrolyte.

In our present SERS spectral data, we again noted that the SERS line shapes of interfacial H_2O remain spectrally narrow and symmetrical after the addition of high concentration CsCl, RbCl, and KCl salts. Furthermore, upon a potential sweep toward more negative values (time D→E), the SERS intensity decreases (see Fig. 4). In contrast, after higher hydration energy cations are added to the electrolyte, the SERS line shapes of interfacial H_2O become spectrally broader and asymmetrical. In addition, upon sweeping toward more negative values (time D→E in Fig. 4), the SERS intensity vs voltage increases. These observations on the SERS line shape and intensity vs potential dependence are consistent with our previous results.[15]

Based on our new measurements on the temporal evolution of the SERS intensity of several adsorbates after oxidation and reduction of the Ag electrode in a low salt electrolyte and after the addition of a high salt concentration solution at a potential hold, a model will be presented which can possibly account for the observed differences in the line shapes, potential dependences, and intensity increases when higher and lower hydration cations are added. We postulate that the Cs^+, Rb^+, and K^+ ions located within an inner Helmholtz layer can also cause disruption of hydrogen bonds between the interfacial H_2O molecules and the rest of the water network,[21] similar to the role of Cl^- ions which are coadsorbed with H_2O.[19] Consequently, hydrogen bond breaking by the cations can result directly or indirectly upon increasing the Cl^- surface density associated with superequivalent adsorption. Furthermore, these lower hydration energy cations in the inner Helmholtz layer do not cause the interfacial H_2O molecules to undergo an orientational change, i.e., the oxygen atom of a H_2O molecule is still bonded to the positively charged Ag electrode. The combination of increased hydrogen bond disruption and no reorientation of the H_2O molecules gives rise to a spectrally narrow SERS line shape, a steeper increase in the SERS intensity between time C→D in Fig. 4, and an intensity vs potential decrease with more negative voltage (time D→E).

We postulate that the higher hydration energy Na^+ and Li^+ ions located outside the inner Helmholtz layer cause the interfacial H_2O molecules to reorient, i.e., the oxygen atoms are coordinated with the Na^+ or Li^+ ions, and a broader, more asymmetric line shape thereby results. By adding high concentration NaCl and LiCl salts, two opposite mechanisms affect the hydrogen bonding with the rest of the water network, one with the increased coadsorbed Cl^- ions and another with the Na^+ or Li^+ cations located in the outer Helmholtz layer. The net effect on the interfacial H_2O molecule is spectral broadening and asymmetry. A slower increase in the SERS intensity between time C→D is partially caused by the spectral broadening of the O–H line shape, as shown in Fig. 4 which plots the peak intensity of the O–H stretching mode, not the total intensity of the O–H stretching mode.

In summary, the cations associated with the supporting electrolyte most definitely affect the SERS characteristics of Ag^O–Cl^-, pyridine, and H_2O modes. Although we have not yet detected cation–Cl^- or cation–H_2O stretching modes, we can infer the cation effect through the temporal evolution of the SERS intensity during a potential hold after

1 M YCl + 0.05 M pyridine is added to an electrolyte originally containing 0.1 M NaCl + 0.05 M pyridine and after the Ag electrode is
oxidized and reduced in this lower salt solution. The growth and decay
rates of the SERS intensity can be grouped according to whether the
cations have lower or higher hydration energy.

A slow growth rate (in the order of several minutes) has been
reported for the 1008 cm^{-1} pyridine mode in 0.1 M KCl + 0.05 M pyridine electrolytes after an anodization pulse.[22] It has been proposed
that changes in the surface morphology or in the adsorbate geometry
can occur during the time interval following the anodization pulse and
before steady-state conditions have been reached.[22] Furthermore, the
observed slow growth and decay rates on Ag electrodes may also be consistent with the recent proposal that SERS is associated with the
cavities or porous nature of the Ag film surface.[23,24] Should these
cavities or pores also exist on the Ag electrode surface, the diffusion rates of Cl⁻, Y⁺, and H$_2$O can then be many orders of magnitude
slower than the diffusion rates of ions and neutral molecules in and
out of the electrode-electrolyte double layer. However, without high
resolution electron microscopy analysis of the surface morphology of
the electrodes at various time intervals after anodization, we are
unable to identify the origin of the slow growth and decay rates, i.e.,
determine whether the slow temporal evolutions are associated with the
diffusion in and out of the pores, changes in the morphology surface,
and/or variations in the adsorbate geometry near the SERS active sites.

ACKNOWLEDGMENTS

We thank Dr. B. L. Laube for many helpful discussions and also gratefully acknowledge the partial support of this work by the Gas Research
Institute (Basic Research Grant No. 5083-260-0795).

REFERENCES

1. M. Fleischmann, P.J. Hendra, and A.L. McQuillan, Chem. Phys. Lett.
   26, 163 (1974).
2. D.L. Jeanmaire and R.P. Van Duyne, J. Electroanal. Chem. 84, 1
   (1977).
3. J.T. Hupp, D. Larkin, and M.J. Weaver, Surf. Sci. 125, 429 (1983).
4. R.K. Chang and T.E. Furtak, eds., Surface Enhanced Raman Scattering (Plenum Press, New York, 1982).
5. A. Otto, in Light Scattering in Solids, Vol. 4, M. Cardona and
   G. Güntherodt, eds. (Springer-Verlag, Berlin, 1984), p. 289.
6. J.F. Owen, T.T. Chen, R.K. Chang, and B.L. Laube, Surf. Sci. 131,
   195 (1983).
7. J.F. Owen and R.K. Chang, Chem. Phys. Lett. 104, 59 (1984).
8. For SERS of interfacial H$_2$O, see: M. Fleischmann, P.J. Hendra,
   I.R. Hill, and M.E. Pemble, J. Electroanal. Chem. 117, 243 (1981);
   B. Pettinger, M.R. Philpott, and J.G. Gordon II, J. Chem. Phys.
   74, 934 (1981); T.T. Chen, J.F. Owen, R.K. Chang, and B.L. Laube,

Chem. Phys. Lett. 89, 356 (1982). •

9. For normal Raman scattering of bulk water, see: G.E. Walrafen, J. Chem. Phys. 55, 768 (1971); C.I. Ratcliffe and D.E. Irish, J. Phys. Chem. 86, 4897 (1982).

10. J.F. Owen and R.K. Chang, Chem. Phys. Lett. 104, 510 (1984).

11. J.F. Owen, T.T. Chen, R.K. Chang, and B.L. Laube, Surf. Sci. 125, 679 (1983); J. Electroanal. Chem. 150, 389 (1983).

12. S.H. Macomber, T.E. Furtak, and T.M. Devine, Surf. Sci. 122, 556 (1982).

13. M. Fleischmann and I. Hill, J. Electroanal. Chem. 146, 367 (1983).

14. B. Pettinger and L. Moerl, J. Electroanal. Chem. 150, 415 (1983).

15. T.T. Chen, K.E. Smith, J.F. Owen, and R.K. Chang, "The Metal Cation Effect on the SERS of Interfacial D_2O and H_2O," Chem. Phys. Lett., in press.

16. The heat of hydration in aqueous solution at 25°C can be found in: D. Dobos, Electrochemical Data (Elsevier, Amsterdam, 1975), p. 99.

17. T.T. Chen, R.K. Chang, and B.L. Laube, "Detection of Interfacial Hydroxyl Ions (OH^- and OD^-) on Ag Electrodes by Surface Enhanced Raman Scattering," Chem. Phys. Lett., in press.

18. J.O'M. Bockris and A.K.N. Reddy, Modern Electrochemistry, Vols. 1 and 2 (Plenum Press, New York, 1970).

19. A.J. Bard and L.R. Faulkner, Electrochemical Methods (John Wiley and Sons, New York, 1980).

20. A. Regis and J. Corset, Chem. Phys. Lett. 70, 305 (1980); G.F. Atkinson, D.A. Guzonas, and D.E. Irish, Chem. Phys. Lett. 75, 557 (1980).

21. D.E. Irish, in Ionic Interactions, Vol. 2, S. Petrucci, ed. (Academic Press, New York, 1971), p. 219; D.E. Irish and M.H. Brooker, in Advances in Infrared and Raman Spectroscopy, Vol. 2, R.J.H. Clark and R.E. Hester, eds. (Heyden, London, 1976), p. 254.

22. A.M. Stacy and R.P. Van Duyne, in Time-Resolved Vibrational Spectroscopy, G.H. Atkinson, ed. (Academic Press, New York, 1983), p. 377.

23. E.V. Albano, S. Daiser, G. Ertl, R. Miranda, K. Wandelt, and N. Garcia, Phys. Rev. Lett. 51, 2314 (1983).

24. H. Seki and T.J. Chuang, Chem. Phys. Lett. 100, 393 (1983).

RAMAN SPECTROSCOPY OF THIN FILMS ON SEMICONDUCTORS

J. C. Tsang
IBM T. J. Watson Research Center
P. O. Box 218
Yorktown Heights, New York 10598

ABSTRACT: We demonstrate the ability of state of the art multichannel photon detectors such as the cooled silicon intensified target vidicon and the microchannel plate photomultiplier with a position sensitive resistive anode to detect Raman scattering from weakly scattering, transparent thin films, reacted metallic layers and disordered surface layers. Vibrational spectra have been obtained for surface layers as thin as 10 A, for frequency shifts as small as 80 cm^{-1} and for substrates which produce background emission orders of magnitude larger than the Raman emission from the surface layer. Measurements of the Raman spectra of thermally grown SiO_2 layers on crystalline Si show that signals from systems with a Raman scattering efficiency below 10^{-15} can be detected in the presence of the strong second order Raman scattering of Si. Since the Raman efficiency of a monolayer of nitrobenzene is about 10^{-15}, this shows we now have the ability to use Raman scattering to study the behavior of adsorbates and surface layers, independent of the identity of the substrate.

INTRODUCTION

The energies, symmetries and lifetimes of the electronic and vibrational states of atoms or molecules at a surface constitute the basic spectroscopic information necessary for the understanding of the interaction of an atom or molecule with a surface. With its energy resolution of typically better than 2 or 3 cm^{-1}, its sensitivity to the symmetry properties of the excitations being studied and its ability to obtain spectroscopic information at gas-solid, liquid-solid and solid-solid interfaces, Raman spectroscopy could strongly complement the electron spectroscopies that are the standard tools for the study of the vibrational and electronic excitations at surfaces.[1] The energy resolution in electron spectroscopy tends to be atleast an order of magnitude worse and it is very difficult for electron spectroscopy to study any interface other than the vacuum-solid interface. However, the small magnitude of the Raman scattering cross sections for most materials, which usually result in signal rates of below 0.1 cps for monolayer coverages, has meant that Raman experiments on surfaces were prohibitively difficult.[2,3]

The recent extensive interest in the surface enhanced Raman effect[4] reflected both intrinsic curiosity about the phenomenon itself as well as the recognition that a significant enhancement of the signal levels obtained in a Raman experiment on a surface species would result in a valuable new tool for the study of the dynamic properties of surfaces and adsorbates. The enormous enhancements of the Raman scattering efficiency observed in the surface enhanced Raman effect made feasible experiments on the vibrational states of certain

B. Pullman et al. (eds.), Dynamics on Surfaces, 379–393.
© 1984 by D. Reidel Publishing Company.

families of molecules adsorbed onto specially prepared noble metal surfaces[4] at the solid-vacuum,[5] solid-gas,[6] solid-liquid [7] and solid-solid interfaces.[8] These enhancements also demonstrated that an enhancement in experimental sensitivity of the order of 10^2-10^3 over that available using a conventional scanning monochromator with photon counting detection would be enough to make Raman scattering a useful tool for the study of surface and adsorbate excitations.[9,10]

In this paper, we demonstrate the improvements in experimental sensitivity that can be achieved through the use of state of the art multichannel photon detectors. Given this objective, we concentrate on results obtained from thin film systems where the Raman scattering efficiencies do not depend on the presence of the surface or interface. This is often not the case for a molecule adsorbed onto a surface where the process of adsorbtion can alter the Raman tensor.[11,12,13] It will certainly not be the case for an isolated atom adsorbed onto a surface since the excitation frequencies and Raman cross sections will be completely determined by the adsorbtion process. We consider both the cooled vidicon detector used by Chang et al.[14] and Campion et al.[9] as well as ourselves[10] and a microchannel plate photomultiplier with a position sensitive resistive anode.[15] There have been a number of discussions of the use of the vidicon detector for low light level light scattering;[6,9,10] we believe this is the first discussion of the use of the microchannel plate photomultiplier with a resistive anode position sensitive detector for such experiments.

We show these detectors can produce enhancements of our experimental sensitivity between ten and one hundred times over that of a conventional single channel detector Raman scattering system. In contrast to the enhancement obtained in the surface enhanced Raman scattering effect which is largely limited to roughened noble metal surfaces,[4] and in the interference technique[16] which is limited to deposited thin film samples with thicknesses of about 50 A, these improvements in detector sensitivity can be applied to arbitrary systems including the surfaces of bulk crystals. We demonstrate the capabilities of these detectors by applying them to problems such as 1) the detection of a submicron thickness oxide layer grown by thermal reaction on a semiconductor surface, 2) the reaction of a metallic layer on a silicon (100) surface and the observation of a thin amorphous layer on Si(100). Our choice of semiconductors as substrates enables us to determine our experimental sensitivity in the presence of a significant signal from the substrate, in this case, the first, second and higher order Raman scattering from the substrates. This can be an important limitation to the useful sensitivity of any experimental system since the noise level will be determined by the background and not the detector.

We first describe our experimental systems including a brief discussion of the vidicon detector and a more detailed discussion of the multiple-microchannel plate photomultiplier. We consider their limitations and compare their capabilities with some nominal data rates for experimental systems of interest. We then describe the results obtained when we apply these experimental systems to the problems mentioned above. We end by considering the current limitations on our sensitivity and possible schemes for its improvement.

TECHNIQUES AND EXPERIMENTAL RESULTS

a. Sensitivity

In Figure 1, we show our Raman scattering system. The Raman spectra were excited using lines from either an Ar^+ or a Kr^+ ion laser. The narrow band filter shown is used to remove non-lasing emission lines from the laser beam. The magnitude of the contribution of these emission lines to the Raman spectrum depends on a variety of factors including the smoothness of the sample surface. In many cases, it was necessary to use more than one narrow band filter inorder to adequately suppress this unwanted contribution to the Raman spectrum. The laser light was focused onto the samples using either a cylindrical lens producing a spot size about 100 microns by 2 mm or a spherical lens producing a spot size of the order of 50 microns. Incident laser power levels were usually between 5 and 300 milliwatts although spectra have been obtained with excitation powers as low as 1 microwatt. All of the spectra presented in this paper were obtained in air at room temperature. Inorder to reduce the intensity of Raman scattering from the low energy excitations of atmospheric molecules near the sample surface, the samples were flushed with flowing He during the measurements. The scattered light was collected by an f/1.2 lens and focused onto the entrance slits of the spectrometer. The elastically scattered light is rejected and the inelastically scattered light dispersed by a commercial three stage spectrometer with a subtractive dispersion double monochromator first stage and a single grating spectrograph second stage. Given the presence in the spectrometer of seven Al coated mirrors, two lenses in the spatial filter of the monochromator, and three Al coated diffraction gratings, the maximum throughput will be of the order of .25 (assuming 80% efficiency for each grating). The results presented in this paper

Fig. 1 An experimental system for Raman spectroscopy using a multichannel detector. Both the cooled SIT vidicon detector and the microchannel plate photomultiplier with a position sensitive resistive anode are shown, though not to scale.

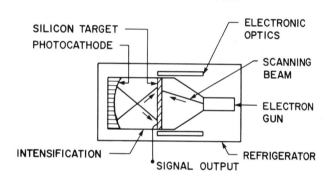

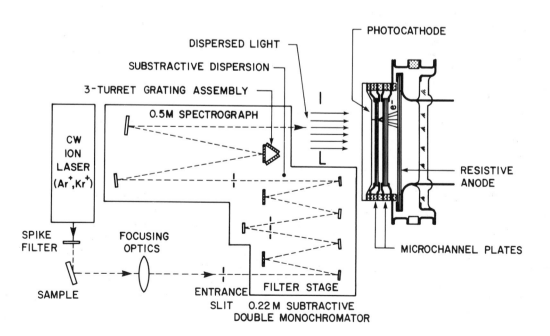

reflect a throughput approximately a factor of two smaller due to somewhat lower efficiencies for our diffraction gratings. The combination of a $f/1.2$ collection lens with our $f/5.6$ monochromator produces a substantial enlargement of the laser spot on the monochromator slits. Depending on the slitwidths used, this could produce a substantial reduction in our optical collection efficiency. The dispersed, inelastically scattered light was detected by either a cooled vidicon detector or a cooled microchannel plate photomultiplier located at the focal plane of the spectrograph. The active area of the vidicon was 0.5 inches in diameter. The active area of the microchannel plate detector was 1 inch in diameter. The ability of both detectors to function in two dimensions rather than as just one dimensional linear arrays, was especially important for the triple monochromator used in this experiment because its optics imaged an input spot on the entrance slits into a long slit at the focal plane of the optical detector. The use of a linear array detector would have resulted in the loss of a significant fraction of the scattered light and a substantial decrease in the experimental sensitivity.

Campion et al.[9] and we[10] have both reported on the use of cooled silicon intensified target vidicons for Raman scattering from surface adsorbates on metals in the absence of any large external enhancement of the Raman signal. The vidicon detector has also been used by Dornhaus et al.[17] in their surface enhanced studies of adsorbates on Ag electrodes and its operation has been thoroughly discussed by Chang et al.[14] The photocathode converts photons into electrons, retaining their spatial distribution, a silicon target stores the photoelectrons and the integrated charge on the silicon target is read out by an electron beam. By cooling the detector, it is possible to reduce the dark count from the photocathode and thermally created carriers in the silicon target to levels where the dispersed spectrum can be integrated on the target for one hour or more. There are limits to this process since at very low temperatures where the dark counts becomes negligible, the silicon intensified target loses efficiency and it becomes increasingly difficult to use. The total background accumulated over a period of an hour even at temperatures near -40 C can exceed 20000 counts per channel where there are significant contributions from the photocathode dark count, the thermally created carriers in the silicon target and the carriers created in the silicon target by the electron beam read out. The minimum detectable signal levels are then about 0.05 cps. Since the quantum efficiency of the semi-transparent photocathode is less than that of the GaAs photocathodes used in most single channel Raman scattering systems and the quantum efficiency of the silicon target is also less than that of the anode chain in a conventional photomultiplier, this corresponds to a signal level of about 0.3 cps for a conventional photomultiplier. This is an order of magnitude better than can be achieved with a conventional single channel detector. Furthermore, the full sensitivity of the detector is not always available. Because the Si target is a charge storage device which is depleted by the flux of photoelectrons, it can be saturated and is not completely linear in its response.[4] This can greatly limit the sensitivity of the detector in the presence of a strong background signal from the substrate, especially if it contains significant structure.

The microchannel plate photomultiplier with a position sensitive resistive plate anode (Fig. 1) represents an adaptation to optical studies of the position sensitive microchannel plate detector used by many groups for electron spectroscopy.[18] The first stage of the ITT F4146M "MEPSICRON" detector consists of a semitransparent bialkali photocathode with a quantum efficiency in excess of 10% for photon energies above 2.2 eV and a quantum efficiency greater than 1% for photon energies above 1.5 eV. The photocathode is 0.75 mm above a stack of 5 microchannel plate electron multipliers. The top surface of the microchannel plate stack is coated with a thin aluminum oxide layer which protects the photocathode from catastrophic rapid degradation due to ion bombardment from the microchannel plates. The close proximity of the photocathode and electron multiplier channel plates when combined with the photocathode to microchannel plate bias voltages between 200 and 400 volts means there is little loss of spatial information as the photoelectrons move from the cathode to the microchannel plates. The amplified electron pulse is detected by the resistive plate anode. The position of the amplified electron pulse on the resistive anode (and due to the proximity of the photocathode to the first microchannel plate, the position of the incident photon on the photocathode) is calculated by a position computer from the voltages induced by the electron pulse at the four corners of the resistive anode. The quantum efficiency of the electron multipliers and detector is limited by the aluminum oxide overlayer and the applied voltage between the photocathode and the first microchannel plate stack. Efficiencies in excess of 70 % are customary with higher values possible.[15] The dark count summed over the full one inch diameter of the detector is about 10 counts per second when the detector is held at -30 C with a total bias voltage of 4200 volts. Given a linear spatial resolution of 400 line pairs for this 2 dimensional detector, this corresponds to a dark count of 0.025 cps per channel or about 90 counts per channel per hour. This is more than two order of magnitude better than that obtained with the cooled vidicon. With this dark count, optical signals with a peak intensity of 0.005 cps can be observed with this detector. Because the output of the microchannel plate photodetector can be stored in a multichannel analyzer and the detector essentially "reset" after each detected photoelectron pulse, this detector should be considerably more linear in its response than the vidicon which integrates its signal as stored charge on the Si target. Inaddition, the microchannel plate detector will not saturate over long integration times, a common problem for the vidicon, even when cooled.

In Fig. 2, we show the Raman spectrum of a (100) oriented crystalline Si wafer obtained using the microchannel plate detector system shown in Fig. 1. The spectrum was excited by 20 mw of p-polarized 5145 A light with the scattered light unpolarized. The spectrum was integrated over 1000 seconds. The spectrometer was used in a low resolution mode where a spectral region covering more than 2400 cm^{-1} was imaged onto the 400 channels of the detector. The dominant line at about 520 cm^{-1} is due to first order Raman scattering while the two weak peaks near 1350 and 1450 cm^{-1} are due to third order phonon scattering.[18] The width of the first order optic phonon is broadened considerably by the low spectral resolution. All of the remaining structures are due to second order Raman scattering. The integrated signal under the 520 cm^{-1} first order Raman line of Si is about 10^6 counts with

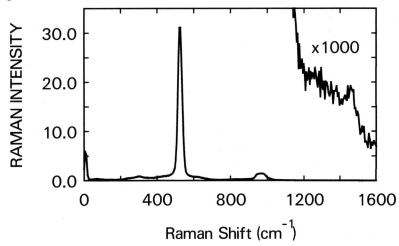

Figure 2. The 1st, 2nd and 3rd order Raman spectrum of Si(100). The scattered light was excited by 20 mw of 5145 A light p-polarized and the scattered light was unpolarized. The spectrum was integrated for 1000 seconds. The resolution of the optical system was about 24 cm^{-1}

a peak intensity of 3×10^5 counts. Integration times a factor of 10 longer have been achieved. As mentioned earlier, unlike the case of the vidicon, there is no intrinsic limit to the integration time. Excitation power levels as high as 500 mw are typical for Raman scattering experiments from opaque crystals. Using the longer integration times and the higher powers and given a total dark signal of about 300 counts over this integration time, we should be able to observe Raman scattering from samples where the strength of the Raman signal is about 6 orders of magnitude weaker than in crystalline Si. We have confirmed this by measuring the first order Raman spectrum of Si excited by 1 microwatt of 5145 A light and integrating for 300 seconds.

The strength of Raman scattering from a number of inorganic solids and molecules is tabulated in Table. 1. For optically opaque solids such as Si and Ge, the strength of the scattering is expressed in terms of a scattering efficiency for an optically thick sample.[19] This is the ratio of the number of photons inelastically scattered into a unit solid angle divided by the number of photons incident on the sample. For an almost transparent thin film of these materials, the scattering efficiency is approximately the value listed in Table 1 multiplied by the ratio of the film thickness and the optical attenuation length. The strengths of the Raman scattering from the optically transparent inorganic[20] and molecular[21] systems are given interms of the scattering efficiency per unit length of the optical path through the sample. In Table 1, we also normalize our results against the Raman efficiency of Si. It is clear from Table 1, that with the exception of SiO_2, we have the ability to observe Raman scattering from all of the materials listed for thicknesses as small as 10 A or less. This of course assumes that the Raman spectra and scattering cross sections are not substantially modified by the presence of the surfaces. For SiO_2, we nominally have the sensitivity to detect Raman signals from layers with thicknesses between 200 and 500 A.

While high detector sensitivity is necessary to observe Raman scattering from adsorbed layers and very thin films on surfaces, it is not sufficient. Most common substrates produce observable Raman scattering and luminescence. The Raman scattering from the surface layer will be superimposed on the optical emission from the substrate. The optically active volume of the substrate, even when strongly absorbing, will be much larger than the scattering volume of any surface layer. It will be necessary to subtract the optical spectrum of the substrate from the Raman spectrum of the surface layer plus emission from the substrate inorder to obtain the spectrum of the surface layer. This will always reduce the sensitivity since the dominant source of noise will be the substrate rather than the detector. The experimentally usable sensitivity will be further reduced if the detector has either a limited dynamic range or is not completely linear in its response as a function of signal intensity. In Fig. 2, we showed the Raman spectrum of crystalline Si. In Fig. 3, we show the Raman spectra of thermally grown oxides of varying thicknesses on Si(100) for Raman shifts between 0 and 700 cm^{-1}.

Table 1. Raman efficiencies of a number of common substrates and overlayers.

material	vibrational frequency (cm^{-1})	absorption coefficient (cm^{-1})	scattering efficiency sr^{-1}	normalized efficiency
Si	522	10^4	3×10^{-9}	1
Ge	303	5×10^5	0.9×10^{-9}	0.3
quartz	466	0	4.6×10^{-9}/cm	1.5/cm
benzene	992	0	3.8×10^{-15}/layer	1.3×10^{-6}
nitro-benzene	1345	0	8×10^{-15}/layer	2.7×10^{-6}
H_2	4161	0	6×10^{-15}/layer	2×10^{-6}

These spectra were excited by s-polarized 5145 A light where the incident field was parallel to the 100 axis of the Si(100) substrate. The scattered light was polarized parallel to the incident light. In this scattering geometry, the Raman selection rules for Si(100) suppress the first order Si Raman scattering at 522 cm^{-1}. This is the only substantial feature in the spectra in Fig. 3 not due to second order Raman scattering from the Si substrate. The spectra in Fig. 3 are largely identical although the sample with 1.7 microns of thermal oxide shows some additional scattering near 500 cm^{-1} In Fig. 4, we show the spectra in Fig. 3 after the Raman spectrum of the Si(100) wafer with 300 A of oxide (Fig. 3c) has been subtracted. Errors in the subtraction are responsible for the "derivative" structure at 522 cm^{-1} in Fig. 4c. The abrupt rise in signal in Fig. 4a for energies above 100 cm^{-1} is due to the monochromator bandpass. Also shown in Fig. 4 is the Raman spectrum of a thin fused quartz wafer. The Raman spectra of the thermally grown oxides are identical to the fused quartz spectrum.

Previous published spectra of the Raman scattering from thermal oxides of Si have been restricted to oxide thicknesses greater than 10 microns and have required the removal of the crystalline substrate underlying the oxide.[22] Fig. 4 shows we can observe Raman scattering from oxide layers as thin as 1300 A. The dominant feature in the oxide spectrum is due to the symmetric stretching modes of oxygen.[23] These modes are readily observed even though they are much weaker than the second order Raman scattering from the Si substrate. The observation in Fig. 4c of a weak feature at 600 cm^{-1} which has been attributed to defects in the SiO_2 lattice shows we can observe scattering from spectral features an order of magnitude weaker still.[22] This shows we can observe very weak Raman scattering even in the presence of a substrate induced background orders of magnitude stronger than the actual signal. This performance is considerably superior to that obtainable using our cooled vidicon detector due the combination of the absence of sharp features in the vitreous SiO_2 spectrum, the smaller dynamic range and weak non-linearity in the response of the vidicon.

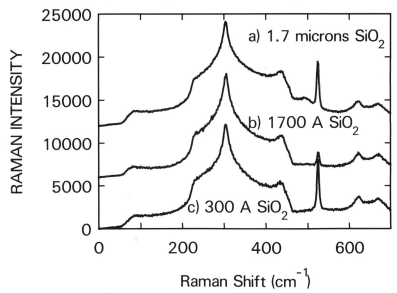

Fig. 3. The Raman spectra of thermally oxidized Si(100). For the scattering geometry used, the first order optic phonon scattering is nominally forbidden. a) An Si wafer with 1.7 microns of thermal oxide. b) A Si(100) wafer with 1700 A of thermal oxide grown on the Si wafer. c) A Si(100) wafer with 300 A of thermal oxide. The spectra were obtained using the microchannel plate photomultiplier.

The spectra in Fig. 4 were obtained by subtracting two successive integrations of the scattered light. The non-SiO$_2$ structures are due to small instabilities in the optical system over the period of time required for each of the integrations. A more sensitive method would involve the summing of many repeated difference spectra, each individual spectrum being obtained over one or two seconds. This can be easily implemented on the microchannel plate photomultiplier. It would be much more difficult to implement on the vidicon because of the non-linearities in the response of the cooled detector and the slow response time of the readout of the cooled Si target. A related advantage of the microchannel plate detector is the fact it reads out in real time. The spectrum can be observed on a multichannel analyser as it is accumulated. This is not the case for the vidicon detector where the spectrum is accumulated in the Si target and can be read out only after the integration is completed. There is a particular limitation of the microchannel plate photomultiplier which arises from its resistive anode. The finite time required for the computation of the position of the electron pulse on the anode means an overall count rate greater than 10^5 counts per second for the whole detector or the arrival of more than two counts within 10 microseconds anywhere on the anode will produce erroneous results.

Figs. 3 and 4 show the strength of the substrate emission can limit our experimental sensitivity. However, the strength of the scattering from a surface layer or an adsorbate relative to the strength of the substrate scattering is not a fixed quantity. It is a strong function of the system studied and the excitation energy used since the Raman scattering efficiency depends on the electronic structure of the scatterers.[24,25] Table 1 showed the Raman scattering efficiency for the optic phonon of Ge under 5145 A excitation is a factor of 3 smaller than that of Si.[19] Studies of the excitation wavelength dependence of the Raman scattering from Ge show that there is a maximum in the scattering efficiency below 2.0 eV with a significant decrease in scattering efficiency for excitation energies above 2.4 eV.[26] The

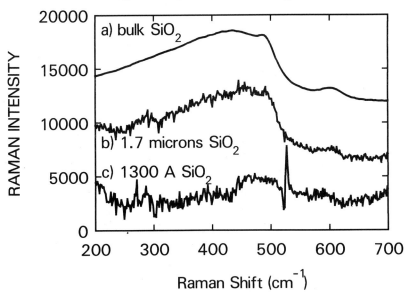

Fig. 4. The Raman spectra of thermally grown oxides of Si and bulk SiO$_2$. a) The Raman spectrum of bulk SiO$_2$. b) The difference spectrum obtained from the subtraction of the Raman spectrum of a sample with 300 A of thermal oxide on Si(100) from the spectrum in Fig. 3a. c) The difference spectrum due to the subtraction of the Raman spectrum shown in Fig. 3c from the Raman spectrum in Fig. 3b.

optical absorption of Ge at these energies is over 10^5 cm^{-1} so the volume of sample probed by the light will be only a few hundred angstroms thick. In contrast, the Raman efficiency of Si rises rapidly between for energies above 2.3 eV and does not begin to decrease until the excitation energy is greater than 3.4 eV.[27] The optical absorption in Si at the lower photon energies is an order of magnitude smaller than in Ge so the sample volume probed by the light can be of 0.5 microns thick. Galeneer[28] has shown that at 5145 A, GeO$_2$ scatters more strongly than SiO$_2$. As a result, for 5145 A excitation, there should be greater sensitivity to the presence of an oxide layer on Ge than on Si. This is shown in Fig. 5. Figure 5a is the Raman spectrum of bulk GeO$_2$. Fig. 5b is obtained from a 400 A layer of GeO$_2$ grown on a Ge(100) surface exposed to O$_2$ at 525 C. The thickness of the oxide was verified by ellipsometry. The spectra in Fig. 5 were obtained using the cooled vidicon detector. The quality of the grown oxide spectrum is limited by the linearity of the detector and optical emission from the oxide as well as the substrate Raman scattering. Nevertheless, it is clear we have the capability of detecting Raman scattering from the thermal oxide of Ge for thicknesses of the order of 100 A. The use of optical excitation energies near the 3.7 eV minimum in the Raman cross section of Si when combined with the much larger absorption coefficient of Si at this energy would enhance our sensitivity to adsorbates and thin films on Si by atleast an order of magnitude.

b. Chemical Reactions and Disorder on Si Surfaces

The Raman spectra shown in Figs 3, 4 and 5 all come from overlayers with relatively low Raman efficiencies. By working with film thicknesses in excess of 100 A, we can neglect any surface effects on the Raman scattering efficiencies. These have allowed us to demonstrate we have the ability to detect Raman scattering from systems where the Raman efficiency is of the order of 10^{-15}. This is the magnitude to be expected from an adsorbed molecular layer. We suggested that this limit can be extended through the careful choice of the excitation wavelength.

In the balance of this paper, we consider a number of examples where the adsorbed or

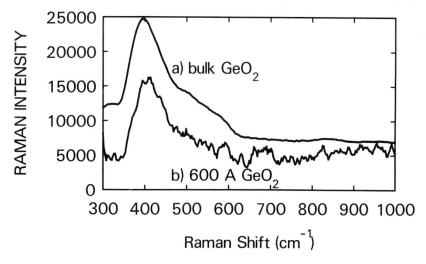

Fig. 5. a) The Raman spectrum of bulk GeO$_2$. b) The Raman spectrum due to the subtraction of the spectrum of Ge(100) from the spectrum of 600 A of a thermal oxide grown at 600 C on Ge(100). The spectra were obtained using a cooled vidicon detector.

reacted surface layer can be considerably thinner than 100 A. We demonstrate the multichannel optical detectors shown in Fig. 1 have the ability to detect Raman scattering from surface layers with average thicknesses of the order of a few atomic layers, an essential requirement if the technique is to be useful in the study of dynamics at surfaces.

In Fig. 6a we show the low energy Raman scattering of a Si(100) wafer. The spectra was excited by 50 mw of 5145 A light in p-polarization with the scattered light unpolarized. Fig. 6b is the Raman spectrum of a Si(100) surface cleaned, covered by 5 A of Pt and reacted at 500 C for 30 minutes in ultra high vacuum.[29] Fig. 6c is the Raman spectrum of Si(100) covered by 15 A of Pt and reacted under the same conditions as 6b. All three spectra are normalized against the intensity of the second order Raman scattering from the Si substrate. The data were obtained using the cooled vidicon detector. There is no significant background emission in the spectral range of interest so we can make full use of the vidicon sensitivity. The Raman spectra of the Si surfaces reacted with 5 and 15 A of Pt show additional Raman scattering which can be attributed to the vibrational modes of PtSi.[29,30] This is done through comparisons made with Raman spectra obtained from thick, reacted films of Pt on Si which were independently characterized as PtSi by electron microscopy and Rutherford Backscatter-

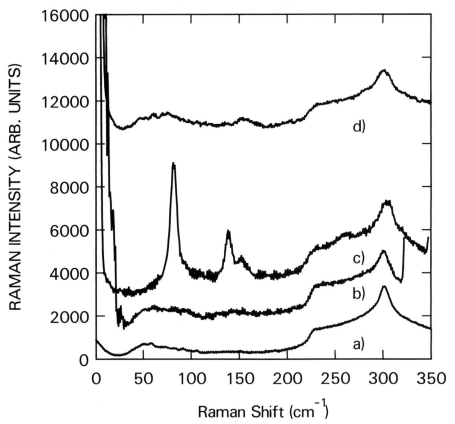

Fig. 6. The Raman spectra obtained with a vidicon detector from Si(100) surfaces covered by thin layers of Pt. a) The Raman spectrum of Si(100). The bands above 200 cm^{-1} are due to two phonon scattering. b) The Raman spectrum of Si(100) with about 10 A of PtSi on the surface. c) The Raman spectrum of Si(100) with about 30 A of PtSi on the surface d) The Raman spectrum of Si(100) after 15 A of Pt was deposited at 300 K.

ing. Fig. 6d is the Raman spectrum of a Si(100) surface prepared in ultra high vacuum and covered by 15 A of Pt at room temperature. The sample was not treated at elevated temperatures. There are broad bands near 80 cm^{-1} and 150 cm^{-1} which are not present in the substrate spectrum. Electron microscopy detects only unreacted Pt while electron spectroscopy observes evidence of an initial reaction between the Pt and the Si.[31] The Raman spectrum in Fig. 6d shows the existence of a poorly crystallized, initial reacted phase, even in the absence of reaction at elevated temperatures.

Fig. 7 shows the Raman spectra obtained from a Si(100) surface covered by 160 A of Pt and processed at different temperatures. The sample used in Fig. 7a was held at 200 C for 30 minutes. Electron microscopy sees no evidence for a well defined reacted phase and this is consistent with the Raman spectrum in Fig. 7a which shows no evidence for any well defined structure similar to that seen for the fully reacted samples in Fig. 6. The sample in Fig. 7b was annealed at 230 C for 480 minutes. Electron microscopy showed a thin 40 A layer of reacted material under a thick layer of unreacted Pt. The Raman spectrum shows structure at 82 and 140 cm^{-1} which are the characteristic long wavelength optic modes of PtSi seen in Fig. 6. Fig. 7c was obtained from a sample held at 300 C for 30 minutes. The 160 A Pt layer was completely reacted to 320A of PtSi by this treatment. The increase in the thickness of the reacted PtSi layer can be gauged by the change with thermal processing in the relative intensities of the PtSi scattering below 150 cm^{-1} and the second order Si substrate scattering between 200 and 350 cm^{-1}. While the substrate scattering is considerably stronger than the PtSi scattering in Fig. 7b, the PtSi scattering dominates the substrate scattering in Fig. 7c. In

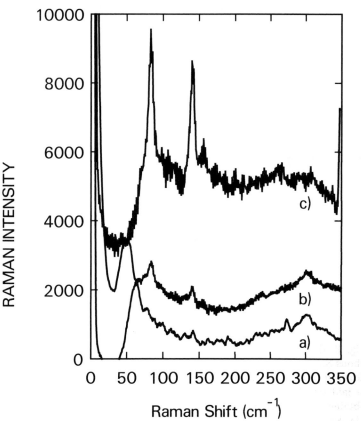

Fig. 7. The Raman spectra of Si(100) covered by 160 A of Pt and reacted by different thermal treatments. a)Sample reacted at 200 C for 30 minutes. Electron microscopy shows Si and unreacted Pt. b)Sample reacted at 230 C for 480 minutes. Electron microscopy shows Si, 40 A of PtSi and unreacted Pt.c)Sample reacted at 300 C for 30 minutes.

this experiment we take advantage of the fact first order Raman scattering is forbidden from metals with simple cubic structures so there is no emission from the unreacted metal. These results show clearly that we have the ability to observe reacted species at a buried semiconductor to metal interface.

Reactions at a surface can produce changes in the surface inaddition to the formation of new chemical species at the surface. They can produce changes in the surface morphology and order. Raman scattering is of course very sensitive to both of these effects.

Although there are many different phonon modes in a crystalline solid, the Raman spectrum usually shows only a few lines. For example, a silicon sample will have 6N vibrational modes where N is the number of atoms but Fig.1 shows only one first order Raman line. This is due to the long wavelength selection rule for Raman scattering in crystalline solids. Because of the requirement for the conservation of momentum, the momentum of the phonon created by Raman scattering must be equal to the difference between the momenta of the incident and scattered photons. Since the photon wavelength is much longer than the atomic spacing, for crystalline solids, only phonons near the zone center can be excited by Raman scattering. If there is no long range order in the solid, this selection rule is inapplicable and the Raman spectrum reflects the full density of states of the vibrational excitations of the scatterer.[32]

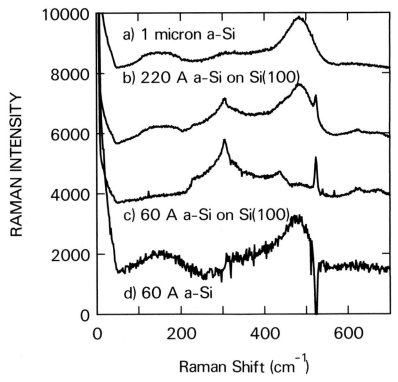

Fig. 8 a) The Raman spectrum for frequency shifts below 700 cm⁻¹ of hydrogenated amorphous Si. b) The Raman spectrum of a sample with 220 A of hydrogenated amorphous Si laid down on Si(100) c) The Raman spectrum of a sample with 60 A of hydrogenated amorphous Si laid down on Si(100). d) The Raman spectrum of 60 A of hydrogenated amorphous Si obtained by subtracting the Raman spectrum of the Si(100) substrate from the spectrum in Fig. 8c.

Fig. 8, demonstrates the sensitive of Raman scattering and the microchannel photo-multiplier detector to this effect when it occurs on a Si surface. A chemically etched Si(100) surface was exposed to a plasma deposition system which deposited thin layers of hydrogenat-ed amorphous Si. Fig. 8a is the Raman spectrum of an optically dense layer of hydrogenated amorphous Si. The Raman spectrum was excited by s-polarized 5145 A light. The polariza-tion of the scattered light was parallel to that of the incident light so that the normal first order phonon scattering from Si could be suppressed. The Raman spectrum of amorphous Si is strikingly different from that of crystalline Si due to the breakdown of the long wavelength Raman selection rule. Instead of a single sharp line at 520 cm^{-1}, the amorphous spectrum shows broad bands which reflect the vibrational density of states of amorphous Si. Fig. 8b is the Raman spectrum of 220 A of amorphous Si on Si(100) and Fig. 8c is the spectrum obtained from 60 A of amorphous Si on Si(100). Fig. 8d shows the spectrum obtained by subtracting Fig. 8c from the Raman spectrum of Si(100). Since our signal to noise in Fig. 8 is better than 10, we have the ability to detect an amorphous Si layer less than 10 A thick on top of a crystalline Si substrate.

In Fig. 9a shows the Raman spectrum for frequency shifts between 1700 and 2300 cm^{-1} of a hydogenated amorphous silicon sample about 400 A thick. The structure near 2000 cm^{-1} is due to the Si-H stretch vibration of the hydrogenated amorphous Si. Fig. 8b was obtained by measuring the Raman spectrum from about 30 A of hydrogenated amorphous Si layed down on Si(100) and subtracting the optical emission due to the uncoated substrate. The H concentration in the 30 A sample is of the order of 15 atomic per cent. Since the 30 A amorphous Si film consists of about 10 atomic layers of Si atoms, this shows we have the ability to observe atomic vibrations of species at the monolayer level.

CONCLUSIONS

We have demonstrated the ability of Raman scattering using either cooled vidicon detectors or microchannel plate photomultipliers to detect the vibrational modes of weakly

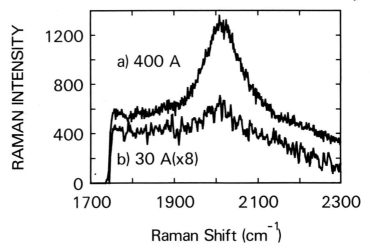

Fig. 9 a) The Raman spectrum due to S-H stretch vibrations of a 400 A thick hydrogenated amorphous Si film. b) The Raman scattering of a 30 A thick film of hydrogenated amorphous Si. The signal due to the Si(100) substrate has been subtracted. The hydrogen concentration is about 15 atomic percent.

scattering thin films on semiconductors. Given the Raman scattering efficiencies of these films, we show that Raman scattering can be used to detect the vibrational modes of adsorbates on surfaces. The principle limitation of the technique is the strength of the optical emission from the substrate which can dwarf the surface scattering. In the cases considered in this paper, the substrate emission was due to symmetry allowed second order phonon scattering and symmetry forbidden first order scattering. The direct subtraction of the substrate emission from the Raman spectra obtained from the surface layer plus the substrate was generally sufficient to produce usable spectra. Substantial improvements in the experimental spectra would be expected if the data were accumulated by obtaining differences over relatively short integration times and then summing the resulting difference spectra.

ACKNOWLEDGEMENT: I wish to acknowledge the assistance of many individuals. D. Checkowski of ITT, Fort Wayne provided essential information on the operation of the MEPSICON detector while Dr. F. Paresce of the Space Telescope Laboratory provided a valuable preprint on this detector. At the IBM Research Center, Yorktown Heights, New York, Ph. Avouris furnished the SIT vidicon which allowed the initial stages of this work while J. R. Kirtley provided the MEPSICON detector with its associated electronics. Both of the above as well as G. Rubloff, R. Matz, A. Reisman and I. Haller furnished samples and many helpful suggestions.

References.

1. J. E. Demuth and Ph. Avouris, Physics Today 36, no.11, 62 (1983).
2. R. G. Greenler and T. L. Slager, Spectrochimica Acta A29, 193 (1973).
3. M. Udegawa, C. C. Chou, J. C. Hemminger and S. Ushioda, Phys. Rev. B23, 6843 (1981).
4. Surface Enhanced Raman Scattering ed. by R. K. Chang and T. E Furtak (Plenum Press, New York 1982).
5. J. E. Rowe, C. V. Shank, D. A. Zwemer and C. A. Murray, Phys. Rev. Lett. 44,1770 (1980).
6. P. B. Dorain, K. U. von Raben, R. K. Chang and B. L. Laube, Chem. Phys. Lett. 84, 405 (1981).
7. D. P. Jeanmaire and R. P. Van Duyne, J. Electroanal. Chem. 84, 1 (1977).
8. J. C. Tsang and J. R. Kirtley, Solid State Commun. 30, 617 (1979).
9. A. Campion, J. K. Brown and V. M. Grizzle, Surf. Sci. 115 L153 (1982).
10. J. C. Tsang, Ph. Avouris and J. R. Kirtley, J. Chem. Phys. 79, 493 (1983).
11. F. W. King, R. P. VanDuyne and G. C. Schatz, J. Chem. Phys. 69, 4472 (1978).
12. S. S. Jha, J. R. Kirtley and J. C. Tsang, Phys. Rev. B.22, 3973 (1980).
13. A. Otto, Surf. Sci. 92, 145 (1980).
14. R. K. Chang and M. B. Long in Topics in Applied Physics- Light Scattering in Solids II ed. by M. Cardona and G. Guntherodt (Springer, Berlin 1982).
15. D. Rees, I. McWhirter, P. A. Rounce and F. E. Barlow J. Phys. E: Sci. Instrum. 14, 229 (1981). C. Firmani, E. Ruiz, C. W. Carlson, M. Lampton and F. Paresce, Rev. Sci. Instrum. 53, 570 (1982).
16. R. J. Nemanich, C. C. Tsai and G. A. N. Connell, Phys. Rev. Lett. 44, 273 (1980).
17. R. J. Dornhaus, M. B. Long, R. E. Brenner and R. K. Chang, Surf. Sci. 93, 240 (1980).
18. F. J. Grunthaner, P. J. Grunthaner, R. P. Vasquez, B. F. Lewis and J. Maserjian, J. Vac. Sci. Technol. 16, 1443 (1979).
19. N. Wada and S. A. Solin, Physica 105B, 353 (1981).
20. P. E. Schoen and H. Z. Cummins in Light Scattering in Solid, ed. by M. Balkanski (Flammarion, Paris 1971).

21. R. P. Van Duyne in Chemical and Biological Applications of Lasers, ed. by C. B. Moore (Academic Press, New York 1979).

22. F. L. Galeener and J. C. Mikkelson, Solid State Commun. 36, 983 (1980)

23. R. M. Martin and F. L. Galeener, Phys. Rev. B23, 3071 (1981).

24. R. Loudon, Advances in Phys. 13, 423 (1964).

25. M. Cardona in Topics in Applied Physics-Light Scattering in Solids II ed. by M. Cardona and G. Guntherodt (Springer, Berlin 1982).

26. M. A. Renucci, J. B. Renucci, R. Zeyher and M. Cardona, Phys. Rev. B10, 4309 (1974).

27. J. B. Renucci, R. N. Tyte and M. Cardona, Phys. Rev. B11, 3885 (1975).

28. F. L. Galeener, A. J. Leadbetter and M. W. Stringfellow, Phys. Rev. B27, 1052 (1983).

29. J. C. Tsang, Y. Yokota, R. Matz and G. Rubloff, Appl. Phys. Lett. 44, 430 (1984).

30. R. J. Nemanich, T. W. Sigmon, N. M. Johnson, M. D. Moyer and S. S. Lau, in Laser and Electron-Beam Solid Interactions and Materials Processing ed. by J. F. Gibbons, L. D. Hess and T. W. Sigmon (North Holland, New York, 1981), p. 541.

31. R. Matz, R. J. Purtell, Y. Yokota, G. W. Rubloff and P. Ho, J. Vac. Sci. Technol. (to be published).

32. R. Shuker and R. W. Gammon, Phys. Rev. Lett. 25, 222 (1970).

RAMAN OPTICAL ACTIVITY OF MOLECULES ADSORBED ON METAL SURFACES

S. Efrima*
Department of Chemistry, Ben-Gurion University of the Negev
Beer-Sheva, Israel

*Bat-Sheva Fellow

ABSTRACT

   The effect of possible large electric field gradients near metal
surfaces on the Raman Optical Activity (ROA) of adsorbed molecules is
discussed. It is shown that large gradients would cause unusually
high values of the measured circular intensity differences, orders of
magnitude higher than those measured in solution. ROA may be used to
investigate the existence of field gradients at surfaces and their
magnitude. Molecular structural information may also be extracted from
the combination of the vibrational spectroscopy and the chiral
sensitivity.

I. INTRODUCTION

   Sass, Neff, Moskovits and Holloway[1a] and Moskovits and DiLella[1b]
have suggested that local large field gradients near a metal surface
may have important effects on the spectroscopical behaviour of adsorbed
molecules. In particular, they discussed the effect on selection rules
in surface enhanced Raman scattering (SERS) and also in infrared measurements. However, at this stage, one cannot unequivocally attribute
the appearance of usually symmetry forbidden bands[1], to this effect
alone. Reduction of symmetry for specific adsorption geometries[1b] is
an alternative explanation. Moskovits and DiLella consider this point
very carefully, comparing the behaviour near silver and lithium, but
could not come up with an unambiguous answer.
   The phenomenon of Raman Optical Activity (ROA) is strongly dependent on electric field gradients (and magnetic fields, as well). Here
chiral molecules exhibit different intensities for the Raman scattering
of left and right circularly polarized light.
   ROA measurements were conducted in several laboratories[2] for molecules in solution and were shown to be feasible and informative in
spite of the small effect measured (~ 0.1% of a normal Raman signal).
   I propose that such measurements are feasible for molecules adsorbed on metal surfaces, as well, in spite of the small number of

B. Pullman et al. (eds.), Dynamics on Surfaces, 395–400.
© *1984 by D. Reidel Publishing Company.*

scattering molecules. Furthermore, I maintain that such measurements
contain information regarding local electric fields, their gradients
and, in general, local dielectric properties of the metal-air (or
metal-solution) interface. In the cases where surface enhancement of
the Raman scattering is apparent it is not more difficult to obtain a
spectrum of an adsorbed molecule than it is to measure Raman scattering
in solution. Therefore, using modern Raman equipment, it should not
be prohibitively difficult to measure surface ROA spectra.

More important, on the metal surface or near it one may expect
large changes in the relative crossections for ROA and regular Raman
scattering which may lead to circular intensity difference values
(c.i.d.'s, Δ) larger than the usual small value of approximately
0.1%, by orders of magnitude. Thus if $\Delta \sim 1$ (instead of 10^{-3}) one
may now-a-days detect the ROA even for small (or no) surface
enhancements[3].

In this paper I discuss one possible mechanism which may lead to
large effects on the measured c.i.d.'s, provided large field gradients
exist near a metal surface.

II. THE EFFECT OF ELECTRIC FIELD GRADIENTS ON ROA

The scattering process of light off a system composed of molecules
adsorbed on a metal surface is a complex one. All components of the
system may scatter the light and their scattering is modified by the
presence of the other components.

We concentrate on the scattering of a single chiral molecule adsorbed on a metal surface. The molecule senses an electric field composed of the "direct" field, that of the incoming light, and a
"reflected" field due to the metal surface. Near the metal surface the
"reflected" field can be quite complex and will in general be composed
of three components: a radiative part, given by the Fresnel reflection
coefficients[4], an evanescent, near-field contribution, which changes
on a range of several nanometers for light in the visible or UV; and
finally there may be a short range field, of microscopic origin and
dying off over molecular distances[5], which may result in extremely
high field gradients.

The total field and field gradients induce a dipole in the molecule, oscillating with the frequency of the light or manifesting the
beating due to internal molecular motion (vibrational for instance).
The dipole is given by[6]

$$\mu = \alpha E + \frac{1}{3} A(VE) + GB; \tag{1}$$

where μ, E and B are the molecular dipole, local electric field and
magnetic field respectively; α, A and G are the normal polarizability, the electric dipole-electric quadrupole polarizability and the
electric dipole-magnetic dipole polarizability tensors, the last two
terms give rise to ROA. We dropped the vectorial directional symbols
as we are interested here only in order of magnitude behaviour.

A is of the order of α multiplied by a molecular size, r_M. Thus

$$\frac{A(\nabla E)}{\alpha \dot{E}} = 0 \left(\frac{(\nabla E)}{E} r_M \right), \tag{2}$$

where expression $= 0(x)$ means that the order of magnitude of the expression on the ℓ.h.s. is x.

For radiative fields

$$\nabla E = 0 \left(\frac{2\pi}{\lambda} E \right), \tag{3}$$

thus usually

$$A(\nabla E)/\alpha E = 0 \ (2\pi r_M/\lambda) \approx 10^{-3}. \tag{4}$$

The third term, GB, is also of the same order of magnitude as for radiative fields, $\underset{\sim}{B} = \underset{\sim}{\nabla} \times \underset{\sim}{E}$.

These two last terms, by interference with the field scattered by the first term (αE), are the origin of ROA in solution[2]. As can be seen they should give c.i.d.'s of the order of 10^{-3}.

On the surface of a metal, if large field gradients occur, this is drastically changed. Here one expects

$$\nabla E = 0 \ (E'/r_M) \tag{5}$$

where the field E' is the short range contribution to the total field and which may be of the same order of magnitude as the long range part, or even stronger.

As a result one immediately finds that the expected c.i.d.'s are of the order of unity

$$\nabla = 0 \ (1) \tag{6}$$

Of course, if the molecule is located in a region where the short range, large gradient, contribution is small, the c.i.d.'s will approach that of the non surface, bulk case, i.e. 10^{-3}.

Unless there is large enhancement of the magnetic fields near the metal surface the $G \cdot B$ terms will be negligible. If the near zone fields are the main contributors to the Raman scattering then usually the magnetic contribution is much smaller than the electric fields[7]. Therefore, for simplicity, we neglect the magnetic terms. For the same reason we chose to overlook, in this paper, the influence of the magnetic moment and quadrupole terms. These terms usually determine the ROA together with the A and G contribution to the induced dipole μ.

II. DISCUSSION

Moskovits and DiLella present results for SERS spectrum of benzene and benzene-d_6 on a silver film[8a] and for benzene-1,3,5-d_3[1b]. Lund,

Smardzewski and Tevault report similar results for sodium[8b]. The former discuss the appearance of bands which are forbidden for the isolated molecule. If indeed field gradients were operative (which is one possibility raised by the authors[1b]) then unusually strong ROA signals would be expected.

In the cases investigated one is not even expected to find the same ratio of ROA of the various active vibrational modes, as in solution. This conclusion stems from the unisotropic nature of the interface, the common "frozen" adsorption configurations, and the large changes in the relative importance of the various terms leading to the ROA. For instance, if indeed large gradients are the dominating feature then the magnetic and quadrupole contributions will be negligible and only the A term in the induced dipole would affect the results. Furthermore, the A and A^2 terms may become comparable. These differences will necessarily change the relative optical activity of the various modes.

Large electric field gradients are not the only possible surface features which can affect the ROA. There are several other mechanisms which can bring about considerable changes in the measured c.i.d.'s.

Consider, for instance, the polarizing effect of the metal surface, which is known to change the ellipticity of the radiation near the surface. In particular, circularly polarized incident radiation, when superimposed with the reflected beam turns into elliptically polarized radiation. Elliptical polarization can be viewed as a combination of linearly polarized light and circularly polarized light of of a smaller amplitude than that of the original circularly polarized light. This effect alone, tends to diminish the ROA c.i.d.'s.

Fortunately such effects can be factored out as it is expected to manifest itself similarly for all the optically active modes of the chiral molecule. Furthermore, the two enantiomers of the same chiral molecule will exhibit the same ROA with the same sign, if this polarizing influence is the only effect.

A trivial, perhaps amusing, surface effect of this kind can be seen by considering a perfect mirror. In this case, a circularly polarized light is changed near the metal surface into plane polarized light leaving no circularity. Thus all the terms proportional to the field itself (electric or magnetic) do not contribute to the ROA. In this respect the system will behave like a racemic mixture composed of the chiral molecule and its "image."

For a non-perfect mirror, such as roughened silver in the visible, partial effects of this kind are expected. They will be global effects acting approximately similarly on all vibrational modes and exactly the same for two enantiomers. These effects may be used to study the local dielectric response of the metal.

Another possible mechanism which may affect the surface ROA is "image" effects, similar to those proposed for the SERS enhancement mechanism[9]. Those effects are controversial[10] but cannot be ruled out as major effects in the SERS problem[11]. A and G may be enhanced in principle in the same way α is increased according this model. Larger ROA is expected if the enhancement of the former two is larger than that of α itself, as can happen in a charge transfer model.

Another possible mechanism for producing large ROA is the local magnetic field enhancement, similar to the electric field enhancement proposed in SERS[12]. As the amplification of the magnetic field is not expected to be larger than that of the electric field, one does not in this case expect unusual Δ values. In fact, for near fields the magnetic field enhancement is much smaller than that of the electric field.

In conclusion, it seems that measurement of ROA of molecules adsorbed on metal surface, expecially in conjunction with SERS, may be a new method to extract molecular information and to study electric field gradients and local dielectric properties near metal surfaces.

REFERENCES

1. a) J.K. Sass, H. Neff, M. Moskovits and S. Holloway, J. Phys. Chem. 85, 621 (1981);
    b) M. Moskovits and D.P. DiLella, in "Surface Enhanced Raman Scattering", R.K. Chang and T.E. Furtak, eds., (Plenum Press, 1982) p. 243;
2. For reviews see:
    a) L.D. Barron and A.D. Buckingham, Mol. Phys. 20, 1111 (1971);
    b) L.D. Barron and A.D. Buckingham, Ann. Rev. Phys. Chem. 26, 381 (1975);
    c) L.D. Barron, in "Molecular Spectroscopy", A Specialist Periodical Report, Vol. 4, R.F. Barrow, D.A. Long and J. Sheridan, Eds., The Chemical Society London, 1976, p. 96;
    d) L.D. Barron, in "Optical Activity and Chiral Discrimination", S.F. Mason, Ed., (Reidel, Dordrecht, 1979);
    e) L.D. Barron, Acc. Chem. Res. 13, 90 (1980);
3. a) A. Campion and D.R. Mullins, Chem. Phys. Lett. 94, 576 (1983);
    b) J.C. Tsang, Ph. Avouris and J.R. Kirtley, Chem. Phys. Lett. 94, 172 (1983);
4. M. Born and E. Wolf, "Principles of Optics" (Pergamon Press, 1975);
5. a) P.J. Feibelman, Phys. Rev. Lett. 34, 1092 (1975); Phys. Rev. B12, 1319, 4282 (1975); Phys. Rev. B14, 762 (1976);
    b) T. Maniv and H. Metiu, Phys. Rev. B22, 4731 (1980); J. Chem. Phys. 76, 696, 2697 (1982);
6. A.D. Buckingham, in "Advances in Chemical Physics", Ed., J.O. Hirschfelder, Vol. XII, 107 (1967);
7. J.D. Jackson, "Classical Electrodynamics", (John Wiley, 1975), p. 395.
8. a) M. Moskovits and D.P. DiLella, J. Chem. Phys. 73, 6068 (1980);
    b) P.A. Lund, R.R. Smardzewski and D.E. Tevault, Chem. Phys. Lett. 89, 508 (1982);
9. a) S. Efrima and H. Metiu, Chem. Phys. Lett 60, 59 (1978); J. Chem. Phys. 70, 1602 (1979); J. Chem. Phys. 70, 1939 (1979); J. Chem. Phys. 70, 2297 (1979); Israel J. of Chemistry 18, 17 (1979); Surf. Sci. 92, 427 (1980);
    b) F.W. King, R.P. Van Duyne and G.C. Schatz, J. Chem. Phys. 69, 4472 (1978); S.G. Schultz, M. Janik-Czachor and R.P. Van Duyne,

Surf. Sci. $\underline{104}$, 419 (1981); G.C. Schatz and R.P. Van Duyne, Surf. Sci. $\underline{101}$, 425 (1980);

10. a) M. Kerker, D.S. Wang and H. Chew, Appl. Opt. $\underline{19}$, 3373, 4159 (1980);

 b) J. Gersten and A. Nitzan, J. Chem. Phys. $\underline{73}$, 3023 (1980); J. Chem. Phys. $\underline{75}$, 1139 (1981);

 c) P.R. Hilton and D. Oxtoby, J. Chem. Phys. $\underline{72}$, 6346 (1980);

11. S. Efrima, in "Modern Aspects of Electrochemistry", J.O'M. Bockris, R. White and B.E. Conway, eds., (Plenum Press), to be published;

12. a) M. Moskovits, J. Chem. Phys. $\underline{69}$, 4159 (1978);

 b) C.Y. Chen and E. Burstein, Phys. Rev. Lett. $\underline{45}$, 1287 (1980);

 c) J. Gersten, J. Chem. Phys. $\underline{72}$, 5779 (1980);

 d) G.W. Ford and W.H. Weber, Surf. Sci. $\underline{109}$, 451 (1981);

 e) F.J. Adrian, Chem. Phys. Lett. $\underline{78}$, 45 (1981).

RAMAN AND INFRARED SPECTROSCOPY
OF MOLECULES ADSORBED ON METAL ELECTRODES

M. R. Philpott, R. Corn,* W. G. Golden, ** K. Kunimatsu† and H. Seki

IBM Research Laboratory, 5600 Cottle Road
San Jose, California 95193 U.S.A.

*Visiting scientist. Present address: Department of Chemistry, Swarthmore College, Swarthmore, Pennsylvania 19801.
**IBM Instruments Inc., 40 West Brokaw Road, San Jose, California 95110.
†Visiting scientist. On leave from Research Institute of Catalysis, Hokkaido University, Sapporo, Japan.

ABSTRACT: The use of Raman and infrared spectroscopy to obtain the vibrational spectra of molecules and ions adsorbed at the metal electrode-aqueous electrolyte interface are described. To obtain Raman spectra, we use surface polaritons and roughness to enhanced spectral intensities. In the infrared, we use Fourier transform and modulated reflection-absorption techniques. Raman results for fatty acids, chloride, and thiocyanate on silver electrodes are presented, together with a combined Raman and infrared study of cyanide adsorbed on silver.

I. INTRODUCTION

The vibrational spectra of molecules and ligands adsorbed at solid-liquid or solid-polymer interfaces can provide valuable information about the structure and functional properties of the interphasial region. In the last few years, techniques have been developed and phenomena discovered that make it possible to detect the IR and Raman of multilayers and even monolayers of adsorbates at these interfaces. Some examples obtained by the IBM San Jose group are described here with particular emphasis on metal electrode-aqueous electrolyte interfaces.

II. SURFACE RAMAN SPECTROSCOPY

In vibrational Raman spectroscopy, a laser is directed at the sample and the scattered light analyzed for changes in frequency corresponding to the subtraction (Stokes scattering) or addition (anti-Stokes scattering) of a vibrational frequency. Ordinary Raman scattering is a very weak process, since

401

B. Pullman et al. (eds.), Dynamics on Surfaces, 401–412.
© 1984 by D. Reidel Publishing Company.

only about one photon in 10^{10} is inelastically scattered. Straightforward calculations indicate that a molecular monolayer should give rise to a very small signal approximating 0.1 photons per cm^{-1} sterradian watt cm^2 at optical wavelengths [1]. For this reason, Raman spectroscopy was largely ignored as a tool for studying interfaces. However, in recent years, this picture has changed because of several developments. The first of these was the discovery of surface enhanced Raman scattering (SERS) from electrochemical interfaces [2-4], a variety of rough metal surfaces in vacuum and from colloidal particles [5]. The second development has been the commercial marketing of multiplex (or array) detectors which allow the recording of large spectral segments of Raman scattered light from unenhanced systems [6]. The increased theoretical understanding of electromagnetic field strengths at metal surfaces resulting from these studies has also lead to the use of surface polaritons to enhance Raman scattering processes on flat surfaces [7]. To increase the rate of Raman scattering from surface species, one can either attempt to increase the electromagnetic field amplitude or increase the cross section by tinkering with the electronic structure (*e.g.*, create a resonance) or bonding (*e.g.*, increase the polarizability) of the adsorbate [8].

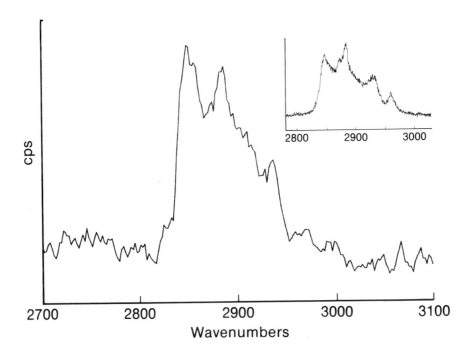

Fig. 1. PSP-enhanced spectra of the CH stretching region of 10 monolayers of cadmium arachidate at a thin silver film. Depicted in the inset for comparison is the Raman spectra from 28 monolayers deposited onto a silver grating (inset spectrum taken from Reference 11).

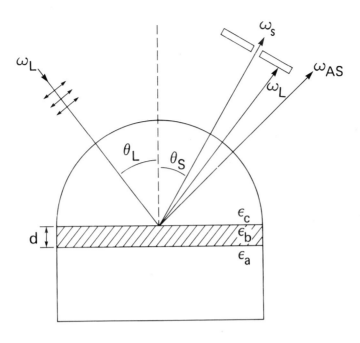

Fig. 2. Schematic diagram showing the plasmon surface polariton scattering geometry.

   We describe first our method of physically enhancing the light intensity at the surface by using plasmon surface polaritons as intermediaries in the scattering process.   These electromagnetic interface modes can have field amplitudes up to an order of magnitude greater than in the case of freely propagating vacuum photons. Consequently, enhancements of 10^4 in the cross section are possible provided surface polaritons are used as input and output channels [9-10]. This can be done by exciting surface waves on Ag, Cu or Au by grating or prism coupling [9-12].   Although grating coupling has many advantages, one drawback is the way the surfaces are made, namely, using photoresist.   The resist materials is a potential source of contamination in electrochemical cells. Consequently, grating electrodes are not too convenient in electrochemical applications. On the other hand, thin films of silver on prisms can be used as working electrodes without fear of contamination since there is no photoresist layer. Figure 1 shows a Raman spectrum of ten monolayers of cadmium arachidate deposited on the flat silver film by the Langmuir-Blodgett dipping technique [9].   Figure 2 shows the experimental arrangement in a schematic fashion. In this experiment, a laser photon excites a surface polariton which undergoes Raman scattering.   The scattered polaritons are then coupled out through the prism. The greater the Raman shift, the smaller is the scattering angle θ_S. In practice, $|\theta_L - \theta_S|$ is approximately a few degrees so that a pin hole or slit must be used to block that part of the specular beam not adsorbed by the metal film. Rough surfaces are a serious source of scattered light at all angles too, so that the collected light must be analyzed using a conventional Raman spectrometer [9] in order to adequately reject scattered laser light.

If one tried to perform the same experiments without a grating or prism coupler, then no Raman spectrum would be detected using a conventional scanning instrument. As mentioned before, the reason for this is the absence of field enhancement due to surface plasmons. However, if one deposits on top of the Langmuir film an island film of silver, then a Raman spectrum is detectable [13,14]. The appearance of Raman scattering in this case is attributed to the presence of high electromagnetic fields around the silver particles due to localized plasma modes that can be driven at optical frequencies. These islands look very rough under SEM and mimic some aspects of electrodes that exhibit SERS.

Surface enhanced Raman scattering has received a lot of attention. As an area of study, it is beginning to mature, i.e., attract lifelong devotees; however, as a phenomenon, it still remains somewhat mysterious because there seem to be several parallel enhancement mechanisms that can operate to different degrees depending on the chemical and physical characteristics of the interface. These mechanisms are roughly divisible into physical (e.g., EM field enhancements at asperities or cavities) and chemical ones due to bonding that strongly couples metal and molecular polarizabilities [5,8]. In addition, there are electronic resonance effects that are voltage and wavelength dependent [15].

For silver electrodes, the metal must be subjected to an oxidation reduction cycle (ORC) that roughens the surface, creating structures capable of supporting localized plasma oscillations with frequencies in the visible. The union of these frequencies spans the visible so that at all wavelengths a SERS process occurs efficiently enough from some point on the surface for signals to be readily detected.

With multiplex detectors, one can record spectra on the millisecond time scale. By way of example, we replot some of the data published in Ref. 16. Figures 3 and 4 show the time development of SERS during an ORC. The system was Ag in contact with 0.1M NaCl and 0.01M NaSCN. The spectra in the fundamental metal-ligand M-L stretch around 250 cm^{-1} and the C-N stretch at 2100 cm^{-1} of the SCN^- moiety are shown separately. Detailed examination of these spectra shows a number of interesting features including displacement of Cl^- by more strongly adsorbing (but present in tenfold less concentration) SCN^- and the narrowing of the active site distribution as shown by the sharpening of the C-N stretching vibration [16].

More recently, we have looked very close to the laser line. At $\Delta\nu \sim 10$ cm^{-1}, there is an intense band of somewhat mysterious origins. One suggestion is that it is a particulate mode, i.e., a vibration due to a surface

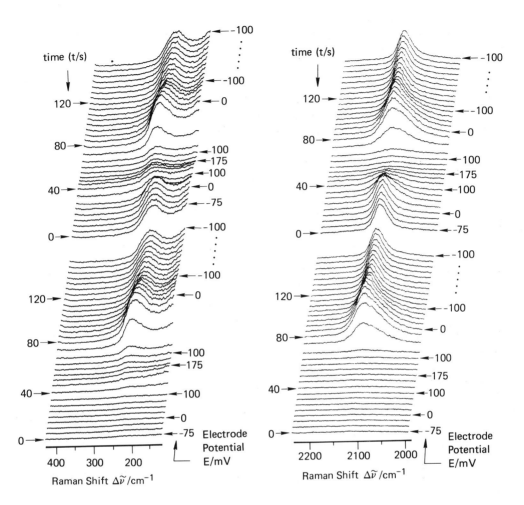

Fig. 3. (left). Time-development of SERS spectrum of a polycrystalline silver electrode in 0.1M NaCl+0.01M NaNCS during two sequential linear scan ORC from −100 mV to 175 mV and back to −100 mV at 5 mV/s. Spectra shown are for the metal-ligand bond stretching mode ν_{ML}. Time at the beginning of each accumulation period shown on left-hand side. Electrode potential at the end of each accumulation period is shown on the right. Laser 530.9 nm, p-polarized, 200 mW. OMA accumulation period was 5 seconds for each spectrum.

Fig. 4. (right). Same electrochemical system as Figure 3. Spectra cover the range 2000 to 2200 cm^{-1} which includes the C-N stretching mode ν_{CN} of the adsorbed thiocyanate ion.

geometrical feature that contains or interacts or supports the localized plasma mode [17-18]. Experiments to understand this mode better have shown that to a first approximation at least it behaves just like the vibrational modes of molecular adsorbates (like pyridine, for example) [19]. This implies that it is located in the vicinity of the active site itself, and is not associated solely with either a particulate or plasma mode as suggested elsewhere [17,18].

III. SURFACE INFRARED SPECTROSCOPY

The theory on which the experimental techniques are based was developed by Greenler and others [20], who were first to note the advantage of using light near grazing incidence and the large difference in the absorption coefficient between light polarized perpendicular to the plane of reflection and that polarized parallel. Both FTIR and dispersive IR at grazing incidence were used to obtain vibrational spectra of thin films [21], Langmuir-Blodgett monolayers [14,22-25] and monolayers adsorbed on metal surfaces in ultrahigh vacuum systems [26-33]. Polarization modulation has also been explored by a number of workers as a way to improve the dynamic range of their experiment [34,35]. The spectra may be normalized to compensate for variation in intensity with wavelength if a double modulation technique is used, *i.e.*, combining polarization modulation with a second modulation scheme. By this technique, spectra of surface species may be obtained even in the presence of gas phase adsorbents. Infrared reflection absorption spectroscopy (IRRAS for short) has been demonstrated for both Fourier transform [36] and dispersion [29] spectrometers.

We have measured by Fourier transform spectroscopy the IRRAS spectrum of several systems including CO on Pt, and CN^- on Ag, Cu, Au and Pt. A thin layer electrochemical cell was used consisting of a cylindrical polycrystalline rod of metal encased in a Kel-F piston. The electrode was polished flat and moved to within a few microns of an optically flat CaF_2 prism through which the IR light was directed. The cell housing was Kel-F, the reference electrode was Ag/AgCl, and the counter electrode a Pt wire. By way of example, we describe here results for CN^- adsorbed on an Ag electrode [37]. The compositions of the electrolytes used were 0.1M KCN and 0.01M KCN in 0.1M K_2SO_4.

The IRRAS spectrum includes the absorption due to the surface species and the thin layer of solution in the immediate vicinity (ca. 30 nm) of the electrode surface as described in a previous paper [38]. Consequently, there is a large background absorption due mainly to the water molecules. It is, therefore, essential to have a high signal-to-noise ratio so that this background can be subtracted out to reveal the contribution from the surface species.

The results shown in Figure 5 were obtained in the following way. First, a background spectrum was recorded for the silver electrode at −1.0V immersed in a K_2SO_4 solution that did not contain any cyanide. Then cyanide was added such that a bulk concentration of 0.1N resulted. This was done with the electrode pulled back from the window. After mixing, the electrode was carefully returned to its original position. Then a series of spectra were taken with potentials between −1.4V to −0.3V and the cyanide free background

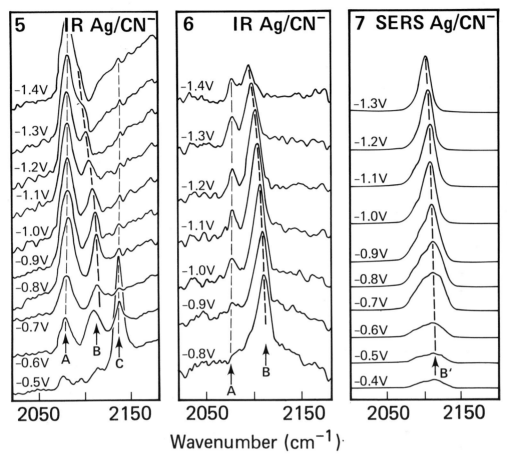

Fig. 5. (left). FT-IRRAS spectra for a solution with 0.1N K_2SO_4 and 0.1N KCN. These spectra are taken starting at −1.4V (*versus* Ag/AgCl). A background spectrum taken at −1.0V before adding the KCN is subtracted.

Fig. 6. (center). FT-IRRAS spectra for the same solution as for Figure 5 but using a spectrum taken at −1.5V (*versus* Ag/AgCl) as the background.

Fig. 7. (right). SERS spectra taken for a solution with 0.1N K_2SO_4 and 0.1N KCN after an ORC from −1.0 to 0.5V and back at 50 mV/sec.

subtracted. Notice that at potentials more negative than −0.8V, there are two distinct bands A and B, and for potentials more positive of −0.7V there is a third band C at 2136 cm^{-1}. For the lowest frequency mode A, at 2080 cm^{-1} the band position and intensity remain relatively constant with potential. In contrast, the mode B continuously shifts to smaller wavenumber continuously losing its intensity with increasingly negative potential. The 2080 cm^{-1} band A is assigned to the C-N stretching mode of the cyanide ions in the solution close to the electrode surface. (For a 0.1N solution, a 1000Å layer has 6×10^{14} solute molecules/cm^2). The prominent band B is assigned as a surface cyanide species since its position depends on the electrode potential and its intensity decreases as the electrode potential is made more negative. This is consistent with the surface ion desorbing as the electrode becomes negatively charged.

The surface species is desorbed at a sufficiently negative potential [39], and this suggests an alternative background subtraction method which was used to prepare the spectra shown in Figure 6. The spectra shown in Figure 6 were obtained by first taking a spectrum at −1.5V and using it as the background to subtract from the spectra taken at increasingly more positive potentials. A large part of the solution cyanide band A is eliminated by this process and the band B assigned to the surface species is more clearly revealed than in Figure 5. Notice that the bulk cyanide band is not completely subtracted. This may be due to the change in cyanide ion concentration in the diffuse layer with potential. As the potential is made more positive, the cyanide ion density increases giving more intensity and reaches a maximum between −1.3 and −1.2V. Thereafter, the band intensity decreases at more positive potential due to the consumption of the cyanide ions near the electrode to produce the adsorbed surface species.

Examination of the spectra displayed in Figure 5 at potentials more positive than −0.6V shows a third band C at 2136 cm^{-1} whose frequency is constant and whose intensity increases with positive potential. This band is assigned to the $Ag(CN)_2^-$ complex ion in aqueous solution. The formation of this species is accompanied by a decrease in the band B due to the surface species and a decrease in the solution cyanide peak C at 2080 cm^{-1}. At increasingly positive potential, around −0.6V, the silver atoms are oxidized consuming the solution cyanide to form $Ag(CN)_2^-$ complex in the solution which is consistent with the cyclic voltammetry of this system.

The peak position of surface cyanide varies linearly with potential with a slope of about 30 cm^{-1}/volt which is essentially identical to that measured for the strongest IR band of CO adsorbed on a platinum electrode [38]. The adsorption of CO on Pt has been the subject of many investigations, both in vacuum and in electrolytes. It is reasonably certain that the band at 2080 cm^{-1} is due to the linearly bonded CO. Comparisons with CO on Pt electrodes raises the following questions for which we currently do not have satisfactory answers: (1) What is the nature of the bonding of CN$^-$ to silver and how is the array of surface cyanide organized? (2) Why does the frequency shift 30 cm^{-1}/volt and is the similarity with isoelectronic CO on Pt significant? (3) How does the oxidation of surface CN$^-$ to $Ag(CN)_2^-$ take place (e.g., randomly, at defects or island edges). We expect that further work with isotopically labelled species and single crystal surfaces will answer some of these questions.

IV. COMPARISONS WITH SERS

We have also carried out Raman measurements in the same cell as used for the IR studies except that a fused quartz window replaced the CaF_2 one [40]. The angle of incidence of the 530.8 nm Kr laser line was about 60° and the collection of the scattered light was normal to the laser line. There is the experimental uncertainty that it is necessary to move the electrode away from the window to carry out the ORC and returning the electrode to the exact same position is not easy. The intensity measured is slightly affected by this factor. Nevertheless, the ORC, which is required for SERS, had little effect on the IRRAS spectra.

In comparing the IRRAS results with the SERS measurements [41,42] the behavior of the C-N stretching band in the negative potential region appears to be most similar. The shift in the peak position with potential in this region has been reported to be 26 or 28 cm^{-1}/volt [43,44]. In order to compare the actual band shape more closely, SERS spectra taken under conditions similar to those for the IRRAS measurements are shown in Figure 7. The ORC consisted of scanning the potential to 0.5V and back to −1.0V at 50 mV/sec. The SERS band is narrower at the more negative potentials and the peak position and peak shift (ca. 28 cm^{-1}/volt) is very close to those observed for the surface species with IRRAS.

Although there are similarities in the SERS and IRRAS spectra as noted above, there are also some significant differences. The intensity of the SERS band decreases more rapidly as the potential becomes more negative than −1.3V and the full width at half maximum is larger for SERS at all potentials (e.g., 20 cm^{-1} versus 10 cm^{-1} at −1.1V). The IR spectra return after polarizing the electrode negative whereas the SERS spectra are irreversibly lost if the potential is more negative than −1.5V. Generally, as the potential is made more positive, the SERS band shape becomes more complex and the overall bandwidth gets larger. At potentials more positive than −0.6V, the SERS band intensity decreases rapidly with increase in potential but there is no indication of bands at 2080 cm^{-1} or 2136 cm^{-1} due to cyanide and $Ag(CN)_2^-$ in solution as seen in the IRRAS spectra. Nor do we observe, as done by Benner et al. [42], the development of a very broad band peaked at 2140 cm^{-1}. The conditions of their experiment were different, however, since the potential was changed rapidly, and the spectra recorded using an optical multichannel analyzer.

Briefly, our interpretation of the data presented here is as follows. The IR experiment looks at all CN^- containing surface species that can project a transition moment perpendicular to the surface. This includes a small subset of CN^- at positions where their Raman cross sections are enhanced many orders of magnitude beyond their concentration. The environment of this SERS subset is more inhomogeneous (chemically and physically) as demonstrated by the Raman linewidths and intensity dependence on potential. However, this inhomogeneity is not so gross that the C-N stretching frequency and its dependence on electrode potential are very different. Detailed discussion of the relative merits of adatom, adcluster and cavity models for SERS *versus* simpler top site, bridged site, *etc.*, models for IRRAS are beyond the scope of the present work.

V. CONCLUSIONS AND OUTLOOK

There are few *in situ*, nondestructive, probes capable of providing information about the molecular structure and dynamical properties of electrochemical interfaces. The excitement following the discovery of SERS was in part connected with the expectation that it could provide a route towards formulating detailed models of the structure of the metal and accompanying Helmholtz double layer. We expect that IRRAS, to be a much more powerful tool for *in situ* probing of the electrode-electrolyte interface with the advantage that it is applicable to flat, unroughened surfaces. Supplementary techniques with some promise include second harmonic generation [45], other nonlinear spectroscopies and photoacoustic spectroscopy. Still lacking are tools that will provide accurate structural information. Even there, however, there are some expectations that surface EXAFs, surface X-ray and scanning tunneling microscopy in liquids will be available during the present decade.

ACKNOWLEDGMENTS

We acknowledge the help and encouragement of J. G. Gordon II. This research has been supported in part by the Office of Naval Research.

REFERENCES

1. R. P. Van Duyne, in *Chemical and Biomedical Applications of Lasers*, Vol. IV, C. Bradley Moore, ed. (Academic Press, New York, 1978), Ch. 5, p. 101.
2. M. Fleischmann, P. J. Hendra and A. J. McQuillan, *Chem. Phys. Lett.* **26**, 163 (1974).
3. M. G. Albrecht and J. A. Creighton, *J. Amer. Chem. Soc.* **99**, 5215 (1977).
4. D. J. Jeanmaire and R. P. Van Duyne, *J. Electroanal. Chem.* **84**, 1 (1977).
5. For a review, see *Surface Enhanced Raman Scattering*, R. K. Chang and T. E. Furtak, eds. (Plenum Press, New York, 1982).
6. A. Campion, J. K. Brown and V. M. Grizzle, *Surf. Sci.* **115**, L153 (1982).
7. For a review of plasmon surface polaritons, see H. Raether, *Surface Plasma Oscillations and Their Applications* (Academic Press, New York-London, 1977); *Physics of Thin Films* **9**, 145-261 (1976), and references therein.
8. M. R. Philpott, *J. Physique (Paris)* **44**, (C10)295 (1983).
9. R. Corn and M. R. Philpott, *J. Chem. Phys.*, in press (1984).
10. S. Ushioda and Y. Sasaki, *Phys. Rev. B* **27**, 1401 (1982).
11. W. Knoll, M. R. Philpott, J. D. Swalen and A. Girlando, *J. Chem. Phys.* **77**, 2254-2260 (1982).
12. J. C. Tsang, J. R. Kirtley and T. N. Theis, *Solid State Commun.* **35**, 667 (1980).

13. C. Y. Chen, I. Davoli, G. Ritchie and E. Burstein, *Surf. Sci.* **101**, 363 (1980).

14. W. Knoll, M. R. Philpott and W. G. Golden, *J. Chem. Phys.* **77**, 219 (1982).

15. A. Otto, "Surface Enhanced Raman Scattering. Classical and Chemical Origins," *Light Scattering in Solids*, Vol. IV, M. Cardona and G. Güntheradt, eds. (Springer, 1983).

16. M. R. Philpott, F. Barz, J. G. Gordon II and M. J. Weaver, *J. Electroanal. Chem.* **150**, 399 (1983).

17. D. A. Weitz, J. I. Gersten and A. Z. Genack, *Phys. Rev. B* **22**, 4562 (1980).

18. A. Z. Genack, D. A. Weitz and T. J. Gramila, *Surf. Sci.* **101**, 381 (1980).

19. R. Corn and M. R. Philpott, *J. Chem. Phys.*, in press (1984).

20. (a) R. G. Greenler, *J. Chem. Phys.* **44**, 310 (1966); (b) S. A. Francis and A. H. Ellison, *J. Opt. Soc. Amer.* **49**, 131 (1959).

21. (a) P. A. Cholett, *Thin Solid Films* **52**, 343 (1978); (b) R. G. Greenler, *J. Chem. Phys.* **50**, 1963 (1968).

22. D. L. Allara and J. D. Swalen, *J. Phys. Chem.* **86**, 2700 (1982).

23. F. A. Burns, N. E. Schlotter, J. F. Rabolt and J. D. Swalen, *IBM Instruments, Inc., Application Note No. 1* (1981)

24. T. Ohnishi, A. Ishitani, H. Ischida, N. Yamamoto and H. Tsubomura, *J. Phys. Chem.* **82**, 1989 (1978).

25. W. G. Golden, C. Snyder and B. Smith, *J. Phys. Chem.* **86**, 4675 (1982).

26. A. Crossley and D. A. King, *Surf. Sci.* **68**, 528 (1977).

27. M. D. Baker and M. A. Chester, in *Vibrations at Surfaces*, R. Caudano *et al.*, eds. (Plenum Press, New York, 1982).

28. H. Pfnür, D. Menzel, F. M. Hoffman, A. Ortega and A. M. Bradshaw, *Surf. Sci.* **119**, 72 (1982).

29. W. G. Golden, D. S. Dunn and J. Overend, *J. Catal.* **71**, 395 (1981).

30. R. Ryberg, in *Vibrations at Surfaces*, R. Caudano *et al.*, eds. (Plenum Press, New York, 1982).

31. P. Hollins and J. Pritchard, in *Vibrational Spectroscopy of Adsorbates*, R. F. Willis, ed. (Springer, New York, 1981).

32. J. C. Campuzano and R. G. Greenler, *Surf. Sci.* **83**, 301 (1979).

33. M. Kawai, T. Ohnishi and K. Tamura, *Appl. Surf. Sci.* **8**, 361 (1981).

34. (a) L. A. Nafie and M. Diem, *Appl. Spectrosc.* **33**, 130 (1979); (b) L. A. Nafie, M. Diem and D. W. Vidrine, *J. Am. Chem. Soc.* **101**, 496 (1979).

35. A. E. Dowry and C. Marcott, *Appl. Spectrosc.* **36**, 414 (1982).

36. W. G. Golden and D. D. Saperstein, *J. Electron. Spec.* **30**, 43 (1983).

37. K. Kunimatsu, H. Seki and W. G. Golden, *Chem. Phys. Lett.*, in press (1984).

38. W. G. Golden, K. Kunimatsu and H. Seki, *J. Phys. Chem.*, in press (1984).

39. N. A. Rogozhnikov and R. Yu. Bek, *Electrokhimiya* **16**, 662 (1980).

40. H. Seki, K. Kunimatsu, K. Binding, W. G. Golden, J. G. Gordon and M. R. Philpott, to be published.

41. J. Billmann, G. Kovacs and A. Otto, *Surf. Sci.* **92**, 153 (1980).

42. R. E. Benner, R. Dornhaus, R. K. Chang and B. L. Laube, *Surf. Sci.* **101**, 341 (1980).

43. R. Kotz and E. Yeager, *J. Electroanal. Chem.* **123**, 335 (1981).

44. M. Fleischmann, I. R. Hill and M. E. Pemble, *J. Electroanal. Chem.* **136**, 361 (1982).

45. R. Corn, M. Romagnoli, M. D. Levenson and M. R. Philpott, *Chem. Phys. Lett.* **106**, 30 (1984).

Electromagnetic Theory Calculations for Spheroids: An Accurate Study of
the Particle Size Dependence of SERS and Hyper-Raman Enhancements

Ellen J. Zeman and George C. Schatz*
Northwestern University
Department of Chemistry
Evanston, IL 60201   USA

The electromagnetic contribution to the enhancements in Raman and
hyper-Raman scattering from molecules adsorbed onto Ag, Cu and Au
spheroids is studied, with an emphasis on the variation of the enhancements with spheroid size and shape. These calculations differ
from several earlier spheroid model studies in that the effects of
surface scattering on the particle dielectric constants, and of radiation damping and dynamic depolarization are included in the electrodynamic description. For particles between 1 and 100 nm in size, we show
that our results reduce to the exact numerical results of Barber, Chang
and Massoudi for the specific shapes which they studied. Our applications to Raman enhancements on Ag, Cu and Au show that the optimum particle size is not strongly dependent on particle shape, with the optimum semi-major axis being 20-30 nm for Ag, 60-80 nm for Cu and 60-80 nm
for Au. Optimum Raman enhancements for a/b = 0.5 are ~10^4 for Ag.
Hyper-Raman scattering shows significant enhancements for particle
shapes where either the input or scattered photons have plasmon resonances, with the relative magnitudes of these resonant enhancements
dependent on particle size.

I.   Introduction

Although the electromagnetic mechanism for surface enhanced Raman
scattering (SERS) and related processes is now widely accepted[1], some
of the important predictions of this mechanism, such as what are the
optimum sizes and shapes of the metal particles involved, and what is
the correct magnitude of the Raman enhancement as a function of wavelength for a given particle size and shape have yet to be worked out
quantitatively for particle shapes other than spheres. One of the
problems associated with treating more general shapes such as spheroids
is that except for one calculation, all spheroid studies to date have
used the LaPlace (small particle) approximation to Maxwell's equa-

*Alfred P. Sloan Fellow and Camille and Henry Dreyfus Teacher-Scholar.

413

B. Pullman et al. (eds.), Dynamics on Surfaces, 413–424.
© 1984 by D. Reidel Publishing Company.

tions[2]. The one exception is a recent numerical solution of Maxwell's equation by Barber et al[3]. This calculation found that the LaPlace equation result is inaccurate even when the spheroid's semimajor axis is only 1/20 the wavelength of light in length because of radiation damping and depolarization effects. Even the Barber et al calculation was not exact, however, since it did not include for the size dependence of the metal dielectric constants associated with scattering of the conduction electrons off the particle surfaces (i.e. quantum size effects). This becomes important for particle sizes smaller than the conduction electron mean free path, which for Ag overlaps with the region region where damping and depolarization are important.

In this paper we present the results of electrodynamic calculations which include all the dominant corrections for size dependent effects in spheroids, and which thereby enable us to determine quantitatively what are the optimum particle size and snapes for different metals and different kinds of surface spectroscopies. We do this by using a recently developed expression for the surface scattering contribution to the plasmon width of a spheroid[4], and by including for damping and depolarization by incorporating the leading k dependent terms from an exact solution of Maxwell's equations for a spheroid[5]. While this calculation still makes several approximations (neglect of adsorbate interactions, neglect of chemical effects) and is strongly dependent on the accuracy of measured dielectric constants in its predictions, the results should be closer to being quantitative for spheroidally shaped particles than in previous work.

II. Theory

We begin with the LaPlace equation solution for a prolate spheroid having major axis 2b and minor axis 2a. The solution for a constant applied field E_0 along the b axis has been discussed in several places[6], so we give only the results here. The quantity of most interest is the square of the electric field E evaluated just outside the spheroid surface, averaged over the surface. This average can be much smaller than the peak value of $|\underset{\sim}{E}|^2$ at the spheroid tips[1c], but it is the average that is more relevant to measurements on randomly distributed molecules. An exact solution of LaPlace's equation gives the following expression for the average of $|\underset{\sim}{E}|^2$ over the surface:

$$<|E|^2> = E_0^2 \left[|1-\zeta|^2 + \left(\frac{\zeta + \zeta^* - 2|\zeta|^2}{Q_1(\xi_0)} + \right.\right.$$

$$\left.\left. + \frac{|\zeta|^2}{Q_1(\xi_0)^2 (\xi_0^2-1)} \right) x \left(\frac{-\sqrt{\xi_0^2-1} + \xi_0^2 \sin^{-1}(1/\xi_0)}{\sqrt{\xi_0^2-1} + \xi_0^2 \sin^{-1}(1/\xi_0)} \right) \right] \quad (1)$$

where

$$\xi_0 = (1 - a^2/b^2)^{1/2} \quad (2)$$

$$Q_1(\xi_0) = \tfrac{1}{2}\xi_0 \; \ell n\left(\frac{\xi_0 + 1}{\xi_0 - 1}\right) \quad -1 \tag{3}$$

and
$$\zeta = (\varepsilon_i - \varepsilon_0) \,/\, (\varepsilon_i + \chi \, \varepsilon_0) \tag{4}$$

with ε_i being the dielectric constant inside the spheroid and ε_0 being that outside the spheroid. The parameter χ is a shape dependent parameter given by

$$\chi = -1 + [Q_1 \, (\xi_0) \, (\xi_0{}^2 - 1)]^{-1} \tag{5}$$

The value of χ is crucial in determining the frequency of the plasmon resonance, since the resonance condition is defined by $Re(\varepsilon_i + \chi \, \varepsilon_0) = 0$.

Corrections to Eq.(1) arising from the finite size of the spheroid have not previously been given. For a sphere, Meier and Wokaun[6] recently developed such corrections by examining the fields responsible for the induced dipole moment μ_{ind} in the sphere caused by an applied electromagnetic field. If α_0 is the sphere polarizability then they write

$$\mu_{ind} = \alpha_0 \, (E_0 + E_{rad}) \tag{6}$$

where E_{rad} is the average field emitted by an ensemble of dipoles distributed uniformly throughout the sphere. Meier and Wokaun find that[6] (in our notation):

$$E_{rad} = \left(\frac{2}{3} \, i \, k^3 + \varepsilon_0 k^2/b\right) \mu_{ind} \tag{7}$$

so that

$$\mu_{ind} = \left[1 - \frac{2}{3} \, i \, k^3 \, \alpha_0 - k^2 \, \alpha_0/b\right]^{-1} \alpha_0 E_0 \tag{8}$$

where $k = \varepsilon_0{}^{1/2} \, \omega/c$ and b is the sphere radius. Eq.(8) indicates that the induced dipole moment μ_{ind} is reduced by the factor in brackets relative to what it would be in the k=0 limit. The $(2/3) \, i \, k^3 \, \alpha_0$ term is identical to what is normally called the radiative damping correction to the LaPlace result, and arises from emission of radiation by μ_{ind}. The $k^2 \, \alpha_0/b$ term arises from dephasing between radiation emitted by different parts of the spheroid and has been called the "dynamic depolarization" term[6].

Although the Meier and Wokaun argument does not generalize directly to spheroids, one suspects that an equation similar to (8) must be valid, with α_0 given by the spheroid polarizability:

$$\alpha_0 = b^3 \frac{(1 + \chi)}{3} \left(\frac{\varepsilon_i - \varepsilon_0}{\varepsilon + \chi\varepsilon_0}\right)\left(\frac{\xi_0^2 - 1}{\xi_0^2}\right) \tag{9}$$

One way to verify if this is true is to compare the expansion of Eq.(8) in powers of k with a corresponding expansion which can be derived from an exact solution to Maxwell's equation. The latter can be obtained from the work of Stevenson et al[5,7] who give expressions for the wave zone electric field expanded up to k^4. The dipole part of this field can be written in terms of an induced dipole moment which is expanded through k^2, which means that the k^2 dependent term in (8) can be compared to the exact result. The lack of an analogous comparison for the k^3 term is not a serious liability, since this term in Eq.8 is expected to be correct for the spheroid on very general grounds because of its relationship to the exact expression for the dipole emission rate of a spheroid.

Considering now the k^2 dependent term, if we start with the results in Ref. 5, then after a great deal of algebraic manipulation, the following expression for μ_{ind} can be derived:

$$\mu_{ind} = \left[1 + \left(\frac{k^2 b^2}{\varepsilon_i + \chi\varepsilon_0}\right)\left(\frac{1}{5}\right)\left\{(1+\chi)(\varepsilon_i - 2\varepsilon_0) + \right.\right.$$

$$\left.\left. + \frac{1}{\xi_0^2}\left(\frac{9}{10}\varepsilon_i - \chi\varepsilon_i + \frac{19}{10}\chi\varepsilon_0\right)\right\}\right]\alpha_0 E_0 \tag{10}$$

The corresponding expansion of (8) through order k^2, using (9) is

$$\mu_{ind} = \left[1 + \frac{k^2 b^2}{(\varepsilon_i + \chi\varepsilon_0)}\left(\frac{1}{3}\right)\left\{(1 + \chi)(\varepsilon_i - \varepsilon_0) + \right.\right.$$

$$\left.\left. + \frac{1}{\xi_0^2}\left(-\varepsilon_i - \chi\varepsilon_i + \chi\varepsilon_0 + \varepsilon_0\right)\right\}\right]\alpha_0 E_0 \tag{11}$$

While it is clear that the two expressions are not the same, they do exhibit the same denominator $\varepsilon_i + \chi\varepsilon_0$ in the k^2 dependent term, which means that the frequency and shape dependence of the depolarization term are the same. In fact, only the multiplicative factor in curly brackets and the factor 1/3 or 1/5 differ between these two formulas. Overall then we conclude that Eq.(8) should be approximately correct for spheroids and we will assume this below. The incorporation of the radiation damping/dynamic depolarization term of Eq.(8) into Eq.(1) for the average of $|E|^2$ can be accomplished by realizing that the ζ factor in (1) arises from polarization effects which in the absence of damping/depolarization are proportional to α_0 in (9). This implies that dividing ζ by the term in square brackets in (8) achieves the appropriate reduction in polarization due to damping/depolarization.

The inclusion of the effects of surface scattering on the field enhancements in (1) is readily incorporated by modifying the metal dielectric constant ε_i by a size dependent factor. The theory of surface scattering for spheroids was recently studied by Kraus and Schatz (4) using both classical and quantum free electron theories. In order to adapt the results of either approach to studies of real metals, we first decompose the metal dielectric constant into the sum of a free electron contribution and a "bound" contribution:

$$\varepsilon_i = \varepsilon_F + \varepsilon_B \tag{12}$$

where ε_F is given by the usual free electron-Drude expression:

$$\varepsilon_F = 1 - \frac{1}{\omega} \frac{\omega_p^2}{\omega + i\gamma} \tag{13}$$

with ω_p being the plasmon frequency and γ the plasmon width. For large metal particles, ω_p and γ are obtained from fits to measured dielectric constants.

The effect of surface scattering on ε_i in lowest order leads to a modification in the width factor γ to

$$\gamma \rightarrow \gamma + \upsilon_F/L_{eff} \tag{14}$$

where υ_F is the Fermi velocity and L_{eff} is an effective length which in the classical treatment is taken to be the average chord length in the spheroid. An expression for L_{eff} is given in Ref. 4a, and in the present notation this is:

$$L_{eff} = \frac{4}{3} b[(\xi_0^2 + 1)/(\xi_0^2 - 1)^{1/2} - \sin^{-1}(1/\xi_0) +$$

$$+ (\xi_0^2 - 1)^{3/2} \ln (1 - 1/\xi_0^2)]$$

$$\times [\xi_0^2 (\xi_0^2 - 1)^{-1}\sin^{-1}(1/\xi_0) + 1/(\xi_0^2 - 1)^{1/2}]^{-1} \tag{15}$$

Note that this expression reduces to $L_{eff} = b$ for $b = a$ (a sphere), to $L_{eff} = 16a/3\pi$ for $b \gg a$ (for needles) and to $L_{eff} = (4/3) b \ln b$ for $b \ll a$ (for pancakes).

Evidently, L_{eff} is determined by the shorter dimension of the spheroid, and υ_F/L_{eff} will be significant whenever this shorter dimension becomes comparable to or smaller than the mean free path υ_F/γ. The corresponding quantum treatment of surface scattering in spheroids[4b] leads to a similar expression for L_{eff} after averaging over spheroid orientations.

III. Applications
IIIA. Average Field Enhancement

We now calculate the enhancement in the average of the square of
the field on the spheroid surface, i.e.,

$$R = <|\underset{\sim}{E}|^2 > / E_0^2 \qquad\qquad (16)$$

using Eq.(1) with ζ therein divided by the term in brackets in Eq.(8),
and using measured metal dielectric constants which have been corrected
according to Eqs.(12)-(15).
We begin by comparing our values of R with those from the exact
calculations of Barber et al[3] for a/b = 0.5 and b = 100 nm. Fig. 1

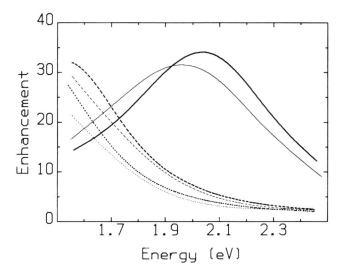

Figure 1. Average field enhancement versus photon energy (eV)
for a silver spheroid having b = 50 nm and a/b = 0.5. Three
different dielectric constants are considered, with the solid
line results for vacuum (ε_0 = 1.0), the dashed line that for
water (ε_0 = 1.77) and the dotted line that for cyclohexane
(ε_0 = 2.04). The thicker line plotted in each case is from the
present calculation while the thinner line is from Barber et al
(Ref. 3b).

plots R versus photon energy $\hbar\omega$ for Ag spheroids using dielectric
constants from Ref. 8 and three different values of ε_0. All three sets
of curves indicate excellent quantitative agreement between our results
and the exact ones. Note that for the size particle considered, surface
scattering effects are unimportant, and thus the fact that Ref. 3 did
not use dielectric constants corrected for surface scattering is unimportant. We have found that the excellent agreement between our results
and the Barber et al results persists for other sizes and shapes except

where surface scattering is important. The quadrupolar resonances seen by Barber et al are not contained in our theory without modifications, but these are generally weak compared to the dipole term which we do describe correctly.

We note from Fig. 1 that R shows a peak near $\hbar\omega$ = 2.05 eV for ε_0 = 1. The position of this peak depends in general on both particle shape and size. The peak value of R (denoted R_p) varies with the size parameter b and with the shape parameter a/b according to what is plotted in Figs. 2 and 3 for Ag particles. Fig. 2 presents the results using Ag dielectric constants from Ref. 9 while Fig. 3 presents the analogous results using dielectric constants from Ref. 8. For each value of a/b, three curves are displayed, including one in which surface scattering is omitted, another in which the depolarization/damping terms have been ignored, and a third in which both corrections have been included.

Figs. 2 and 3 both show that R_p optimizes at a particle size between 20 and 30 nm for both sets of dielectric constants, with the higher magnitudes of R_p associated with the more prolate objects. Figs. 2 and 3 are qualitatively similar in appearance, but the magnitude of the peak enhancement is significantly larger when the dielectric constants from Ref. 8 are used. Since the dielectric constants from Ref. 9 are consistent with a Kramers-Kronig analysis while those from Ref. 8 are not (9), it could be argued that the results in Fig. 2 are more realistic.

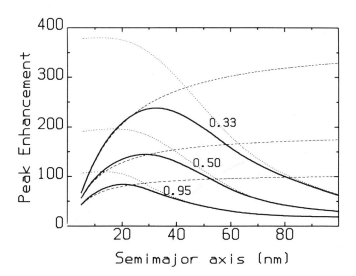

Figure 2. Peak enhancement versus b for Ag spheroids in vacuum having a/b = 0.95, 0.50 and 0.33. Dielectric constants were taken from Ref. 9. Dashed line in each case shows the result including surface scattering, dotted line including radiation damping and dynamic depolarization, and solid line including all corrections.

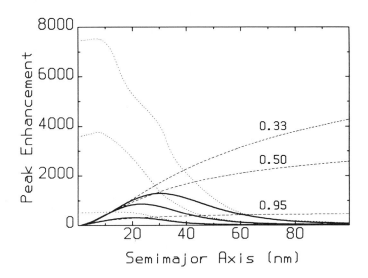

Figure 3. Enhancement versus b as in Fig. 2 but using the Ag
dielectric constants from Ref. 8.

    Note that in both Figs. 2 and 3, the effect of surface scattering
is mainly important for particles smaller than 30 nm while damping/de-
polarization effects are important above 30 nm. The influence of surface scattering is most pronounced for the a/b = 0.95 case and using the
dielectric constants from Ref. 8. Fig. 3 indicates that there is a
factor of 5 drop in the peak R_p value upon inclusion of surface scattering.
    Figs. 4 and 5 show the variation of R_p with b for Cu and Au,
respectively, using dielectric constants from Ref. 9. These two figures
are very similar to each other (with the Au enhancements smaller than Cu
by typically a factor of 2), and they are quite a bit different in b
dependence from the results in Fig. 2 (although the peak R_p value for Ag
is only 50% higher than that for Cu). The peaks in the solid curves for
both Cu and Au are between b = 60 and 80 nm, implying that the optimum
Cu and Au particles are 2-3 times larger than the optimum Ag particle.
    Note that surface scattering is relatively unimportant for the more
spherical particles (a/b = 0.95, 0.50) and that damping/depolarization
sometimes causes the peak enhancements to increase. The reason for the
latter result is that the damping/depolarization term causes the resonance energies to shift downward. For Au and Cu, such a shift downward
in energy leads to substantially smaller values of the imaginary part of
the dielectric constant for particles with resonances near $\hbar\omega = 2$ eV.

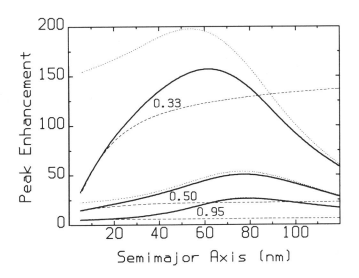

Figure 4. Enhancement versus b as in Fig. 2 but using the Cu
dielectric constants from Ref. 9.

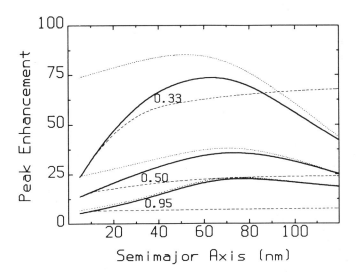

Figure 5. Enhancement versus b as in Fig. 2 but using the Au
dielectric constants from Ref. 9.

IIIB.Raman and Hyper-Raman Enhancements

The Raman and hyper-Raman enhancements may be approximated by multiplying the appropriate powers of R at the incident and outgoing frequencies ($R(\omega)R(\omega')$ for Raman and $R(\omega)^2 R(\omega')$ for hyper-Raman). Better treatments of the Raman emitted photon field are only available for spheres[10], and would be quite difficult to derive for spheroids.

Fig. 6 presents the Raman enhancement for Ag as a function of $\hbar\omega$ for three different frequency shifts $\Delta\omega$ using a spheroid size which is close to the optimum for a/b = 0.5 in Fig. 2. The figure indicates a peak enhancement of nearly 9000 for $\Delta\omega$ = 1000 cm^{-1}. The plasmon resonance width is large enough so that the separate resonances for the input and scattered photons are not distinguishable in the figure.

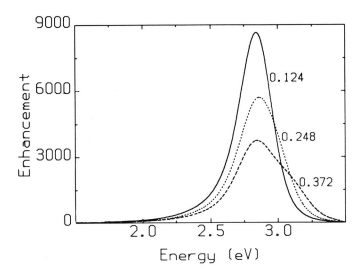

Figure 6. Raman enhancement versus incident photon energy for Ag spheroid having b = 50 nm and a/b = 0.5. The Stokes energy shift is taken as 0.124 eV (1000 cm^{-1}) in the solid curve, 0.248 eV in the dotted curve and 0.372 eV in the dashed curve.

Fig. 7 presents hyper-Raman enhancements for Ag particles having the same size and shape as in Fig. 6, and for the same frequency shifts. Here the plasmon resonance at 2.7 eV shows up twice, once when the incident frequency ω is 2.7 eV, and once when frequency $\omega' = 2\omega - \Delta\omega$ is at 2.7 eV (i.e., ω = 1.35 eV + $\Delta\omega/2$). The resonant value of ω in the second case increases with increasing $\Delta\omega$ as is observed in the figure. Note that for the particle shape and size chosen the more intense resonance occurs when ω = 2.7 eV, and the enhancement in this case is 2 x 10^4. For more prolate particles, we often find that the lower frequency peak gives more intense scattering. This indicates that hyper-Raman excitation spectra are quite sensitive to particle shape distributions.

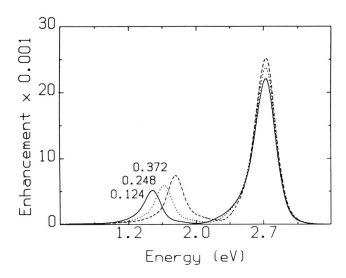

Figure 7. Hyper-Raman enhancement versus incident photon energy for the same spheroid as in Fig. 6 and for the same Stokes shift.

IV. Conclusions

By correcting the spheroid LaPlace equation solution for the dominant size dependent effects, we have made a detailed study of the size and shape dependence of enhanced fields relevant to SERS and surfaced enhanced hyper-Raman scattering. The results indicate that the optimum enhancements are obtained from spheroids having b = 20-30 nm for Ag and 60-80 nm for Cu and Au. Particles with smaller b values are attenuated by surface scattering effects while larger particles are subject to significant radiation damping and dynamic depolarization effects.

Optimal SERS enhancements for a/b = 0.5 are about 9×10^3 for $\Delta\omega$ = 1000 cm^{-1} and the optimal hyper-Raman enhancement is about 2×10^4. The fact that the SERS enhancement is less than 10^6 is evidence for the importance of chemical mechanisms in SERS.

V. Acknowledgement

This work was supported by the Air Force Office of Scientific Research, grant no. AFOSR-83-0350.

References

1. Recent reviews of electromagnetic theory include:
    (a) Surface Enhanced Raman Scattering, edited by R.K. Chang and
        T.E. Furtak (Plenum, New York, 1982);
    (b) Metiu, H., Progress in Surf. Sci., 1984 in press;
    (c) G.C. Schatz, Accts. Chem. Res., to be published;
    (d) M. Kerker, Accts. Chem. Res., to be published.
2. (a) F.J. Adrian, Chem. Phys. Lett. 78, 45 (1981);
    (b) J.I. Gersten and A. Nitzan, J. Chem. Phys. 73, 302 (1980);
    (c) M. Kerker, D.S. Wang and H. Chew, Appl. Opt. 19, 2256, 3373,
        4159 (1980);
    (d) U. Laor and G.C. Schatz, Chem. Phys. Lett. 83, 566 (1981);
    (e) U. Laor and G.C. Schatz, J. Chem. Phys. 76, 2888 (1982);
    (f) A. Wokaun, J.P. Gordon, P.F. Liao, Phys. Rev. Lett. 48, 957
        (1982).
3. (a) P.W. Barber, R.K. Chang, H. Massoudi, Phys. Rev. Lett. 50, 997
        (1983);
    (b) P.W. Barber, R.K. Chang, H. Massoudi, Phys. Rev. B27, 7251
        (1983).
4. (a) W.A. Kraus and G.C. Schatz, Chem. Phys. Lett. 99, 353 (1983);
    (b) W.A. Kraus and G.C. Schatz, J. Chem. Phys. 79, 6130 (1983).
5.   A.F. Stevenson, J. Appl. Phys. 24, 1143 (1953).
6.   M. Meier and A. Wokaun, Opt. Lett. 8, 581 (1983).
7.   A.F. Stevenson, J. Appl. Phys. 24, 1134 (1953).
8.   P.B. Johnson and R.W. Christy, Phys. Rev. B6, 4370 (1968).
9.   H.J. Hagemann, W. Gudat and C. Kunz, DESY Report No. SR-74/7
        (May, 1974).
10.   H. Chew, M. Kerker and D.D. Cooke, Phys. Rev. A16, 320 (1977).

SURFACE ARRANGEMENTS ON MULTILAYER ICOSAHEDRA

J. FARGES, M.F. de FERAUDY, B. RAOULT, G. TORCHET
Laboratoire de Physique des Solides – Université Paris Sud
91405 ORSAY (France)

ABSTRACT

An atomic cluster with icosahedral external shape must contain a well-defined –"magic"– number of atoms, $13 - 55 - 147 - ...$etc, in order to have a complete outer layer. The question then arises to know what should be the surface arrangement of the cluster when the number of atoms is not a "magic" one. Our purpose is to show that the answer depends on whether the icosahedral cluster shape is due to a better growth rate or to a better stability. In the former case, the external surface layer is always icosahedral while in the latter a transition may occur in the surface atom arrangement of small-sized icosahedra from twin to icosahedral positions.

I. INTRODUCTION : MACKAY ICOSAHEDRA

In 1962, Mackay[1] introduced some amazing structures built with hard spheres as possible models for atomic clusters. Their external shape was that of a regular icosahedron and their size could be extended from the atomic scale to infinity. Despite the perfect five-fold symmetry of these structures, their internal local order was nearly that of a crystal. In effect, Mackay described the icosahedron as made of 20 identical tetrahedra possessing a common vertex and connected with each other through adjacent faces, each of which forms twinning planes, the icosahedron and the tetrahedra from which it is composed possessing the same number of atoms on each edge. To a first approximation the tetrahedra represent a face-centered cubic structure having thus four (111) faces ; in reality however this structure is slightly deformed, because the tetrahedra are not absolutely regular. In fact the three radial edges derived from the common vertex are equal, just as the three surface edges are, but the latter are longer than the former by approximately 5%.
The smallest icosahedron contains 13 atoms. It has only one layer (n=1) and 2 atoms on each edge. In a general way, an icosahedron with n layers has (n + 1) atoms on each edge, like the tetrahedra which it is made of. It is easy to calculate the total number of atoms N corresponding to

425

B. Pullman et al. (eds.), Dynamics on Surfaces, 425–430.

a n layer icosahedron knowing that the layer of index m contains
(10 m^2 + 2) atoms. Thus the well-known sequence 13,55,147, etc.. is
found.
Mackay calculated the effective density for the case of an icosahedron
of hard spheres. The radial interatomic distances are thus put equal to
the diameter of the spheres while the lateral distances are 5% larger.
Because of this, he found an effective density somewhat less than that
of the face-centered cubic structure (0.69 as against 0.74). It therefore appeared to him "somewhat unlikely that a large number of atoms
would naturally adopt a multilayer icosahedral arrangement".

II. "CRYSTAL GROWTH" OF ICOSAHEDRA

These studies of models with pentagonal symmetry remained almost unno-
ticed until in 1966 [2] [3] [4] the discovery by electron microscopy of
icosahedral particles in thin gold films caused a revival of interest.
The question was then what could favour the growing of such particles.
It has been noticed that they were usually obtained by condensation of
a metallic vapour onto a cold support, that is under conditions of high
supersaturation and out of thermodynamical equilibrium. Such experimental conditions lead to the following consequences : first the stable
critical nuleus is reduced to a single atom and, second, the structure
and shape of the particles mainly depend on the growth kinetics.
Let us examine how may grow an icosahedron with n complete layers. A critical nucleus for the layer of index (n + 1) must first appear. It will
consist of a single atom sticking to one face of the layer of index n.
Now this atom can choose among two kinds of site leading to different
growth sequences :
A. The first possibility is to choose a f.c.c. site with respect to the
tetrahedron on which it sticks and then adatoms coming afterwards will
form a part of a regular (111) layer. The considered face will be gradually covered and as soon as one atom will stand over an edge, the adjacent face will be covered too without the necessity of any new critical nucleus. Step by step, neighbouring faces will be covered, thus forming the icosahedral layer of index (n + 1). During this growth sequence,
one can see that the surface of the particle consists in an incomplete
but regularly close-packed layer. Gillet[5] has shown that this distinctive feature of needing one single nucleation for each layer makes the
growth rate of icosahedra much larger than that of standard f.c.c. particles which need one nucleation for each face.
B. The other possibility is to choose a non f.c.c. site with respect to
the tetrahedron. In this case, adatoms will form a part of a (111) plane
in twin fault position and adatoms will never stand above one edge. Consequently, the growth of the layer is limited to the considered face and
will never extend to adjacent faces. The growth of an icosahedral layer
will then take place only if a nucleation as described in A occurs on one
of the remaining faces.

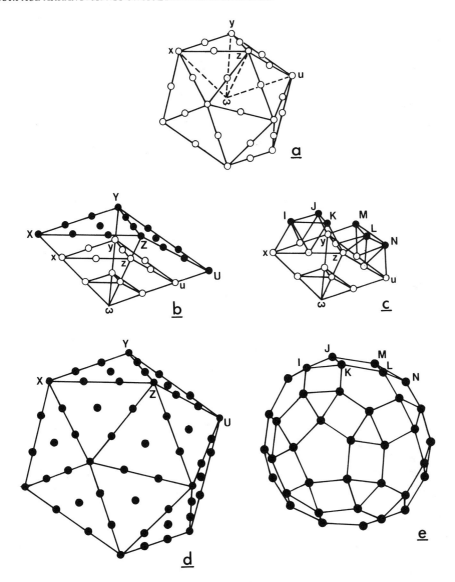

Fig. 1

a) The atom position on the 2^d layer of an icosahedron.

b) Two of the 20 tetrahedra which constitute the previous icosahedron have their external face covered with a third regular f.c.c. layer. Note that the two faces share the atoms on the edge xy. It is the A arrangement.

c) The same tetrahedra are now covered with third layer in twin fault position. The edge yz is not covered. It is the B arrangement.

c) Appearance of the surface when the second layer is completely covered in A arrangement. It is a three layer icosahedron.

e) The second layer is now completely covered in B arrangement. It is a rhombicosidodecahedron.

III. THE ICOSAHEDRON AS A STABLE PARTICLE

More recently, it has been shown that [6] [7] rare gas clusters, obtained by homogeneous nucleation in an adiabatic expansion, also present the structure of Mackay icosahedra when they contain from several tens to several hundreds of atoms. It must be pointed out that in these experiments, clusters are in the state of highest thermodynamical stability due to the formation process and to the relatively high cluster temperature which is about 2/3 of their melting temperature.

The question of knowing how adatoms are located on the surface of an icosahedron with n complete layers should no more be stated in terms of growth rate, but in terms of best energetic stability. From this point of view both possibilities described previously are likely to be competitive but it is not easy to decide *a priori* which one will prevail : the surface density is higher in the A arrangement while in the B arrangement, the absence of atom over the edges may constitute an energetic advantage.

Both situations are presented in Fig.1 in the case of an icosahedron with 2 complete layers covered with atoms in A and B arrangements. In that case where the number of atoms is not too large, it is possible to calculate the energy of the models assuming a two-body potential, for instance Lennard-Jones, in order to account for atomic interactions. In this way, the expected results will be valid for rare gas clusters as well as for some simple molecular clusters. We have done this calculation for several models with the same total number of atoms N but different surface arrangements (A or B), so that we can compare cluster energies at a given cluster size. The calculation has been performed with relaxed models.

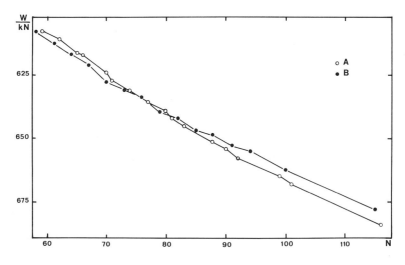

Fig. 2
Potential energy of models with A and B surface arrangement versus the number N of atoms in the model.

Lines in Fig.2 join together points which represent the potential energy of models with the same arrangement, either A or B, and thus give
the energy variation versus N. For each arrangement and each N value,
several models have been used when they seemed to be *a priori* equivalent. The reported energy value is that of the model providing the
minimum energy value. It can be observed on Fig.2 that the B arrangement prevails for N < 71, a number corresponding to 16 adatoms on the
second complete layer. In this case, the surface arrangement is that of
Fig. 1c and 1e. For N > 85, that is 30 adatoms, the A arrangement prevails and corresponds to Fig. 1b and 1d. Between these N values, a
transition occurs in the local order and no surface arrangement prevails
for a given cluster size.

IV. DISCUSSION
1. Experimental observations. The existence of a transition in the surface local order which extends over a well-defined size range makes it
possible to interpretate some data provided by two experiments on rare
gas clusters formed in a free jet expansion.
The first already mentioned experiment concerns electron diffraction
studies that we have performed in Orsay [6] [7]. Diffraction patterns
are sensitive to the local order and B → A transition is actually observed between N = 70 and N = 90.
The second experiment refers to mass spectroscopy on Xe clusters [8].
Peaks on these spectra at precise N values reveal a greater relative
abundance of several ionized clusters. In the size range considered in
the present study, one can note peaks at N = 55 and 147 corresponding to
clusters with complete layers and, more surprising N = 71, 87 ... The
particularly low abundance of clusters in the range 72 < N < 85 should
be due to the fact that adatoms are then experiencing a transition in
their local order which makes them move and facilitates their evaporation.
2. Multilayers icosahedra with n ⩾ 3 (Fig. 3)
First of all, it can be noted that in the case examined precedently,
when third layer begins to grow, the A arrangement is unstable because
most of the corresponding sites stand over edges. Adatoms then need a
sufficient number of neighbours to be stabilized in such sites, a situation which occurs near N ∿ 85.

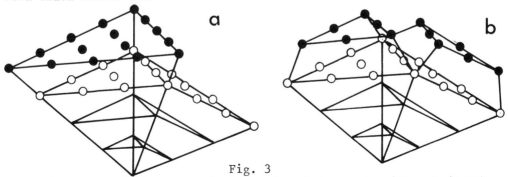

Fig. 3
Three layer tetrahedra covered with a fourth layer in either A (left)
or B (right) arrangement.

When increasing the number of layers, the number of sites over edges increases as n, that is less rapidly than the number of sites on the faces which increases as n^2. The competition between arrangement A and B will then rapidly turn to the advantage of the former. This is confirmed by our experiment since the B arrangement has never been observed in electron diffraction pattern of rare gas clusters with n > 3.

REFERENCES

1 -   A. MACKAY, Acta Cryst., 15, 916-918,(1962).
2 -   S. INO, J. Phys. Soc. Jpn, 21, 346-362, (1966).
3 -   J.G. ALLPRESS & J.V. SANDERS, Surf. Sci., 7, 1-25, (1967).
4 -   E. GILLET & M. GILLET, J. Cryst. Growth, 13-14, 212-216, (1972).
5 -   M. GILLET, J. Cryst. Growth, 36, 239, (1976).
6 -   J. FARGES, M.F. de FERAUDY, B. RAOULT, G. TORCHET, J. de Physique
      C3 n°4 , Tome 41, (1980).
7 -   J. FARGES, M.F. de FERAUDY, B. RAOULT, G. TORCHET, J. Chem. Phys.
      78 (8), 5067, (1983).
8 -   O. ECHT, K. SATTLER, E. RECKNAGEL, Phys. Rev. Lett., 47, 1121,(1981).

QUANTUM SIZE EFFECTS IN THE ELECTRONIC PROPERTIES OF SMALL SEMICONDUCTOR CRYSTALLITES

L. E. Brus
AT&T Bell Laboratories
600 Mountain Avenue
Murray Hill, NJ, 07974

1. Introduction

Metals and semiconductors have strong interatomic chemical bonds. The valence electrons determine the bonding energies, crystal structure, and excited electronic states of the infinite crystal; these electrons are extensively delocalized. Delocalization implies that the electronic properties of these materials develop gradually as a small cluster or crystallite grows in size. In the small size regime where bulk electronic properties have not yet been achieved, such clusters are partially molecular in nature in that their properties depend upon the size and shape of the crystallite. Small metal clusters in this range (perhaps 5-100 atoms) are effective heterogeneous catalysts, and intense interest has recently developed in understanding their transition from molecules to bulk metals as a function of size.

Semiconductor crystallites of $\geqslant 80\text{Å}$ diameter have long been employed as colloidal photosensitizers for redox chemical reactions. Particles of this size appear to behave as bulk semiconductors. Optical excitation of the crystallite produces mobile electrons and holes which may reduce or oxidize surface adsorbed molecules. Such colloidal systems provide elementary prototypes for electrochemical processes at macroscopic electrodes. We have recently applied the technique of time resolved resonance Raman spectroscopy to these surface reactions, in order to chemically identify, and probe the vibrational structures of, the initial surface adsorbed reaction products.[1] We find that simple one electron and hole transfer processes occur with very minor distortion of vibrational structure due to molecular adsorption. Strong chemisorption does not seem to be a prerequisite for charge transfer across the semiconductor: liquid interface, in the few examples that have been studied to date.

As part of this effort, we developed colloidal chemical syntheses for small ZnS and CdS crystallites in the 20-50Å diameter range. We discovered that the crystallite absorption, fluorescence, and phonon resonance Raman excitation spectra become a strong function of size.[2] These observations imply that the bulk band gap has not yet formed; we interpret the shift between the crystallite lowest excited state and the bulk band gap as essentially a quantum size effect.

B. Pullman et al. (eds.), Dynamics on Surfaces, 431–435.
© *1984 by D. Reidel Publishing Company.*

A 30Å diameter crystallite contains $\sim$ 750 atoms, and a metal cluster of this size is commonly considered to be a bulk metal. Semiconductors, however, appear to be especially sensitive to size effects because carriers (electrons and holes) at the band edges have crystal momenta k $\simeq$ 0, very long de Broglie wavelengths λ, and hence are especially sensitive to localization. In this manuscript we review the elementary model of quantum size effects in 3 dimensional semiconductor crystallites, and then discuss the experimental evidence.

2. The Elementary Model

2.1 Excited Electronic states

The Wannier Hamiltonian describing bound electronic states formed by an electron and hole in an infinite crystal is

$$\hat{H} = \frac{-\hbar^2}{2m_h} \nabla_h^2 - \frac{\hbar^2}{2m_e} \nabla_e^2 - \frac{e^2}{\epsilon_2|r_e-r_h|} \tag{1}$$

m_e and m_h are the effective masses of the electron and hole; in Cds these values are m_e = .19 and m_h = .8 in units of the actual electron mass. In the standard band theory of solids, these low numbers are caused by extensive delocalization of conduction and valence band orbitals. ϵ_2 is the dielectric coefficient, principally due to atomic core polarizabilities, and is $\simeq$ 5.7 for CdS at optical frequencies. Equation 1 describes hydrogenic states, weakly bound due to the small masses and the screening of the Coulomb potential.

We solve Schrödinger's equation for an electron and hole in a small spherical crystallite, at the same level of approximation as eqn. (1). The fact that the Coulomb potential is significantly screened in (1) implies that the electron and hole have substantial dielectric solvation energies in the bulk crystal. In small crystals there should be a partial loss of solvation energy. In order to correctly incorporate these electrostatic effects in small crystallites, where there is a dielectric discontinuity at the surface, we obtain the exact electrostatic interaction energy (for S wavefunctions):[3]

$$V_0(r_e,r_h) = \frac{-e^2}{\epsilon_2|r_e-r_h|R} + \frac{e^2}{2R} \sum_{n=1}^{\infty} \alpha_n \left[\left(\frac{r_e}{R}\right)^{2n} + \left(\frac{r_h}{R}\right)^{2n} \right] \tag{2}$$

Here R is the sphere radius, and the α_n are expansion coefficients dependent only upon ϵ_2 (the sphere dielectric coefficient) and ϵ_1 (the coefficient of the external medium). Schrödinger's equation for an electron and hole localized in a spherical crystallite becomes:

$$\left[\frac{-\hbar^2}{2m_e} \nabla_e^2 + \frac{-\hbar^2}{2m_h} \nabla_h^2 + V_0(r_e,r_h) \right] \Phi(r_e,r_h) = E\Phi(r_e,r_h) \tag{3}$$

There are no adjustable parameters. The potential is taken to be infinite outside the crystallite. The solution depends upon R and upon the bulk electronic properties m_e, m_h, and ϵ_2. The principal physical assumption is that the lattice structure inside the crystallite is the same as in the bulk material.

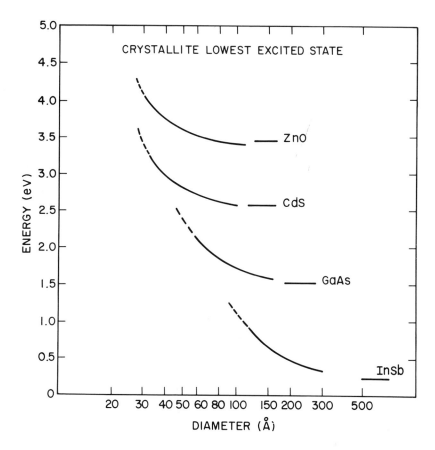

Figure 1.   Predicted dependence of the crystallite lowest excited state energy upon
            diameter. Short solid lines are bulk band gap energies.
            (from Ref. 3, with permission of Am. Inst. of Physics)

Equation 3 can be solved for the lowest eigenstate (i.e., the lowest excited state
of the crystallite) by standard variational procedures. The size dependence of the lowest
excited state thus calculated appears in Figure 1 for ZnO, CdS, GaAs, and InSb.[3] An
analytical approximation for this dependence is:

$$E = E_g + \frac{\hbar^2\pi^2}{2R^2}\left[\frac{1}{m_e} + \frac{1}{m_h}\right] - \frac{1.8e^2}{\epsilon_2 R} + \frac{e^2}{R}\left[\sum_{n=1}^{\infty}\alpha_n\left(\frac{r}{R}\right)^{2n}\right] \qquad (4)$$

There are three corrections to the bulk band gap E_g. The first is a quantum localization
energy for the electron and hole, varying as R^{-2}. The second is the screened Coulomb
attraction varying as R^{-1}; the electron and hole are physically closer to each other in
smaller crystallites. The third term can be interpreted as a loss of solvation energy: it is
numerically the smallest. The localization term increases E while the Coulomb term
decreases E. The two terms have different R dependencies, and E always shifts blue for

sufficiently small R as seen in the Figure. At very small R the model fails as the effective mass approximation is no longer valid.

Bulk semiconductors, and large crystallites, are characterized by continuous optical absorption above E_g. The model predicts that the apparent band gap (i.e., the lowest excited electronic state) will shift blue in small quantum size regime crystallites. A detailed consideration of oscillator strengths indicates that the intensity of absorption at the edge should increase, and a resolved peak (the 1S exciton) should appear.[3]

2.2 Ionization and Redox Potentials

Using these same methods and physical assumptions, one can model ionization potentials in vacuum, and chemical redox potentials in aqueous solutions.[4] For example, the electrostatic potential energy for a charged sphere with one extra hole or electron at r is

$$V_1(r) = \frac{(\epsilon-1)e^2}{2\epsilon_2 R} + \frac{e^2}{2R}\left[\sum_{n=1}^{\infty}\alpha_n\left(\frac{r}{R}\right)^{2n}\right] \tag{5}$$

Here $\epsilon = \epsilon_2/\epsilon_1$. The first term is the classical charging energy of a dielectric sphere. This particular term increases as the square of the number of excess electrons or holes. The second term describes a relatively weak radial electric field pulling the charge to $r = o$, the point of greatest dielectric stabilization. Solutions to the appropriate Schrodinger's equation indicate that a conduction band electron should be a more potent reducing agent, and the a hole should be a stronger oxidizing agent, in smaller crystallites than in bulk crystals. Alternatively stated, carriers are destabilized in small crystallites due to the loss of dielectric solvation energy, and due to the quantum energy of localization.

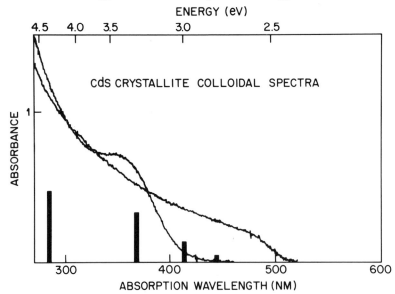

Figure 2.   Comparison of absorbance spectra of small CdS crystallites (28Å average diameter) in an acetonitrile colloid, with larger CdS crystallites (over 100Å average size) in an aqueous colloid.

(from Ref. 3, with permission of Am. Inst. of Physics)

3. Experimental Optical Consequences

In principle, the size dependence of the lowest excited state is observable in absorption spectra, fluorescence spectra, and the resonance Raman excitation spectra of crystallite phonon modes. In an aqueous colloid of $\simeq$ 45Å diameter CdS crystallites, we originally observed that the apparent absorption edge was modestly shifted $\simeq$ 0.2 eV to higher energy than the bulk band gap.[2] In this same colloid, the resonance Raman spectrum of the CdS LO phonon ($\simeq$ 305 cm^{-1}) was observable. The excitation spectrum of this Raman line was blue shifted and broadened somewhat as compared with large crystals. The resonant intermediate state in the LO phonon Raman scattering is the 1S exciton, the state described by eqn. 1 in bulk material and eqn. 3 in crystallites. The Raman spectra confirm a shift in this state in crystallites $\simeq$ 45Å diameter. We had somewhat earlier observed a shift in fluorescence spectra; this shift we subsequently attributed to size effects.[5]

The crystallite sizes are determined by transmission electron microscopy. The electron microscope, with $\simeq$ 2.5Å resolution, is also able to directly image the lattice potential variation along the (111) crystal (face centered cubic) axis.[6] This result demonstrates that the chemical colloidal syntheses employed yield crystalline material.

Figure 2 compares the observed spectrum in a $\simeq$ 28Å average diameter CdS colloid (in acetonitrile) with the spectrum of large crystallites.[6] Bulk crystalline CdS has a band gap E_g near 500 nm. The two absorbance spectra are mass normalized so that we essential compare a few larger crystals with many smaller ones. There is a large shift in the absorption edge, with partial resolution of a peak near 360 nm (shifted $\simeq$ 0.8 eV from the bulk E_g). The particle size distribution function is moderately broad, and we interpret this spectrum as inhomogeneously broadened due to the distribution in sizes and shapes. The solid bars give an indication of the predicted relative intensities and exciton wavelengths of major components in the size distribution.

These data confirm the two major qualitative predictions of the elementary model: A) A large shift to higher energy by the lowest excited state (exciton), and B) an increase in exciton oscillator strength relative to above gap absorption. Quantitative tests of the model will require further experimental work now in progress.

Acknowledgments

We collaborated with S. Beck and R. Rossetti in the optical and Raman studies, and with S. Nakahara, M. Gibson, and R. Hull in the TEM characterizations. We had extensive discussions with and useful suggestions from D. A. Kleinman and B. Miller. Figures 1 and 2 are reprinted with permission of the American Institute of Physics.

References:

1) R. Rossetti, S. M. Beck, and L. E. Brus, *J. Am. Chem. Soc.* **106**, 980 (1984).

2) R. Rossetti, S. Nakahara, and L. E. Brus, *J. Chem. Phys.* **79**, 1086 (1983).

3) L. E. Brus, *J. Chem. Phy.* **80** (May 1984).

4) L. E. Brus, *J. Chem. Phys.* **79**, 5566 (1983).

5) R. Rossetti and L. Brus, *J. Phys. Chem.* **86**, 4470 (1982).

6) R. Rossetti, J. L. Ellison, J. M. Gibson, and L. E. Brus, *J. Chem. Phys.* **80**, (May 1984).

LASER SPECTROSCOPY OF Li$_3$ ISOLATED IN RARE GAS MATRICES, COMPARISON
WITH OTHER METAL TRIATOMICS

M. Moskovits and W. Limm
Department of Chemistry and Erindale College, University of
Toronto, Canada M5S 1A1.
T. Mejean
Laboratoire de la Spectroscopie Infrarouge, Université de
Bordeaux I, Talence, France.

The laser induced resonance fluorescence (Raman) of three lithium
containing species isolated in solid Xe and Kr are presented. In Xe
one finds in addition to a progression assignable to Li$_2$ two others
both of which are assigned to Li$_3$, one, a fluxional molecule is
characterized by a number of progressions in the symmetric stretch of
the molecule (310 cm^{-1}) built upon several vibronic terms, the other is
ascribed to a rigid <u>acute</u> angle form of Li$_3$ characterized by a
symmetric frequency of 353 cm^{-1} and a bending frequency of 231 cm^{-1}.
With yellow (R6G dye laser) excitation yet another species is excited
which is tentatively assigned to Li$_4$. Comparison is made with other
metal triatomics such as Cu$_3$, Ni$_3$, Cr$_3$ and Sc$_3$.

    The properties of very small metal molecules–diatomics, triatomics
and larger aggregates – has captured the interest of a large number of
researchers (1) partly because one may, in the fullness of time, apply
what one has learned about these particles to practical problems such
as the design of new heterogeneous catalysts and to the understanding
of such phenomena as homogeneous nucleation and partly as a result of
the discovery, as more data came to light that these small clusters
possess some remarkable physical and spectroscopic properties in
their own right. Precise structural and vibrational information about
these species has come almost entirely from two sources: luminescence
and ESR studies of matrix isolated species (2) and laser induced
fluorescence and two or multiphoton ionization spectroscopy of clusters
produced in molecular beams (3). In this manner a great deal of data
have become available for metal diatomics some of which has recently
been summarized (4). In this article we concentrate on the few
triatomics that have been studied and recent results with the Li$_3$
molecule will be presented in some detail.
    The experimental technique has been thoroughly presented elsewhere
(4) and will 'not be discussed further.
    The dependence on apical angle (ϕ) of the three vibrational
frequencies of an ordinary homonuclear triatomic is shown in fig. 1.
At 60° the bending and asymmetric stretching vibration are degenerate.
An accidental degeneracy also occurs for the asymmetric and symmetric
stretching vibrations near $\phi = 100°$. The curves in fig. 1 were

B. Pullman et al. (eds.), Dynamics on Surfaces, 437–446.
© 1984 by D. Reidel Publishing Company.

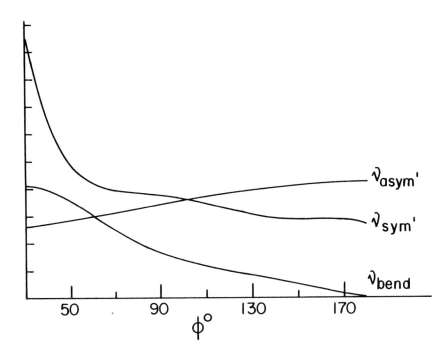

*Fig. 1. Calculated fundamental frequencies of an arbitrary triatomic
          as a function of apical angle.*

obtained by assuming equal force constants for the three bonds at
ϕ = 60° and allowing the force constant of the "odd bond" to obey the
relation kr^3 = constant, for other values of ϕ in the spirit of
Badger's rule, k and r being the force constant and bond length,
respectively. The molecules Ni_3, Cr_3 and Sc_3 illustrate molecules
of various structure. Ni_3 (5) for example was shown based on the
isotopic fine structure of its resonance Raman spectrum to be a bent
triatomic with an apical angle near the crossing of the symmetric and
asymmetric stretching frequencies i.e. of about 100°. It was argued
that in the gas phase this molecule might in fact be a linear molecule
subject to Renner–Teller instability and that its bent geometry was
imposed upon it by the matrix in its attempt to occupy matrix sites
whose geometries did not vary greatly from linear.

A Raman spectroscopic study of Cr_3 (2a) isolated in solid Ar
revealed all three of its fundamentals. The symmetric stretching
vibration was shown to have the highest frequency and the frequency
spacing between adjacent pairs of its three fundamentals was about
equal implying, on the basis of figure 1, that Cr_3 is a bent triatomic
with apical angle in the neighborhood of 75°. Finally the two low
frequency fundamentals of Sc_3 (4) were found to have almost equal
frequencies while the symmetric stretch was considerably higher in
frequency implying that this molecule is essentially equilateral
triangular in shape. This shape was also determined by ESR (6).

The spectroscopy of some triatomic molecules such as Cu_3 and Li_3

cannot be easily understood in terms of the familiar vibrational
fundamentals.

The case of Li_3 illustrates the point best since it has been the
subject of meticulous computational analysis by Schumacher and Gerber
(7) and by Martins et al. (8). The D_{3h} form of this molecule
would have a $^2E'$ ground state which according to the Jahn-Teller
theorem is unstable to a linear distortion along one of the normal
coordinates belonging to the direct product e⊗e to form an isoscelese
triangle with an apical angle which is either more obtuse
or more acute than 60°. The adiabatic surface calculated for this
molecule (7) (8) reveals the obtuse form to be the preferred one, with
such low barriers to pseudorotation, however, that the Born–Oppenheimer
approximation (assumed in calculating the adiabatic surface) cannot in
principle be made. The molecule is therefore fluxional with a true
vibration only along its symmetric stretching coordinate. The bending
and asymmetric stretching nuclear coordinates blend with electronic
coordinates to form vibronic states whose terms are not separated by
simple intervals as in the case of vibrations. An illustration is
provided in fig. 2 which shows a resonance Raman or fluorescence
spectrum attributed to Cu_3 (9). The spectrum is dominated by a progression with 354 cm^{-1} spacing which is assigned to the symmetric
stretching frequency. There are, moreover, four other progressions in
the 354 cm^{-1} frequency built upon other frequencies, however, these
are 404 or 50, 478 or 124, 579 and 295 cm^{-1}. (the uncertainty arises

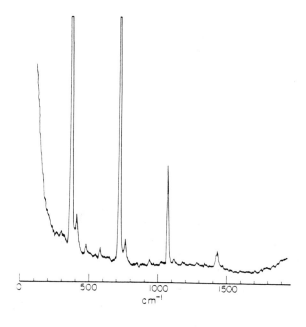

Fig. 2. *Resonance Raman spectrum attributed to Cu_3 excited with 594 nm
laser radiation.*

from the unavailability of data close to the exciting frequency). Any
attempt to account for these extra frequencies in terms of two other
fundamentals fails. These frequencies are assigned, therefore, to
vibronic levels of a pseudorotating Cu_3 molecule (even at 15K and in
an argon matrix). This interpretation is challenged by two recent
reports. The first, an ESR study of Cu_3 isolated in solid adamantane
(10) concludes on the basis of spin density distribution that Cu_3 is
not fluxional. The second, an elegant beam study of Cu_3 (11) prefers
the interpretation that Cu_3 is in a deep well and hence not fluxional
in its ground state but fluxional in the excited $^2E''$ state. In the
latter study the conclusion regarding the ground state is based on two
hot bands and cannot therefore be construed to be final (parenthetically
the excited state data of that study is accounted for quite well by
means of two fundamentals, of a rigid molecule: the symmetric stretch
and the bend); the results of the former study may be specific to the
adamantane matrix used. In Ar the molecule may be fluxional. That
such a state of affairs is possible is nicely illustrated below in
the case of Li_3 which is fluxional or rigid according to the matrix or
even matrix site.

   Matrix isolated lithium has been the subject of numerous studies
(12 (13). In one of the most recent, Welker and Martin (13) report
four bands in addition to those due to atomic lithium in the absorption
spectrum of lithium in xenon. These are centered at 429, 476, 510 and
554 nm, of which three are fortuitously near Ar^+ laser exciting lines.
The 510 nm feature was assigned by the authors to Li_2 and indeed

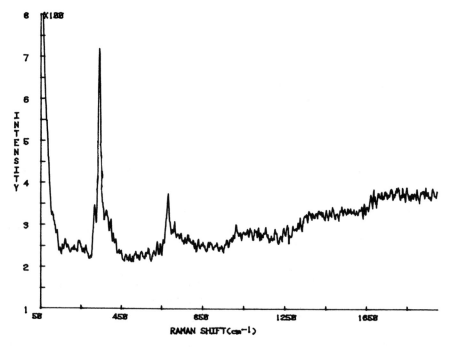

*Fig. 3. Resonance fluorescence spectrum of Li_2 in solid Xe excited
with 514.5 nm Ar^+ laser radiation.*

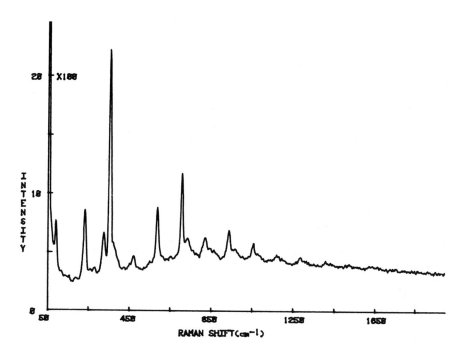

Fig. 4.   Spectrum attributed to "rigid" Li₃ in solid Xe excited with 457.9 nm Ar⁺ radiation.

exciting into that band with 514.5 nm Ar⁺ laser radiation produces the resonance fluorescence progression shown in fig. 3 with a vibrational spacing close to that of gas-phase Li₂ (14). Welker and Martin could not make a precise assignment of the other bands but suggested that they belong either to Li₃ or Li₄. Exciting with 457.9 and 476.7 nm respectively one obtains the spectra shown in figs. 4 and 5. The observations of Welker and Martin (13) together with the behavior of the intensity of these emissions upon matrix warming convince us that they belong to clusters of the same nuclearity. We assign them to Li₃ isolated in different matrix sites. The spectrum obtained with 476.7 nm excitation contains the characteristic hallmark of the spectrum of a fluxional molecule, a single "good" frequency, 310 cm⁻¹, which is assigned to the symmetric stretch and a large number of other frequencies upon which progressions of the symmetric stretching frequency are built. These frequencies are assigned to vibronic states of pseudorotating Li₃ in table 1 where the calculated frequencies of Schumacher and Gerber are also included. The comparison both for the vibronic state frequencies and the symmetric stretching frequency is quite acceptable, suggesting that the surface calculated by those authors is equally acceptable.

With 457.9 nm Ar⁺ laser excitation one obtains a much simpler spectrum characteristic of a rigid molecule. The symmetric stretching frequency for Li₃ in this site is, moreover, increased noticably to 353 cm⁻¹ from 310 cm⁻¹ in the "fluxional" site. The data in fig. 1

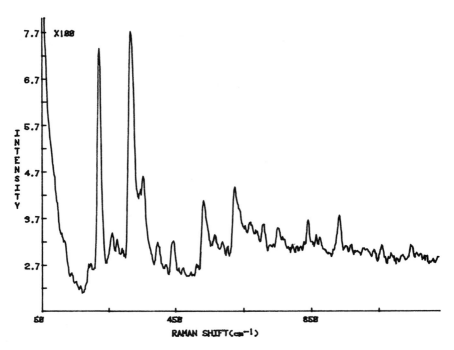

Fig. 5. Spectrum attributed to "fluxional" Li$_3$ in solid Xe excited with 476.7 nm Ar$^+$ radiation.

imply that such an increase in frequency suggest a rigid molecule with
an apical angle <u>less</u> than 60°. This conclusion is also borne out by
the isotopic fine structure. Lithium occurs naturally in two isotopes
^{6}Li (7.4%) and ^{7}Li (92.6%) which form four isotopic variants of Li$_3$
of which only two ^{7}Li$_3$ and ^{7}Li$_2$ ^{6}Li are of significant abundance:
79% and 19% respectively. Using the same technique employed in
generating fig. 1 one can calculate the expected separation between the
symmetric vibrations of these two molecules. A best fit was obtained
with φ approximately equal to 45°. This is shown in fig. 6 but for
Li in Kr. In krypton only a single form of Li$_3$ was obtained, which
corresponded closely to the rigid form obtained in Xe except that the
symmetric frequency observed was 326 cm^{-1}. The propensity for
fluxional Li$_3$ to adopt the acute form when it becomes rigid, a form
which in the gas phase is predicted to correspond to saddle points in
the potential energy surface (7) of the molecule indicates the
influence of the rare gas matrix upon the molecule and therefore the
care one should use in drawing conclusions about such delicately bonded
species as pseudorotating molecules from matrix data.

     With 579 nm (argon pumped R6G dye laser) excitation of Li in Xe
one obtains the spectrum shown in fig. 7. The spectrum consists of
two progressions built upon 375 and 298 cm^{-1}. Two progressions imply
two "good" frequencies which would not be inconsistent with Li$_4$ as
being the carrier. Beckmann et al. (15) predict Li$_4$ to be a square
species Jahn-Teller distorted into a lozenge shaped molecule for which

TABLE 1

Comparison of observed and computed vibronic term frequencies for Li_3.

Term energy above the ground state (cm^{-1})

representation	calculated*	observed
E'	0	
A_2'	47	
A_1'	113	110
E'	163	
E'	255	221
A_2'	259	259, 276
E'	369	
E'	394	396
A_1'	406	441
A_2'	499	472
A_1'	539	
A_2'	540	
E'	557	

symmetric stretch frequency   303   310

*calculated from data of ref. 7.

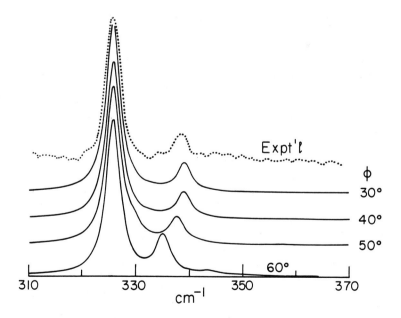

Fig. 6. Observed and calculated isotopic fine structure of fundamental attributed to "rigid" Li_3 · in solid Kr.

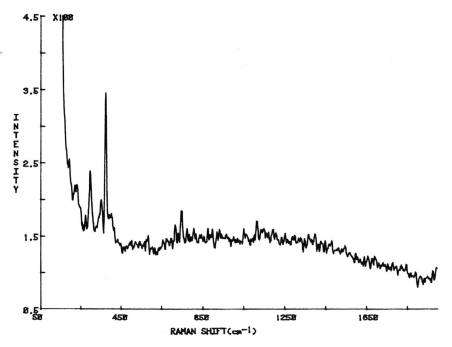

Fig. 7. *Spectrum tentatively attributed to Li$_4$ in solid Xe excited with 579 nm excitation.*

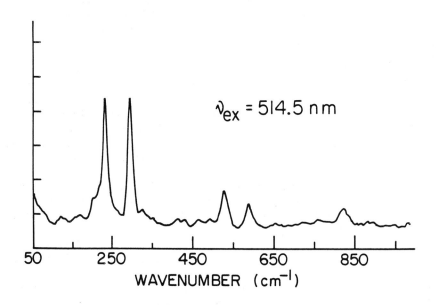

Fig. 8. *Spectrum of unassigned Li species isolated in argon excited with 514.5 nm radiation.*

those authors predict two symmetric frequencies at 334 and 224 cm^{-1}. Although one is edified by the agreement the assignment of this particular resonance fluorescence spectrum must remain tentative.

The emission spectrum of Li isolated in argon must also be left unassigned. The spectrum fig. 8 consists essentially of two progressions in 295 cm^{-1} built upon the ground state and 235 cm^{-1}. Although such a spectrum is not inconsistent with Li$_3$ we prefer to leave it unassigned. Based on ESR data Garland and Lindsay (16) report Li$_7$ to be the dominant species in Li containing argon matrices with Li$_3$ appearing only upon photolysis. The spectrum shown in fig. 8 was obtained immediately upon deposition, while prolonged photolysis with a D$_2$ lamp uv source produced neither a substantial decrease in the observed spectrum nor the appearance of new species.

Acknowledgement

We are indebted to the Natural Science and Engineering Research Council of Canada for financial support. The authors wish to thank NATO for a travel grant.

References

1. Metal Bonding and Interactions in High Temperature Systems, edited by J.L. Gole and W. Stwalley (Amer. Chem. Soc. Symposium Series No. 179, Washington, 1982).
2. a) D.P. DiLella, W. Limm, R.H. Lipson, M. Moskovits and K.V. Taylor, J. Chem. Phys. 77, 5263 (1983) and references therein; b) D.M. Lindsay, D.R. Herschbach and A.L. Kwiram, Mol. Phys., 32, 1199 (1976); c) R.J. Van Zee, C.A. Baumann, S.V. Bhat and W. Weltner Jr., J. Chem. Phys., 76, 5636 (1982).
3. A. Hernmann, E. Schumacher and L. Wöste J. Chem. Phys. 68, 2327 (1978); D.R. Preuss, S.A. Pace, and J.L. Gole, J. Chem. Phys. 71 3553 (1979); D.E. Powers, S.G. Hausen, M.E. Geusic, A.C. Puiu, J.B. Hopkins, T.G. Dietz, M.A. Duncan, P.R. Langridge-Smith, and R.E. Smalley, J. Phys. Chem. 86, 2556 (1982); J.L. Gole, and J.H. English, and V.E. Bondybey, ibid. 86, 2560 (1982); S.J. Riley, E.K. Parks, C.R. Mao, L.G. Pobo, and S. Wexler, ibid. 86, 3911 (1982).
4. M. Moskovits, D.P. DiLella and W. Limm, J. Chem. Phys. 80, 626 (1984); Metal Bonding and Interactions in High Temperature Systems, by J.L. Gole and W. Stwalley (Amer. Chem. Soc. Symposium Series Nol. 179, Washington, 1982).
5. M. Moskovits and D.P. DiLella, J. Chem. Phys. 72, 2267 (1980).
6. L.B. Knight Jr., R.W. Woodward, R.J. Van Zee and W. Weltner Jr. ibid. 79, 5820 (1983).
7. W.H. Gerber and E. Shumacher, J. Chem. Phys., 69, 1692 (1978); W.H. Gerber, Ph.D. Thesis, University of Berne 1980.
8. J.L. Martins, R. Car and J. Buttet, J. Chem. Phys. 78, 5646 (1983).

9.   D.P. DiLella, K.V. Taylor and M. Moskovits, J. Phys. Chem. 87, 524 (1983).

10. J.A. Howard, K.F. Preston, R. Sutcliffe and B. Mile, J. Phys. Chem. 87, 536 (1983).

11. M.D. Morse, J.B. Hopkins, P.R.R. Langridge-Smith and R.E. Smalley, J. Chem. Phys. 79, 5316 (1983).

12. L. Andrews and G.C. Pimentel, ibid. 47, 2905 (1967).

13. T. Welker and T.P. Martin, ibid. 70, 5683 (1979).

14. K.P. Huber and G. Hertzberg, Molecular Spectra and Molecular Structure IV: Constants of Diatomic Molecules, (Van Nostrand, New York 1979).

15. H.O. Beckmann, J. Koutecky, P. Botschwina and W. Meyer, Chem. Phys. Lett., 67, 119 (1979).

16. D.A. Garland and D.M. Lindsay, J. Chem. Phys. 78, 2813 (1983).

ELECTROMAGNETIC RESONANCES IN SYSTEMS OF SMALL METALLIC PARTICLES AND SERS

UWE KREIBIG
Fachbereich 11 – Physik – Universität des Saarlandes
D-66oo Saarbrücken/Germany

ABSTRACT

SERS of molecules deposited at/near the surface of small noble metal particles is due to several origins, one of which is the electromagnetic (Mie-) resonance of the particles. This effect has been calculated to be extremely strong for proper systems of particles. Realistic particle systems as used for experiments, however, differ from these model systems. Changes resulting therefrom are qualitatively discussed by making use of well-known results of the optical plasma resonance extinction of such particle systems, which is caused by the same resonance of the inner field. In general, the enhancement is expected to be markedly reduced compared to the model calculations.

1. INTRODUCTION

It is now accepted that SERS, i.e. Surface Enhancement of Raman Scattering of molecules at or near noble metal surfaces, is the result of a multitude of effects which mainly may be divided into
. electromagnetic (long range) effects
. chemical bonding effects and
. electron-tunneling effects.

They all have been investigated (see (1) and literature cited therein) to explain the enhancement of the Raman scattered intensity which has been observed to amount up to one million, compared to the free molecule, in systems of various substances and sample geometries.

Up to now, however, only rough estimates can be made for the contributions of these effects to the total observed enhancement, and it is not my purpose to join the discussion concerning their relative importance. Instead, I shall restrict myself to consider some experimental aspects of the electromagnetic resonances underlying the first of the effects mentioned above, which has been shown (2 and lit. in 1) to be the most important one in samples which consist of small metallic particles, the scattering molecules being located at or near the interface between metal and embedding medium.

447

B. Pullman et al. (eds.), Dynamics on Surfaces, 447–460.
© *1984 by D. Reidel Publishing Company.*

2. SMALL PARTICLE SYSTEMS

If a such system is described by very simple models, it appears to be
the best defined geometrical sample structure for SERS experiments, since, then, no statistically uncertain quantities have to be considered,
as is the case for the plane surface geometry including surface rough-
nesses.

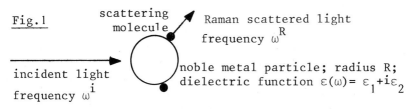

Fig.1   scattering
         molecule   Raman scattered light
                     frequency ω^R

incident light   noble metal particle; radius R;
frequency ω^i dielectric function $\varepsilon(\omega) = \varepsilon_1 + i\varepsilon_2$

embedding medium ; dielectric constant ε_m

Referring approximations to such models are:
(1) spherical metal particles with atomically smooth interface and uni-
    form sizes, the latter being extremely small compared to the wave-
    length of the incident light,
(2) well defined, bulk-like mean dielectric function of the particle
    material, $\varepsilon(\omega)$
(3) homogeneous embedding medium, which only acts via its dielectric
    constant ε_m
(4) large distances between the particles, which are assumed to be ran-
    domly arranged in the embedding matrix,
(5) low coverages of the scattering molecules at the interfaces.

        In this approach, enhancement factors due to the excitation of
spherical surface plasmon polaritons in the particles have been calculated (2) which amount up to the order of one million. Simply, the metal
particles act as microamplifiers for the scattering molecules because
of this resonance behavior of the inner field in the particle (3, 4).

        Yet, it is very difficult, if not impossible at all, to pre-
pare samples which meet even approximately all these demands. As a general consequence, it can be stated that the corresponding enhancement factors which are effective in the experiment, will be seriously lower
than calculated. It cannot be stated, however, to which amount this may
happen in a special sample, though several attempts have been made (5)
to calculate some effects in detail.

        As well, it is not yet possible to extract -vice versa- effec-
tive electromagnetic enhancement factors from experimental SERS results
because of the complex concurrence of the different enhancement mechanisms mentioned above.

        So, in the following, I shall, instead, discuss the optical
absorption of such small metal particle systems, which is better known
today and, being also a consequence of the electromagnetic resonance
phenomena in the metallic particles, is related to the electromagnetic

enhancement.

3. OPTICAL ABSORPTION AND SERS

The correspondence between SERS and optical absorption for small particle systems has been emphazised in several basic papers by Wang and Kerker (2, 5). Restricting the following discussion to a scale of orders of magnitude of the enhancement, details of this relationship are out of its scope, as, e.g., there are influences of radial field components, special orientations of the Raman active molecules, bimodal enhancement maxima in the case of Ag particles etc.

Let us consider   one molecule at the interface of a spherical metal particle which is very small compared to the wavelength of light. Then, the Raman radiation is emitted in a two-step process (2,1):

1) Excitation of the molecule. The exciting field can exceed the in-cident field due to the resonance behavior of the inner field in the particle; its increase is given (2,4) by the factor $3 \cdot \varepsilon(\omega^i)/(\varepsilon(\omega^i) + 2 \cdot \varepsilon_m)$ which reflects the spherical geometry of the particle.

2) Emission from the molecule. The emitted field, again, is enhanced by the metal particle resonance; the factor is now $3 \cdot \varepsilon(\omega^R)/(\varepsilon(\omega^R)+2 \cdot \varepsilon_m)$.

Hence, the total enhancement factor for the scattered inten-sity due to the electromagnetic resonance of the small metal particle amounts to

$$G = A \cdot \left| \frac{\varepsilon(\omega^i)}{\varepsilon(\omega^i)+2 \cdot \varepsilon_m} \cdot \frac{\varepsilon(\omega^R)}{\varepsilon(\omega^R)+2 \cdot \varepsilon_m} \right|^2 \tag{1}$$

or, if $\varepsilon(\omega^i) \approx \varepsilon(\omega^R)$,

$$G = A \cdot \left| \frac{\varepsilon(\omega^i)}{\varepsilon(\omega^i)+2 \cdot \varepsilon_m} \right|^4 = A \cdot \left| \frac{\varepsilon_1^2 + 2\varepsilon_1 \cdot \varepsilon_m + \varepsilon_2^2 + i \cdot 2 \cdot \varepsilon_m \cdot \varepsilon_2}{(\varepsilon_1 + 2 \cdot \varepsilon_m)^2 + \varepsilon_2^2} \right|^4 \tag{2}$$

Introducing the dipole polarizability of the spherical particle in the li-mit $R \ll \lambda$, $\alpha_{sphere} = 4\pi\varepsilon_0 \cdot \varepsilon_m \cdot R^3 \cdot (\varepsilon-\varepsilon_m)/(\varepsilon+2 \cdot \varepsilon_m)$ we obtain for the lea-ding term $G \propto |1+B \cdot \alpha_{sphere}(\omega^i)|^4$ with $B = (2 \cdot \pi \cdot \varepsilon_0 \cdot \varepsilon_m \cdot R^3)^{-1}$. \tag{3}

The resonant enhancement near the spherical plasmon polariton frequency ω_s follows mainly from the resonant behavior of the denomina-tor of eq. (2):
Around the frequency ω_s, we have $\varepsilon_1 \approx -2 \cdot \varepsilon_m$ and, hence, G reaches a maxi-mum.

This resonance behavior is reflected in the optical absorp-tion spectrum $K(\omega)$ of such spherical metallic particles (without the Raman active molecule), which is given by $K(\omega) \propto \text{Im}\{\alpha_{sphere}\}$ i.e.

$$K(\omega) \propto \frac{\varepsilon_2(\omega)}{(\varepsilon_1(\omega)+2 \cdot \varepsilon_m)^2 + \varepsilon_2^2(\omega)} \quad , \text{ and, hence, } \quad G \underset{\sim}{\propto} K^4 . \tag{4}$$

Fig. (2) shows, as an example, the spectrum of silver particles and, for comparison, the spectrum of a plane silver film, which does not show this resonance , the absorption extending, instead, to $\omega \sim 0$.

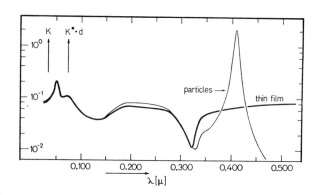

Fig.2: Calculated absorp-
tion spectra of spherical
of Ag particles (2R<<λ),
compared to a thin Ag
film (6).
There are strong diffe-
rences due to the plasma
resonance at 400 nm.

Hence, by investigating the well known optical absorption spectra of
small noble metal particles in some detail , we will, in the following,
obtain information about changes of the enhancement factor in samples
which do not fulfill the assumptions listed above.

4. PROPERTIES OF SMALL PARTICLE SYSTEMS
(1a) Non-uniform particle sizes:

Within the limits of eq(4), both, peak position and width of the absorption spectra do not depend on R and, hence, also G does not. In fact,
however, they depend on R in all particle size regions. In small particles, a size dependence enters via $\varepsilon(R)$ as shown in the next section,
whereas the inner field itself depends on R in large particles (see, e.g.
fig. 3).

The experimentally observed spectra of samples with a distribution of particle sizes therefore are, in general, broadened and shifted, compared to those following from eq(4) for samples with uniform
sizes.

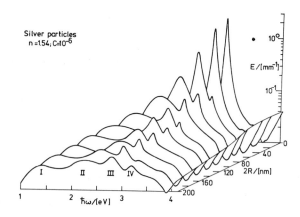

Fig.3:
Extinction spectra of spherical Ag particles of various sizes (0-200 nm), calculated from Mie theory (3)
with $\varepsilon(\omega)$ of the bulk.
(I,II, ... denote the dipo-
lar, quadrupolar, ... resonance modes).

Since roughly, a sum rule, $\int K(\omega) \cdot d\omega \approx$ const, holds for this absorption
(6), the broadening is accompanied by a decrease of the peak height.
Passing over to SERS: the resonance condition of the inner field is ful-

filled at different frequencies in particles of different sizes and the
measured maximum enhancement of the whole sample, thus, is decreased.
          (1b) Non-uniform particle shapes:
Most simply, such particles are approximated by spheroids with varying
eccentricities. Then, in the small particle limit, α of eq. (3) is replaced by

$$\alpha_{spheroid} \propto \frac{\varepsilon(\omega) - \varepsilon_m}{\varepsilon(\omega)+(\varepsilon(\omega)-\varepsilon_m)\cdot N_i} \; , \qquad (5)$$

N_i being the depolarization factors (7,5).
          For a many particle system with arbitrary orientations of the
particles relative to the incident light, α takes an average value with
respect to N_i. In the case of noble metal particles, the resulting absorption peak is shifted strongly with varying eccentricity and a second, weaker, peak occurs (fig. 4).

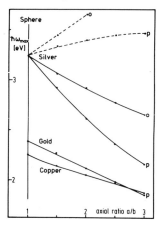

Fig. 4:
Calculated absorption peak positions
of ellipsoidal noble metal particles
(size $\ll\lambda$) with various eccentricities a/b. There is a second peak,
which is strong for Ag but weak for
Cu und Au.
(o: oblate; p: prolate spheroids).

In contrast to Ag particles, the peak height increases in the case of
Au particles, both, for oblate and prolate shapes. In general, the arguments of Sect.(1a)   hold analogously, and as a consequence, the resonance is broadened and -except, perhaps, for Au particles-, is lowered
in a sample containing particles with varying particle shapes and orientations.
          (2) Size dependent mean dielectric function $\varepsilon(\omega,R)$ of the par-
ticle material:
In small metal particles, below, say, 15 nm diameter ε deviates from the
bulk dielectric function in the region of the plasma resonance absorption, due to changes of, both, the "conduction" electron susceptibility
(6,8) and   the interband transition susceptibility (9).
These two contributions add up in ε:

$$\varepsilon(\omega) = 1 + \chi^{cond}(\omega) + \chi^{interb}(\omega) \qquad (6)$$

χ^{cond} depends on size in small particles since the extension of the
electron wave functions is limited by the particle surface. This size
effect (8) affects the relaxation frequency $1/\tau$ as

$$(1/\tau)^{\text{particle}} = (1/\tau)^{\text{bulk}} + A \cdot v_{\text{Fermi}}/R, \tag{7}$$

A being of the order of unity. v_{Fermi} is the Fermi velocity, R the particle radius. The resulting $\varepsilon(\omega, R)$ is shown in Fig. 5 and compared to experimental results of Ag particles.

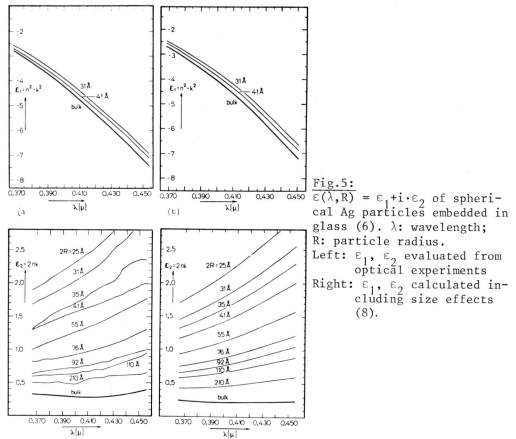

Fig.5:
$\varepsilon(\lambda, R) = \varepsilon_1 + i \cdot \varepsilon_2$ of spherical Ag particles embedded in glass (6). λ: wavelength; R: particle radius.
Left: ε_1, ε_2 evaluated from optical experiments
Right: ε_1, ε_2 calculated including size effects (8).

The bulk contribution in eq (7) is due to relaxation processes, which are as well present in the infinite material. It has been stated (10), that in silver hydrosol particles, as frequently used for SERS experiments, the contributing grain boundary scattering may be markedly increased compared to the bulk, giving rise to an additional change of $1/\tau$.

Fig.6 shows for Au particles that this size effect causes marked broadening of the plasma resonance absorption, i.e., an increased damping of the inner field if the particle size is reduced from 2o nm to 2 nm. Again: as a consequence, the electromagnetic enhancement factor for SERS experiments is expected to be reduced as well.

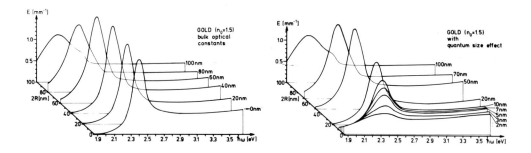

<u>Fig.6:</u> Extinction spectra of spherical Au particles of
various sizes, calculated from Mie's theory (3) with $\varepsilon(\omega)$
of the bulk (left) and with size dependent $\varepsilon(\omega)$ (right) (4).

Changes of χ^{interb} were only observed in extremely minute particles (9)
of less than, say, 3nm diameter, indicating structural transitions from
the molecular to the crystalline solid state (fig.7).

From both of these size effects it follows that particles with
radii of less, say, 5nm are less well suited for SERS experiments because of increased damping of their inner field and, hence, decreased
electromagnetic enhancement.

On the other hand, if R is varied in otherwise unchanged
systems between, say, lo and 1 nm, the electromagnetic enhancement is
suppressed by let's guess, between one and two orders of magnitude, thus
making   feasible to check the relative importance of other enhancement contributions which are not or less size dependent.

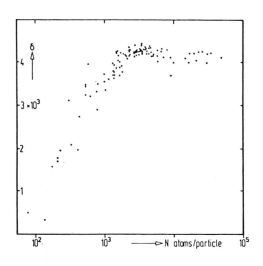

<u>Fig.7:</u> Size dependence of the
slope of the optical interband
transition edge, χ_2(interband),
of small Au particles, embedded
in glass (9).

(3) Large spherical particles:

The validity of eq (1) is limited to noble particlesof less than, say, 2o nm. In large particles, retardation effects have to be included, and also scattering contributions to the extinction (beside the consumptive absorption) and higher multipole excitations (beside the dipole excitation) have to be considered. Then, the leading term for the multipole polarizabilities is

$$\alpha_{\nu}^{\text{sphere}} \propto u_{\nu} \cdot \frac{\varepsilon - v_{\nu} \cdot \varepsilon_m}{\varepsilon + ((\nu+1)/\nu) \cdot w_{\nu} \cdot \varepsilon_m} \,, \tag{8}$$

where $\nu = 1$ denotes the dipolar, $\nu = 2$ the quadrupolar, $\nu = 3$ the octupolar excitations etc.

Figs.3 and 8 show the   structure of the resulting extinction spectrum of spherical 2oo nm Ag particles (11). Avoiding again, detailed discussion, we draw the conclusions therefrom, that for 2oo nm particles

. the extinction resonance is extremely broadened, extending, in the case of Ag particles, over the whole visible and NIR region, and

. the maximum is lowered for almost two orders of magnitude, compared to particles of 2o nm diameter.

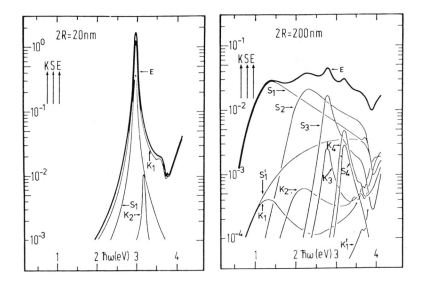

Fig.8: Extinction spectra E (ω) of 2o nm and 2oo nm Ag particles due to Mie theory(3) and their deconvolution into absorption (K(ω)) and scattering (S(ω)) contributions of the first four multipole modes (Index 1: dipole, 2: quadrupole, etc.)

Also in this case, the extinction spectrum reflects the resonance of the inner field in the particles, and, hence, the SERS-enhancement of large particles is expected to be markedly lowered compared to small particles. This has been shown explicitely by Wang and Kerker (2,5).

Summarizing the above mentioned effects occurring in the SINGLE particles, we may state that -perhaps with the exception of some ellipticity effects- all of them show a tendency to diminish the inner field resonances in the particles and therefore, also the electromagnetic enhancement factor.

(4) Particle systems with interactions:

The fourth of the model assumptions listed above was that particles in a sample should be well separated to ascertain that the inner field resonance of one particle is not influenced by the presence of other particles.

However, a multitude of preparation methods yields samples with more or less densely packed particles (evaporation of island films, gas-evaporation of particle-"blacks", etc.), and even in hydrosols of low MEAN particle concentrations one frequently observes clusters of aggregated particles, as particle pairs, tiplets, linear chains etc. (fig.9).

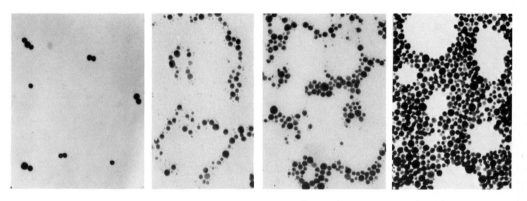

Fig.9: Electron micrographs of two-dimensional systems of Au particles (mean size: 1o nm) with various amounts of aggregation.

In the last section we will discuss some consequences of the formation of such aggregates on the optical absorption and on SERS enhancement.

(A) If the particles of a densely packed sample are randomly arranged, effective medium theories (see,e.g.(4)) may be applied to calculate their resulting absorption spectra. Here, the dominating parameter is the filling factor f, i.e. the relative volume concentration of the particles in the sample.

Fig.1o shows, again for the example of Ag particles, results of the Maxwell-Garnett effective medium theory which predicts the peak position to shift with increasing f, in quantitative agreement with experiments (12). The considered experiments also show an increase of the resonance widths; additional assumptions have to be made to explain this phenomenon (13).

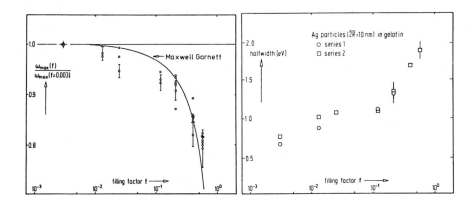

Fig.1o:Effective medium systems containing spherical Ag particles (mean size: 1o nm)(12): Absorption peak energies (left)
and widths (right) for varying filling factor f.
Points: experimental; line : Maxwell-Garnett theory

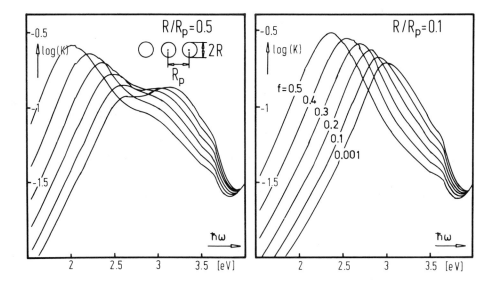

Fig.11: Calculated absorption spectra of effective medium systems containing linear Ag particle triplets with
filling factor f (4,15).
Left: touching particles (R_p=2R); Right: particle distance
R_p = 1oR (i.e. Maxwell-Garnett case).
(The spectra of both figures belong to the same sequence of f's.)

(B) The effective medium theories base upon the assumption that the single particle polarizability is not influenced by the existence of neighboring partners. This assumption does, however, not hold any longer, if neighboring particles approach to such an extent that direct multipolar coupling becomes effective. These coupling effects are large, if the surface to surface distance of two neighboring particles is below, say, the particle diameter. Such clustered particles form aggregates, which have to be treated as coupled oscillators.

As a consequence of the coupling, the plasma resonance modes of the single particles are split up, and the splitting increases with decreasing neighbor distance as shown by the calculated spectra of fig.11 for linear Ag particle triplets, which are packed in a sample of many particle aggregates with varying filling factor (4).

The splitting depends on the number of particles and their geometrical arrangement in such aggregates. This effect has been calculated (14) and is shown in fig. 12 for particles in direct contact which form aggregates of various symmetries.

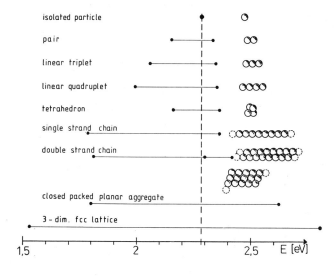

Fig.12:
Absorption peak positions of aggregates of Au particles (2R<<λ), according to the theory of Clippe et al. (14).

In real systems, however, as, e.g. aqueous colloidal systems, the formation of aggregates is a statistical process, driven by v.d. Waals forces among the particles and, hence, highly symmetric aggregates as shown in fig. 12 are rather rare. So, comparisons of such calculations with experiments can be performed only to limited extent. Within this frame, existing experimental results confirm the theoretical predictions (13a,12). In figs. 13,14 recent unpublished results on Au particle systems with the amount of aggregation as the ONLY varied parameter are shown (16).

Clearly, there are two well separated modes in the extinction spectra which are due to longitudinal and transverse modes of anisometric particle aggregates, examples of which are shown in fig.9 . Fig.13 demonstrates that by this kind of interaction it is possible to suppress in the overall extinction spectrum completely the selective resonance of the inner field occurring in the single Au particle. This does not mean, that the resonance of the inner field in ONE special particle has vanished (in the case of the longitudinal mode of Au it even increases), but that the resonance frequencies of different particles in the sample are distributed over the whole visible and NIR region.

Since SERS experiments give results which are averages over all particles in the sample, as does the optical extinction measurement, it is expected that the effective enhancement factors as may be evaluated from experimental results of samples with coagulation aggregates will also be much smaller than those which may be obtained from samples with well separated single particles, IF relation (4) holds, at least, roughly.

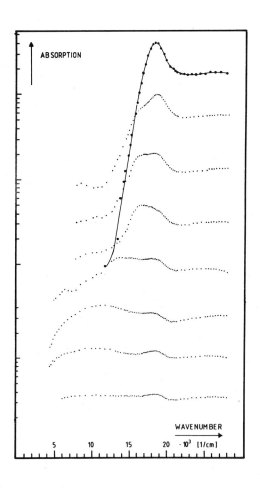

Fig.13:
Measured extinction spectra
of two-dimensional Au particle systems with the amount
of aggregation increasing
from top to bottom (see fig.9).
Mean particle diameter: 1o nm.
(16).

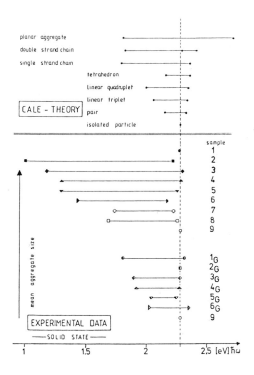

Fig.14: Peak positions of the absorption spectra of Au particle systems with varying amount of particle aggregate formation (16).

Upper part: after Clippe et al.(14)
Lower part: experimental results of particle systems embedded in gelatin
$2\bar{R} \approx 10$ nm.

So, it may be possible to obtain information about other enhancement effects which are not or less dependent of the formation of aggregates by varying the measurable electromagnetic enhancement of a sample via variation of the amount of aggregation of the particles.

At this point the list of effects which influence the inner field resonance in systems of small particles, and, hence, the electromagnetic enhancement of Raman scattering, shall be closed, though being far from complete.

My purpose was to give some impression of difficulties that may arise when SERS experiments are evaluated numerically. After all, it appears more probable for results which are obtained from different kinds of small particle samples to differ markedly than to give correspondence, both, in the spectral behavior of the SERS spectra and in the evaluated overall enhancements. If, however, SERS experiments are accompanied, both, by optical extinction measurements and electron microscopic registration of the SAME samples, it should be possible to define the sample properties properly enough to calculate corresponding electromagnetic contributions to the scattering enhancement from existing theoretical models for the inner field resonance in small particles to, at least, a correct order of magnitude.

REFERENCES

(1) A. Otto in: <u>Light Scattering in Solids</u> Vol.IV, Ed.by M. Cardona,
        G. Güntherodt, Springer, Berlin, 1984

(2) M. Kerker, D.S. Wang, H. Chew, Applied Optics <u>19</u>, 4159 (1980)

(3) G.Mie, Ann.Physik <u>25</u>, 377 (1908),P.Debye,Ann.Phys. <u>30</u>, 57 (1909)

(4) U. Kreibig in: <u>Impact of Cluster Physics in Materials Science and</u>
        <u>Technology</u>, Nato-ASI, Agde/France; in press

(5) D.S. Wang, M. Kerker, Phys. Rev.B <u>24</u>, 1777 (1981)

(6) U. Kreibig, Z.Physik <u>234</u>, 307 (1970)

(7) R. Gans, Ann.Physik <u>37</u>,881 (1912)

(8) U. Kreibig, C.v.Fragstein, Z.Physik <u>224</u>, 307 (1969), J.Physics <u>F.4</u>,
        999 (1974) J.Physique <u>38</u>, C2 (1977)
    L.Genzel, P.Martin, U.Kreibig Z.Physik <u>21</u>, 339 (1975)
    R. Ruppin, H. Yatom , phys.stat.solidi <u>74</u>, 647 (1976)
    W.A.Kraus, G.C.Schatz J.Chem.Phys. <u>79</u>, 6130 (1983)

(9) U. Kreibig, Sol.State Commun. <u>28</u>, 767 (1978)
    in: <u>Growth and Properties of Metal Clusters,</u> Ed.by I.Bourdon, Else-
        vier, Amsterdam, 1979
    T.Yamaguchi,M.Ogawa,H.Takahashi,N.Saito,E.Anno, Surf.Science (1982)

(1o) U. Kreibig, Z.Physik <u>B 31</u>, 39 (1978)

(11) U. Kreibig, P.Zacharias, Z. Physik <u>231</u>, 128 (1970)

(12) U. Kreibig, A.Althoff, H.Pressmann,Surf.Science <u>106</u>, 308 (1981)

(13) S.Yoshida,T.Yamaguchi,A.Kinbara,J.O.S.A. <u>61</u>, 62 (1971);<u>62</u>, 1415(1972)
        <u>64</u>, 1563 (1974)
    B.Persson,A.Liebsch, Sol.State Commun. <u>44</u>, 1637 (1982)

(14) D.Bedeaux,J.Vlieger, Physica <u>73</u>, 287 (1974)
    P.Clippe,L.Evrard,A.Lucas, Phys.Rev. <u>B14</u>, 1715 (1976)
    M.Ausloos,P.Clippe,A.Lucas,Phys.Rev. <u>B18</u>, 7176 (1978)
    Further papers: <u>B22</u>, 4950 (1980), <u>B26</u> (1982)

(15) D.Schoenauer, U.Kreibig,L. Genzel, to be published

(16) D. Schoenauer, U. Kreibig , to be published.

DYNAMICS ON SURFACES - CONCLUDING REMARKS

R.B. Gerber
Department of Physical Chemistry
and the Fritz Haber Research Center of Molecular Dynamics
Hebrew University of Jerusalem, Jerusalem, Israel

A. Nitzan
Department of Chemistry, Tel-Aviv University, Tel-Aviv, Israel

Surface science and in particular the study of dynamical processes on surfaces has advanced in about fifteen years from relative obscurity to become one of the most rapidly growing fields in the forefront of physics and chemistry research. Some of the most interesting subjects from the mainstream of this field were represented in this meeting. In broad overview these were:

(a) Atom surface scattering as a tool for surface structure and dynamics.

(b) Translational, rotational vibrational and electronic energy transfer processes involving molecules adsorbed on or colliding with the surface.

(c) Adsorption desorption and diffusion processes.

(d) Electron and photon interactions with pure and adsorbate covered surfaces.

(e) Reactive processes on surfaces.

(f) Structure and dynamics of metal, semiconductor and molecular clusters.

1.   ATOM DIFFRACTION AND PHONON PROCESSES

Until just a few years ago the emphasis in this subject has been on relating scattering intensities to surface structure, using essentially a rigid surface framework for interpretation. The emphasis has now shifted to the ability to measure phonon processes, and to sophisticated analysis of their participation in diffraction. Contributions at the Symposium indicate that the understanding of thermal attenuation of diffraction scattering is progressing now into a new quantitative phase that transcends the simplistic, old Debye-Waller treatments (lectures of Celli, Manson, Armand). One of the most interesting features of the results reported is that strong deviations may occur from the conventional linear dependence of the logarithmic scattering intensity upon surface temperature. At the level of the new generation theories presented, it seems no longer possible <u>in general</u> to discuss thermal

461

B. Pullman et al. (eds.), Dynamics on Surfaces, 461–466.
© *1984 by D. Reidel Publishing Company.*

attenuation in terms of simple Debye-Waller factors.

Measurements of individual phonon creation and annihilation events in atom scattering have been an exciting highlight of the field in recent years (Toennies, Sibener). Toennies has shown that phonon-assisted trapping and desorption processes in atom-surface collisions can be studied in detail by time-of-flight measurements. Quantitative theoretical interpretation of individual phonon transitions is now possible (Toennies, Benedek). A remarkable development is that the phonon structure of well-defined overlayer systems can now be studied by He scattering (Sibener). Results were obtained for monolayer, double layer, etc. up to higher order multilayer covering of Xe on an Ag surface, and it was shown that the scattering reveals an Einstein vibrational spectrum for the monolayer, but double and higher over-layers show dispersive phonons (Sibener).

The study of scattering intensities as related to surface structure and corrugation remains, of course, extremely important. Lapujoulade studied collisions of several projectiles (He, H_2, Ne) from different faces of Cu, finding a surprising order for the corrugation strengths pertinent to these three colliders. This raises an interesting question for interpretation in terms of electron structure theory. Rabitz in his contribution discussed scattering intensities from disordered clusters adsorbed on smooth surfaces. Rabitz and his collaborators developed a new quantum-mechanical approximation for calculating the scattering intensities from such disordered targets in the low-energy range. Scattering from disordered surfaces is likely to emerge as an important future direction in studies of surface structure.

2. MOLECULAR ENERGY TRANSFER

This also is one of the most intensely active areas in current studies of surface processes. Experimental studies of vibrational energy trans-fer in collisions of small molecules with metal surfaces were described by Houston and by Asscher. Houston studied CO_2 deactivation in colli-sions at moderate energies. Houston concluded that energy transfer occurs via trapping, and that electron-hole pairs are the most likely receiving modes. Asscher discussed a system of much stronger molecule-surface coupling (NO/Pt, where chemical interactions are present). His interpretation also involves vibrational energy transfer to electron-hole pairs.

The study of rotational transitions in molecule-surface collisions exhibits two interesting effects in two opposite regimes: trapping resonances and rotational rainbows. Schinke presented a theoretical study of both effects (for HD/Ag and for NO/Ag) focussing on the extrac-tion of potential functions from the data. Kouri's article treats the related question of molecular reorientation upon collision with the surface, as reflected in the occurrence of transitions in Δm_J, the mag-netic quantum number. Levine, in his presentation, considered the impli-cation of the principle of detailed balance for the energy transfer processes in molecule-surface collisions. Excitation of surface vibra-tions upon molecular impact was a central topic in several presentations. Newns discussed several versions of the quasi-classical trajectory method (treating the translational motion as classical) as a useful tool in the regime of multiphonon excitation. Amirav described experiments in which large transfer of collision energy into solid vibrations was observed in cases of heavy colliders (I ,Hg) in high energy impact upon light-atom solids. In collaboration with Elber and Gerber the large energy transfer was found to be due to a compression-wave excitation of the target solid.

Although major progress was made in recent years in studying energy transfer in molecule-surface collisions at the state-selected level, the quantitative importance of the possible mechanisms (e.g. electron-hole vs. phonon excitations) remains, in most cases, to be explored. One may expect more definite answers within the next few years.

3. REACTION DYNAMICS

This topic is obviously of outstanding chemical interest. However, well-defined experiments are notoriously hard to carry out. Progress along several novel directions was reported at the meeting, which seems to hold much promise for the future. Polanyi described interesting experiments in which H_2 and F atoms were trapped on a (CF_4) surface, with a reaction taking place in the adsorbed state.

Vibrational and rotational state distributions of the reaction products were measured, yielding insight into the reaction dynamics. Amirav and Kolodney studied dissociation of I_2 molecules in high-energy collisions with chemically inert surfaces (MgO, sapphire). The experiments combined with theoretical simulations yield detailed insight into the dissociation dynamics. Bernasek described studies of molecular reactions at iron surfaces, with results that suggested that large effective corrugation exists and plays a major role in the dynamics of the process. Shapiro explored theoretically the interactions of negative ions at the vicinity of alkali halide surfaces and suggested dissociative electron transfer as a mechanism of photodesorption in such systems. Dissociation of small molecules adsorbed in tungsten planes was investigated by Folman, using FEM and TDS studies. Insight into the interaction mechanisms was inferred from changes in the work function.

4. ADSORPTION, DESORPTION AND DIFFUSION

It has long been recognized that chemical kinetics at interfaces is often determined by the transport processes associated with the flows of reactants and products towards, out of and on the surface. The investigation of surface transport phenomena has therefore been for a long time in the mainstream of surface dynamics research. In recent years kinetic studies of both diffusion and adsorption-desorption have been increasingly coupled to molecular level investigation of these processes.

On this level, this much studied subject may yield some unexpected and surprising results. Gomer, in his presentation described the latest

results of his investigation of the diffusion dynamics of H, D and T on
W surfaces. Combined studies of the temperature dependence, of the variation with coverage and of the anisotropy of the diffusion process,
span a tantalizing and fascinating landscape of facts that at present
still escape complete interpretation. Landman, examined the mechanism
of Pb_2 diffusion on a metal surface by detailed molecular dynamics simulations. One interesting result stressed was that rotation of the Pb
dimer plays a major role in enhancing the diffusion, this being effective in view of favorable matching between phonon and pertinent rotational frequencies.

Recent studies of adsorption desorption and diffusion processes
suggest that transition state theory provides the main features of
these activated processes. The dynamic contributions may be incorporated e.g. by introducing the sticking coefficient. These aspects were
discussed by Cardillo who also stressed the importance of special
desorption sites and the role of diffusion into them.

A different category of atom migration is involved in the work of
Greene, who discussed a variety of processes that follow the deposition
from molecular beams of alkali atoms on semiconductor surfaces. He
found noticeable diffusion of large alkali atoms into the bulk solid.
Also phase transitions on the surface layer were found to take place.
Kinetics of diffusion was pursued by Rabitz in the framework of a model
that involves random reaction sites, with diffusion between them, and
has shown the possible occurrence of cooperation effects between the
reaction sites. Finally, it should be mentioned that laser induced
desorption processes (Heidberg, Chuang) open new possibilities in this
field both because laser surface heating make it possible to probe the
desorption process on a very short timescale and because laser excitation raises the possibility of non-thermal selective desorption.

5.   ELECTROMAGNETIC PROCESSES

The use of electron and photon beams as probes of surface structure and
dynamics has long proved to be of crucial importance in surface research, with tools like UPS, XPS, Auger spectroscopy and LEED now
becoming widely spread. The last few years have seen rapid progress in
electron energy loss spectroscopy. A theoretical treatment of the principles involved in this technique was presented by Persson and experimental applications in studies of molecule-metal surface interactions
were described by Avouris. Another rapidly developing technique is the
use of adsorbate selective low energy photon probes such as Raman and
infrared spectroscopies. The two latter techniques have the important
advantage of providing convenient tools for studying the solid-liquid
interface. Surface enhanced Raman scattering, even though it is still
a subject of some theoretical debate (Otto, Schatz, Kreibig) is rapidly
becoming such a tool (Van Duyne, Chang, Philpott). Its main limitation -
of being rather substrate specific - is rapidly overcome by the increasing sensitivity of avilable detection systems which make it now possible to observe unenhanced Raman scattering from adsorbates in monolayer
quantities as described by Tsang. In the solid liquid interface we

evidence a similarly promising progress in the use of resonance Raman scattering (Van Duyne) and infrared spectroscopy (Philpott) on metal electrodes. Interesting new possibilities are opened also by the use of time resolved Raman scattering whereby dynamics may be followed down to time scales as short as picoseconds (Van Duyne). Finally, detailed studies of surface enhanced Raman scattering of adsorbed water which until recently eluded workers in this field were reported by Chang.

A new and very different type of enhancement has been discussed by Naaman who showed that the photoionization yield of molecules is enhanced by the proximity to metal and semiconductor surfaces.

In addition to the rapid development of these tools the investigation of electronic and optical properties of adsorbed molecules continues to be a subject of much experimental and theoretical interest. Properties such as absorption and fluorescence cross-sections as well as quantum yields and lifetimes have been shown in recent years to be strongly modified by the surface. The possibility of observing surface enhanced Raman optical activity associated with the strong electric field gradient near the surface suggested by Efrima, represents one more example for this class of effects. More specific effects of surfaces on molecular optical properties have been demonstrated in the presentations of Koch and of Avouris. Such specific interactions are related to the configuration of chemisorbed systems and are encountered in excitations involving orbitals associated with the chemisorbed bond. Core level spectroscopy may thus reveal such cases where adsorbate core levels participate in chemisorption.

Another aspect of radiative interactions involving adsorbed molecules is photon induced desorption. Electron and photon induced desorption involving adsorbate electronic transitions is one much studied process. Progress in infrared induced desorption and in attempts to affect selective photodesorption were reported by Chuang and Heidberg. Such studies carry an important analogy to gas phase vibrational predissociation phenomena, though more work is needed for a further clarification of the role played by the different possible routes: direct photodesorption, resonance surface heating with the adsorbate as the light absorber and direct surface heating. A different type of light induced desorption was discussed by Shapiro who proposed a model for the radiative degradation of silver halides propelled by electron transfer.

It may be safely assumed that radiative interactions on surfaces will continue to be a major direction in surface dynamics research. Similarly electron interactions, in particular EELS will provide a tool of increasing importance for studies of adsorbates on metals and semiconductors.

6. THE STUDY OF CLUSTERS

Cluster structure and dynamics is rapidly becoming a focus for much experimental and theoretical research effort. Situated between the single molecule and matter in bulk, small clusters possess some unique features which are associated with their large surface to volume ratio.

The unique optical properties of small dielectric clusters have been known since the turn of the century. The relation of these properties to surface enhanced Raman scattering has been discussed in this meeting by Kreibig. A microscopic look at the modification of these phenomena due to finite size effects on the cluster dielectric function was provided by Schatz, using a quantum mechanical study of electrons confined in ellipsoidal particles. On the other extreme end are the small molecular clusters, whose structure and dynamics can be probed in detail by laser spectroscopy of such clusters formed in supersonic beams or in rare gas matrices. Spectroscopic studies of organic molecules bound in Van der Waals clusters were described by Even. Studies of small metal clusters in rare gas matrices were reported by Moskovits.

Very interesting models of the development of cluster structure have been introduced by Farges based on geometric arguments and diffraction data.

A repeatedly raised question concerning clusters is how do physical and chemical properties of small clusters depend on their size and how do they approach their bulk values as the cluster size increases. Evidently different cluster size dependence is associated with different properties. A demonstration of this dependence for one of the lesser studied phenomena - the optical response of small (colloidal) semiconductor particles has been provided by Brus.

Clearly, our understanding of the physics and chemistry of small cluster is at its initial stages. The progress in experimental techniques make it increasingly easy to perform controlled studies of clusters in solid matrix, liquid and vacuum environments. We expect that strong interest and rapid advance of our knowledge will continue in this direction.

7. CONCLUSIONS

It is safe to assume that the next few years will see a very rapid increase in our knowledge and understanding of dynamical processes on surfaces. With an increased understanding of the physical phenomena we expect to see a substantial progress in understanding chemical and photochemical processes involving well characterized solid-vacuum interfaces as well as solid-gas, solid-liquid and small cluster interfaces. This meeting has given us an overview of the ground work done towards this goal.

INDEX OF SUBJECTS